2nd International Symposium on Marine and Fisheries Research 2017

IOP Conference Series: Earth and Environmental Science
Volume 139

Yogyakarta, Indonesia
24 - 25 July 2017

Editors:

Alim Isnansetyo
Indun Dewi Puspita
Noranizan Mohd. Adzahan
Suadi
Dini Wahyu Kartika Sari
Anes Dwi Jayanti

ISBN: 978-1-5108-6153-4
ISSN: 1755-1307

Printed from e-media with permission by:

Curran Associates, Inc.
57 Morehouse Lane
Red Hook, NY 12571

Some format issues inherent in the e-media version may also appear in this print version.

Printed by Curran Associates, Inc. (2018)

For permission requests, please contact the Institute of Physics at the address below.

Institute of Physics
Dirac House, Temple Back
Bristol BS1 6BE UK

Phone: 44 1 17 929 7481
Fax: 44 1 17 920 0979

techtracking@iop.org

Additional copies of this publication are available from:

Curran Associates, Inc.
57 Morehouse Lane
Red Hook, NY 12571 USA
Phone: 845-758-0400
Fax: 845-758-2633
Email: curran@proceedings.com
Web: www.proceedings.com

Table of contents

Volume 139

The 2nd International Symposium on Marine and Fisheries Research

24–25 July 2017, Yogyakarta, Indonesia

Accepted papers received: 16 March 2018
Published online: 12 April 2018

Preface

Papers

Aquaculture

Fisheries Management and Oceanography

Fisheries Socio-Economics

Fish Processing Technology

Preface

The role and significance of fisheries and marine research become more crucial in the future around the world as the ocean covers almost 2/3 of our planet and produces huge resources for human needs. Fisheries and marine resources have crucial roles in providing not only world energy, seafood productions, significance in climate change, marine aquaculture, but also in the world coastal socio-economic dynamics. Fisheries and marine resources serve as main income generating sector for people in coastal area of tropical countries especially in the archipelagic countries such as Indonesia. Department of Fisheries, Faculty of Agriculture, Universitas Gadjah Mada Indonesia organizes a symposium biannually, International Symposium on Marine and Fisheries Research (ISMFR). This symposium is a forum to be the primary focal assembly to share research outcomes and discussions on the role of fisheries and marine scientists in sustaining ocean ecosystems. This 2nd ISFMR was participated by more than 150 participants with a total of 106 presented papers. The symposium also invited the prominent scientists in fisheries and marine science mainly in the topics of marine sciences, fisheries management, aquaculture and fish processing technology. The symposium theme was "tropical marine and fisheries resources in a changing environment" covered broad topics in fisheries and marine science such as aquaculture, fish diseases, fish genetics, biotechnology, oceanography, seafood processing and safety, fisheries biology, climate science, marine ecotoxicology, fisheries management, marine natural product, and fisheries socio-economics.

This proceeding provides an opportunity for readers to engage with refereed papers presented on 24-25 July 2017 of the 2nd ISFMR. The papers published in this proceeding were selected from a total of 106 presented papers. This proceeding is divided into 5 sections, namely aquaculture, marine science, fisheries management and oceanography, fish processing technologies, and fisheries socio-economics. Therefore, in this proceeding, readers might discover the recent issue and results of research in the broad topics of fisheries and marine sciences.

We would like to deeply appreciate all of parties for the successful of the 2nd ISFMR. We deeply express our gratitude for organizing committee, keynote and invited speakers, reviewers, editors, and editing staffs for fully dedication, tireless efforts, and continuing hard work along the conference events and the process of this proceeding publication. We express special acknowledgement for Rector, Dean of Faculty of Agriculture, and Publication Agency of Universitas Gadjah Mada for their consistent and solid support for this conference. We also deeply appreciate all of participants and authors for the excellent participating and taking the best opportunity for disseminating, discussing and publishing the papers. A broad parties and large number of people have to be acknowledged for their crucial contribution for this successful ISFMR. Finally, we hope this proceeding provides the readers up to date and prominent information in fisheries and marine science from various points of view.

Chief Editor,

Dr. Alim Isnansetyo

EDITORS

Alim Isnansetyo
Universitas Gadjah Mada, Indonesia
isnansetyo@ugm.ac.id

Indun Dewi Puspita
Universitas Gadjah Mada, Indonesia
indun_dp@ugm.ac.id

Noranizan Mohd. Adzahan
Universiti Putra Malaysia, Malaysia
noraadzahan@upm.edu.my

Suadi
Universitas Gadjah Mada, Indonesia
suadi@ugm.ac.id

Dini Wahyu Kartika Sari
Universitas Gadjah Mada, Indonesia
dini.sari@ugm.ac.id

Anes Dwi Jayanti
Universitas Gadjah Mada, Indonesia
anes.dwijayanti@ugm.ac.id

IOP Conf. Series: Earth and Environmental Science **139** (2018) 011003 doi:10.1088/1755-1315/139/1/011003

REVIEWERS

1. Alim Isnansetyo (Universitas Gadjah Mada, Indonesia)
2. Amir Husni (Universitas Gadjah Mada, Indonesia)
3. Anes Dwi Jayanti (Universitas Gadjah Mada, Indonesia)
4. Anindya Wirasatriya (Diponegoro University, Indonesia)
5. Arini Wahyu Utami (Universitas Gadjah Mada, Indonesia)
6. Bambang Triyatmo (Universitas Gadjah Mada, Indonesia)
7. Dian Wijayanto (Universitas Diponegoro, Indonesia)
8. Dini Wahyu Kartika Sari (Universitas Gadjah Mada, Indonesia)
9. Djumanto (Universitas Gadjah Mada, Indonesia)
10. Donghwa Chung (Seoul National University, Korea)
11. Eko Setyobudi (Universitas Gadjah Mada, Indonesia)
12. Faizal Rachman (Universitas Gadjah Mada, Indonesia)
13. Hamdan Syakuri (Universitas Jenderal Soedirman, Indonesia)
14. Ign. Hardaningsih (Universitas Gadjah Mada, Indonesia)
15. Indah Istiqomah (Universitas Gadjah Mada, Indonesia)
16. Indun Dewi Puspita (Universitas Gadjah Mada, Indonesia)
17. Ingrid van Putten (CSIRO, Australia)
18. Jeong-Ho Kim (Gangneung-Wonju National University, Korea)
19. Julie Ekasari (Institut Pertanian Bogor)
20. Latif Sahubawa (Universitas Gadjah Mada, Indonesia)
21. M. Saifur Rohman (Universitas Gadjah Mada, Indonesia)
22. Mala Nurimala (Institut Pertanian Bogor)
23. Mark Tamplin (Tasmania Institute of Agriculture, Australia)
24. Murwantoko (Universitas Gadjah Mada, Indonesia)
25. Namastra Probosunu (Universitas Gadjah Mada, Indonesia)
26. Noer Khasanah (Universitas Gadjah Mada, Indonesia)
27. Noranizan Mohd. Adzahan (Universiti Putra Malaysia, Malaysia)
28. Nurfitri Ekantari (Universitas Gadjah Mada, Indonesia)
29. Paul McShane (Monash University, Australia)
30. Ratih Ida Adharini (Universitas Gadjah Mada, Indonesia),
31. Riza Yuliratno Setiawan (Universitas Gadjah Mada, Indonesia)
32. Rustadi (Universitas Gadjah Mada, Indonesia)
33. Siti Ari Budhiyanti (Universitas Gadjah Mada, Indonesia)
34. Suadi (Universitas Gadjah Mada, Indonesia)
35. Susumu Ohtsuka (Hiroshima University, Japan)
36. Triyanto (Universitas Gadjah Mada, Indonesia)
37. Ustadi (Universitas Gadjah Mada, Indonesia)
38. Zuprizal (Universitas Gadjah Mada, Indonesia)

IOP Conf. Series: Earth and Environmental Science **139** (2018) 011004 doi:10.1088/1755-1315/139/1/011004

KEYNOTE SPEAKER ABSTRACTS

BIOPROSPECTING OF MICROBIAL SYMBIONTS OF MARINE INVERTEBRATES FOR SUSTAINABLE USE OF CORAL REEFS

Ocky Karna Radjasa
Department of Marine Science, Diponegoro University, Indonesia

The marine environment is the largest habitat on Earth, representing more than 70% of the surface of our planet. These diverse marine environments still remain largely unexplored, understudied and underexploited in comparison with terrestrial ecosystems and organisms The tropical marine environment is well-known for housing unique and diverse marine invertebrates. Natural products from marine invertebrates greatly expand the chemical diversity available for biotechnological exploitation. One of the most serious bottlenecks in developing natural products from marine sources during the past decades has been the availability of biomass and/or of optimized cultivation conditions to gain sufficient amounts of compound for preclinical and clinical studies. Among the unusual niches for novel microbes are marine invertebrates, which host hundreds of different bacterial groups and contain diverse symbionts. Marine invertebrates are sources of a diverse array of bioactive metabolites with great potential for development as drugs and research tools. Marine invertebrates are the target of the search for marine natural products because their associated microorganism communities occupy a unique niche in the ocean's biota. Since marine diversity also reflects chemical diversity, the isolation of the under-exploited symbiotic bacteria from marine invertebrates offers a great opportunity for discovering novel bioactive compounds based on screening against various disease targets with significant impact and strong potential for treatment of diseases. Therefore, it is reasonable to expect that exploration of untapped marine microbial diversity and resources will improve the rates at which new classes of secondary metabolites are discovered.

IOP Conf. Series: Earth and Environmental Science **139** (2018) 011004 doi:10.1088/1755-1315/139/1/011004

SYMBIOTIC BIOLOGY AND FISHERIES OF JELLYFISH IN SOUTHEASTERN ASIA

Susumu Ohtsuka
Graduate School of Biosphere Science, Hiroshima University, Japan

Large-sized jellyfishes can harbor a wide variety of eukaryotic symbionts, including protists, cnidarians, platyhelminths, nematodes, nudibranchs, polychaetes, barnacles, copepods, crabs, shrimps, ophiuroids, and fishes. These interspecific interactions can be ordinarily observed in the world's oceans and include dynamic changes in populations of the host and symbiotic groups, often in conjunction with developmental stages of the latter. These symbioses may be classified into two basic types based on the presence or absence of a trophic interaction between them. This can be confirmed through simple gut or stomach-content analyses of the symbionts, or can sometimes be supported by stable isotope analyses. In general, early life-stages of the symbionts tend to utilize host jellyfishes in a complex manner, although some caridean shrimps use the host mainly for breeding. In the case of the former, the host jellyfish plays a role in one or more combinations of vector, vehicle, food collector, food source, or predator shield. For instance, in Indonesia, Malaysia, Thailand and the Philippines, large scyphomedusae were frequently associated with megalopa larvae and juveniles of crucifix crab *Charybdis feriata*, young ophiuroids (e.g., *Ophiocnemis marmorata*), 0-year-old juveniles of shrimp scad *Alepes djedaba*, and juveniles and adults of caridean shrimps (e.g., *Latreutes anoplonyx*). In Southeast Asia, large rhizostome jellyfish are often harvested commercially as a component of Chinese cuisine. The main targeted species are *Rhopilema hispidum*, *R. esculentum*, *Lobonemoides robustus*, *Acromitus hardenbergi*, and *Crambionella helmbiru*. According to recent FAO data, the annual catch of jellyfishes can reach approximately 160,000 metric tons. Although the edible species may also harbor symbionts, fishermen have largely ignored the existence of these organisms, other than fish juveniles. Our data on the prevalence and mean intensity of symbionts in host edible jellyfishes off Thailand suggest they may severely inhibit normal recruitment, particularly the symbiotic ophiuroids due to their firm attachment to the host. In the case of ophiuroids, almost all individuals in jellyfish fishery captures seem to be killed. In contrast, juveniles of shrimp scad can easily escape from large-meshed nets, although relatively large-sized juveniles may enter the interstices of the oral arms of jellyfish when scared, and will be finally killed upon capture. Thus, jellyfish fisheries seem to reduce the numbers of protective hosts available for the symbionts. Therefore, we propose that jellyfish fisheries in tropical waters may greatly influence the recruitment of planktonic, benthic, and nektonic animals, and even the content of whole ecosystems.

IOP Conf. Series: Earth and Environmental Science **139** (2018) 011004 doi:10.1088/1755-1315/139/1/011004

AUSTRALIAN RECREATIONAL FISHERIES AND INDONESIAN SMALL SCALE FISHERIES: CHALLENGES AND SIMILARITIES

Paul E. McShane
School of Social Sciences, Faculty of Arts, Monash University, Clayton, Australia

As an archipelagic state, Indonesia has vast aquatic natural resources. Opportunities abound for its 250 million people. However, Indonesia faces three major challenges to sustainably develop its fisheries so as to provide economic and social wellbeing. It must eliminate destructive fishing practices (bombing and poisoning) and address other impacts on coastal ecosystems (e.g. pollution). It must improve cold chain management so that safe high quality seafood can be provided to meet domestic and international demand. It must find ways of effectively regulating fisheries. Indonesia has about 2 to 3 million people directly dependent on fishing for livelihoods. Typically, these fishers are small family-operated operations for subsistence or modest livelihoods. Australia has about 5 million persons who enjoy recreational fishing: fishing for sport or to get some food for their families. Both the Indonesian subsistence fishery and the Australian recreational fishery are, in effect, unregulated. In Australia, daily catch limits and size limits apply to certain popular recreational fish species. However, participation rates are unlimited which means that constraints on total catches are few. As for Indonesia, information on total catches and the effects of fishing on the biomass of targeted species, or on the ecosystem more generally (given increasing participation rates for recreational fishers) is scant. Yet Australia has made progress, largely through coordinated awareness raising campaigns, in encouraging conservative practices among the recreational fishing community. Similarly, education and awareness raising, harnessing traditional wisdom (where this aligns with conservative fishing practices) are key to eliminating destructive fishing practices and conserving vitally important fish stocks in Indonesia.

IOP Conf. Series: Earth and Environmental Science **139** (2018) 011004 doi:10.1088/1755-1315/139/1/011004

EQUIPPING COASTAL COMMUNITIES TO DEAL WITH CHANGE IN THE MARINE ENVIRONMENT

Ingrid van Putten
CSIRO Oceans & Atmosphere, Australia

People in coastal communities depend on the marine environment for many aspects of well-being such as food security, employment, but also less tangible benefits such as tradition and culture. To develop enduring information that will lead to suitable fisheries management approaches, the integration of natural, social and economic studies, together with stakeholder participation is needed. A more fragmented approach will not achieve the same level of information on feedback between oceanographic, biological change and coastal community impacts which is needed for effective management. For this purpose a comparison was undertaken of coastal communities and stakeholders dependent on marine resources. Survey data was collected from over 1,200 people in coastal communities in six different countries that are hotspots for marine change to uncover how socio-cultural and economic factors relate to sensitivity and how observational, attitudinal and experiential factors influence exposure. The links between exposure and vulnerability to change in the physical and biological environment is considered in different cultural contexts and marine resource management implications are drawn.

IOP Conf. Series: Earth and Environmental Science **139** (2018) 011004 doi:10.1088/1755-1315/139/1/011004

CHALLENGES ON FISH DISEASES STUDY IN INDONESIA

Murwantoko
Department of Fisheries, Faculty of Agriculture, Universitas Gadjah Mada, Indonesia

As an archipelago country, freshwater, brackish water and marine aquaculture in Indonesia is one of important main income. Increasing aquaculture production efforts usually faced problems of quality and quantity of the seeds, feeds, market potential, price and the most serious problem is fish diseases as well as decreased in the quality of aquaculture condition. Several fish diseases had led to serious economic and ecological losses to the national aquaculture business. Those diseases were due to parasites, bacteria, viruses infections. The challenges of disease in Indonesia are on the detail understanding of the disease that already exist in Indonesia and the threat of new disease. A more comprehensive study needs to be done to trace the origins of the white tail disease, to determine the mechanism of *Temnosewellia* infection to freshwater giant prawn, *Macrogyrodactylus* infection on African catfish. Some important diseases already exist in the country around Indonesia. Early mortality syndrome is an acute disease in shrimp that is known to spread in many countries including Malaysia and Thailand. Tilapia lake Virus has caused mass death in fingerling *Tilapia* has been known caused ocutbreak recently in Thailand in 2017. Those diseases may spread and threat to aquaculture as new diseases in Indonesia. Cooperation and a more comprehensive approach are needed to uncover the scientific facts of these diseases and of course for its control.

IOP Conf. Series: Earth and Environmental Science **139** (2018) 011004 doi:10.1088/1755-1315/139/1/011004

ANISAKID NEMATODES OF MARINE FISHES IN KOREA: PREVALENCE AND THEIR PATHOGENIC POTENTIALS

Jeong-Ho Kim
Gangneung-Wonju National University, Korea, Republic of South Korea

Anisakid nematodes belong to Family Anisakidae, comprising 10 genera with more than 60 species. They have indirect, complex life cycles involving various whale species as definitive hosts while marine crustaceans as first intermediate hosts, and fish and cephalopods as second intermediate or transport hosts. Humans can be accidentally infected with these worms by ingesting raw or undercooked fish or squids containing the larvae and the particular source of human infection varies according to the countries. Symptoms include mainly epigastric pain, nausea and vomiting, as well as allergic reactions. Morphological identification of these larvae is difficult because the morphological keys for larval species identification are lacking. Thus, most of the worms from human infection have been conventionally diagnosed as *Anisakis* spp. or *Anisakis simplex* (sensu lato). However, molecular approach has revealed that there are at least 9 species in Genus *Anisakis.* Of these, *A. simplex* (sensu stricto) is widely distributed in the northern hemisphere and the main cause of human infection, particularly in Asia. A sibling species, *A. pegreffii* sympatrically occurs with *A. simplex* in the northern hemisphere, and it is considered another important source of human *Anisakis* infection, particularly in Mediterranean Europe. We have been investigating anisakid nematodes fauna of marine fishes caught around Korean peninsula since 2011. The collected nematodes were identified by PCR-RFLP with subsequent sequencing, and the infection status was evaluated. Interestingly, *A. pegreffii* has been dominantly found from most of the fish species in this study. When these 2 different *Anisakis* species were examined for comparing physical invasiveness *in vitro*, these 2 species had no considerable difference. Moreover, their similar physical invasiveness was confirmed by *in vivo* experimental infection. We recently obtained trancriptomes of these 2 species by high throughput sequencing and employed bioinformatics to characterize them. When their potential allergen and protease gene profiles were compiled and compared, we found different number of putative allergens in each species, but also several novel *Anisakis* allergen genes in both species. For protease profiles, there was no considerable difference in each 2 species. All of these results suggest *A. simplex* and *A. pegreffii* may have no significant difference in terms of their pathogenicity against humans, although the different transcription levels of target genes should be validated at post-transcriptional level. This information will be helpful for epidemiological survey of human *Anisakis* infection.

IOP Conf. Series: Earth and Environmental Science **139** (2018) 011004 doi:10.1088/1755-1315/139/1/011004

LOVING THE UNDERLOVED SPECIES THROUGH PRODUCT INNOVATION

Noranizan Mohd. Azhadan
Universiti Putra Malaysia, Malaysia

Underloved fish or trash fish are often seen as low-value catch. The difference between a low-value catch and high-value catch is demand for the type and how the fish species is perceived. The list of what consumers like in food is endless, but there are certain trends that usually stands out. Consumers now want fresh, less-processed and semi-prepared raw materials to land in their kitchen as the home-cooking and brown bag lunch trend is back. The changing population demographics plays a big role in the shifting preference for source of protein in the diet. Taste and texture of a fish product can be manipulated using various techniques in product innovation. These techniques involve development of formulation, refining process selection and innovating packaging for the product. Incorporation of functional ingredients and application of suitable processing methods can be beneficial in making the underloved fish more palatable and attractive. Another critical aspect to consider is product promotion and consumer education. This aspect can result in a positive impact on the supply-demand cycle. Thus, the possibilities of transforming trash fish into treasure.

IOP Conf. Series: Earth and Environmental Science **139** (2018) 011004 doi:10.1088/1755-1315/139/1/011004

USE OF PREDICTIVE MODELS AND SENSORS TO EVALUATE THE SAFETY AND QUALITY OF SEAFOOD IN SUPPLY CHAINS

Mark Lewis Tamplin
Tasmanian Institute of Agriculture, University of Tasmania, Australia

Access to domestic and international markets highly depends on the performance of supply chains that must deliver safe and high quality seafood. This demand is driven by market forces that require companies to implement effective food safety systems across the whole of the supply chain, from production to consumer, and provide greater transparency about supply chain performance and inputs. Companies must also proactively manage expectations of customers, who are increasingly aware of safety. Sensors provide an efficient means to monitor and evaluate supply chain performance, and communicate information about food product characteristics and environments, all of which may have a positive influence on purchasing behaviour. For example, sensors that transmit information about supply chain temperature can influence product acceptance by wholesalers, food safety authorities, and even consumers. Also, many consumers increasingly prefer seafood produced with environmentally-sustainable practices and consideration for animal welfare. Research projects in Tasmania, Australia have demonstrated the value of predictive models and sensors in supply chains for live Pacific oysters (*Crassostrea gigas*) and fresh Atlantic salmon (*Salmo salar*). For oysters, certain importing countries, such as Japan and USA, impose market access limits for the bacterial pathogen, *Vibrio parahaemolyticus*. To help oyster companies better manage supply chains, models for *V. parahaemolyticus* and Total Viable Counts (TVC) growth rates were used to design and monitor supply chain performance. Results showed that *V. parahaemolyticus* replicated above 15°C, and was inactivated below 15°C. The resulting predictive models were extended across Tasmanian oyster companies through a quantitative microbial risk assessment, demonstrating 'sensitive' points in supply chains that maximize quality and safety. A second research project addressed growing markets for Atlantic salmon, and the benefits of predictive models and sensors to ensure optimum shelf-life and safety, including managing potential hazards caused by *Listeria monocytogenes*. Predictive models were successfully produced for fresh whole, head-on, gutted Atlantic salmon, specifically for TVC, *L. monocytogenes* and Quality Index Metrics (QIM), over a temperature range of 0-15°C. These research products provide oyster and salmon industries with tools to increase access to domestic and international seafood markets. In the near future, predictive models and sensors will be coupled with real-time reporting systems, providing seafood companies with important decision-making capabilities that expand their commercial markets.

IOP Conf. Series: Earth and Environmental Science **139** (2018) 011004 doi:10.1088/1755-1315/139/1/011004

MOLECULAR INTERACTIONS OF FISH GELATIN WITH POLYSACCHARIDES

Donghwa Chung
Graduate School of International Agricultural Technology, Institutes of Green Bio Science and Technology, Seoul National University, Korea

The interactions between proteins and polysaccharides are of increasing interest because they have significant relevance in important biological processes, including elastogenesis, enzyme channeling, and cytoplasm organization, and also have numerous applications in food, pharmaceutical, and cosmetic industries, including encapsulation, texturization, interfacial stabilization, gelation, fat replacement, meat mimesis, and protein recovery. Protein–polysaccharide interactions are either attractive or repulsive in aqueous solutions, depending on the molecular characteristics and mixing conditions. Attractive interactions occur mostly via electrostatic interactions when the two biopolymers are oppositely charged, leading to the formation of soluble or insoluble biopolymer complexes. The formation of insoluble complexes and their subsequent phase separation is called complex coacervation or precipitation for the separation of liquid-state insoluble complexes or solid-state insoluble complexes as fractal aggregates or electrostatic gel, respectively. Fish gelatin is a promising alternative to mammalian gelatin because it can be produced from fish processing by products and has no consumer concerns regarding bovine spongiform encephalopathy or ethnic/religious dietary restrictions. This presentation will discuss the basic features of molecular interactions between fish gelatin and polysaccharides and their applications in food systems.

IOP Conf. Series: Earth and Environmental Science **139** (2018) 011005 doi:10.1088/1755-1315/139/1/011005

INDEX

Peer review statement

All papers published in this volume of *IOP Conference Series: Earth and Environmental Science* have been peer reviewed through processes administered by the proceedings Editors. Reviews were conducted by expert referees to the professional and scientific standards expected of a proceedings journal published by IOP Publishing.

IOP Conf. Series: Earth and Environmental Science **139** (2018) 012001 doi:10.1088/1755-1315/139/1/012001

Effect of water irrigation volume on *Capsicum frutescens* growth and plankton abundance in aquaponics system

Y Andriani[1], Y Dhahiyat[1], Zahidah[1], U Subhan[1], Iskandar[1], I Zidni[1], T Mawardiani[1]

[1] Faculty of Fisheries and Marine Science, Universitas Padjadjaran, Jl. Raya Bandung–Sumedang KM 21, Jatinangor, Sumedang 45363, Jawa Barat, Indonesia
*Email: yuliyusep@yahoo.com

Abstract. This study aimed to understand *Capsicum frutescens* growth and plankton abundance in aquaponics culture. A Completely Randomized Design (CRD) with six treatments in triplicates comprising of treatment A (positive control using organic liquid fertilizer), B (negative control without fertilizer), C (drip irrigation aquaponics with a water debit of 100 ml/day/plant), D (drip irrigation aquaponics with a water debit of 150 ml/day/plant), E (drip irrigation with a water debit of 200 ml/day/plant), and F (drip irrigation aquaponics with a water debit of 250 ml/day/plant) was applied. The water used in treatments C, D, E, and F contained comet fish feces as fertilizer. *C. frutescens* growth and plankton abundance were observed. Analysis was conducted using analysis of variance for plant productivity and descriptive analysis for plankton abundance and water quality. The results of this study showed that the highest plant growth was seen in plants receiving F treatment with 50 ml/day drip irrigation. However, no significant difference was found when compared to the positive control with organic artificial fertilizer. Eleven types of phytoplankton and six types of zooplankton were found, with *Stanieria* sp. as the most abundant phytoplankton and *Brachionus* sp. and *Epistylis* sp. as the most abundant zooplanktons.

1. Introduction

Aquaponics system is one of the culture technologies that combine fish and plant cultures. This system excels because of its nature as an environmentally friendly system with more productive culture environment, producing fish and plants with a higher quality without using chemicals as fertilizer, pesticide, or herbicide [1].

One of the well-known aquaponics systems is the drip irrigation system. This system is known as a system that can save water through minimization of water loss that enables the system to be used in agricultural areas with limited water source. This creates an opportunity to broaden the market of the plants cultured using this system, including *Capsicum frutescens,* to include household, domestic, and export markets [2]. Currently, the demand for *C. frutescens* is quite high, i.e. around 4 kg/capita/year [3].

To meet the demand of the market for *C. frutescens* is not easy, especially with the low production of chili in Indonesia. Challenges faced in increasing the chili production in this country includes pest and disease attacks exaggerated by damaged soil due to toxic contamination of chemicals from the use of inorganic fertilizers or excessive use of pesticides. Hence, proper fertilizing process that pays more attention to soil conservation is needed [4].

Nutritional needs of plants can be satisfied using fish metabolism waste, in addition to the use of fertilizers. This can be attained by combining fish culture and plant culture in an aquaponics system. Both fish for consumption and ornamental fish can be used for this purpose.

In an aquaponics system, the waste produced by the fish is used as the fertilizer for the plants [5]. Fish produces 80–90 % of ammonia through osmoregulation process. Fish feces and urine contain around 10–20 % of total nitrogen [6]. The nutrients contained in the fish waste will be different, depending on the feed used. A preliminary study has shown that comet fish fed with feed containing 35 % of protein and a feeding rate of 5 % of its body weight produces feces with 3.40 % of N, 6.08 % of P_2O_5 and 0.05 % K_20, *C. frutescens* absorbs 70 kg·ha^{-1} N, 16 kg·ha P_2O_5, and 92 kg·ha^{-1} K_2O [7]. By this in mind, this study focused on the relationship between the nutrients produced in comet fish feces and the growth of *C. frutescens* in a drip irrigation aquaponics system.

2. Materials and Methods

2.1. Time and place of study
This study was performed in the period of March to June 2016 at the Aquaculture Wet Laboratory, Ciparanje, Faculty of Fishery and Marine Science, Universitas Padjadjaran, Indonesia. The water quality measurement was performed at the Aquatic Resources Management Laboratory of Universitas Padjajaran and a preliminary testing was performed in the Soil Fertility and Plant Nutrition Laboratory of the Faculty of Agriculture, Universitas Padjadjaran, Indonesia.

2.2. Materials and instruments
The main instruments used during this study were aerator, concrete pond, plankton net, and water quality measurement tool (Lutron pH-207). The materials used were (a) fish, i.e. 5-month-old of *Carassius* sp. with an average length of 9 cm and average weight of 6 g, (b) three-week-old *C. frutescens*'s seeds, (c) cocopeat as the growing media (JM Tani), (d) *FFF 999* commercial feed (PT Central Pangan Pertiwi) with a crude protein, fat, fiber, ash, and maximum water contents of 35 %, 2 %, 3 %, 13 %, and 12 %, respectively (e) *Cobra* organic liquid fertilizer which was used every 10 days (Bharti Bio Chem) with the following compositions: 5.60 % N, 2.70 % P_2O_5, 3.23 % K_2O, 0.3 % C-Org, 0.18 % CaO, 0.20 % MgO, 0.70 % SO_4, 0.21 % NH_4, 0.03 % B, 0.04 % Cl, 33.17 ppm Zn, 40.38 ppm Cu, 0.2 ppm Mo, 67.34 ppm Mn, and 75.27 ppm Fe, (f) reagent solutions comprising of $MnSO_4$ and H_2SO_4 reagents for dissolved oxygen titration, 0.5 Nestler and 1 ml Siegnet_for ammonia testing, SnCl and NH_4 molybdate for phosphate testing, and phenol and NH_4OH 10 % for nitrate testing (Sigma).

2.3. Procedure

2.3.1. Aquaculture pond preparation
The culture pond used in this study was a concrete pond of 3 m × 2 m × 1.1 m. The preparation of the ponds included pond cleaning after drying and then water filling.

2.3.2. C. Frutescens planting
The planting of *C. frutescens* was started by seed selection and seeding. *C. frutescens* seeds were soaked in warm water for 30 minutes to clean the seed and the seeding was done using trays. Seeding was performed to select seeds, separating normal from abnormal or diseased ones. In addition, seeding was also used to prepare the seeds until they were strong enough to be planted in a bigger site. The *C. frutescens* seed was considered ready to be planted in a bigger growing media when it grew four or more leaves.

2.3.3. Drip irrigation aquaponics system
In this study, the culture pond was a concrete pond of 3 m × 2 m × 1.1 m with a total density of comet fish of 430 fishes, and a water depth of 60 cm. The pond was equipped with aerators for oxygen supply and a clear plastic roof to prevent rain water falling into the aquaponics media. Polybags were filled with coco peat and *C. frutescens* plants were planted in the bags that were then arranged around the pond with a space of 20 cm between each polybag. The water from the culture media was pumped and distributed through a PVC pipe that was installed to match the form of the pond (figure 1).

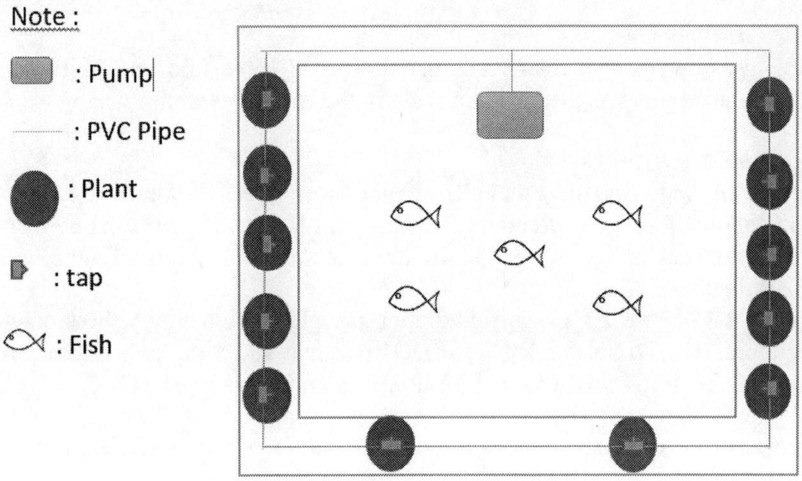

Figure 1. Schematic aquaponics design used in this study.

2.4. Steps of study
A three month period was assigned for this study. Fishes were fed three times a day with a feeding rate of 5 % of the total fish biomass. Fish weighing was done every two weeks. Water quality measurement that included Dissolved Oxygen (DO), temperature, pH, phosphate, nitrate, ammonia, and plankton abundance measurement was conducted once every 7 days. This study used an outdoor location; hence, the sun was used as the source of light. Data on *C. frustescens* were collected by measuring the height of the plant and the number of leaves throughout the study period.

2.5. Design experiment
This study was an experimental study using Completely Randomized Design (CRD) that consisted of six triplicate treatments. The treatments in this study were:
Treatment A = positive control (use of organic liquid fertilizer)
Treatment B = negative control (without organic liquid fertilizer)
Treatment C = irrigation volume of 100 ml/day/plant
Treatment D = irrigation volume of 150 ml/day/plant
Treatment E = irrigation volume of 200 ml/day/plant
Treatment F = irrigation volume of 250 ml/day/plant

2.6. Observed parameters

2.6.1. Comet fish feces content analysis
Feces analyzed in this study was the feces of Comet fish that fed with 5 % of feeding rate. Analysis of total Nitrogen, P_2O_5 content, and K_2O content in feces samples were conducted. The nitrogen in the feces was measured using Kjehdal method while the P_2O_5 and K_2O contents were measured using

Olsen method and spectrophotometry, respectively [8]. The results were then compared to the nutrient requirement of the plant using the basic formula of fertilization [9].

$$Recommended\ dose\ of\ fertilizer\ plot = Plot\ size\ x\ 1\ ha\ size^{-1}\ x\ Dose \qquad (1)$$

Notes:
Per plot = per plant
Plot size = polybag size (30 × 30)
1 ha size = soil/cocopeat mash

2.6.2. C. frustescens Plant Growth
Plant height was measured using a ruler while the number of leaves, flowers, and fruits were counted manually. Fruit produced was weighed using a scale in the end of research.

2.6.3. Water Quality
The measurement of water quality included temperature, pH, dissolved oxygen (DO), nitrate, phosphate, and ammonia level measurements. The parameter measurement was performed once every seven days since the beginning of the study until the end of the study through titration and spectrophotometry methods.

A pH meter (LUTRON PH-207) was used to measure pH and a mercury thermometer was used to measure temperature. The Dissolved Oxygen (DO) was measured through titration by applying 21 drops of O_2 reagent, 21 drops $MNSO_4$, and 42 drops of concentrated H_2SO_4 on the sample that was then placed into the Erlenmeyer flask (50 ml) for titration until the color changed. Nitrate, phosphate, and ammonia contents were measured using a spectrophotometer (Thermo Scientific Genesys 20).

2.6.4. Phytoplankton and Zooplankton Densities
Sampling was done using a plankton net 25 mikron size. Phytoplankton samples was stored at around 5 °C to slow down physical and chemical reactions that will cause deterioration of species cell structure. Before counting, samples was acclimatized at room temperature for 12 hours for obtaining random distribution of species in counting chamber as well as for preventing air bubble formation. Sample volume of 1 ml was observed using Sedgewick-Rafter Counting Chamber method [10]. Sample was observed under microscope to determine plankton density using the following formula [11]:

$$\mathrm{N} = \mathrm{n}\ x\ Vr\ Vo^{-1}\ x\ Vs^{-1} \qquad (2)$$

Notes:
N = number (cell·m^3)
n = number of identified cells
Vr = volume of filtered water (50 ml)
Vo = volume of water observed in *counting chamber* (1 ml)
Vs = volume of filtered water (25 ml)

3. Results and Discussion

3.1. Chemical content of Comet fish feces
The Comet fish feces contained 3.40 % N, 6.08 % P_2O_5, and 0.05 % K_2O. Based on the values, the N and P levels in the water were calculated and adjusted to the plant's requirement.

Table 1. Comparison of comet fish chemical contents and the nutrient requirement for *C. frutescens.*

Content	Feces (%)	Calculation result (ppm)	Plant's Requirement (ppm)
N	3.4	1.19	0.303
P	6.08	2.128	0.263
K	0.05	0.0175	0.169

Comet fish cultured in this study was 2.5 kg with an average weight of 6.72 ± 3.23 g. Based on that stocking density, the disposed feces in the water was estimated to be 110 g. Hence, it was estimated that the nitrogen and phosphate levels in the water were 1.19 ppm and 2.128 ppm, respectively (table 1) The nutrient requirement was estimated to be 0.303 nitrogen, 0.263 phosphate, and 0.169 ppm potassium.

However, based on the water quality test performed once every seven days, the nutrient content in the comet fish culture media could not meet the total need of *C. frutescens*. This has inhibited the plant to grow well. The difference in the water quality test result and calculation result was suspected as the result from the challenge in the nitrogen cycle in the culture pond or the small amount of feces in the body of water, making the dissolved organic materials became less. The inadequate amount of nutrients received by the plants has led to low and abnormal-like growth. Furthermore, according to [12], plants that do not receive enough nitrogen will slow growth with pale leaves. If the deficiency is very severe, the leaves will turn color into light green and yellow [9]. This is in line with the results of this study where the *C. frutescens* plant grew slowly and had yellowish leaves (figure 2).

Figure 2. *C. frutescens*' leaves that lacks of nutrient (left) and well-nourished leaves (right).

Furthermore, the limited amount of nutrients in the *C. frutescens* planting media was also influenced by the use of coco peat as the planting substrate. When cocopeat is used, the water distributed through the irrigation drops cannot not hold for a long time because of higher porosity than that of soil. This, eventually, will reduce the amount of nutrients that can be used by the plant.

3.2. C. Frutescens growth

Plant growth can be identified from the increase in the stem length (plant height), number of leaves, number of flowers, and fruit weight. Chili plant has two phases of growth, i.e. vegetative phase and generative phase. The vegetative phase is marked by increased plant height and increased number of leaves while the generative phase is marked by flower and fruit growth.

3.2.1. Plant height

The vertical growth is identified from the increased height of plant's stem. This growth is also considered as the characteristics of the vegetative phase. The increase in plant height for each treatment ranged between 15–27 cm during the three months of study (figure 3).

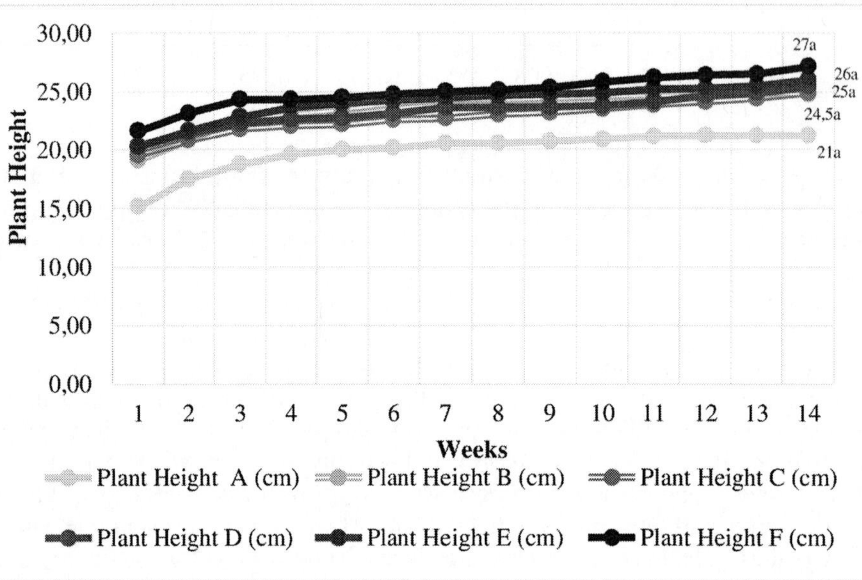

Figure 3. Vertical growth of *C. frutescens* planted in aquaphonics system with various irrigation volume. (A) Positive control; (B) Negative control; (C) Irrigation volume of 100 ml/day/plant; (D) Irrigation volume of 150 ml/day/plant; (E) Irrigation volume of 200 ml/day/plant; (F) Irrigation volume of 250 ml/day/plant.

Based on the above chart, the lowest vertical growth was seen in treatment A, which used organic fertilizers, as the positive control. This was because the plants in treatment A were attacked by aphids more often than those in other treatment groups. The plants in treatment B, which was the negative control without any fertilizer, were higher than the plants in treatment A, which was around 25 cm high. It was suggested that the water in Treatment B group was mostly used to increase the plant height.

Treatment F group, which was the group that used the highest amount of dripped water produced the highest height of 27 cm. It was suggested that the amount of nutrient absorbed by the plants in this group was higher due to the bigger volume of water. The followings are the results of the statistical testing using analysis of variance (table 3).

Based on the statistical analysis, the plant growth did not show any significant difference between treatments. The nutrient with the most influence towards vertical growth is phosphate. In plants, phosphate is used to form roots, accelerate fruit maturation, and strengthen the plant stem [13]. This study showed that the phosphate content in the pond and in the irrigation drip tap during the study ranged between 0.025–2.187 $mg \cdot L^{-1}$ and 0.077–0.797 $mg \cdot L^{-1}$, respectively. The phosphate content in the comet fish culture pond was adequate for plant growth.

3.2.2. Leaf growth

During this study, the leaf growth ranged between 8–66 leaves. The lowest leaf growth was seen in treatment B and the highest leaf growth was seen in treatment A, where Treatment B showed a higher vertical growth compared to the Treatment A. This was caused by the priority in water use which, in

Treatment B, was prioritized more on vertical growth than on adding leaves. Below is the chart depicting leaf growth during the study (figure 4).

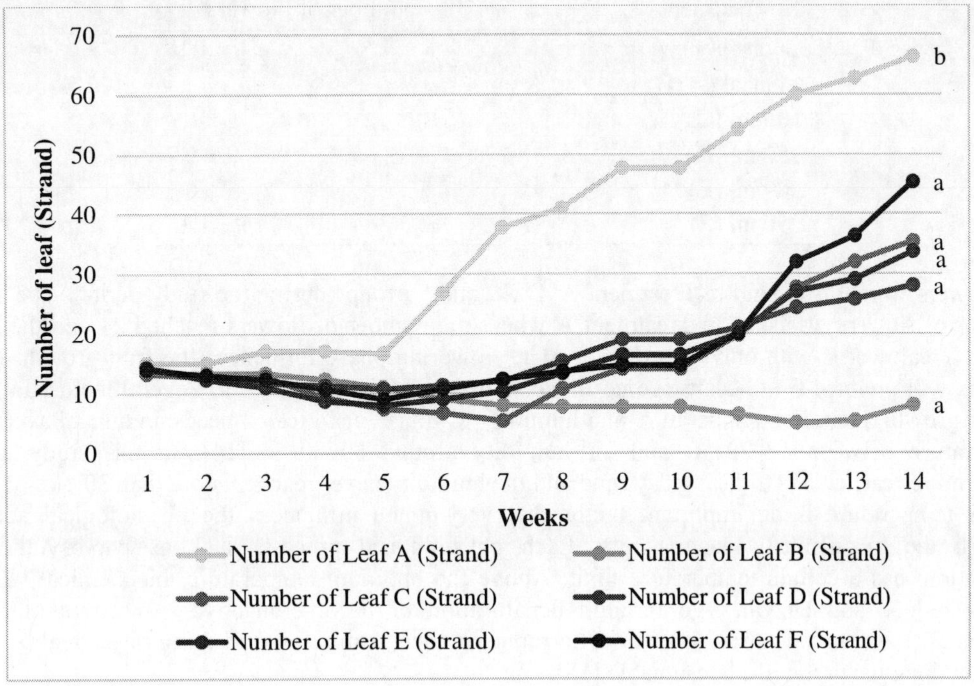

Figure 4. Average number of leaf of *C. frutescens* planted in aquaphonics system with various irrigation volume. (A) Positive control; (B) Negative control; (C) Irrigation volume of 100 ml/day/plant; (D) Irrigation volume of 150 ml/day/plant; (E) Irrigation volume of 200 ml/day/plant; (F) Irrigation volume of 250 ml/day/plant.

Based on the chart, it is obvious that the heighest leaf growth was seen in treatment A group. In Treatment B group, the leaf growth was reduced because the number of leaves shed was higher due to inadequate nutrition. Among the treatment groups, Treatment F group was found to be the group with the most leaves, which was proportional to the vertical growth of this group.

The statistical analysis results showed that the leaf growth in Treatment A was not significantly different from Treatment F but significantly different from treatment B, E, and D while treatment F was not significantly different from treatment B, E, and D. This was apparent from the fact that treatment A had the highest average leaf growth, which was followed by treatment F. Hence, control treatment showed a better result compared to other treatments because the plants in treatment A group received adequate nutrition from the artificial fertilizer while the plants in treatment F received more nutrition compared to other treatments because the number of drips and water volume were higher. The most influential compound for leaf growth is nitrogen, which is in line with the result of this study where treatment F group had the biggest nutrient-containing water volume, meaning that more N was provided and more leaves were produced. Lack of nitrogen will make the leaves turn yellowish and slow down the growth, making the leaves weak and easily shed [13].

3.2.3. Flower growth

Flower is one of the characteristics of the generative (reproductive) phase in a plant. In this study, not all treatment groups produced flowers. The number of flower is presented in the following table (table 2).

Table 2. Average flower number of *C. frutecens* planted in aquaphonics system with various irrigation volume.

Treatment	Number of Flowers
Control (+) (A)	24 ± 1.18
Control (-) (B)	-
100 ml (C)	-
150 ml (D)	9 ± 0.7
200 ml (E)	5 ± 0.7
250 ml (F)	15 ± 1.15

Flowers were only found in Treatment A, D, E, and F groups during the study period. The highest number of flowers was seen in Treatment A where the number of flowers reached 24 and the lowest was in Treatment E with only five flowers. The flowering phase influences the fruit growth. Not all flowers will produce fruit because some flowers shed and fail to form fruit. Several factors influence flower growth, including temperature and humidity. Chili (*C. frutescens*) needs an annual average air temperature between 18–30 °C and air humidity around 60–80 % [14]. In this study, the air temperature reached 32 °C during day time and the humidity never reached more than 30 %.

Air temperature is an important factor that very much influences the physiological activities through the biochemical reaction rate. Each physiological process, such as photosynthesis or respiration, has a certain temperature limit. Above the optimum temperature, biochemical reactions slowly reduce and enzymes go through denaturation or become inactive. Air temperature also influences flower growth, fruit formation, hormone balance, maturation and aging rates, quality, result, and how long the product is consumable [15].

3.2.4. Fruit growth

The main product component of *C. frutescens* culture is the fruit. The fruit of *C. frutescens* can be harvested after 80–90 days or two or three months after planting. The fruit is initially green then turns into red in several weeks. The following table lists the fruit growth during the study (table 3).

Table 3. Fruit growth of *C. frutescens* planted in aquaphonics system with various irrigation volume.

Treatment	**Number of Fruits**	**Weight (gr)**
Control (+) (A)	6 ± 0.63	2,3
Control (-) (B)	-	-
100 ml (C)	-	-
150 ml (D)	3 ± 0.43	1,4
200 ml (E)	-	-
250 ml (F)	2 ± 0.24	1,2

From the data in table 3, it is apparent that the highest number of fruits was seen in treatment A group with six chilies, followed by treatment D with three chilies and treatment F with two chilies. The fruit growth may also influence the final weight of the plant. The more fruit produced, the higher the possibility that the fruit weight is higher. The plants in treatment A and D groups experienced faster fruit growth than the plants in treatment F; therefore, the number of fruits and weight were higher than those in treatment F.

The difference in flower and fruit growth was also seen in treatment D, which was suspected to had undergone a stress phase, or a phase where the plant feels threatened by the environment. This might be due to the lack of water. Lack of water makes plants divert their energy into flower growth and, eventually, produce fruits. This happens when the amount of water absorbed is not enough for photosynthesis process because of the closing of stomata [16]. Treatment F received more water but the nutrition was not fully adequate that the amount nutrition absorbed by the plants was used to prolong the vegetative phase of the plants, making them entered the generative phase later than the plants in treatment D.

3.3. Plankton abundance

Besides the artificial feed, natural feed was also available for the comet fish in the form of planktons. Planktons grew in the water because the water contains phosphate. Phosphate is one of the factors that influence plankton growth in the water because planktons absorb nutrition in the water, and may make the nutrition not sufficient for the plant growth. Nitrate and phosphate are important nutrients in plankton growth and metabolism. In general, plankton is one of the indicators to evaluate the quality and fertility level of water [11]. Results of the plankton observation are shown below (table 4).

Table 4. Phytoplankton abundance in comet fish culture media (cell·ml^{-1}).

PHYTOPLANKTON	CLASS	TOTAL NUMBER OF INDIVIDUALS
Synendra sp	*Bacillariophyceac*	4
Phormidium sp	*Cyanophyceae*	16
Paramechium sp	*Oligohymenophoreae*	2
Chlorella sp	*Chlorophyceae*	8
Scenedesmus sp	*Chlorophyceae*	26
Eudorina sp	*Chlorophyceae*	26
Oscillatoria sp	*Cyanophyceae*	2
Tribonema sp	*Xanthophyceae*	4
Chlosterium sp	*Chlorophyceae*	4
Euglena sp	*Euglenaphyceae*	10
Stanieria sp	*Cyanophyceae*	1,240

Table 5. Zooplankton abundance in comet fish culture media (cell·ml^{-1}).

ZOOPLANKTON	CLASS	TOTAL NUMBER OF INDIVIDUALS
Brachiounus sp	*Eurotatoria*	10
Tetrahymena sp	*Ciliphora*	2
Plumatella sp	*Phylatolaeata*	6
Monostyla sp	*Monogononta*	2
Rotifera sp	*Eutrotatoria*	2
Epistylis sp	*Ciliata*	8

Tabel 4 shows that most of the phytoplanktons seen was the *Stanieria* sp. This phytoplankton is a *Cyanophyta* plankton, which is also categorized as cyanobacteria. The cyanobacteria planktons also play a role in adding organic materials so that other organisms can grow. In water ecosystem, cyanobacteria act as producers for other organisms such as zooplanktons, small fishes, or small shrimps due to their ability in photosynthesis.

The most abundant zooplanktons were *Brachionus* sp and *Epistylis* sp. The later one is known as one of the diseases in fish. The presence of planktons can be beneficial if the plankton in the water by adding nutrients for the plants through the microbial decomposition of dead planktons, producing compost that is beneficial for plants.

4. Conclusion

1. The nutrient contents produced in this drip aquaponics system include 0.032–1.133 ppm of nitrogen and 0.077–0.736 ppm of phosphate. This shows that the nutrient components in this system are not adequate to meet the nutrient requirement of *C. frutescens*, making the growth represented by plant height, number of leaves, number of flowers, and number of fruits lower than the positive control treatment.
2. The best result is gained through 250 ml/day/plant of drip water; however, the result is not better than the result in the positive control treatment.
3. The most frequently found phytoplankton is *Stanieria* sp., which is a *Cyanophyta,* while the most abundant zooplanktons are *Brachionus* sp. and *Epistylis* sp.

5. Acknowledgement

This manuscript is based on a study supported by the Academic Leadership (ALG) Program of Universitas Padjadjaran 2016.

References

[1] Diver S 2006 *Aquaponic-Integration of hydroponic with aquaculture* (Montana: National Centre for Appropriate Technology Agriculture Specialist) p 28
[2] Rukmana R 2002 *Usaha Tani Cabai Rawit* (Yogyakarta: Kanisius) p 88
[3] Warisno, K Dahana 2010 *Peluang Usaha dan Budidaya Cabai* (Jakarta: Gramedia Pustaka Utama) p 128
[4] Poerwanto, R, A D Susila 2014 *Teknologi Hortikultura* (Bogor: PT Penerbit IPB Press) p 380
[5] Wahap N, A Estim, A Y S Kian, S Senoo and S Mustafa 2010 *Aquaponics Journal* **58** 28–32
[6] Rakocy J E, T M Losordo and M P Masser 1992 *Southern Regional Aquaculture Center Pub.* **454** 1–7
[7] Halliday D J, M E Trenkel 1992 *IFA World Fertilizer Use Manual* (Paris: International Fertlizer Industry Association Cop.) p 632
[8] Iatrou M, A Papadopoulos, F Papadopoulos, O Dichala, P Psoma, and A Bountia 2014 *Journal Communication in Soil Science and Plant Analysis* **45** 2207–2214
[9] Wijayani A, D Muljanto, Soenoeadji 1998 *Ilmu Pertanian* **6** 8–13
[10] Erdogan S and M B Yerli 2014 *Phytoplankton Counting and Identification Methods* (Turkey: Technical Assistance for Capacity Building on Water Quality Monitoring) p 26
[11] Fachrul F M, H Haeruman and L C Sitepu 2005 *Komunitas Fitoplankton sebagai Bio-indikator Kualitas Perairan Teluk Jakarta* (Jakarta: Seminar Nasional FMIPA–Universitas Indonesia) 24-26 November
[12] Marschner H 1986 *Mineral Nutrition of Higher Plants* (London: Academic Press Inc) p 889
[13] Hardojiwigeno S 2007 *Ilmu Tanah* (Jakarta: Akademika Pressindo) p 296
[14] Cahyono B 2003 *Cabai Rawit Teknik Budidaya dan Analisis Usahatani* (Yogyakarta: Kanisius) p 112
[15] Knott J E and J R Deanon Jr 1967 *Vegetable production in Southeast Asia* (Phillipines: Univ. of Phillipines College of Agricultural College) 97–133
[16] Akinci S, Losel D M 2012 Plant water-stress response mechanism *Water Stress* ed I M Rahman (Croatia: InTech) chapter 1 pp 1-42

IOP Conf. Series: Earth and Environmental Science **139** (2018) 012002 doi:10.1088/1755-1315/139/1/012002

The performance analysis of hybrid seeds between catfish (*Clarias gariepinus* Burchell) semarang and sangkuriang strains

F Basuki[1]*, T Susilowati[1] and D Harwanto[1]

[1] Aquaculture Department, Faculty of Fisheries and Marine Science Faculty, Diponegoro University, Jl. Prof. H. Soedharto S.H, Tembalang, Semarang, Indonesia
*Email: fajar.basuki@live.undip.ac.id

Abstract. This study aimed to analyze the performance of hybrid seeds between catfish (*Clarias gariepinus* Burchell) semarang and sangkuriang strains by parameters of fecundity, length and weight gain and seeds survival rate. The material used in this study was the parent of catfish semarang and sangkuriang strains. The research method used was experimental method using Completely Randomized Design (RAL), consists of four treatments and three replications. Those were Treatment A (semarang/HH ♀ × semarang/HH ♂), Treatment B (semarang/HH ♀ × sangkuriang/SS ♂), Treatment C (sangkuriang/SS ♀ × sangkuriang/SS ♂) and Treatment D (sangkuriang/SS ♀ × sangkuriang/SS ♂). The results showed that fecundity of most catfish eggs were obtained from treatment A, B, C and D. The hatching rate of catfish eggs from various treatments was not significantly different. Performance of seeds with length and weight parameters showed that newly hatched larvae until the age of one-week-feeding catfish are not significantly, the difference in seed performance began to appear on the 14th day of feeding. The order of the best treatment result were C, B, D and A. The difference was more significant at 28 days of feeding.

1. Introduction

Catfish is one of the superior products of freshwater aquaculture because of its good taste, source of protein and preferred by all levels of society, especially in Central Java. This fish is suitable to be developed to increase food security as food demand will continue to increase along with the increase in population especially in Central Java [1]. The production of *Clarias gariepinus* catfish (Burchell, 1822) in Central Java is also continuously improved. The production of catfish in Central Java reached 62,686 tons and 75,236 in 2012 and 2013, respectively or increased by 20.02 % [2]. The production of catfish in Central Java in 2016 amounted to 122,292.16 tons [3]. The high demand for seeds has not been followed by improved seed quality. This condition is indicated by increase in length of fish harvesting time, as has been illustrated that in 1985 in the first year the developed catfish showed that seeds measuring 3–5 cm were stocked then within 60 days it was harvested. However, gradually there has been a decrease in growth, on 2000 when the 3–5 cm seeds size were stocked then the harvest time has reached 90 days [4].

One effort that has been made to improve the performance and increase the ability of catfish production is hybridization program. Hybrid excellence is showing a faster rate of growth than either parent, or has other characteristics, which are often economically significant to a better quality than that of parent [5].

IOP Conf. Series: Earth and Environmental Science **139** (2018) 012002 doi:10.1088/1755-1315/139/1/012002

For hybridization, cross-breeding is conducted between strains. This study used the parent of catfish semarang and sangkuriang strains reciprocally. The purpose of this research was to analyze the performance of cross-breeding seeds between the catfish of semarang and sangkuriang strains with the parameters consisting of egg fecundity, the length and weight gain and the seeds survival rate.

2. Materials and Methods

2.1. The test fish

The parent of catfish (*C. gariepinus* Burchell) sangkuriang strain. Sangkuriang strain was a result of catfish selection of individual and backcrossing so its growth was very well and have been released by the Ministry of Maritime Affairs and Fisheries, Republic of Indonesia through the Ministry of Marine and Fishery No. KP.26/MEN/2004 dated July 21, 2004. Whereas, the catfish of semarang strain has been located in semarang more than 27 years and done the simple selection: every time of spawning, it was taken the biggest in size. The body weight of catfish semarang and sangkuriang strains range between 1.10–1.45 kg.

2.2. Spawning pond

A 2 m^2 × 3 m^2 pond with a height of 0.60 m made of tarpaulin with 12 pieces of bamboo.

2.3. Experimental design

The experimental design used was Completely Randomized Design (RAL) with four treatments and three replications i.e. A (semarang/HH ♀ × semarang/HH ♂), Treatment B (semarang/HH ♀ × sangkuriang/SS ♂), Treatment C (sangkuriang/SS ♀ × sangkuriang/SS ♂) and Treatment D (sangkuriang/SS ♀ × semarang/SS ♂). In order know the growth of larvae to be a seed, every one week the length and weight of the body was measured. Food given to larvae was egg yolks for 3–5 days old larvae, silk worm for aged 5–15 days of larvae, fine pellets for seed aged 15–28 days.

Gonadosomatic Index (GSI) calculation fecundity, hatching rate (HR) and weight and length gain are measured as follows:

Gonadosomatic Index (GSI) = gonad mass / total body mass × 100 %
Fecundity = No. of eggs in sub sample × Gonad weight Fecundity (F1) / Weight of sub sample
Hatching Rate (HR) = (Number of eggs hatched / Total number of eggs in a batch) × 100 %

Growth in weight and length were measured for 28 days of breeding, while the growing size of the fish by using the formula of [8]:

- *Growth of Individual Weights / Absolute Weight (g/each) = the average final weight (g) – the average initial weight (g)*
- *Growth of Individual Length / Absolute Length (mm/each) = the average final length (mm) – the average initial length (mm)*

2.4. Data analysis

The data obtained as fecundity of obtained eggs, hatching rate (%), and performance hybrid seeds development were analyzed using the analysis of variance, as for the analysis tool of variance using the program MSUSTAT.

3. Results and Discussion

3.1. Fecundity of obtained eggs

As mentioned in the experimental material that the parent catfish used in this research is the parent catfish semarang strain with weight ranged from 1.1–1.45 kg, while the sangkuriang strain ranged between 1.1–1.35 kg.

The egg weights obtained from treatment A, B, C and D were 209.02 ± 30.45; 202.43 ± 19.49; 188.77 ± 17.14 g and 168.55 ± 19.81 g, respectively. Meanwhile, the fecundity were 146.314 ± 21.315; 141.701 ± 13.643; 132.139 ± 11.998 and 117.985 ± 13.867 eggs, respectively (table 1).

Table 1. Eggs Weight /Parents used for breeding in this study (g).

Replication	Treatment			
	A	B	C	D
1	218.7	201.25	174.8	183.6
2	174.9	182.85	183.6	175.95
3	233.45	223.2	207.9	146.1
Total	627.05	607.3	566.3	505.65
Average	209.02 ± 30.45^a	202.43 ± 19.49^a	188.77 ± 17.14^a	168.55 ± 19.81^a

The result of gonadosomatic index calculation at treatment A, B, C and D were 16.07 ± 0.15; 16.03 ± 0.12; 15.30 ± 0.10 and 15.23 ± 0.12 %, respectively (table 2). The analysis showed no significant differences between treatments.

Table 2. GSI (Gonadosomatic Index) of brood stock used in this study.

Replication	Treatment			
	A	B	C	D
1	16.2	16.1	15.2	15.3
2	15.9	15.9	15.3	15.3
3	16.1	16.1	15.4	15.1
Total	48.2	48.1	45.9	45.7
Average	16.07 ± 0.15^a	16.03 ± 0.12^a	15.30 ± 0.10^a	15.23 ± 0.12^a

The results showed that the catfish both the semarang strain and the sangkuriang strain are sufficiently feed and healthy, *C. gariepinus* Burchell has gonadosomatic index (GSI) of 14.68 ± 4.86 indicating that the fish was healthy, feeding and the water temperature were quite sufficient [9]. In general, catfish *C. gariepinus* Burchell has a gonadosomatic index (GSI) of 15–20 % [10].

3.2. Hatching rate (%)

The percentage of hatching rate of cross-breeding eggs is shown in table 3. A treatment produced hatching rate of 80.48 ± 1.12 %, treatment B produced hatching rate of 80.55 ± 0.45 %, treatment C produced hatching rate of 82.26 ± 1.16 % and treatment D produced hatching rate of 79.93 ± 1.58 %.

Table 3. Hatching Rate eggs produced by hybridization of semarang and sangkuriang catish.

Replication	Treatment			
	A	B	C	D
1	79.21	80.23	81.12	79.34
2	81.32	81.11	82.24	80.22
3	80.91	80.32	83.43	80.22
Total	241.44	241.66	246.79	239.78
Average	80.48 ± 1.12^a	80.55 ± 0.45^a	82.26 ± 1.16^a	79.93 ± 1.58^a

The results showed that fertilization and hatching results were not significant with high hatching rate more than 80 % due to enough parents' weight, health and suitable environment. It is in accordance to an opinion [11], who stated that the heavier the parent catfish spawning produces the higher hatching rate of eggs. High hatching rate of eggs indicates that nutritional content of oocytes in the gonads is enough for physiological processes of embryonic development for hatching. The time of oocyte growth, there will be an accumulation of the substance which is necessary for the early development of the embryo [12]. VTG absorption by the embryo explained that the yellow protein in the embryo covered by a *synsytial* layer or *periblast*, two *coelemic* mesoderm leaves, *splanchnopleure* and *somatopleure* and epidermis. Viteline vascular tissue develops in *splanchnopleure* and by using these networks, embryo can use protein stored in the egg yolks period. To break up the yolk globule, on the surface of *periblast*, there are *viteolisis* zones by the activity of *thiol* proteinase that is *cathepsin L*. The increase in activity of the enzyme causes yolk *globula* torn from the yolk period, so that the protein can be absorbed [13], and that has happened to yolk metabolism in embryos and larvae [14].

3.3. Performance hybrid seeds development

There was no significantly different of weight and length of newly-hatched larvae among treatments (figure 1). The absorption of the yolk by the embryo in very short time could not provide the significant different of larvae weight although there is some genetic differences.

Results of a study on the newly-hatched larvae found that larvae length ranged from 5.73 ± 0.71 mm to 6.47 ± 0.61 mm from various treatments. The newly hatched-larvae in this study is longer than that of reported in the previous study [15] which has total length of 3.5–4.0 mm.

The observation result of weight and length gains on the 3rd day or at the time the larvae begin to eat can be seen in figure 1 and 2. The weight and length of the larvae were not significantly different at the 3 days old. It is assumed egg yolk has not been used for weight gain but it has only used to develop the larvae for the completeness of its digestive system.

The observation of weight and length gains on the 7th day of feeding or the 10th day of larvae can be seen in Figures 1 and 2, it can be shown that the weight and length of the larvae of the various treatments showed no significant difference. It is assumed that the use of feed from the outside has not been able to significantly accelerate the growth, or the larvae still adapt from the intake of feed from the outside. Consequently the growth of the larvae of the various treatments is not significantly different.

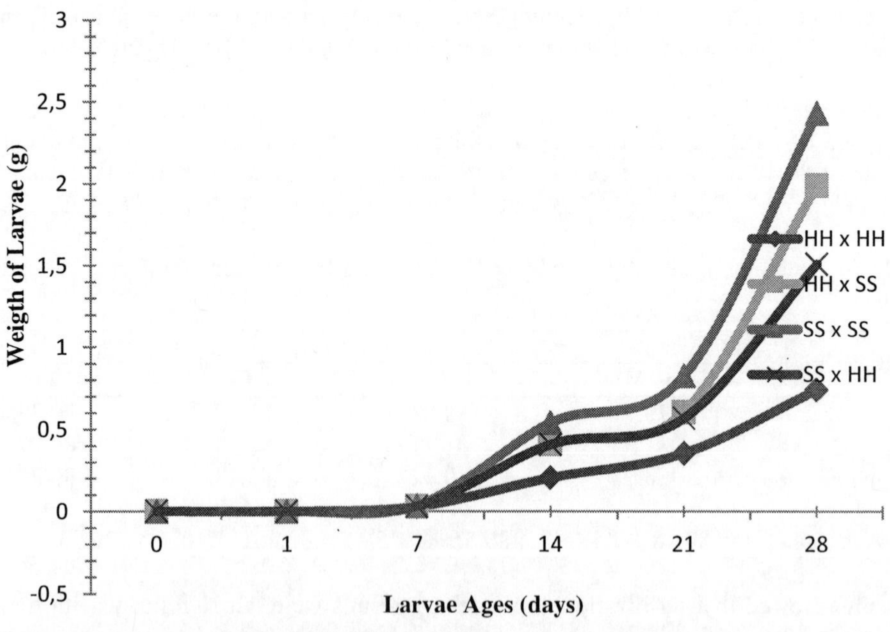

Figure 1. Weight gain and larvae and fingerling of catfish produced from the breeding of semarang and sangkuriang strains.

The observation result of weight and length gain on the 14th and 21st day of feeding or the 17th and 23rd days of larvae age can be seen in figures 1 and 2. The weight and length of the larvae showed significantly different among treatments. It is assumed that the exogenous feed has been used for growth and the genetic factor has worked resulting a difference in growth among the treatments. The similar reason is also suitable for the weight and length of 28th old fingerling. The observation of weight and length gain on the 28th day of feeding or day 31 can be seen in figure 1 and 2. Weight and length of fingerling in various treatments showed significantly different.

The result of hybrid breeding between the female and the male catfish of sangkuriang strain showed the same result between female sangkuriang breeding with male sangkuriang. This result indicated that hybridization was able to improve the quality of the catfish of semarang strain. Hybrids can play an important role in increasing aquaculture production of several types of freshwater and marine fish [5, 6]. For example, hybrid catfish in Thailand, striped bass hybrids in the United States, hybrid tilapia in Israel and tilapia hybrids in Venezuela. Although the hybridization of catfish has shown an increase in production [17], but it is suggested that further research is required for large scale catfish hybrid production which can be used for commercial catfish culture.

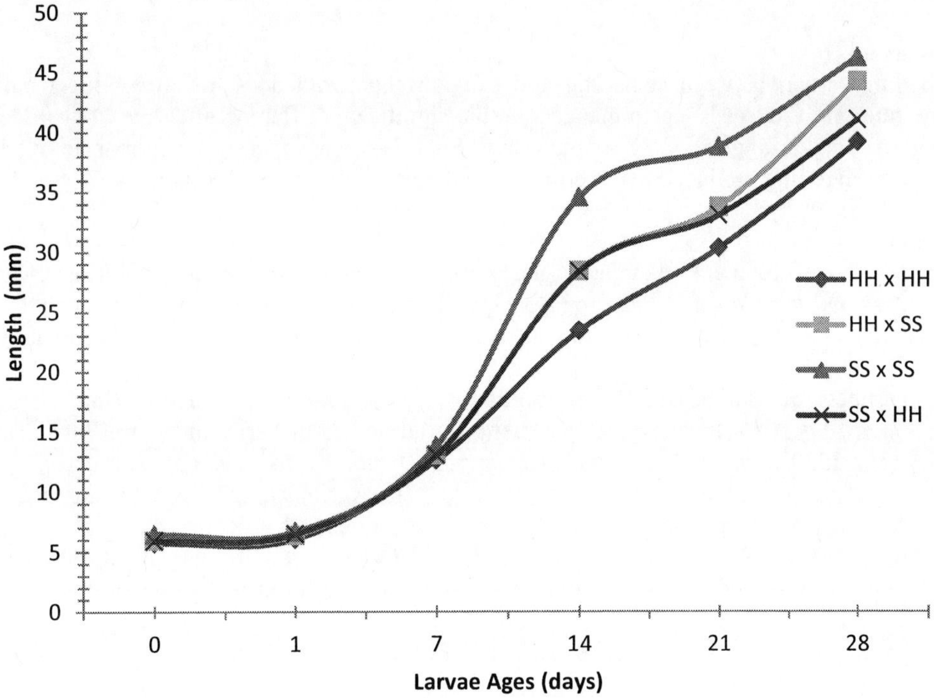

Figure 2. Length gain and larvae and fingerling of catfish produced from the breeding of semarang and sangkuriang strains.

The result of the observation on the survival rate at 28th day of feeding or the 31st day of larvae age is presented in table 4. It can be shown that the best survival rate was obtained in treatment C successively followed by treatment B, treatment D and treatment A.

Table 4. Survival Rate of fingerling of catfish produced from the breeding of semarang and sangkuriang strains.

Replication	Treatment			
	A	B	C	D
1	45.36	49.56	53.45	50.34
2	46.42	50.32	54.43	48.56
3	47.32	51.25	56.39	49.18
Total	139.1	151.13	110.82	148.08
Average	46.37	50.38	55.41	49.36

The low percentage of survival rate is assumed because of cannibalism as a result of uneven seed growth. Cannibalism begun to appear on the 14th day of feeding or 17-day-old seeds, and increased along with the age of seeds and the occurrence of difference in seed size. Therefore size grading was performed periodically from the 17th days-old. This results indicate that the grading of catfish size was crucial to avoid cannibalism, since adequate feeding was not appropriate to decrease the cannibalism [18, 19].

4. Conclusion

The breeding catfish between semarang and sangkuriang strain does not affect the fecundity and hatching rate. Hybrid seeds performance exhibit significant different after day 14 old of larvae. Breeding of sangkuriang/SS ♀ × sangkuriang/SS ♂) give a superior performance of fingerling, followed the fingerling produced by breeding of semarang/HH ♀ × sangkuriang/SS ♂.

References

[1] National Development Planning Agency 2013 *Indonesia Population Projection 2010–2035 National Development Planning Agency*, (Indonesia: Central Bureau of Statistics Jakarta)

[2] Director General of Aquaculture 2013 *Data-Statistics-Series-Production-Fishing-Culture Indonesia*

[3] Department of Marine and Fisheries 2017 *Main Commodity Production Data Prov. Central Java Year 2016* (Indonesia: Department of Marine and Fisheries of Central Java Province)

[4] Rustidja 1999 *Genetic Quality Improvement of Dumbo Catfish and Cryopreservation.* Paper at the Agreement Meeting of Seed Technology of Agribusiness of Freshwater, Brackish and Marine Fish (Jakarta: Directorate General of Fisheries)

[5] Childers W F 1971 *Seminar/Study Tour in The U.S.S.R.* (Rome: Genetic Selection And Hybridization of Cultivated Fishes Food And Agriculture Organization of The United Nations)

[6] Bithy K, Miah M I, Haquel M S, Hasan K R and Islam M F 2012 *J. Environ Sci & Natural Resources* **5** 295 – 300

[7] Ndimele P E and Owodeinde F G 2012 *Turkish Journal of Fisheries and Aquatic Sciences* **12** 619–26

[8] Islam M Z, Sarder M R I and Akhand M R I 2015 *IJFAS* **2** 186–91

[9] Al-Deghayem W A, Al-Balawi H F, Kandeal S A and Suliman El A M 2017 *Braz. Arch. Boil. Technol.* **60** ISSN 1678–4324

[10] de Graaf G and Janssen H 1996 *Artificial reproduction and pond rearing of the African Catfish Clarias gariepinus in sub Sahara Africa-A Hand Book FAO Fisheries Technical Paper no.* 362 (Rome: FAO) p 73

[11] Ataguba G A, Solomon S G and Onwuka M C 2012 *Livestock Research for Rural Development* **24** 2012

[12] Yamashita M 2000 *Zooll. Sci.* **17** 841–51

[13] Sire M F, Babin P J and Verner J M 1994 *J. Exp. Zool.* **269** 69–83

[14] Miyares, Rosa L, de Rezende, Victor B, Steven A and Farber 2014 *Dis. Model Mech.* **9** 15–27

[15] Peteri A, Shibabrata N and Samarendra N 1994 *Manual on seed production of african catfish (Clarias gariepinus)* (Food and Agriculture Organization of The United Nations)
[16] Rahman MA, Arshad A, Marimuthu K, Ara R and Amin S M N 2013 *Asian Journal of Animal and Veterinary Advances* **8** 139–53
[17] Adah P M, Onyia L U and Obande R A 2014 *Biotechnology* **13** 248–251
[18] Onwuteaka J and Prince C O 2015 *Journal of Environmental Science, Toxicology and Food Technology* **9** 46–49
[19] Obirikorang K A, Madkour H A and Adjei-Boateng 2014 *International Journal of Science and Technology* **3**

IOP Conf. Series: Earth and Environmental Science **139** (2018) 012003 doi:10.1088/1755-1315/139/1/012003

Evaluation of traditional plant extracts for innate immune mechanisms and disease resistance against fish bacterial *Aeromonas hydrophila* and *Pseudomonas* sp.

E H Hardi[1], G Saptiani[1], I W Kusuma[1], W Suwinarti[1], R A Nugroho[1]

[1] Department of Aquaculture, Faculty of Agriculture, Mulawarman University, Jl. Gunung Tabur, Gunung Kelua, Samarinda Ulu, Samarinda, East Kalimantan, Indonesia

*Email: estieriyadi2011@gmail.com

Abstract. The purposes of this study were to evaluate effect of ethanol herbal extracts of *Boesenbergia pandurata*, *Solanum ferox* and *Zingimber zerumbet* on Tilapia (*Oreochromis nilaticus*) innate immune mechanisms and disease resistance against *Aeromonas hydrophila* and *Pseudomonas* sp. Fish were intramuscularly injected with 0.1 mL/fish (10^{10} CFU·mL^{-1}) of each bacterium on the day 6th of post treatment using extract by several methods (injection, oral administration and immersion). The doses of extract were 600 ppm of *B. pandurata*, 900 ppm *S. ferox* and 200 ppm of *Z. zerumbet*. The percentage mortality, Relative Percent Survival (RPS) and innate immune response were assessed on weeks 1, 2, 3 and 4. All the methods were effective to enhance the immune parameters after 2 weeks application and the RPS of treatment reached more than 90 %. The results showed that the injection method of extracts was the most effective method to control *A. hydrophila* and *Pseudomonas* sp. The result indicated that all the doses of extracts could be significantly influence the immune response and protect the health status of tilapia against *A. hydrophila* and *Pseudomonas* sp. infections.

1. Introduction

Aeromonas hydrophila and *Pseudomonas* sp. are two bacterial pathogens that infect tilapia fish in aquaculture. *Aeromonas hydrophila* causes exophthalmia, fin root, darkened and ulcerative lesions on the body even severe bleeding [1, 2]. *Pseudomonas* sp. causes injury to internal organs such as changing the consistency of the kidneys and heart [3]. Herbal plants extract for disease control is highly recommended because besides safe for fish and the environment, the plant extract is also cheap and easy to application in aquaculture [4].

The plant extract effectiveness for disease control is very diverse. Feed application of *Azadirachta indica*, *Ocimum sanctum* and *Curcuma longa* ethanol or methanol extract increase the immune response and resistance of goldfish infected with *A. hydrophila* [5, 6]. Mangrove (*Avicennia marina*) leaves extract demonstrated inhibit the growth of bacteria *Pseudomonas fluorescens*, *Pseudomonas aeruginosa*, *Vibrio parahaemolyticus* and *Vibrio anguillarum* isolated from sea water ornamental fish [7]. Furthermore, methanol extract of leaves of *A. marina* is capable to increase the survival of ornamental fish and also inhibits the growth of pathogenic bacteria in the fish body.

Extract plants that contain levamisole, flavonoids, steroids, carbohydrates have ability to inhibit the growth of pathogenic bacteria in fish [8], saponin [9] and glycyrrhizin exhibit as natural antibacterial properties. Some plant extracts have a antibacterial activity to inhibit the pathogen bacteria and increase the fish nonspecific and specific, such as *O. sanctum* [10, 11], azadirachtin [12], *Viscum album*, *Urtica dioica* and *Zingiber officinale* [13], *Radix astragalin seuhedysari* and *Radix angelicae sinensis* [14, 15], *Astragalus radix* and *Scutellari radix* [16] and *Achyranthes aspera* [17, 18].

Administration of plant extracts through injection enable to stimulate immune system faster than oral administration, because the extract is slowly absorbed by the fish in the later administration [5]. Some plant extracts application through fed in fish were able to enhance the total leukocytes and phagocytic activity [7, 19, 20, 21, 22, 23]. Immersion method of *Azadirachta indica* leaf extract was effective to control the *A. hydrophila* infection in goldfish (*Cyprinus carpio*) [24]. Show that use of garlic extract (1,000, 2,000 and 3,000 ppm) to prevent *A. hydrophila* infection was effective with survival rate ranged 83 to 100 % [25].

In this research, we observed the immunostimulant activity of tree plant extracts *Boesenbergia pandurata*, *Solanum ferox* and *Zingimber zerumbet* to control *A. hydrophila* and *Pseudomonas* sp. bacteria pathogens in tilapia through three different methods: injection, immersion and oral administration.

2. Materials and Methods

This research was done in April to December 2016 at The Laboratory of Environmental Microbiology, Faculty of Fisheries and Marine Science and Laboratory of Forest Products Chemistry, Faculty of Forestry, Mulawarman University, East Kalimantan, Indonesia.

2.1. Fish and bacteria

Oreochromis nilaticus were purchased from Kutai Kartanegara fresh water hatchery. The fish with 10.1 ± 3.2 g (mean ± SD) in weight were used in trials. The *A. hydrophila* (EA-01) and *Pseudomonas* sp. (EP-01) isolates were obtained from Laboratory of Environmental Microbiology, Faculty of Fisheries and Marine Science. The bacteria was cultured in BHI (Brain Heart Infusion Broth, DIFCO®) and BHIA (Brain Heart Infusion Agar, DIFCO®) media for 24 h at 30 °C and the density of bacteria for injection test was 1010 CFU·mL^{-1} [26].

2.2. Preparation of plants extract B. pandurata, Z. zerumbet and S. ferox

Herbal materials were collected from traditional market in Samarinda (figure 1). The ethanol extract of plants were prepared according to previous study by [8, 27]. For the treatment, extract concentrations 600 mg·L^{-1} of *B. pandurata*; 900 mg·L^{-1} of *S. ferox* and 200 mg·L^{-1} of *Z. zerumbet* with three replicates and three administration methods i.e. injection, immersion and oral administration [8 26]. The extracts dosages in this research were the best dosages as antibacterial and imunostimulant in tilapia against *A. hydrophila* and *Pseudomonas* sp. infection.

2.3. Experimental design

The tilapia fish were injected intramuscularly with *A. hydrophila* and *Pseudomonas* sp. (1.6 10^{10} CFU·mL^{-1}) and then at 6th days post injection, tilapia ware treated by using the extract through several methods (injection, immersion and oral). The fish group infected with *A. hydrophila* was treated with *B. pandurata* and *Z. zerumbet*. While, fish group infected with *Pseudomonas* sp. was treated with *S. ferox* extract.

Administration of extract by injection was done by injecting the tilapia with extract intraperitoneally (0.1 mL/fish) and the fish is reared for four weeks.

For oral administration, the extracts were incorporated in pellet feed by mixing 500 mL herb extracts with 1–2 % yolk egg and adding to 1 kg of commercial fish feed. Tilapia was fed with the pellet fed twice a day for 14 days, and reared for four weeks. Immersion administration was done by immersing tilapia in extract solution for 30 min and rearing for four weeks.

This experiment consists of six groups: 1) negative control, the fish was not treated with extract and not injected with bacteria; 2) the fish was injected with *A. hydrophila* and treated with *B. pandurata* extract; 3) the fish was injected with *Pseudomonas* sp. and treated with *S. ferox* extract; 4) the fish was injected with *A. hydrophila* and treated with *Z. zerumbet* extract; 5) the fish was injected with *A. hydrophila* and not treated with extract; and 6) the fish was injected with *Pseudomonas* sp. and not treated with extract.

2.4. Observation
The observation parameters in this research were abnormality swimming response (gasping, weakness and aggressive); anatomy pathology (fin root and darkness exophthalmia) and total bacteria count in the fish blood and RPS (relative presentation survival). Total of leukocyte and phagocyte index were observed every week until 4th weeks.

3. Results and Discussion
The result showed that the three extracts effectively enhance the fish immune system and increasing the recovery after bacterial infection in the fourth week.

(a) (b) (c)

Figure 1. Herb plants used in the research, a) *Boesenbergia pandurata*, b) *Zingimber zerumbet* dan c) *Solanum ferox*.

3.1. Fish infection
Aeromonas hydropila and *Pseudomonas* sp. infection caused upnormal swimming such us gasping, weakness and aggressive, while the fish were given extracts *B. pandurata* and *Z. zerumbet* showed normal swimming begun in the 2nd weeks until fourth weeks post-treatment through three methods. The fish treated with extract through the injection method recovered quickly, and also exhibited normal pattern of fish swimming and gross pathology. In detail can be seen in figure 2 and 3.

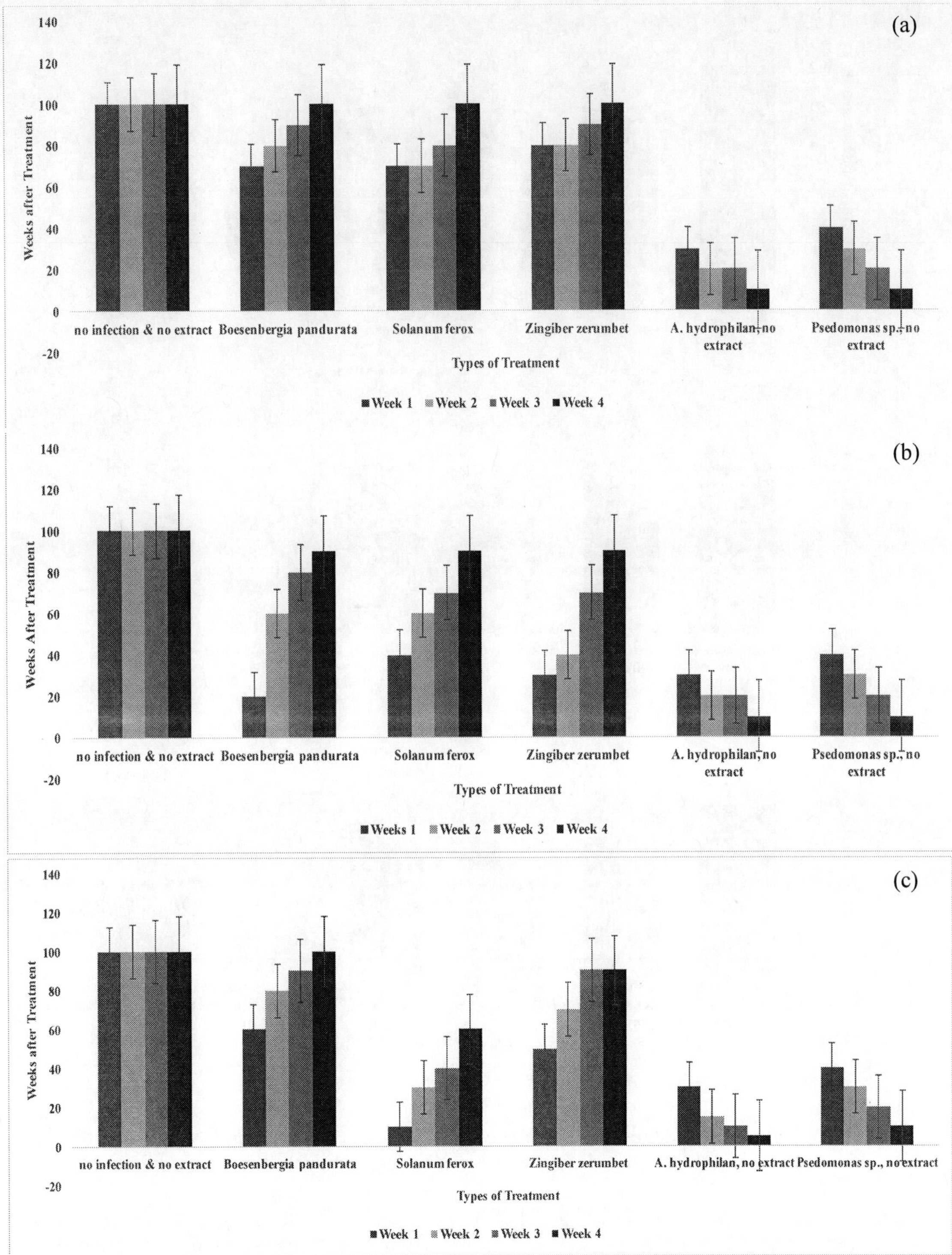

Figure 2. Fish swimming recovery (gasping, weakness and aggressive) of treated fish with three extract and three methods. a) Immersion, b) oral administration and c) injection.

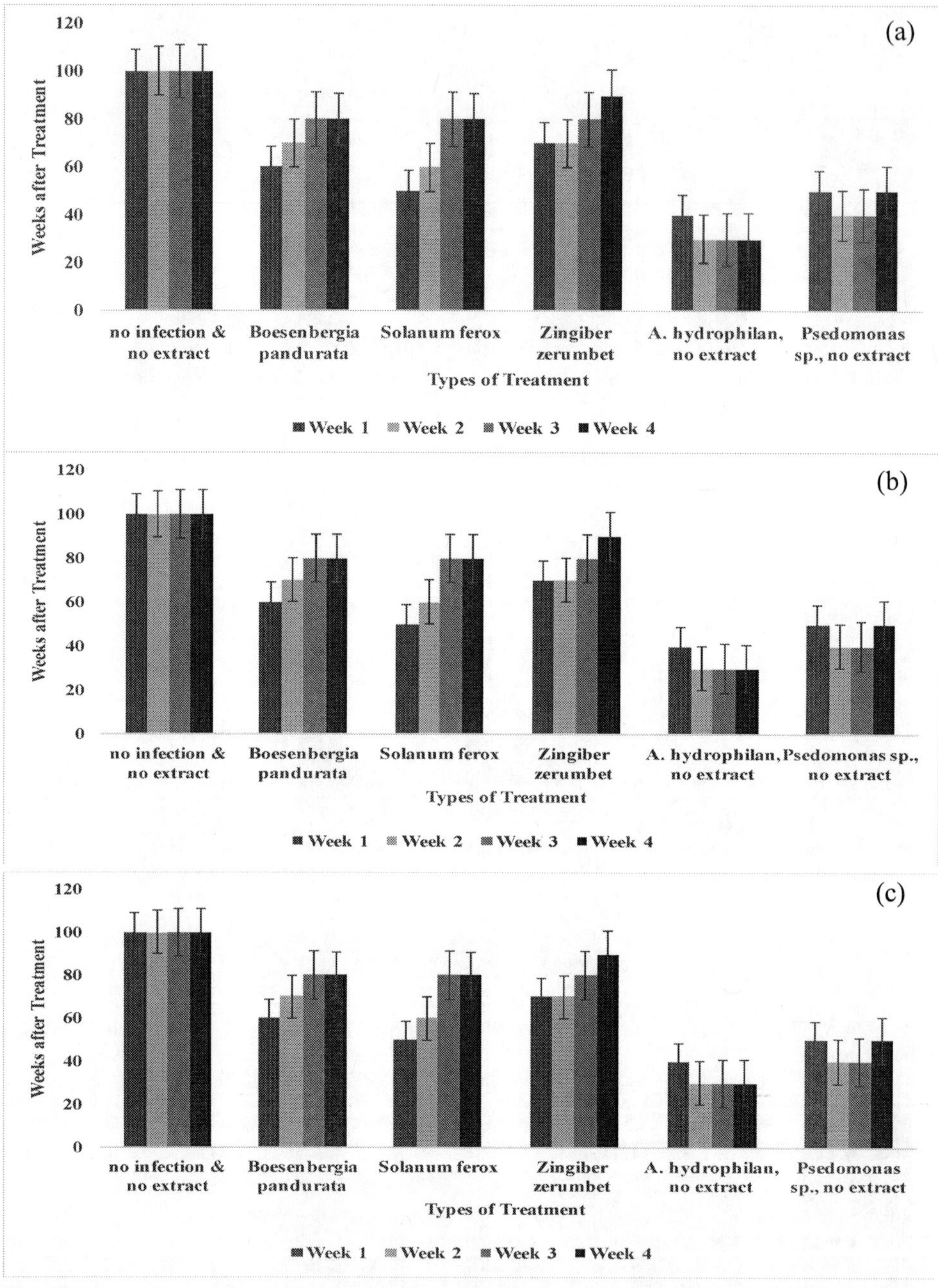

Figure 3. Anatomy pathology (darkness, exophthalmia and fin rot) of treated fish with three extracts and three methods. a) Immersion, b) oral administration and c) injection.

Treatment using extract through feed method showed increase in fish recovery after being infected. The recovery occurred as early as in 1-week to the 4-week reaching 90 % recovery rate with *Z. zerumbet*, and 80 % with *S. ferox* and *B. pandurata*. Immersion method was effective, but the recovery process was slower than by injection and oral administration. This might be having correlation with the absorption process of extract [5]

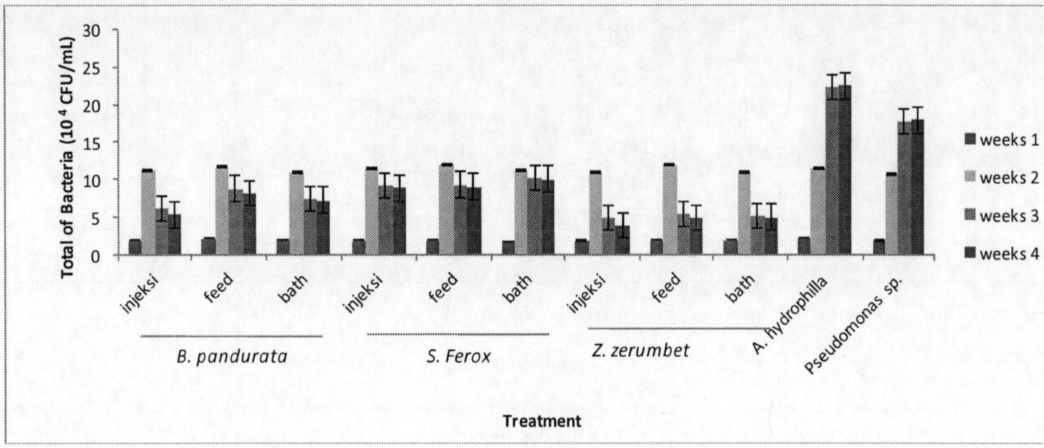

Figure 4. Total bacteria count of treated fish with three extract and three administration methods.

Total bacteria count in treated tilapia was almost lower than control group. Generally, application of the *Z. zerumbet* extract was able to reduce the number of *A. hydrophila* at week 2 to week 4 either through injection, oral administration or immersion and resulted the total bacteria count lower than that of fish treated with extract *B. pandurata*. Treatment of *S. ferox* extract through oral administration and injection were able to suppress the growth of bacteria *Pseudomonas* sp. up to 5.5 10^4 CFU·mL^{-1} while through immersion, TPC decreased to 10^5 CFU·mL^{-1}.

The phagocytic index (leucocyte) of extract treatment have increased since week 2. The *Z. zerumbet* treatment were highest increase compared *B. pandurata* and *S. ferox*.

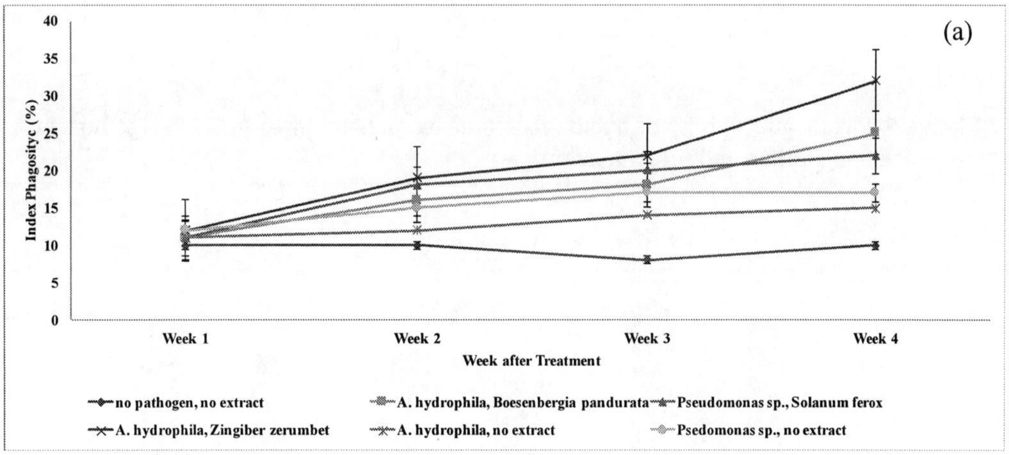

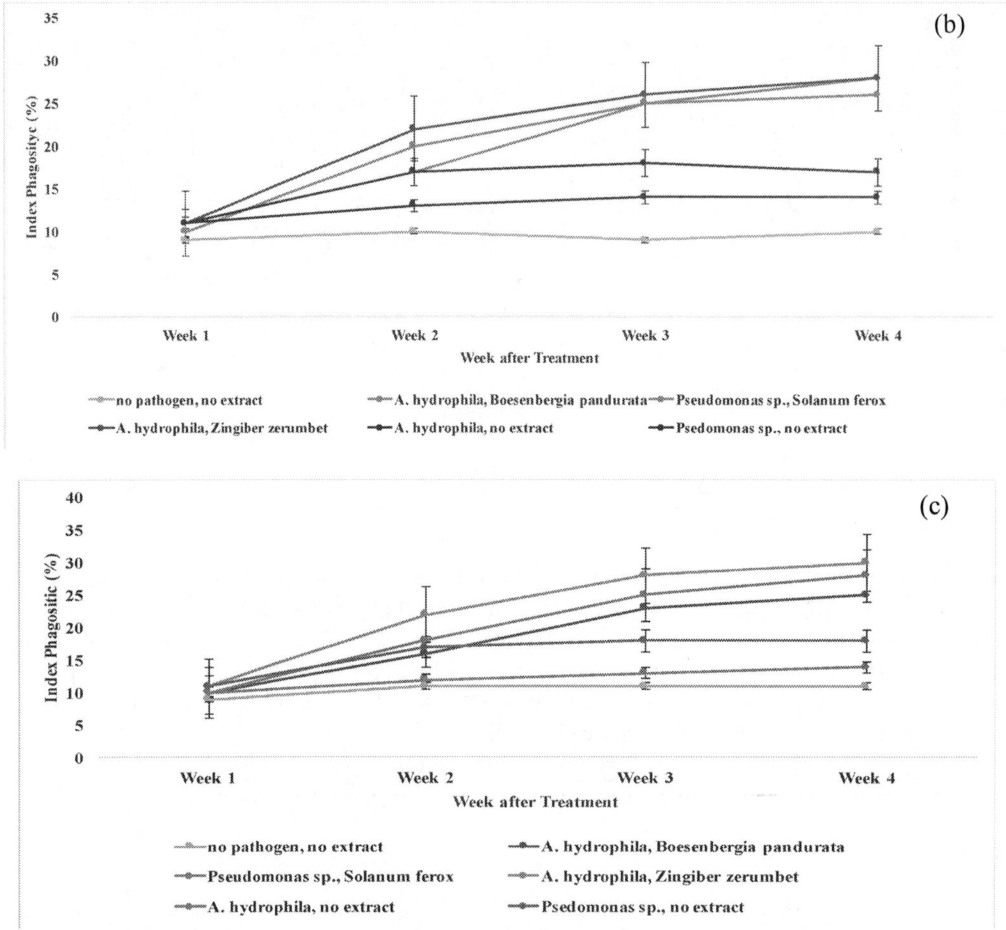

Figure 5. Phagocytic activity of treated fish three extract and three methods: (a) Injection, (b) oral administration and (c) immersion.

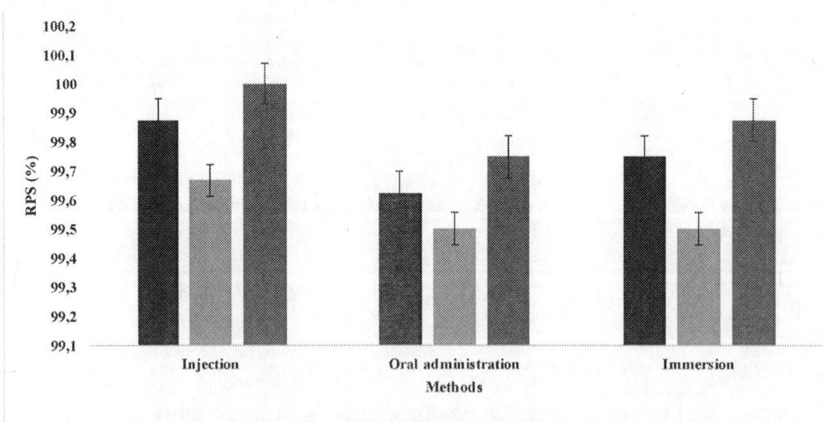

Figure 6. Effect of different extract administered on the Relative Percent Survival (RPS) (%) in Tilapia.

After challenge with *A. hydrophila* and *Pseudomonas* sp., all treated groups showed lower mortality compared to the control treatment. The best survival rate was observed in the group treated with *Z. zerumbet* extract (100 %) through injection. All extract were effective to control diseases caused *A. hydrophila* dan *Pseudomonas* sp. The extract administration as a immunostimulan increase the fish nonspecific immune system [23, 24, 28]. Fish innate immune response enhance the ability to eliminate the pathogen, and increase the fish survival rate after infection [29, 30, 31].

Administration of plant extracts as an immunostimulatory agent in fish maight be through injection, feed (oral administration) and immersion. The later method has advantages and disadvantages, but still effective [23]. Utilization plant extracts as antibacterial and immunostimulant in fish have been developed, but still limited information in the most effective method, dose and mechanisms in vivo are available. The results showed that the ethanol extract of *Z. zerumbet*, *S. ferox* and *B. pandurata* were able to increase fish phagocytic activity after two weeks of immersion and oral administration. Injection gave faster effect in one week.

The oral administration of *Azadirachta indica*, *Ocimum sanctum* and *Curcuma longa* administration in goldfish (*Carassius auratus*) increase the leucocyte phagocytic activity in two weeks administration until fourth week [5]. Ginger extract in fish feed increased the total protein level in blood plasma of fish and the highest level of plasma proteins was observed in the group fed with 1 % ginger extract [13]. The plant extract contains components that increase the phagocytic activity [13, 21, 32]. Phagocytic cells are important cells that play a role in the defense mechanism of fish [33].

After challenging with *A. hydrophila* and *Pseudomonas* sp. untreated tilapia showed high mortality (80 %) in week 4, while fish treated with the extract showed low mortality (10–20 %) in the end of experiment in the injection and oral administration methods [32, 34]. Injection of *O. sanctum* and *Nyctanthes arbortristis* in *Oreochromis mossambicus* reduced the mortality after *A. hydrophila* infection [5, 12]. The survival rate of tilapia fed with *Rosmarinus officinalis* extract in combination with *Astragalus membranaceus* and *Lonicera japonica* extract increase after *A. hydrophila* infection in 4th weeks [35, 36].

4. Conclusion

Extracts of *B. pandurata, S. ferox* and *Z. zerumbet* with the concentrations 600, 900 and 200 ppm can be used in the treatment of *A. hydrophila* and *Pseudomonas* sp. infection in tilapia. This extracts improve the phagocytic index, recovery process from the infection. Injection method is more effective method to treat *A. hydrophila* infection by using *Z. zerumbet* and *B. pandurata* than oral administration and immersion method. Similar conclusion is also obtained for *S. ferox* extract in the treatment of *Pseudomonas* sp. infection.

Acknowledgments

The authors are grateful to RISTEK DIKTI, Indonesia for the award of Research, STRANAS with contract number 370/UN17.41/KL/2017 which made this work possible. All authors are grateful to the Department of Aquaculture, Fisheries and Marine Science, for the facilities. The authors would like to thank Mr. Ardians, from marine and fisheries department, Kutai Kartanegara, East Kalimantan province.

Reference

[1] Austin B and Austin D A 2007 *Bacterial Fish Diseases of Farmed and Wild Fish. Clinical Picture of Pseudomonas septicemia Characterized* (United Kingdom: Springer-Praxis Publishing Ltd)

[2] Hardi E H, Kusuma I W, Suwinarti W, Agustina and Nugroho R A 2016 *Nusantara Bioscience* **8** 18–21

[3] Hardi E H and Pebrianto C A 2012 *Jurnal Ilmu Perikanan Tropis* **16**(2) 35–39

[4] Raa J, Roerstad G, Ingested R and Robertson B 1992 *Asian Fisheries Society* 39–50.

[5] Harikrishnan R, Rani M N and Balasundaram C 2009 *Fish & Shellfish Immunology* **2211**(4) 41–

50
[6] Harikrishnan R, Balasundaram C and Heo S M 2010 *Fish & Shellfish Immunology* **28** 354–361
[7] Dhayanithi N B, Kumar T T A, Balasubramanian T and Tissera K 2013 *Journal of Coastal Life Medicine* **1**(3) 217–224
[8] Hardi E H, Kusuma I W, Suwinarti W, Agustina, Abbas I and Nugroho R A 2016 *AACL Bioflux* **9** 638–646
[9] Ninomiya Y and Imoto T 1995 *American Journal Physiology* 268 1029–1035
[10] Logambal S M, Venkatalakshmi S and Michael R D 2000 *Hydrobiologia* **430** 113–120
[11] Venkatalakshmi S and Michael R D 2001 *Journal Aquaculture Tropic* **16** 1–10
[12] Logambal S M and Michael R D 2001 *Journal Aquaculture Tropic* **16** 339–347
[13] Dugenci S K, Arda N and Candan A 2003 **88** 99–106
[14] Jian J and Wu Z 2003 Aquaculture **218** 1–9
[15] Jian J and Wu Z 2004 *Fish Shellfish Immunology* **16** 185–191
[16] Yin G, Jeney G, Racz T, Xu P, Jun X and Jeney Z 2006 Aquaculture **253** 39–47
[17] Rao Y V and Chakrabarti R 2005 *Fish Shellfish Immunology* **18** 327–34
[18] Rao Y V, Romesh M, Singh A and Chakrabarti R 2004 *Aquaculture* **238** 67–73
[19] Chitmanat C, Tongdonmuan K and Nunsong W 2005 *Journal Science Technology* **27**(1) 359–364
[20] Elkamel A A and Mosaad G M 2012 *Journal of Aquaculture Research & Development* **3** 147
[21] Alambra J R, Alenton R R R, Gulpeo C P R, Mecenas C L, Miranda A P, Thomas R C, Velando M K S, Vitug L D and Maningas M B B 2012 **AACL Bioflux 5** 13–17
[22] Yin Q Y, de Groot P W, Dekker H L, de Jong L, Klis F M and de Koster CG 2005 *J. Biol. Chem.* **280**(21) 20894-20901
[23] Yin G, Ardó L, Thompson K D, Adams A, Jeney Z and Jeney G 2009 *J.* **26**(1) 140–145
[24] Harikrishnan R, Nisha Rani M and Balasundaram C 2003 *Aquaculture* **221** 41–50
[25] Lukistyowati I, Windarti, Morina, Isnansetyo A and Kurniasih 2008 *J. Fish Science* **10**(1) 11–19
[26] Hardi E H, Pebrianto C A, Hidayanti T and Handayani R T 2014 *Journal of Veterinary Sciences* **8**(2) 130–134
[27] Limsuwan S and Voravuthikunchai S P 2008 *FEMS Immunology & Medical Microbiology* **53** 429–436
[28] Galina J, Yin G, Ardo L and Jeney Z 2009 *Fish Physiol. Biochem.* **35** 669–676
[29] Gopalakannan A and Venkatesan A 2006 *Aquaculture* **255**(1–4) 179–187
[30] Lin S, Yu Pan, Lin Luo and Li Luo 2011 *Fish & Shellfish Immunology* **31**(6) 788–794
[31] Bricknell I, Roy A and Dalmo 2005 *Fish & Shellfish Immunology* **19**(5) 457–472
[32] Wang E, Chen X, Wang K, Wang J, Chen D, Geng Y, Lai W and Wei X 2016 *Fish & Shellfish Immunology* **59** 196–202
[33] Zhang G, Gong S, Yu D and Yuan H *Fish Shellfish Immunol.* **26** 467–472
[34] Cho S H, Jeon G H, Kim H S, Kim D S and Kim C 2013 *Asian-Australas J. Anim. Sci.* **26**(1) 90–96
[35] Abutbul S, Golan-Goldhirsh A, Barazani O, Ofir R and Zilberg D 2005 *J. Aquacult. Bamid* **57** 71–80
[36] Ardo L, Yin G, Pao Xu, Váradi L, Szigeti G, Jeney Z and Jeney G 2008 *Aquaculture* **275**(1–4) 26–33

IOP Conf. Series: Earth and Environmental Science **139** (2018) 012004 doi:10.1088/1755-1315/139/1/012004

Comparison of three inert markers in measuring apparent nutrient digestibility of juvenile abalone under different culture condition and temperature regimes

K U Nur[1*], L Adams[2], D Stone[3], N Savva[4] and M Adams[2]

[1] Muhammadiyah University of Parepare, Faculty of Agribusiness, Veterinary and Fisheries, Jl. Jend. Ahmad Yani Km 6, Bukit Harapan, Soreang, Parepare, South Sulawesi 91113, Indonesia
[2] IMAS (Institute for Marine and Antarctic Studies), Private Bag 1370, Launceston, Tasmania, Australia
[3] SARDI Aquatic Science Centre, 2 Hamra Ave, West Beach, South Australia
[4] AbTas (Abalone Tasmania), 17 Bevic Rd, Clarence Point, Tasmania 7270, Australia

*E-mail: kurniati.umrah@gmail.com

Abstract. A comparative research using three inert markers, chromic oxide, yttrium and ytterbium to measure the apparent nutrient digestibility of experimental feed in juvenile Hybrid abalone (*Haliotis rubra* X *H. laevigata*) and Greenlip abalone (*H.laevigata*) revealed that apparent digestibility of crude protein (AD_{CP}) measured using yttrium and ytterbium in hybrid abalone were significantly different across the treatments. Protein digestibility measured in experimental tanks was higher than those measured in indoor and outdoor commercial tanks, regardless of inert marker used. Chromic oxide led to overestimated AD_{CP} compared to when measured using yttrium and ytterbium. There were no significant interactions between temperature and inert markers when measuring AD_{CP} and apparent digestibility of gross energy (AD_{GE}). However, there was a significant difference of AD_{CP} amongst inert markers when measured in greenlip abalone cultured at two temperatures. While measurements of AD_{GE} calculated using three inert markers shared the same value.

1. Introduction

Determining the nutrient digestibilities of certain feed ingredients is pivotal to provide credible information to formulate a cost effective formulated feed. Measuring nutrient digestibility directly in Haliotids is difficult due to the problems related to the accurate measurements of feed consumed and faeces voided [1]. Digestibility determination can be affected by time between the production of faeces and their collection [2]. If faeces are exposed to water over time, nutrients may leach from the feed and faeces, which may lead to overestimation of AD. Additionally, experimental feeds tend to disintegrate in water, causing collection of whole uneaten feed to be difficult with potential contamination problem between feed and faeces [3]. Therefore, it is requisite to use indirect method by using inert markers that cannot be absorbed by the animal, should pass through the digestive system at a similar rate as the other ingredients, it must be inert and not impede the digestive processes, consistently included into the feed and can be easily and accurately analysed even at very low concentrations, and be nontoxic to people and environment [4, 5].

There are various possible markers for use to measure digestibility of particular nutrients in Haliotids, In a study conducted by [1] using blacklip abalone and greenlip abalone, it was concluded that markers including insoluble mineral components determined as acid insoluble ash (AIA), hydrolysis resistant organic matter (HROM) and crude fiber content of the feed were inappropriate due to the abalone capability to digest complex carbohydrates, which create a component of these markers. However, amongst all of the observed markers in this study, AIA and chromic oxide (Cr_2O_3) were found to be the promising markers.

Chromic oxide found to be reliable in determining protein digestibility of greenlip abalone [3, 6, 7]. However, it was not accurate to determine carbohydrate or cholesterol digestibility due to a low correlation between the movement of the marker to carbohydrate and cholesterol [8]. Trivalent oxides of yttrium and lanthanides (rare earth metals) have been investigated as inert markers for protein digestibility experiments since the beginning of 1990's [4]. Those inert markers appear to satisfy the desirable attributes of inert markers. They are moderately soluble in weak acid, can be detected by plasma emission spectroscopy and atomic absorption, therefore they can be analyzed in feed and feces even at low concentrations (in the $mg \cdot kg^{-1}$ range) [4, 9].

Another rare earth metal that can be used as inert marker is ytterbium. Ytterbium has not been used as an inert marker in mollusc, particularly in haliotids. However, this inert marker has been successfully used to determine nutrient digestibility in humans, salmonid fish and crustaceans [4, 10, 11]. Ytterbium is useful since it is unabsorbed across the digestive system and does not interfere with the metabolic processes or react with digestive enzymes [12]. These markers have replaced the use of chromic oxide in most aquaculture nutrition research in recent years due to more accurate recovery of marker, better homogeneity in feces and more accurate measurement techniques at low concentration. Thus, evaluation of ytterbium as inert markers in determining digestibility of abalone feed is of significance.

Inert markers provide the opportunity to measure apparent digestibility in both experimental systems but also in commercial tank setting. The majority of the data used in feed formulation is derived from experimental measurements conducted in laboratory systems [13–21], whereas in commercial scale abalone also have access to biofilms on tank surfaces and are exposed to different light levels, temperature and water flows according to the farm design. Justifying the results gained from experimental systems and digestibility measurements made on farm is of interest to know if in-tank commercial measurements might be valid during times of rapid commercial growth or during adverse culture conditions.

Providing fundamental data about nutrient digestibility of greenlip and hybrid abalone at different temperature regimes in different culture systems is of important in order to formulate cost-effective formulated feed that can be adjusted to the temperature seasonal changes. Moreover, a reliable inert marker to measure digestibility of feed ingredients should be investigated to provide accurate estimation of digestibility of the feed. This will assure optimal growth rates and reducing grow out period through the formulation of feeds specific to species, size and culture conditions.

The aims of this study were to compare yttrium, ytterbium and chromic oxide as inert digestibility markers to measure apparent protein digestibility in subadult hybrid abalone (*H. rubra x H. laevigata)* and greenlip abalone (*H.laevigata)*, in experimental and commercial tanks.The second one was to compare yttrium, ytterbium and chromic oxide as inert digestibility markers to measure apparent protein and energy digestibility at two temperatures in subadult greenlip abalone *(H. laevigata)*.

2. Materials and method

2.1. *Experiment 1 (protein digestibility and inert markers comparison in subadult hybrid abalone)*

2.1.1. *Experimental animals*

Subadult hybrid abalone with initial body weight 33.72 ± 5.7 g and initial shell length 60.18 ± 3.2 mm (n = 600 animals) were used in this experiment. The abalone were provided by AbTas, Clarence Point, Tasmania.

2.1.2. *Experimental system*

The experimental systems consisted of 4 square outdoor tanks, 4 square indoor tanks (780 L Solid Nally Megabins, MS7800; Viscount Plastics Pty Ltd, Hawthorn East, Victoria, Australia), and 20 rectangular white plastic tanks (41 cm L × 27 cm W × 16 cm H). Outdoor and indoor tanks were cleaned but biofilm was not removed, while the rectangular white plastic tanks were installed in a controlled room to resemble the experimental tanks used in laboratory based nutrition research. Concrete hides were placed into each indoor and outdoor tank to provide more space, while there were no hides placed in experimental tanks. To avoid abalone escaping, 40 mm wide black stick on hook tape (Velcro Australia Pty Ltd., Hallam, Victoria, Australia) was installed around the top inside perimeter of experimental tanks. The experiment was conducted in a flow-through sea water system with ambient temperature conditions (13.6 °C ± 0.046). The tanks were cleaned every day at 8 AM to remove overnight feces and uneaten feed.

2.1.3. *Experimental protocol*

The abalone were gently removed from a commercial tank surface using a scraper. The animals were weighed and measured using a digital balance and a caliper. Each indoor and outdoor commercial tank was randomly stocked with 60 animals, while six animals were randomly stocked into each experimental tank. Animals were acclimated to the systems and diet for one week. During the acclimation period, one mortality was replaced by a new animal from the same population with similar weight and size.

2.2. *Experiment 2 (Protein, energy digestibility and inert markers comparison in subadult greenlip abalone)*

2.2.1. Experimental animals. Subadult greenlip abalone (initial shell length 51.3 mm ± 0.075, initial body weight 20.450 g ± 0.032, Mean ± SE) were used in this experiment. The juvenile were obtained from commercial abalone farms and pre-conditioned in SARDI Aquatic Science Centre, Adelaide and fed commercial diet prior to stocking in the experimental system.

2.2.2. Experimental system. The experiment was carried out in a temperature controlled flow-through UV treated salt water system (Model 025120-2, 120w, Emperor Aquatics, Pottstown, PA, USA). The experimental system consisted of six circular shaped tanks, assigned in pairs to two temperature levels, 14 °C and 18 °C. The 14 °C temperature water was the ambient temperature of the flow through system, while the 18°C temperature water was controlled using immersion heater (3 kw, 240 V, 3 kw, JQ20; Austin & Cridland, Carlton, NSW, Australia). The water flow was maintained at a rate of 2 L $\cdot min^{-1}$ and the water level was maintained at 14.5 cm. Water temperature and dissolved oxygen (ppm and % saturation) were measured on a daily basis using an OxyGuard^{TM} Handygamma dissolved oxygen meter (OxyGuard International A/S, Birkerod, Denmark). pH was measured using a pH meter (Oakton pHtester 20; Oakton Instruments, Vernon Hills, IL, USA). Salinity (ppt) was measured daily using a portable refractometer, model RF20 (Extech Instruments, Nashua, NH, USA).

IOP Conf. Series: Earth and Environmental Science **139** (2018) 012004 doi:10.1088/1755-1315/139/1/012004

2.2.3. Experimental feeds. Feeds for experiment 1 and 2 were manufactured using the SARDI formulation, contained 33% crude protein. Dry ingredients were initially mixed (Brice, Model: VFM-2 °C, Brice Australia Pty. Ltd). The dough mixture was then pelleted using a commercial pasta machine (ITALPAST, Food Machinery International Pty.Ltd, Australia), and pellets steamed for 15 mins using a domestic food steamer (TEFAL Steam Cuisine, Series S04, China). Feed was then dried to 10% moisture using fan forced fumehood for 4 days and stored at -20^{0}C prior to the experiment.

2.2.4. Fecal collection. Collection of feces from experiment 1 was conducted 2 times a day at 2 h intervals (1100 hours to 1500 hours) for 2 weeks. Feces were collected using plastic pipette onto 63 µm screen before transferring into plastic tubes. The samples were then stored frozen at -20°C prior to analysis.

Feces from experiment 2 were collected using plastic pipette into labeled 1.5 ml eppendorf tubes. All eppendorf tubes were kept on ice during collection. Collection was conducted 3 times a day in 2 h intervals (1000 hours to 1900 hours) for 4 weeks. The feces samples then centrifuged for 30 s at 1000 rpm and excess water was removed by plastic pipette. Samples were stored frozen at -20°C prior to analysis.

2.2.5. Feeding and cleaning. Animals were fed to excess in all experiments. Feeding was conducted at 4:00 PM daily. Diets were randomly allocated to each tank. Cleaning and collection of uneaten feed was accomplished by screening the tank contents using either a fine meshed net or a feeding tray to retain pellets.

2.2.6. Chemical analysis. Prior to sample analysis, all samples were freeze dried until no loss of moisture was detected. All of the samples were then pooled and ground to fine homogenous powder. Gross energy (GE) was determined by complete combustion using a bomb calorimeter (Gallenkamp Autobomb). Due to the silica contamination, gross energy of samples from experiment 2 could not be analysed. Crude protein was determined as N × 6.25 where nitrogen (N) was measured after sulphuric acid digestion, auto Kjeldahl (Kjeltec™ 8100, Foss, Sweden) distillation, and titration against hydrochloric acid as described in the Association of Official Analytical Chemists (AOAC) guidelines for chemical testing [22]. Lipid was determined by chloroform methanol extraction as described by [23]. The inert markers (yttrium, ytterbium and chromic oxide) concentration in the diet and feces were digested in 3:1 nitric:hydrochloric acid solution for 4 days then centrifuged for 10 mins at 3000 rpm. Centrifuged samples were diluted volumetrically in distilled water, then transferred into acid-washed test tubes before analysis using inductively coupled plasma atomic emission spectrophotometry (ICP-AES) based on method described by [24]. Total nitrogen, carbon, hydrogen and sulfur was determined using elemental analysis (Thermo Finnigan EA 1112 Series Flash Elemental Analyser). Ash was determined gravimetrically by incineration in a muffle furnace at 600^0 C for 2 h NFE was calculated by difference.

2.2.7. Digestibility determination. Animals were fed to excess in all experiments. Feeding was conducted at 4:00 PM daily. Diets were randomly allocated to each tank. Cleaning and collection of uneaten feed was accomplished by screening the tank contents using either a fine meshed net or a feeding tray to retain pellets. The apparent digestibility of protein and energy of diets was calculated using the following equation [25]:

$$Digestibility = 100 - \left(100\ \frac{I\,(feed)}{I\,(feces)} \times \frac{N\,(feces)}{N\,(feed)}\right)$$

Where N = nutrient (protein or energy), I = indicator (yttrium, ytterbium and chromic oxide)

2.2.8. Statistical Analysis.. Residual plots were used to examine the homogeneity and normality of the data. A two way ANOVA was performed to determine the interaction between three different inert markers and applied treatments. Tukey's Post-hoc test was used to distinguish significant differences between means. All statistical analysis were done using IBM SPSS version 20 for Macintosh (IBM SPSS Inc., Chicago, IL). Data are presented as mean ± SE.

3. Results and Discussion

3.1. *Results*

3.1.1. Experiment 1 (Protein digestibility and inert markers comparison in subadult hybrid abalone). There were significant interactions between treatments and marker ($P < 0.05$). Chromic oxide in outdoor and indoor tanks displayed higher AD_{CP} than calculated according to digestibility of yttrium and ytterbium (Figure 1).

Results calculated using ytterbium and yttrium were the same across all treatments ($P > 0.05$). Chromic oxide results were higher than yttrium and ytterbium in all experimental tanks. There were significant differences between AD_{CP} of chromic oxide in outdoor and indoor tanks with AD_{CP} of yttrium and ytterbium under the same treatments, however there were no significant differences between AD_{CP} of all inert markers in experimental tanks.

There was no significant difference for AD_{CP} of chromium oxide in all treatments ($P > 0.05$). However, AD_{CP} measured using yttrium and ytterbium were significantly different across the treatments. AD_{CP} of yttrium and ytterbium was higher in experimental tanks compared to AD_{CP} by these two inert markers in outdoor and indoor tanks. Outdoor tanks exhibited the lowest apparent crude protein digestibility.

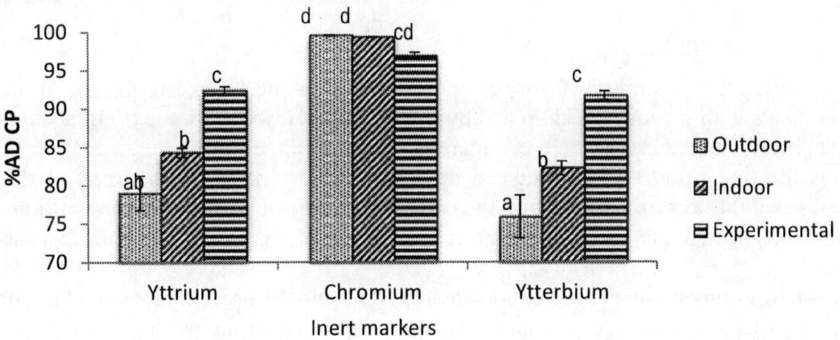

Figure 1. Apparent crude protein digestibility of subadult hybrid abalone (*H. laevigata* x *H. rubra*) fed 33% dietary protein levels with 3 inert markers at three treatments (outdoor, indoor and experimental tanks). Values that share the same superscript are not significantly different ($P > 0.05$).

3.1.2. Experiment 2 (Protein, energy digestibility and inert markers comparison in subadult greenlip abalone). There were no significant interactions by two way ANOVA between temperature and inert markers ($P > 0.05$). There were also no significant differences between the temperature ($P > 0.05$) but there were significant differences amongst inert markers ($P < 0.05$) (Figure 2)

AD_{CP} when measured using chromic oxide was higher (81.9 ± 1.6%) than yttrium (76.9 ± 0.9%) but not significantly different to AD_{CP} of ytterbium (77.2 ± 0.8%).

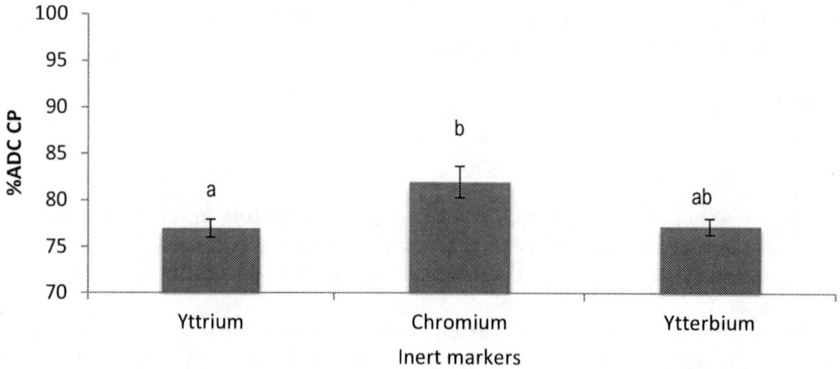

Figure 2. Apparent crude protein digestibility of subadult greenlip abalone (*H. laevigata*) fed 33% dietary protein levels at two different temperatures (14°C and 18°C). Values that share the same superscript are not significantly different (P > 0.05).

There were no significant interactions by two way ANOVA between temperature and markers for apparent gross energy digestibility (P> 0.05). Moreover, AD_{GE} calculated using the three inert markers was the same (P > 0.05).

3.2. *Discussion*

3.2.1. Inert markers digestibility comparison

The recent study verified the feasibility of yttrium and ytterbium as inert markers for use in protein digestibility experiments with juvenile greenlip and hybrid abalone. However, some caution should be given to results of AD_{CP} obtained using Cr_2O_3 as a marker.

This study was the first study that investigated the application of ytterbium as inert marker on abalone nutrient digestibility determination, and the observed significant AD_{CP} between yttrium and ytterbium across all treatments suggest a possibility to use ytterbium as inert marker in digestibility study on abalone.

The AD_{CP} measured using chromic oxide overestimated AD_{CP} and AD_{GE} when compared to yttrium and ytterbium when using standard digestion and analytical technique for inductively coupled plasma atomic emission spectrophotometry (ICP-AES). These three inert markers were digested using the hydrochloric:nitric acid method while [4] suggested the digestion of chromic oxide should use phosphoric acid since solubilisation of chromic oxide could be incomplete if using hydrochloric:nitric acid method [4]. Using standard faecal digestion protocols provide unsatisfactory chromic oxide recovery and results.

Chromic oxide has been widely used as inert digestibility markers of aquatic animals [26], Nevertheless, its reliability as an inert marker cannot be validated due to the analytical variability and lower recovery rate [27]. [28] measured feed digestibility in green abalone (*H. fulgens*), found that chromic oxide resulted in lower digestibility coefficient values compared to acid insoluble ash. However, it was digested using perchloric acid and not phosphoric acid as [4] have suggested. Moreover, negative apparent digestibility coefficients were observed when using chromic oxide as inert marker in digestibility study of *H. midae* [29]. The recovery rate value of chromic oxide was over 100% and apparent digestibility coefficient were higher than the other two inert markers demonstrating the limitation of using chromic oxide using standard digestion techniques. Moreover, to avoid analytical variation, the inclusion of chromic oxide into the feed should be in high concentration (5–10 g kg^{-1} feed) [4], while the amount of the chromic oxide that included into the formulated feed in this study was 0.02 g $100g^{-1}$. The inclusion of chromic oxide in high concentrations is to be avoided as it interferes with the absorption and metabolism nutrients and may affect the feed consumption[4, 30].

Compared to chromic oxide, yttrium and ytterbium provided consistent and accurate estimates of apparent digestibility at lower concentrations (100 $mg.kg^{-1}$) than chromic oxide [4].

3.2.2. Experimental approach

Results from this study assessed the possibility to obtain an accurate estimation of nutrient digestibility of feed in a commercial system. Conducting digestibility research in commercial systems seemed to be inappropriate due to several limiting factors. Contamination of silica in the feces samples occurred when conducting this study. This contamination was likely caused by unfiltered sea water used on farm that brings suspended solids and sediments from the sea, particularly during storm events. This contamination led to an inability to determine AD_{GE} of all samples collected from the experiment conducted on abalone farm due to imperfect combustion of the samples when determining gross energy using the bomb calorimeter. A different result was obtained when determining gross energy of samples from experimental tanks that had no contamination where the samples ignited perfectly.

Culture conditions including water temperature, water flow, tank design and the presence of microflora in tanks might have influenced the nutrient digestibility. Microflora that grows in the tanks is a food source for abalone containing nitrogen and energy. Therefore, instead of consuming the feed provided, the animals might preferentially to graze on microflora bed. Visual observation during the sample collection showed the growth of microflora bed in outdoor tanks, however further investigation is required to confirm this event was occurred.

On farm management was also a limiting factor in doing on farm digestibility research. The workers might accidentally feed the experimental animals with their commercial feed. During this study, despite clear instructions and training, on one occasion farm workers fed experimental tanks with incorrect feeds. Therefore, the samples from the contaminated tank were not used, reducing the samples available to be analysed. This highlights potential limitation of commercial-based trials where feeding regime is critical to results.

4. Conclusion

Ytterbium and yttrium are reliable inert markers in determining digestibility of feed ingredients. An appropriate method in chromic oxide digestion should be applied to obtain a more accurate estimation of nutrient digestibility. It is recommended to conduct nutrient digestibility research under experimental conditions to avoid sample contamination. Doing on-farm digestibility research should be well considered and all limiting factors should be mitigated.

Acknowledgement

I would like to express my greatest thanks to my valuable supervisor Dr. Louise Adams, Dr. Mark Adams of IMAS Launceston, University of Tasmania, Dr. David Stone of South Australia Research Development Institute for his excellent supervise, restless support and endless patience. Also my sincere thanks to Nick Savva and his Abalone Tasmania for facilitating my research. Daniel Pountney, Tan Duc Nguyen and Mean Torchaeron for their help during my lab work.

References

[1] Wee K L, Maguire G B and Hindrum S M 1994 Methodology for digestibility studies with abalone. 2. Comparisons of markers in an artificial diet for blacklip abalone (Haliotis rubra) and greenlip abalone (H. laevigata). *Third Asian Fisheries Forum* **1** 152
[2] Anderson T A 1988 *J. Fish Biol.* **32** 911
[3] Fleming A E, Van Barneveld R J and Hone P W 1996 *Aquaculture* **140** 50
[4] Austreng E, Storebakken T, Thomassen M S, Refstie S and Thomassen Y 2000 *Aquaculture* **188**

65
[5] De Silva S S 1989 Digestibility evaluations of natural and artificial diets *Asian Fish Soc. Spec* **4** 36
[6] Barneveld R J van, Fleming A E, Vandepeer M E, Kruk J A and Hone P W 1998 *J. Shellfish Res.* **17** 649
[7] Vandepeer M E, Van Barneveld R J, Hone P W and Havenhand J N 1999 *Proceedings of the 6th Annual Abalone Aquaculture Workshop* pp 17
[8] Ishikawa M, Teshima S, Kanazawa A and Koshio S 1996 *Fish. Sci.* **62** 229
[9] Combs D K and Satter L D 1992 *J. Dairy Sci.* **75** 2176
[10] Katersky R S and Carter C G 2010 *Comp. Biochem. Physiol. Part A Mol. Integr. Physiol.* **156** 529
[11] Ringø E 1993 *Can. J. Microbiol.* **39** 1169
[12] Ringø E 1994 *Aquac. Res.* **25** 341
[13] Bautista-Teruel M N and Millamena O M 1999 *Aquaculture* **178** 117
[14] Bautista-Teruel M N, Koshio S S and Ishikawa M 2011 *Aquaculture* **312** 172
[15] Britz P J 1996 *Aquaculture* **140** 55
[16] Cho S H, Park J, Kim C and YOO J 2008 *Aquac. Nutr.* **14** 61
[17] Ismail I, Jones C L W, Britz P J and Esterhuizen A J 2009 *African J. Mar. Sci.* **31** 103
[18] Mai K, Mercer J P and Donlon J 1995 *Aquaculture* **136** 165
[19] Viana M T, Lopez L M and Salas A 1993 *Aquaculture* **117** 149
[20] Wang W, Mai K, Zhang W, Xu W, Ai Q, Yao C and Li H 2009 *J. Ocean Univ. China (English Ed.* **8** 254
[21] Wang W, Mai K, Zhang W, Xu W, Ai Q, Liufu Z and Li H 2012 *Aquaculture* **330** 42
[22] AOAC 1990 *Official methods of analysis* (Virginia: Association of Official Analytical Chemists Inc)
[23] Bligh E G and Dyer W J 1959 *Can. J. Biochem. Physiol.* **37** 911
[24] McQuaker N R, Brown D F and Kluckner P D 1979 *Anal. Chem.* **51** 1082
[25] Maynard L A and Loosli J K 1969 Animal nutrition. *Anim. Nutr.*
[26] Vandenberg G W and De La Noüe J 2001 *Aquac. Nutr.* **7** 237
[27] Riche M, White M R and Brown P B 1995 *Nutr. Res.* **15** 1323
[28] Montaño-Vargas J, Shimada A, Vásquez C and Viana M T 2002 *Aquaculture* **213** 339
[29] Sales J and Britz P J 2001 *Aquaculture* **202** 113
[30] Browman H I and Marcotte B M 1987 *Progress. Fish-Culturist* **49** 140

Effect of inorganic fertilizer on the growth of freshwater *Chlorella* sp.

R A Rahardini, S Helmiati* and B Triyatmo

Department of Fisheries, Faculty of Agriculture, Universitas Gadjah Mada, Jl. Flora Bulaksumur, Yogyakarta, Indonesia

*e-mail: senny@ugm.ac.id

Abstract. One of the alternative nutrient sources in the culture of *Chlorella* sp. is inorganic fertilizer. The aim of this study was to determine the effect of inorganic fertilizer composition in the culture medium on the growth of *Chlorella* sp. This research was conducted by using the 2.5 L transparent containers containing 2 L freshwater. The culture was carried out in laboratory facilitated with an air conditioner, fluorescent light and continuous aeration for 14 days of culture. The treatments consisted of fertilizer composition of Urea:ZA:TSP i.e. P1 (1:1:0,50), P2 (2:2:0,70), P3 (3:3:1), P4 (4:4:1,25) and P0 (Walne medium) as a positive control. Each treatment was performed in triplicates and arranged based on the Completely Randomized Design. *Chlorella* sp. was inoculated with the initial density of 30×10^4 cells·mL^{-1}. The results of this study showed that the inorganic fertilizer composition of 3:3:1 provided the highest growth rate of 25.9 %/day, and the highest population density reached 2.348×10^4 cells·mL^{-1}. The doubling time was not affected by the inorganic fertilizer treatments.

1. Introduction

Plankton is natural feed for fish larvae, either zooplankton or phytoplankton. Plankton should be high nutrient, not harmful to fish, not pollute the environment, suitable with the size of fish's mouth and has a high tolerance to environmental change [1]. One of the phytoplankton used as natural feed is *Chlorella* sp.

The factors affecting the growth of *Chlorella* in the cultivation include light intensity, dissolved oxygen, temperature and nutrients. Indoor culture generally use 80 watt fluorescent bulb, with the light intensity of 5,000 lux–10,000 lux [2]. Good dissolved oxygen content for the growth of *Chlorella* sp. is > 2 mg L^{-1} [3]. A common temperature for the growth of *Chlorella* sp. ranges from 26 to 33 °C [4]. The nutrients needed for *Chlorella* consist of macro and micro nutrients. Macro nutrients are C, H, N, P, K, S, Mg and Ca, meanwhile micro nutrients consist of Fe, Cu, Mn, Zn, Co, Mo, Bo, Vn and Si [3]. Among these nutrients N and P are often limiting factors for the growth of microalgae. In addition to macro and micro nutrients, algae also need vitamins in supporting its growth. The vitamins needed by algae consist of cobalamin (vitamin B_{12}), thiamin (vitamin B_1) and biotin.

The nutrient required by Chlorella in the cultivation is obtained from culture medium. Water as a growing medium of *Chlorella* sp. only has limited mineral content. Thus, the addition of minerals from outside is absolutely necessary. The additional nutrient on laboratory scale cultivation is usually in the form of Walne medium while on a mass scale is in the form of inorganic fertilizers. The inorganic fertilizer commonly used is urea, TSP and ZA. Urea is an artificial fertilizer containing primary nutrient which is nitrogen with the chemical formula $CO(NH_2)_2$. Urea contains 46 % element of nitrogen [5].

Ammonium sulfate (ZA) has the chemical formula $(NH_4)_2SO_4$. ZA in a pure state contains 21 % element of nitrogen (N) and 23 % nutrient of sulphur (S). Phosphorus (P) in the fertilizer is indicated in P_2O_5. TSP (Triple Super Phosphate) contains phosphorus of 44 % P_2O_5. The dose of inorganic fertilizer given to the cultivation media will affect the growth of *Chlorella* population, hence this study aim was to determine the effect of the different doses of inorganic fertilizer on the growth of *Chlorella* sp.

2. Materials and methods

The research was conducted from February to May 2016 in the Laboratory of Aquaculture Department of Fisheries, Faculty of Agriculture, Universitas Gadjah Mada. This research was conducted by the experimental method using Completely Randomized Design with four treatments and one control, each treatment consisted of three replications. The factors observed were the effect of inorganic fertilizers with ratios of urea: ZA: TSP as follows P1= 1:1:0.05; P2 = 2:2:0.70; P3 = 3:3:1; P4 = 4:4:1.25.

Chlorella sp. was obtained from a pure culture developed by the Technology Development Center of Marine and Fisheries Yogyakarta. The culture medium at the control treatment used in the study was Walne medium (Table 1). Cultures of *Chlorella* sp. was conducted in a jar containing 2 L fresh water media for 2 weeks. Observation was performed on the growth of *Chlorella* sp. population and water quality. The water quality consist of temperature, pH, carbon dioxide-free, nitrate, phosphate and ammonia.

The research was carried out by maintaining a pure culture of *Chlorella* sp. in the jar. Water culture medium was freshwater which came from groundwater. The culture medium in the jar was enriched with inorganic fertilizers and performed the initial stocking density of *Chlorella* sp. at 30×10^4 cells·mL^{-1} which was determined by using the formula [6]:

$$V1 \times N1 = V2 \times N2$$

Description:
V1 = the volume of the desired seed in the stocking
V2 = the volume of media culture
N1 = stocking density
N2 = the desired initial stocking density

The density of *Chlorella* sp. was observed daily by haemocytometer periodically at 10:00 am. The density observation was started on the second day until the death phase. The density of *Chlorella* sp. was enumerated by following [7].

$$\text{Cell density (cells·mL}^{-1}\text{)} = \frac{\text{The cell total in 4 block (=4)}}{\text{The sum of block (=4)}} \times 10^4$$

The growth of *Chlorella* sp. was determined based on the specific growth rate which was the change in the number of cell mass per unit of time. The specific growth rate was determined by following [8]:

$$K = \frac{\ln Nt - \ln N0}{t} \times 100\ \%$$

Description:
K = Specific growth rate (cells/day)
Nt = the density of *Chlorella* sp. in the peak population
No = the density of *Chlorella* sp. in the initial population
t = the time required to reach the maximum population

Table 1. Composition of Walne Medium.

	Content	In 1000 mL H_2O
1	$NaNO_3$	100.00 mg
	Na_2EDTA	45.00 mg
	H_3BO_3	33.60 mg
	NaH_2PO_4. 2 H_2O	20.00 mg
	$FeCl_3$. 6 H_2O	1.30 mg
	$MnCl_2$. 4 H_2O	0.36 mg
2	Vitamin	
	B_1	0.100 mg
	B_{12}	0.005 mg
3	Micro Metallic Solution	
	$ZnCl_2$	0.021 mg
	$CoCl_2$. 6 H_2O	0.020 mg
	$(NH_4)_6$. Mo_7O_{24}. 4 H_2O	0.009 mg
	$CuSO_4$. 5 H_2O	0.020 mg

Source: [6]

The doubling time was calculated by using the formula [9]:

$$k=\frac{t\log 4}{\log b-\log a}$$

Description:
k = the doubling time (day)
t = the time to reach maximum population (day)
a = the initial density of *Chlorella* sp. ($\times 10^4$ cells·mL^{-1})
b = the maximum density *Chlorella* sp. ($\times 10^4$ cells·mL^{-1})

Water quality including water temperature, pH, light intensity, content of ammonia, phosphate, nitrate and free carbon dioxide were observed. The temperature measurement was carried out everyday. The measurements of free carbon dioxide, water pH, contents of ammonia, phosphate, nitrate, and the light intensity were performed at the beginning, middle and end of the culture.

3. Results and Discussion

3.1. The growth rate of Chlorella sp.

Cell density, population peak, doubling time, specific growth rate and growth rate are parameters that commonly used to estimate the growth of cells in the culture medium [10]. The cell density was measured every day during the culture until the cell growth reached the death phase.

Daily population density of *Chlorella* sp. in all treatments increased from the first day. The treatment with the composition of fertilizers 3:3:1 resulted the greatest growth rate which was 25.9 %/day. Based on the Dunnet analysis of the growth rate of *Chlorella* sp., it was found that a composition of inorganic fertilizer 3:3:1 resulted the similar growth rate with the control treatment.

The differences of population growth rate of *Chlorella* sp. in all treatments was resulted from the difference dosage of fertilizer [11]. The differences in the growth rate in this study is considered because of the variation of nutrient ratio in treatments. The composition of inorganic fertilizers 3:3:1 produced the highest population growth rate, because the dose was similar with the control and was not excessive dose for growth of *Chlorella* sp.

Extra doses of inorganic fertilizer in the culture medium of *Chlorella* sp. does not always produce the higher growth rate. The excessive doses of the nutrient to the treatment of inorganic fertilizer composition 4:4:1.25 produced the lowest population growth rate among all treatments. Nutrient deficiencies in the media will inhibit the growth rate seen in the treatment of inorganic fertilizer composition 1:1:0.5.

3.2. Population Peak

Daily cell density showed an increase since the first day of observation, in which the initial density was 30×10^4 cells·mL^{-1} and reached the maximum population at a certain time. In the first day of inoculation, the population of *Chlorella* sp. grew in the exponential growth phase without going through a stage of adaptation. The exponential growth phase generally occurs at the 7th day to the 9th day. The cell density began to decline after the culture of *Chlorella* sp. reached a maximum population and the observation was stopped at the 14th day (Figure 1).

The results showed that *Chlorella* sp. has high adaptability to the new environment. It could be seen from the value of the growth rate from the first day which was quite high at 49.2 %/day to 51.8 %/day. This happened because the inoculum of algae and the water media used were obtained from asimilar laboratory environment. Another supporting factor was that *Chlorella* sp. cells has been adapted to the research environment before the experiment was conducted.

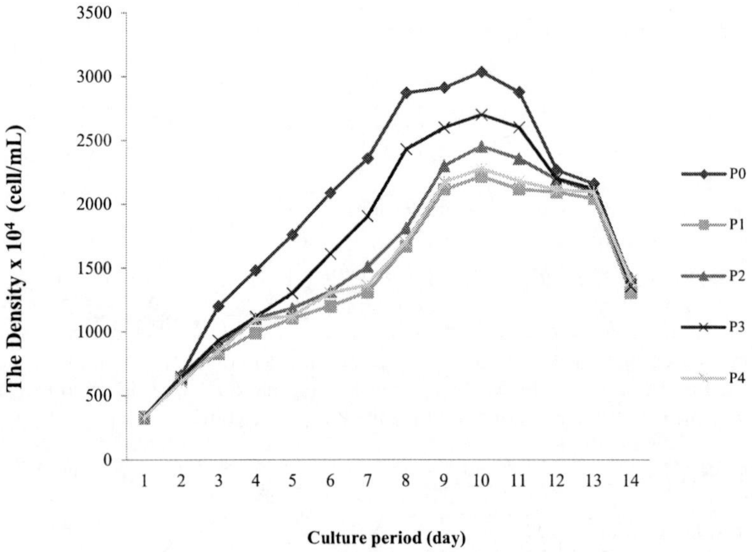

Figure 1. The density of *Chlorella* sp. with various ratio of urea:ZA:TSP in culture medium.

The peak population is the final process of the exponential phase of growth in the cycles of phytoplankton growth. The peak population is usually used to find out when the culture of microalgae can be harvested because at that time the population density reaches its maximum. The highest peak population density of *Chlorella* sp. was found in inorganic fertilizer with the composition of 3:3:1 at $2{,}348 \times 10^4$ cells·mL^{-1} which was achieved at the day of 9.7, while the lowest peak population density for the inorganic fertilizer with the composition 1:1:0.50 at $1{,}924 \times 10^4$ cells·mL^{-1} which was achieved at day of 11.2. The maximum population density of *Chlorella* sp. during the maintenance along with the treatment of a different inorganic fertilizer composition was presented in the Table 2.

Table 2. The maximum cells density of *Chlorella* sp. at the various ratio of urea:ZA:TSP in culture medium.

Fertilizer composition	Time to reach peak (day)	Density (x10^4 cells·mL^{-1})
0.3:0.3:0.2	9.1	2,733 $\pm$ 857.018
1:1:0.50	11.2	1,924 $\pm$ 621.157
2:2:0.70	11.0	2,088 $\pm$ 678.043
3:3:1.00	9.7	2,348 $\pm$ 775.449
4:4:1.25	11.2	1,989 $\pm$ 635.349

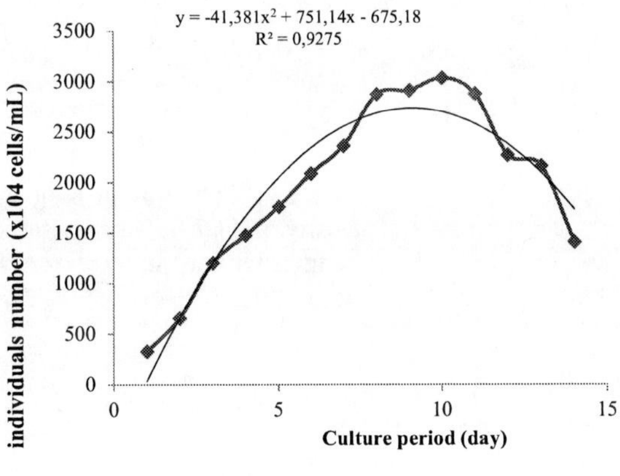

Figure 2. The curve of population density of *Chlorella* sp. cultured in Walne medium 0.3:0.3:0.2.

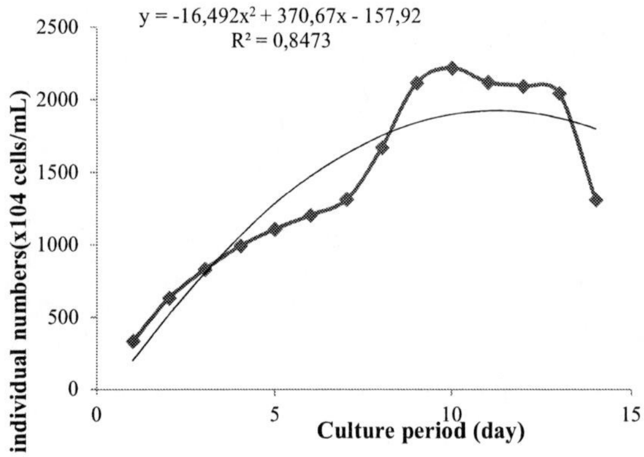

Figure 3. The curve of population density of *Chlorella* sp. cultured in medium with a ratio urea:ZA:TSP (1:1:0).

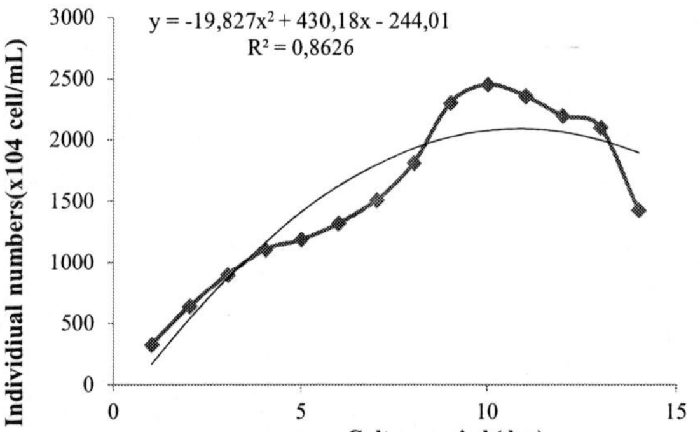

Figure 4. The curve of population density of *Chlorella* sp. cultured in medium with a ratio urea:ZA:TSP (2:2:0.70).

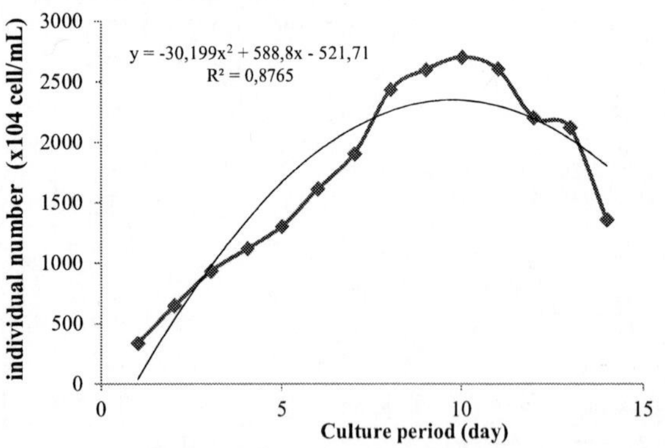

Figure 5. The curve of population density of *Chlorella* sp. cultured in medium with a ratio urea:ZA:TSP (3:3:1).

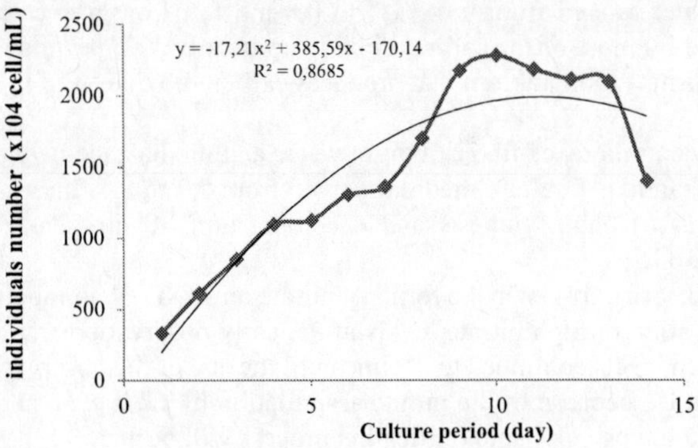

Figure 6. The curve of population density of *Chlorella* sp. cultured in medium with a ratio urea:ZA:TSP (4:4:1.25).

Population of *Chlorella* sp. decreased after reaching its maximum population. The decline was caused by the availability of nutrients in the culture medium decreased and eventually loss so that the cells deprived nutrients for growth. The growing population was occurred when nutrients concentration was enough, so that the cell division occurs rapidly (exponential phase), followed by the formation of autosphora and enlargement of cells. Population growth decreased (stationer phase) after the nutrients have been in limited concentration or run out. Dead phase occurred when the nutrients run out, so the population growth stopped and there were many deaths cells in a population of *Chlorella* sp.

3.3. Doubling time

Doubling time is time required for microalgae to be double in density, while the growth rate is the rate of growth per unit time during the last exponential phase. Analysis of variance of the doubling time indicated that inorganic fertilizers did not affect significantly the population doubling time of *Chlorella* sp.

Inorganic fertilizer application in medium provided nutrients for the process of cell division that caused the population continue to increasing. Based on the observations it can be concluded that the inorganic fertilizers with specific composition are not always able to short the cell doubling time of *Chlorella* sp.

3.4. Water quality

Phytoplankton growth quietly related to the macro and micronutrients availability, and also it was affected by the environmental condition in the culture media. Environment factors that affect the phytoplankton growth are light, temperature, pH, water and carbon dioxide-free, which can stimulate or inhibit the growth of phytoplankton [12]. The light intensity during the study ranged from 3,020 lux to 5,770 lux. Optimum light intensity for growth of *Chlorella* sp. ranged between 2,000 lux to 8,000 lux [13].

Proper temperature for *Chlorella* sp. growth is 25 to 32 °C [6]. Water temperature of media during research was in the range of 26 to 29°C. Water temperature of media during research was in optimal range and stable because during the culture the laboratory was facilitated with the lamp and in indoors room with the air conditioner. Temperature is an important limiting factor for organism life, because every organism has limiting ability to tolerate temperature change in environment. Instead of that, temperature affects the stability of the enzyme catalyst in the process of photosynthesis by phytoplankton. Algal cell metabolism increases with rise in temperature to achieve optimal temperature and will fall back if the temperature has exceeded the optimum point.

Carbon dioxide is limiting factor to all of living creatures in aquatic [14]. Lack of free carbon dioxide would harm producers organisms (phytoplankton and water plants) so stunted. Too high carbon dioxide content also has detrimental effects on aquatic organisms. Free carbon dioxide is important factor for *Chlorella* sp. growth because it is directly necessary for organic molecules synthesis through photosynthesis. Free carbon dioxide in water ranged from 0 mg L^{-1} to 0.8 mg L^{-1}. Low free carbon dioxide content in water media was caused by most of the carbon dioxide up took by *Chlorella* sp. in photosynthesis, and no other aquatic organisms (zooplankton) that produces carbon dioxide lived in the culture media.

pH plays a role in determining the concentrations of free carbon in water and the balance between bicarbonate and carbonate. During this research, pH culture medium ranged from 7.7 to 8.1. Increased in medium pH was caused by the process of photosynthesis and decomposition of nitrogen from fertilizers that produced acid CO_2 and NH_4^+ [15].

Phytoplankton including *Chlorella* sp. absorb nitrogen in the forms of nitrate ions (NO_3^-), ammonium (NH_4^+) and nitrite (NO_2^-) [16]. The observation of high nitrate levels in the early observations ranged from 1.17 mg L^{-1} up to 1.99 mg L^{-1}. Nitrogen levels continued to decline until the end of the observation ranging from 0.253 mg L^{-1} up to 0.609 mg L^{-1} because of the nitrate assimilation by *Chlorella* sp. for growth. When organic materials are aerated intensively, ammonification process will occur.

Result indicated that phosphate concentration in water at the beginning of observation was high, but decrease until the end of observation around 0.009–0.146 mg L^{-1}. Orthophosphate (PO_4^{3-}) is dissolved in the form of inorganic phosphates, which can directly be used by *Chlorella* sp. The concentration of orthophosphate can be reduced more quickly in alkaline water media than in acidic water as the reduced levels of phosphorus are directly affected by pH and Ca concentration [15].

4. Conclusion

Culture of *Chlorella* sp. with inorganic fertilizer composition of 1:1:0.50; 2:2:0.70; 3:3:1; 4:4:1.25 produce highest population density of 1,924, 2,088, 2,348, 1,989 × 10^4 cells·mL^{-1} and achieved at day of 11.2, 11, 9.7 and 11.2, respectively. Walne medium produces highest population density of 2,733 × 10^4 cells·mL^{-1} at day of 9.1. The composition of inorganic fertilizers 3:3:1 produces the highest population density of 2,348 × 10^4 cells·mL^{-1} at day of 9.7. The composition of inorganic fertilizers 3:3:1 generates the highest growth rate of 25.9 %/day.

Suggestion

Inorganic fertilizer application provide nutrients to sustain life and reproduction of *Chlorella* sp. in organic fertilizers might be used as a substitute for Walne medium, since it is cheaper and sufficient nutritional content for the growth of *Chlorella* sp. inorganic fertilizer with the composition of 3:3:1 might be applied for the cultivation of *Chlorella* sp. in a mass scale as it is able to generate the highest rate of population growth.

References

[1] Becker W 2004 *Handbook of Microalgal Culture* ed A Richmond (Oxford: Blackwell) p 315
[2] Fulks W and Main K L 1992 *Rotifer and Microalgae Culture Systems Proceeding of a US-Asia Workshop* (Hawai: The Oceanic Institute Honolulu)
[3] Round F E 1973 *The Biology of Algae* (London: Edward Amold Ltd)
[4] Fogg G E 1975 *Algae Culture and Phytoplankton Ecology* (London: The University of Wisconsin Press)
[5] Sutedjo M M 2000 *Pupuk dan Cara Pemupukan* (Jakarta: Rineka Cipta)
[6] Isnansetyo A and Kurniastuty 1995 Teknik Kultur Phytoplankton dan Zooplankton *Pakan Alami untuk Pembenihan Organisme Laut* (Yogyakarta: Kanisius)
[7] Taw 1990 *Petunjuk Kultur Murni dan Massal Mikroalga* (USA: UNDP, FAO)
[8] Banerjee S, Wei W E, Helena K, Mohamed S and Fatimah M Y 2011 *Growth and Proximate Composition of Tropical Marine Chaetoceros calcitrans and Nannochloropsis oculata Cultured Outdoors and Under Laboratory Conditions*. African Journal of Biotechnology **10** 1375–1383

[9] Wirosapoetro S 1997 *Chlorella, Budidaya, Manfaat, dan Prospek Pasar* (Yogyakarta: Jurusan Perikanan. Fakultas Pertanian Universitas Gadjah Mada)
[10] Sostaric J, Golob M, Bricelj D, Klinar, A and Pivec 2009 *Chem. Biochem. Eng.* **23** 471–477
[11] Jayasankar R and Valsala K K 2008 *J. Mar. Biol. Assoc* **50** 74–78
[12] Yani and Yosar 2009 *Pemanfaatan Algae Chlorella sp. dan Eceng Gondok Untuk Menurunkan Tambaga (Cu) Pada Industri Pelapisan Logam* (Semarang: Jurusan Teknik Kimia, Fakultas Teknik. Universitas Diponegoro)
[13] Sylvester B, Nelvy D and Sudjiharno 2002 *Persyaratan Budidaya Fitoplankton dan Zooplankton. Balai Budidaya Lampung* (Lampung: Departemen Kelautan dan Perikanan)
[14] Goldman C R and Horne J 1983 *Limnology* (New York: McGraw-Hill)
[15] Boyd C E 1982 *Water Quality in Warm Water Fish Pond* (Auburn Alabama: Auburn University Agricultural Experimental Station) p 359
[16] Chen F and Michael R J 1991 *Journal of Applied Phycology* **3** 203–209

IOP Conf. Series: Earth and Environmental Science **139** (2018) 012006 doi:10.1088/1755-1315/139/1/012006

The effectiveness of crude papain enzyme supplement for tilapia's (*Oreochromis niloticus*) growth at the floating nets of Cirata Reservoir

R Rostika[1]*, Sunarto[1], H N Sugiyanto[1] and L P Dewanti[1]

Fisheries Study Program–Fisheries and Marine Science Faculty, Universitas Padjadjaran, Jl. Raya Bandung Sumedang, Hergamanah, Jatinangor, West Java, Indonesia

*Email: ritarostika_unpad@yahoo.com

Abstract. Papain is an enzyme capable of hydrolyzing protease into a more simple elements i.e. the peptide to amino acids. The enzyme in the feed can increase the absorption of protein and digestion rate in the digestive tract of fish. This research examined the effective level of enzyme papain to increase the Feed Utilization Efficiency (FUE), Protein Efficiency Ratio (PER) and Average Daily Gain (ADG). This research used Completely Randomized Design (CRD) with five treatments i.e. treatment A (control), treatment B (1.5 %), treatment C (2.25 %), treatment D (3 %) and treatment E (3.75 %) in triplicate. Tilapia (*Oreochromis niloticus*) with the average initial weight of 17 g, and initial total lenght of 8–10 cm was fed three times daily at feeding rate of 5 % of the total body weight. The results showed that supplementation of papain in the feed significantly increased the activity of protease, FUE, PER and ADG. The optimal dose of the enzyme papain at 3.75 % was able to increase 48.31 % of FUE, 2.13 % of PER and 2.07 % of ADG.

1. Introduction

Cirata Reservoir is one of sites to perform aquaculture activities in West Java. It is categorized as a multipurpose dam. Nowadays, people are highly enthusiastic to culture fish at Floating Net Cages (FNC) in Cirata Reservoir. However, those have already exceeded the reservoir carrying capacity that, at some point or when ignored, mass mortality of fish may occur due to its severe eutrophication. Eutrophication occurs when there are too many inputs coming from Citarum River upstream's from various activities, such as agriculture, livestock, industry, household and aquaculture itself.

One of the efforts to reduce eutrophication is by improving the feed quality for aquaculture through papain enzyme addition. Papain is a protease enzyme found in papaya sap. The enzyme is used to break down perfectly peptide bonds in protein that are decomposed into simpler peptide bonds as papain may catalyze the hydrolysis reaction [1].

The role of enzyme is greatly essential in protein digestive processes. In addition to endogenous enzymes, the exogenous ones are also required to speed up the protein hydrolysis processes. One exogenous enzyme which has the ability to rapidly hydrolyze proteins is papain enzyme. Papain is expected to improve protein feed utilization for tilapia growth. Efficient feed can improve body deposition as well as to reduce the total amount of dissolved solid in the waters. It means that the quality of water media in the Floating Net Cages is expected to be well maintained in accordance with the criteria for aquaculture.

The aims of this research was to examine the most appropriate dosage of papain enzyme to improve the value of Feed Efficiency Utilization (FEU), Protein Efficiency Ratio (PER) and Daily Growth Rate (DGR).

2. Materials and Methods

2.1. Research place and time

The research place was in Cirata Reservoir which was located in Cianjur Regency, West Java. Fish identification was performed at Animal Physiology Laboratory of Faculty of Fisheries and Marine Sciences, Padjadjaran University, Jatinangor. The enzyme activity test was performed in the Laboratory of Researches and Services, Chemistry Department, Faculty of Mathematics and Natural Sciences, Padjadjaran University. The research was conducted from August 2015 to January 2016.

2.2. Methods

This research was experimentally conducted with five treatments and three replications. Observations on tilapia weight were conducted every 15 days for 2 months. The cultivated tilapia size was about 10–12 cm with the initial weight of 16–18 g, fed with 5 % of its body weight, and then differently treated with papain enzyme addition in feed. The inclusion level used in this research is as follows:

A: Feed without enzyme (Control)
B: Feed hydrolyzed with papain with the dosage of 1.5 %
C: Feed hydrolyzed with papain with the dosage of 2.25 %
D: Feed hydrolyzed with papain with the dosage of 3 %
E: Feed hydrolyzed with papain with the dosage of 3.75 %

The variables studied were Survival Rate (SR), Feed Efficiency Utilization (FEU), Protein Efficiency Ratio (PER), Daily Growth Rate (DGR) and water quality.

2.3. Research parameters

2.3.1. Survival Rate (SR). Survival Rate was calculated by using the following formula [2]:

$$SR = \frac{Nt}{N0} \times 100\ \%$$

Description:
SR = Survival Rate (%)
Nt = fish number day-t (fish)
N0 = fish number day-0 (fish)

2.3.2. Feed Efficiency Utilization (FEU). Feed Efficiency Utilization (FEU) was calculated by using the following formula [3]:

$$FEU = \frac{(Wt - W0)}{F} \times 100\ \%$$

Description:
FEU = Feed Efficiency utilization (%)
W0 = fish biomass day-0 (g)
Wt = fish biomass day-t (g)
F = fish feed amount consumed during the study (g)

2.3.3. Protein Efficiency Ratio (PER). Protein Efficiency Ratio was calculated by using the following formula [3]:

$$PER = \frac{(Wt - W0)}{Pi} \times 100\ \%$$

Description:
PER = Protein Efficiency Ratio (%)
W_0 = fish biomass day-0 (g)
W_t = fish biomass day-t (g)
Pi = the consumed feed protein weight (g)

2.3.4. Daily Growth Rate (DGR). Daily Growth Rate was calculated by using the following formula [4]:

$$DGR = \frac{(lnWt - lnW0)}{t} \times 100\ \%$$

Description:
DGR = Daily Growth Rate
lnW0 = fish biomass day-0 (g)
lnWt = fish biomass day-t (g)
t = Length of Cultivation (day)

2.4. Data analysis
Data were analyzed using variant investigation analysis (F-test) with confidence level of 95 %. If the treatment had a significant influence on variant analysis (ANOVA), then Duncan's multiple range test was conducted to determine the difference between treatments [5]. Water quality data were descriptively analyzed and used as the research supporting data.

3. Results and Discussion

3.1. Protease enzyme activities on papain fish digestion
The tested treatments are control treatment A and treatment B which was given the lowest papain enzyme inclusion in the feed by only 1.5 %. Protease enzyme activities in treatment B was higher than those in treatment A with the difference of 0.95 activity U·mg^{-1} Protein. It is assumed that in treatment B, the exogenous papain enzyme which was added to the fish feed might improve the activities of fish protein break down within fish intestine that amino acid absorption is much better and might result in the improving feed digestibility level that the body might utilize for its metabolism and growth.

3.2. Survival rate
Tilapia's survival rate was not significantly different in each treatment and still within the most favorable range due to the fish positive reactions to the enzyme- added feed characterized by the eaten enzyme-added feed. In addition to feeding factors, the environmental conditions could also influence tilapia's survival rate.

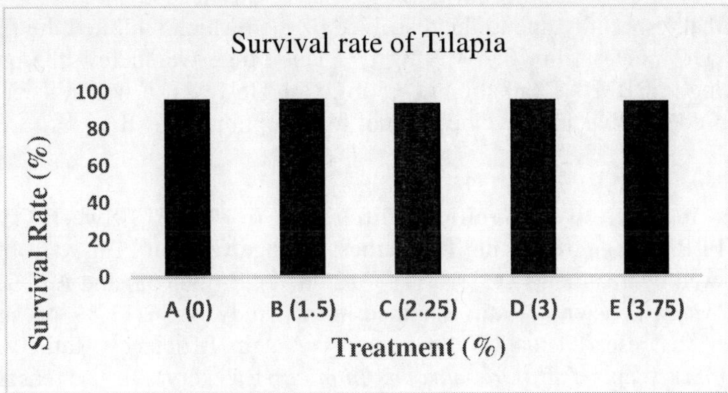

Figure 1. Survival rate of tilapia.

Feed was not only serves as energy sources, but also to maintain fish's survival, growth and reproduction. The protein content in merkartha feed given was ± 25 %, then supplementation of papain protease enzyme accelerate the process of complex protein break down into simple protein up to amino acid that protein might be easily digested as energy sources, growth and reproduction.

3.3. Feed Efficiency Utilization (FEU)

Feed Efficiency Utilization value showed the significant difference and tended to improve at higher dosages. When treatment A (control) was compared with the lowest papain enzyme inclusion (treatment B), the difference was about 5.77 %.

Papain was protease enzyme which might hydrolyze proteins into simpler elements, such as peptides up to amino acids. Papain addition as exogenous enzyme into feed might improve its protein hydrolysis which might result in the improving feed digestibility level. Higher digestibility level might improve its amino acid absorption ability to its body to grow [6]. In addition, feed quality is one factor which could influence protein efficiency value.

Table 1. Duncan test feed efficiency utilization fed with papin-supplemented feed at various dose.

Treatment of Papain Enzyme	Feed Efficiency Utilization (%)
A (0 %)	40.26 ± 0.07^{a}
B (1.5 %)	46.03 ± 0.05^{b}
C (2.25 %)	47.58 ± 0.03^{bc}
D (3 %)	48.01 ± 0.04^{cd}
E (3.75 %)	48.31 ± 0.09^{cd}

Mean value followed by different superscript indicates the significant different ($p < 0.05$).

Results of this study were also in line with those conducted by previous researchers [6] who reported that feed efficiency utilization is directly proportional to the exogenous enzyme dosage added to the feed. Their study used catfish fed with enzyme-supplemented feed in doses of A (0 %), B (0.75 %), C (1.5 %) and D (2,25 %) exhibiting Feed Efficiency Utilization value of 44.63 %; 52.71 %; 53.85 % and 62.83 %, respectively.

The obtained Feed efficiency utilization value ranges from 40.26–48.31 % that this could not be yet considered in a good category. Feed might be considered in good category when the feed efficiency value is more than 50 % or even close to 100 %. It was assumed that the water quality at the research

IOP Conf. Series: Earth and Environmental Science **139** (2018) 012006 doi:10.1088/1755-1315/139/1/012006

sites was less optimal, especially due to the dissolved oxygen which inhibited the fish growth [7]. The cultivation pools' DO ranges from 0.7–3.9 mg·L^{-1}. This range was below the quality standard for cultivation according to BBPBAT Sukabumi (> 5 mg·L^{-1}). However, it was still within the reasonable limit that tilapia could still live and develop, yet not for its optimal growth.

3.4. Protein Efficiency Ratio (PER)

This study showed that there were significant differences ($p \leq 0.05$) between the values of Protein Efficiency Ratio (PER) among treatments The highest average value of PER was obtained in treatment E by 2.13 %, followed by treatments D (2.10 %), C (2.06 %), B (2.03 %) and A (1.75 %) (table 2).

The value of Protein Efficiency Ratio obtained in this study was 2.13 %. Meanwhile, in research conducted by previous researchers [8], the value of Protein Efficiency Ratio was 2.37 % which examined young black tilapia fish (*Oreochromis niloticus* Bleekery). In the reasearch conducted by previous researchers [6] with the tested catfish, the highest Protein Efficiency Ratio value was 1.97 %. Similarly, the research conducted by previous researchers [9] with the tested carp fish shows that the highest Protein Efficiency Ratio value was 2.24 %.

Table 2. Test of Protein Efficiency Ratio (PER) fed with papin-supplemented feed at various dose.

Treatment of Papain Enzyme	PER (%)
A (0 %)	1.75 ± 0.02^{a}
B (1.5 %)	2.03 ± 0.01^{b}
C (2.25 %)	2.06 ± 0.00^{c}
D (3 %)	2.10 ± 0.01^{cd}
E (3.75 %)	2.13 ± 0.02^{cd}

Mean value followed by different superscript indicates the significant different ($p < 0.05$).

PER values was significantly difference among treatment ($P \leq 0.05$). This result might be caused by the different concentrations of protease in intestines of tilapia. This results supported by a research conducted by previous researchers [10], on the development of tilapia digestive enzymes concluding that all digestive enzymes, including protease enzymes are already exist in fish intestines. Due to the papain exogenous enzyme addition to the feed, it was expected that the addition might accelerate the intestinal digestive processes to increase the breakdown proteins to amino acids.

3.5. Daily Growth Rate (DGR)

Results indicated that the different papain enzyme supplemented in the feed generated various daily growth rate. Papain enzyme-containing feed given to tilapia had positive response to the fish growth indicated by increasing tilapia weight average in each sampling time (15 days). The average weight of tilapia in each treatment increased during the culture period.

This study provided the Daily Growth Rate averages that might be considered in good category, as it was above the minimum good growth rate value of 1 % [11]. Variant analysis indicated that the treatment influenced the growth rate of tilapia. The results of Duncan's multiple range test showed that daily growth rate of tilapia fed with papain supplemented feed was significantly different ($p \leq 0.05$) comparing to the control (table 3).

Tabel 3. Result of Daily Growth Rate (DGR) fed with papin-supplemented feed at various dose.

Treatment	DGR (%)
A (0 %)	2.11 ± 0.01^a
B (1.5 %)	2.44 ± 0.02^b
C (2.25 %)	2.54 ± 0.02^{bc}
D (3 %)	2.68 ± 0.01^d
E (3.75 %)	2.67 ± 0.02^d

The administration of papain enzyme 3 % in treatment D of this study showed the highest daily growth rate by 2.68 %. The results of research conducted by previous researchers [6], showed that the supplementation of papain up to 2.25 % resulted in daily growth of 2.89 % for catfish. In addition, the results of research conducted by previous researchers [8] showed that the supplementation of papain enzyme up to 2.25 % resulted in daily growth of 1.83 % for young black tilapia (*Oreochromis niloticus* Bleeker). Thus, it was then assumed that papain might hydrolyze proteins contained in feed that the feed given to the fish might be efficiently utilized and increase its daily growth rate.

4. Conclusion

Based on the research results, it could be concluded that the administration of feed-containing papain enzyme significantly improves protease enzyme activity in tilapia's digestive system. The optimal dose of 3.75 % papain enzyme improves Feed Utilization Efficiency value by 48.31 %, Protein Efficiency Ratio value by 2.13 % and Daily Growth Rate by 2.67 % of tilapia.

References

[1] Muchtadi D, Palupi N S and Astawan M 1992 *Enzim dalam Industri Pangan* (Bogor: Pusat Antar Universitas Pangan dan Gizi ITB)
[2] Effendi M I 2002 *Biologi Perikanan* (Yogyakarta: Yayasan *Pustaka* Nusatama)
[3] Tacon A G 1987 *The Nutrition and Feeding of Farmed Fish and Shrimp* (Brazil: FAO) pp 106–09
[4] Steffens W 1989 *Principles of Fish Nutrition Ellis* (West Sussex: Horwood Limited) pp 384
[5] Srigandono 1992 *Rancangan Percobaan Design* (Semarang: Universitas Diponegoro) p 23–36
[6] Amalia R, Subandiyono and Arini E 2013 *Jurnal of Aquaculture Management and Technology* **2** 136–43
[7] Craig S and Helfrich L A 2002 *Understanding fish nutrition, feeds, and feeding* (Virginia: Cooperative Extension, Publication) 420–256
[8] Rachmawati D, Irawati D and Pinandoyo 2015 *Journal of Aquaculture Management and Technology* **4** 1–9
[9] Singh P, Maqsood S, Samoon M H, Phulia V, Danish M and Chalal R S 2011 *J. Int. Aquat. Res.* **3** 1–9
[10] Tengjaroenkul B, Smith B J, Caceci T and Smith S A 2000 *Aquaculture* **211** 241–51
[11] Retnosari D 2007 *Skripsi* Universitas Padjajaran Jatinangor

Characteristics and environmental carrying capacities of coastal area in Yogyakarta Special Region for aquaculture

B Triyatmo[1], Rustadi[1], S B Priyono[1]

[1] Department of Fisheries, Faculty of Agriculture, Universitas Gadjah Mada
Jl. Flora Bulaksumur, Sleman, Yogyakarta, Indonesia
*Email: btriyatmo@ugm.ac.id; btriyatmo@yahoo.com

Abstract. The purpose of this study were to determine characteristics and environmental carrying capacities of coastal area in Yogyakarta Special Region for aquaculture. This study was conducted in 2015 by characterizing land and water dynamics, land use, and the suitability of coastal environments for aquaculture. Evaluation on the coastal environments suitability for aquaculture ponds was based on the landforms, soil properties, water quality and land. Selection of coastal locations for aquaculture development was based on the level of suitability of coastal environment. The results showed that the coastal in Kulon Progo and Bantul Regencies were characterized by sand dune and beach ridge with sandy soil texture, while in Gunungkidul Regency was characterized by limestone hill with rocky texture. Water sources of the coastal area were the sea, river, and ground water with the salinity of 31–37, 7–11, 7–31 ppt and pH of 7.4–8.4; 7.0–8.2 and 7.4–9.9, respectively. The coastal lands were used for seasonal/annual planting, ponds, fish landing sites, tourism areas and conservation areas. The coastal carrying capacity was rather suitable for aquaculture, especially in the sandy soil area. Aquaculture in that area can be done intensively for shrimp (*Litopenaeus vannamei*), using biocrete (biological material) or plastic sheet.

1. Introduction

Carrying capacity is an important concept for ecosystem-based management, which helps set the upper limits of aquaculture production given the environmental limits and social acceptability of aquaculture, thus avoiding "unacceptable change" to both the natural ecosystem and the social functions and structures. Carrying capacity for aquaculture is based solely on production, they have been developed further into a more comprehensive four-category approach based on physical, production, ecological and social carrying capacities [1].

Coastal areas have diverse natural resources with various functions, i.e. agriculture, fisheries, settlements, ports, tourism and/or industry. Some activities in the region often exceed the carrying capacity of the environment [2]. Responsible coastal management should be used as a basis for coastal development, both for fish farming and other sector activities. The fish/shrimp ponds were developed in order to optimize the utilization of coastal environment [3].

Production of aquaculture in some places within few years after the intensification has decreased, since the carrying capacity of the ponds become lower. This problem happened in ponds production centers such as the north coast of Java and the east coast of Lampung Province. About 80 % of the land on the north coast was used for shrimp farms [4]. The coastal area has been damaged and affected by various diseases and pollution problems. The waste of residual feed and shrimp faeces into the aquaculture environment contributed the addition of nutrients to the aquatic environment [5]. On the other hand, coastal area of southern part of Java Island is a potential area for ponds development. Pond constrution in the area is one of alternatives to increase the land productivity, especially with regards to the good water quality and opening to the sea.

Coastal areas with diverse landforms and dynamic waters generally have a varied aquaculture fishery model. The landform units reflect the similar morphological and soil properties [6]. The dynamics of coastal waters throughout the year are determined by tidal sea water, freshwater supply from rivers and rain. The relation of land character and coastal waters, and their use has been used as the basis for determining the suitability (carrying capacity) of the coastal environment for aquaculture. The suitability of a particular land unit is a natural ability to support a specific use for aquaculture. The dynamics of waters throughout the year is also a factor that determines the model of aquaculture in relation to time (season) [7, 8].

The coastal area of Yogyakarta Special Region (Daerah Istimewa Yogyakarta) has a karst hill land, sand dune, swale and alluvial plains. The rivers that flows to the coastline have a water dynamic throughout the year. The current use of coastal sand is also varies. In the last four years, shrimp farming conducted by local communities has been growing rapidly on the coastal sandy land. Salt/brackish water source for shrimp ponds was pumped from coastal groundwater (wells). The coastal area of DIY have certain environmental characteristics (abiotic, biotic and culture). Therefore, shrimp farming in the coastal areas should be developed based on the specific location model [8]. The purpose of this study were to determine the characteristics and environmental carrying capacities of DIY coastal area for aquaculture ponds. In this study, especially on the physical carrying capacity is based on the suitability of aquaculture development by considering the physical factors of the environment and the farming system. The result of the present study is important for development of sustainable and eco-friendly coastal aquaculture fisheries along the coastline throughout the year.

2. Materials and Methods

Research has been conducted to assess the characteristics and carrying capacities of the land and coastal waters. This study was conducted in June–August 2015. Research location was coastal area of Yogyakarta Special Region (Daerah Istimewa Yogyakarta). The study area included rural/subdistrict along the coast of Kulon Progo Regency, Bantul Regency, and Gunungkidul Regency. The coastal area covered the area approximately 1 km from the coastline. In the coastal area there were river estuaries. The general condition of the coastal area were studied from secondary data, which included climatic conditions (rainfall), river water discharge, sea tides, landforms and coastal land uses. The research was conducted on spatial and temporal coastal physical factors including the characteristics of land and waters dynamics, land use, and suitability of coastal environment for aquaculture. The land parameters used were land shapes, soil texture and soil pH. Parameters of aquatic dynamics used included the fluctuations of water surface, water salinity and water pH. Land parameter data were collected from along the coast, while the water parameter data were collected from the coastal estuaries. The data of land, soil and water parameters were obtained from secondary data as well as direct observation on the coast of Yogyakarta Special Region (DIY).

The physical data for aquaculture was analyzed descriptively, both spatial pattern and temporal pattern. The coastal land uses was for aquaculture ponds. The research was conducted by survey, interview and observation methods. The surveys and interviews were conducted on the farmers of aquaculture ponds. The parameters in questionair sheet were the farm physical properties, distribution, type, season and shrimp production. The data obtained were presented descriptively in relation to environmental characteristics along the coast for a year. The suitability (carrying capacity) of the coastal environment for aquaculture was based on the land characteristics and land uses. Evaluation of conformity (carrying capacity) of coastal environments using land quality parameters covering: landform, slope, soil texture, soil pH, and land use, while water quality parameters include: water salinity, and water pH. The degree of conformity of the coastal environment for aquaculture were calculated based on the quality parameters of land and water.

3. Results and Discussion

3.1. State of territory
DIY region was bordered directly by the south Indian Ocean; Purworejo Regency in the west; Magelang, Klaten and Sukoharjo regencies in the north; and Wonogiri regency in the east. Along the coast of DIY there were rivers that flow into the Indian Ocean that included Bogowonto River, Serang River, Progo River, Opak-Oya River. Coastal area of DIY with coastline of approximately 120 km covering the coastal of Kulon Progo Regency (± 30 km), Bantul Regency (± 20 km) and Gunungkidul Regency (± 70 km).

3.2. Landforms and soil properties
Coastal landforms of DIY (table 1 and figure 1) included the landforms of marine origin, the origin of the wind, and fluvial processes. The landforms of the marine processes consist of whisper (beach) and swale (watery/wet and dry). The land form of the wind process is in the form of sand dunes. The fluvial processed forms are from alluvial plains and flood plains. The coastal area of Kulon Progo Regency and Bantul Regency were dominated by sand dunes, while the coastal area of Gunungkidul Regency was dominated by conical hills karst plain and doline complex (Figure 1). The coastal sand dunes in Kulon Progo and Bantul Regency were 2,022 ha (87.1 %) and 752 ha (49.6 %), respectively.

Slopes of the lower sand dunes ranged from 0.7 to 3.0 %, while the upper sandbanks and swale (wet or dry) were 3.0–9.3 % and 0–0.5 %, respectively. Areas with sand dune and beach ridge generally have a sandy soil texture, or textured loamy sand in some places. The degree of acidities (pHs) of soil in the coastal area were varied. The soil pHs in the sand dunes area were ranging from 6.6–7.4 (7.2 in average) [7]. The soil in sand dunes was dominated by coarse texture (granulated) with a neutral to slightly alkaline pH [9].

Table 1. Landforms of DIY Coastal Area.

Landforms	Land Area (Ha)			
	Kulon Progo Regency	Bantul Regency	Gunungkidul Regency	DIY Region
Alluvial Plain	72.71	335.72	0	408.43
Flood Alluvial Plain	119.11	140.38	0	259.49
Beach Ridge	60.59	86.36	0	146.95
Wet Swale	46.74	0	0	46.74
Dry Swale	0.36	89.45	0	89.81
Sand Dunes	2,021.93	751.58	0	2,773.51
Breccia Hills	0	112.2	0	112.20
Conical Hills Complex Karst and Doline Plains	0	0	6,618.89	6,618.89
Total	2,321.44	1,515.69	6,618.89	10,456.02

Source: Primary data based on Landforms Maps.

3.3. Climate, river water discharge and sea water tides
According to he climatic classification by Koppen, the climatic types found in the coastal area of Gunungkidul Regency was included into the Awa type, while the coastal areas of Bantul Regency and Kulon Progo Regency were included into the climate type Am. The Awa type has characteristics that the amount of rain in wet months (rainfall > 100 mm) can not offset the rainfall in dry months (rainfall < 60 mm), whereas the climate type Am has the characteristics of the amount of rain in the wet months that able to offset the shortage of rain in the dry months. The average annual air temperature in the

study area was 26.50 °C with a minimum average temperature of 23.45 °C and a maximum of 31.31 °C. Wind direction in the rainy season blow from southwest to northwest, while in the dry season, the wind blows from south to southeast [10]. Most winds are less than 10 knots (5 $m \cdot sec^{-1}$). In the western seasons (northwest monsoon winds), the flow of currents in the waters of the Indian Ocean to the west spreads between latitude 10–20 °LS. In the east season (southeast monsoon wind), the flow of currents in the waters of the Indian Ocean to the east extends beyond the latitude of 10 °LS [11]. Wind direction and current in the Indian Ocean affect the sedimentation and closing of estuary of rivers in coastal of DIY which generally occur in August until November [7].

The Bogowonto River water debit in 2008–2009 has a monthly value ranging from 11,436–2,253,287 $m^3 \cdot sec^{-1}$ with the daily average ranges of 369–78,231 $m^3 \cdot sec^{-1}$. In the dry season, low rainfall and small river water discharge, wind direction and speed as well as tides of the sea water caused sand sedimentation and closed the mouth of Bogowonto and Jali Rivers. The water flow at the mouth of the Bogowonto River turns westward through the Pasir-Jati River to the mouth of the Jali River [7] River water discharge was significantly affected by the rainfall. During the rainy season, the rainfall was high causing the high water flow rate in the river. In the rainy season, high rainfall and large river water discharges as well as the sea tides naturally gave a flashing effect the sediments in the mouth of Bogowonto River (estuary).

Tidal data of the Indian Ocean was obtained from the Observation Station at Cilacap Port under the Hydro-Oceanographic Service [12]. The monthly observations in 2008–2009 found a mean monthly tidal average of 1.1 m (ranging from 0.0 to 2.2 m), a mean full moon tidal of 1.08–1.20 m (min. 0.1 m and max. 2.2 m) and a 2nd Quater Month tidal of 1.06–1.12 m (min. 0.3 m and max. 2.0 m) [7]. Sea tide on the Full Moon had a larger water level fluctuation compared to the second quarter. The monthly tide on the Full Moon and the 2nd Quarter of 2008–2009, has a mean of 1.1 m, including a low one, but was still appropriate as a pond water resource (with extensive system). The tides at that time ranged from 0.0 to 2.2 m including very low-high enough. Coastal areas with maximum tidal water of 2–3 m and average amplitude of 1.1–2.1 m were included as suitable for management of shrimp farming ponds in the intertidal area [13].

3.4. Coastal water salinity and pH

Coastal areas of Bantul Regency (from 30 locations of ground water and ponds) had an average groundwater salinity of 19 ppt (ranging between 7–31 ppt) and pond water 19 ppt (12–27 ppt), degrees of acidity (pH) of 8.8 (7.4–9.9) or 9.0 (8.0–10.2) for ground water or pond water. Observation of groundwater and ponds (24 plots) for 90 days (in November 2014–May 2015) in the Kuwaru beach area of Poncosari Village, Srandakan Subdistrict showed an average salinity of 20 ppt (ranging from 8–27 ppt) and 17 ppt (13–26 ppt), and the average acidity (pH) degree of 8.7 (7.6–9.8 ppt). White shrimp is able to live in water salinity of 4–45 ppt. White shrimp can live well (optimal) at water salinity of 10–30 ppt and pH 7–8.5 [14]. Water salinity might decrease because of freshwater supply from groundwater and run off of water rain. High acidity (pH) of water was found in open water during the day because photosynthesis reduced free CO_2 content.

Bogowonto River and Pasir River (Jangkaran Village, Temon District, Kulon Progo Regency), and the sea in FullMoon and the 2nd Quarter Moon started from August 2008 to July 2009 showed the water salinity of 7 ppt (ranged from 0–22 ppt), 10 ppt (1–28 ppt), and 34 ppt (31–37 ppt), respectively. Adequate water salinity for shrimp aquaculture ranges from 10–35 ppt (optimal 15–25 ppt) [13]. Pasir Jati River has relatively low water salinities of 2–11 ppt for aquaculture [7].

(a)

(b)

(c)

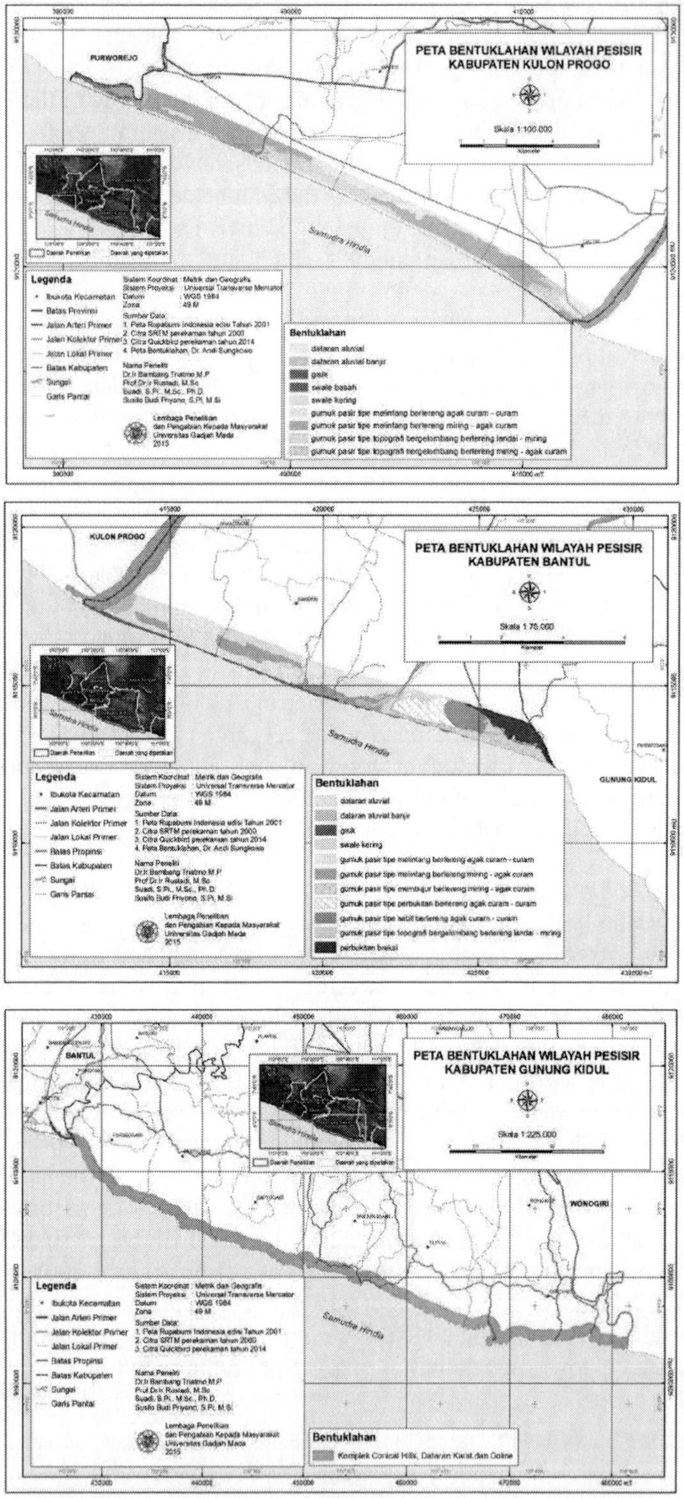

Figure 1. Landform maps of several regencies area in Yogyakarta Special Region [8]. (a) Kulon Progo Regency; (b) Bantul Regency; and (c) Gunungkidul Regency.

Bogowonto River and Pasir River (Jangkaran Village, Temon District, Kulon Progo Regency), and the sea in Full Moon and the 2nd Quarter Moon started from August 2008 to July 2009 showed the average water pH of 7.5 (ranged from 7.0 to 8.2), 7.5 (6.3–8.5), and 7.5 (7.0–8.4), respectively. The degree of acidity (pH) of water suitable for aquaculture ponds ranges from 7.5–8.7 (8.0–8.5) [13]. The water pH of the Pasir River ranged pH 7.2–7.7 that is categorized to be suitable for aquaculture. The degree of acidity (pH) of seawater was relatively stable ranging from 7.4–8.4 (alkaline–alkaline), while pH of Bogowonto estuarine water was 7.0–8.2 (neutral–alkaline). The degree of acidity (pH) of water along the Pasir River throughout the year fluctuated between 6.2–8.5 (slightly acid–alkaline). The degree of acidity (pH) of water along the river is influenced by water sources from river mouths, springs and rain. Besides, the degree of acidity (pH) of water along the river is also strongly influenced by soil pH around each location of observation.

3.5. Landuse in Coastal Areas

Landuse of coastal area (table 2 and figure 2) were generally for rice fields, moor, buildings/yards, dry soil, ponds, forests and others (shrubs, meadows, swamps, deserts/vacant). Coastal villages of Kulon Progo, Bantul and Gunungkidul Regencies covered the area of 6,094.5; 3,943.0 and 31,878.8 Ha, respectively, while the wetland area covered 1,357.8; 1,156.0 and 144.6 Ha, respecteively. Dryland in the area were 2,243.4; 1,156.0 and 23,704.5 Ha [15, 16, 17], respectively. Rice fields in the coastal villages of Kulon Progo Regency was the most widespread, whereas in Kabupaten Gunungkidul was the least. In contrast, dry land in the coastal villages of Gunungkidul Regency was the most widespread, whereas in Kabupaten Bantul was at least.

Table 2. Landuse of DIY Coastal Area.

Landuse	Land Area (Ha)			
	Kulon Progo Regency	Bantul Regency	Gunungkidul Regency	DIY Region
Irrigated rice fields	59.27	272.73	0	332.00
Moor	1,279.39	344.16	0	1,623.55
Garden	363.18	99.44	0	462.62
Forest	0	85.76	6,068.61	6,154.37
Open field	119.78	40.23	547.41	707.42
Open land and natural vegetation	245.42	313.24	0	558.66
Settlement	65.34	162.94	2.87	231.15
River	77.11	160.17	0	237.28
Ponds	111.95	37.01	0	148.96
Total	2,321.44	1,515.68	6,618.89	10,456.01

Source: Primary data based on Landuse Maps.

Land use in coastal areas of DIY included Irrigated rice fields, fields, gardens, forests, open land, and vegetation, settlements, rivers/bodies of water, and ponds. Land use of coastal area in Kulon Progo, Bantul and Gunungkidul Regencies are presented in table 2. Ponds are in DIY were estimated 148.96 Ha, located in coastal area of Kulon Progo Regency of 111.95 Ha and Bantul Regency of 111.95 Ha, while in Gunungkidul was not recorded in the map.

The pattern of land use for cultivation of crops and aquaculture were largely determined by the state of coastal landform and availability of water resources. The availability of spatial water resources (along the coast) as well as temporal (year-round or seasonal) fluctuated. Land use for wetland farming depends on water sources mainly from rainfall and springs, whereas for dry land agriculture depends on rainfall and groundwater (bore or open wells). The use of land for ponds depends on the sea/brackish water source of the river mouth and the tides. The brackish ponds developed in the coastal area used the groundwater water sources.

3.6. Coastal carrying capacity for aquaculture

Aquaculture development has consequencies to increase in the pressure on aquatic and terrestrial resources, impact on environment and biodiversity, as well as impact on the society. The application of ecosystem approaches to aquaculture (EAA) is a sustainable aquaculture development effort while avoiding negative impacts. Carrying capacity is a major component of the ecosystem approach used to implement site selection for well–planned aquaculture [1]. Site selection for aquaculture development based on the level of suitability of coastal environment is crucial. The suitability of coastal environments for aquaculture (in sand dunes) which was included as suitable (S2) is only 10 % and somewhat appropriate (S3) as much as 90 % [7]. Pond construction on coastal sand land needs technology input using plastic (mulch) or biocrete (permanent).

The sand dunes are ± 3–5 m above sea level, so water should be pumped. Pond water resources can be directed pumped from sea water, river, or groundwater (drill). Sea water has a stable salinity and pH which are suitable for shrimp, but pumpimg process faced a difficulty due to the high and strong current and sea waves. River water has a fluctuated widely in salinity and pH, so sometimes they were not suitable for shrimp. In addition, the river water is easily polluted. Groundwater near the coast has a stable salinity and pH which is suitable for shrimp, but sometimes the salinity may decrease.

Based on International Principles for Responsible Shrimp Farming and good fish farming method (CBIB), the ponds should be located in the suitable environmental conditions, so that land and water uses are efficient with optimaum ecosystem function [18] The development of ponds is performed in areas that have not reached the maximal environmental carrying capacity. Ponds are designed and constructed to fulfill the adequate water quality for cultivated fish with integrated pond management. Disturbance to environmental ecosystem is kept as small as possible.

(a)

(b)

(c)

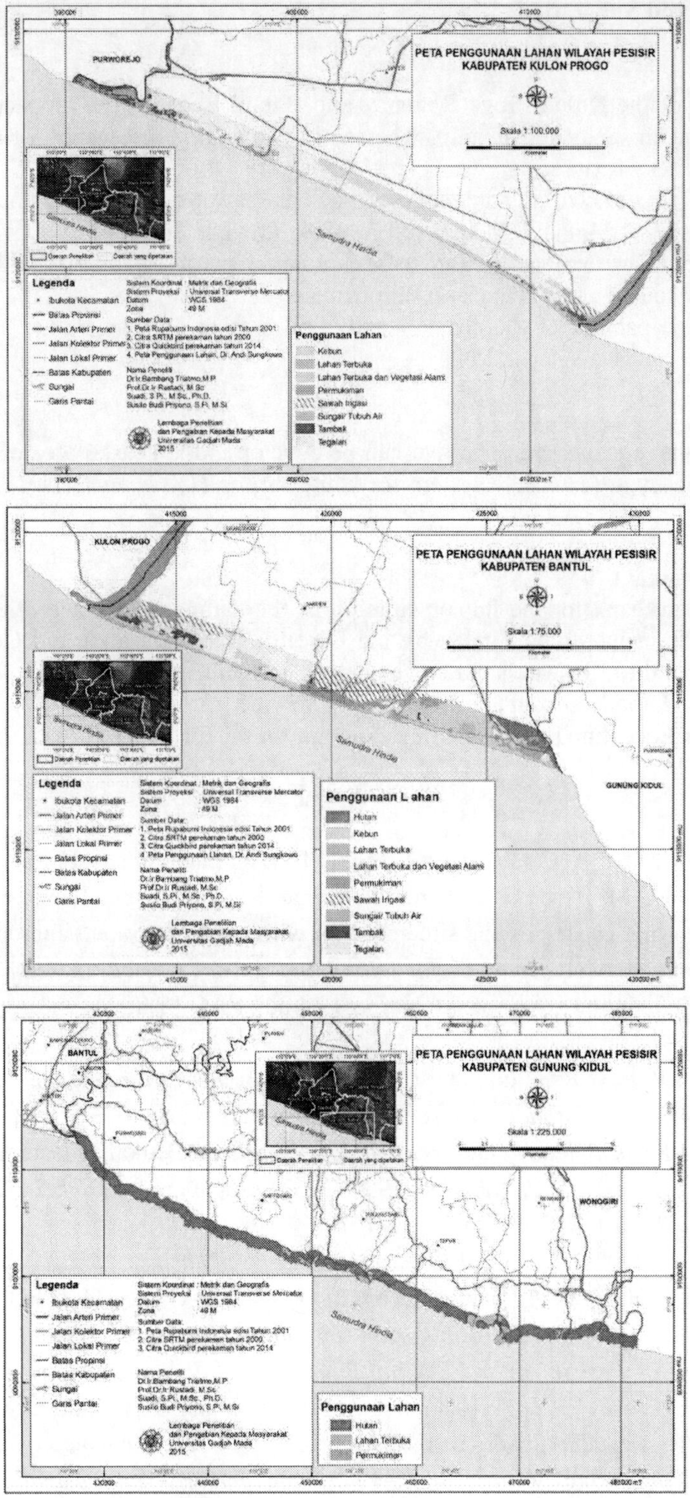

Figure 2. Landuse maps of several regencies area in Yogyakarta Special Region [8]. (a) Kulon Progo Regency; (b) Bantul Regency; and (c) Gunungkidul Regency.

4. Conclusion and Suggestion

4.1. Conclusion

1. Coastal area of the Kulon Progo Regency and Bantul Regency are charactized by sand land and beach ridge with sandy soil texture, whereas in Gunungkidul Regency is generally limestone hill (karst) with rocky texture.
2. Salinities and pHs of coastal water sources, such as seawater, river water and ground water (drilling well) are 31–37, 7–11 and 7–31 ppt, and 7.4–8.4; 7.0–8.2, 7.4–9.9, respectively.
3. The coastal lands were generally used as vacant land (bush), seasonal/ annual plantation, ponds, fish landing sites, tourist areas and conservation areas.
4. The carrying capacities of the coastal sandy soil area of DIY are rather suitable for aquaculture ponds.

4.2. Suggestion

It is suggested that aquaculture pond that can be develop in the coastal area of Kulon Progo Regency and Bantul Regency are intensive system for white shrimp (*L. vannamei*) by using plastic or biocrete ponds.

5. Acknowledgement

We would like to thank for the implementation of the main research activities on the Pond Culture Fisheries Development Model funded by the Directorate of Research and Community Service, the Directorate General for Research and Development, the Ministry of Research, Technology and Higher Education through the Research and Community Service Institute of Universitas Gadjah Mada, In the Superior Higher Education Research Grant Program for the Fiscal Year 2015, 2016 and 2017.

References

[1] Ross L G, Telfer T C, Falconer L, Soto D, Aguilar-Manjarrez J, Asmah R, Bermúdez J, Beveridge M C M, Byron C J, Clémen A, Corner R, Costa-Pierce B A, Cross S, De Wit M, Dong S 2013 Carrying capacities and site selection within the ecosystem approach to aquaculture *Site selection and carrying capacities for inland and coastal aquaculture* ed L G Ross, T C Telfer, L Falconer, D Soto and J Aguilar-Manjarrez (Rome: FAO Fisheries and Aquaculture Proceedings) pp 19–46

[2] Dahuri R, J Rais, S P Ginting and M J Sitepu 2001 *Pengelolaan Sumber Daya Wilayah Pesisir dan Lautan Secara Terpadu* (Jakarta: PT Pradnya Paramita)

[3] DJPB 2004 *Profil Budidaya* (Jakarta: Direktorat Jenderal Perikanan Budidaya KKP)

[4] Stroethoff K H and A P H M Hovers 1996 *INFOFISH International* **5** 24–29

[5] Sukadi M F 2010 *J. Ris. Aquaculture* **5** 1–12

[6] Sartohadi J 2007 *Geomorfologi Tanah dan Aplikasinya untuk Pembangunan Nasional* (Yogyakarta: Dies Natalis Fakultas Geografi Universitas Gadjah Mada) Oral Presentation

[7] Triyatmo B 2012 *Pola Pengembangan Perikanan Budidaya Tambak Berbasis Karakteristik Lingkungan di Pesisir Antara Sungai Bogowonto Kabupaten Kulon Progo dan Sungai Jali Kabupaten Purworejo* (Yogyakarta: Sekolah Pascasarjana Universitas Gadjah Mada)

[8] Triyatmo B, Rustadi, Suadi, and Sukardi 2015 *Model Pengembangan Perikanan Budidaya Tambak Ramah Lingkungan dan Berkelanjutan pada Lahan Pasir Pesisir Daerah Istimewa Yogyakarta* (Yogyakarta: PUPT Universitas Gadjah Mada)

[9] BAKOSURTANAL 2000 Pedoman Survei Cepat Terintegrasi (Jakarta: Badan Koordinasi Survei dan Pemetaan Nasional)

[10] Khakhim N 2009 *Kajian Tipologi Fisik Pesisir Daerah Istimewa Yogyakarta untuk Mendukung Pengembangan dan Pengelolaan Wilayah Pesisir* (Bogor: Sekolah Pascasarjana IPB)

[11] Wilopo M D 2005 *Karakteristik Fisik Oseanografi di Perairan Barat Sumatera dan Selatan Jawa-Sumbawa dari Data Satelit Multi Sensor* (Bogor: Institut Pertanian Bogor)

[12] DHO 2009 *Daftar Pasang Surut Kepulauan Indonesia* (Jakarta: Dinas Hidrografi TNI Angkatan Laut)

[13] Poernomo A 1992 *Pemilihan Lokasi Tambak Udang Berwawasan Lingkungan Seri Pengembangan Hasil Penelitian* Series No. PHP/KAN/PATEK/004/1992 p 40

[14] BBAP 2013 *Budidaya Udang Vaname di Tambak* (Situbondo: Balai Budidaya Air Payau)

[15] BPS 2014[a] *Kabupaten Bantul dalam Angka* (Bantul: BPS Kabupaten Bantul)

[16] BPS 2014[b] *Kabupaten Gunungkidul dalam Angka* (Gunungkidul: BPS Kabupaten Gunungkidul)

[17] BPS 2014[c] *Kabupaten Kulonprogo dalam Angka* (Kulonprogo: BPS Kabupaten Kulonprogo)

[18] FAO/NACA/UNEP/WB/WWF 2006 *International Principles for Responsible Shrimp Farming* (Thailand: Network of Aquaculture Centres in Asia-Pasific (NACA))

IOP Conf. Series: Earth and Environmental Science **139** (2018) 012008 doi:10.1088/1755-1315/139/1/012008

Masculinization of nile tilapia (*Oreochromis niloticus*) using extract of bull testes

A Yustiati[1]*, I Bangkit[1] and I Zidni[1]*

[1] Study Program of Fisheries Faculty of Fisheries and Marine Sciences
Universitas Padjadjaran, Jl. Raya Bandung- Sumedang KM 21, Jatinangor 40600 West Java Indonesia. Telp. 081314551573
*E-mail : yustiati@yahoo.com and irfan.zidni12@gmail.com

Abstract. Tilapia males grow faster to increase the production of the fish, there is a necessity to produce all males by using extract of bull testes. The research method used in this study was the experimental Completely Randomized Design (CRD) with five treatments and three replications. The treatments were A: without immersion and without oral treatment (control); B; immersion 600 $\mu g \cdot L^{-1}$ without oral treatment; C: immersion 600 $\mu g \cdot L^{-1}$ and oral at a dose of 30 $mg \cdot kg^{-1}$ feeds, D: immersion 600 $\mu g \cdot L^{-1}$ and oral at a dose of 40 $mg \cdot kg^{-1}$ feeds, E: immersion 600 $\mu g \cdot L^{-1}$ and oral at a dose of 50 $mg \cdot kg^{-1}$ feeds. Results showed that the use of bull testicle extract mixed to media at concentration of 3 ml L^{-1} produced red male nile of 69.07 %. In addition, the immersion of the extract at 500 $\mu g \cdot L^{-1}$ resulted the male of 86.71 %. Sex reversal by dipping at 600 $\mu g \cdot L^{-1}$ 17α-*methyltestosterone* combined with oral administration at 40 $mg \cdot kg^{-1}$ produced the highest male of Nile Tilapia Java carp (86.67 %) compared with dipping, which produced 65.56 % of male. These results suggested that combination of dipping and oral is more effective for sex reversal.

1. Introduction

Nile Tilapia is one of freshwater commodities that has a high economic value, easy to culture and has a thick body flesh so it is suitable to be used as refined products, one of them is fillet. Tilapia fillet has very wide export market, especially in the United State [1].

Innovation and technology are needed to increase production and productivity of tilapia culture to fulfill market demand. One of technology innovations which can be conducted is selective breeding to obtain good strain, one of the good tilapia strains is Nirwana (Nila Ras Wanayasa).

Nile Nirwana is the strain resulted from the Development Center for Tilapia and Carp Wanayasa, Purakarta. The advantage of nirwana is faster growth than other strains. Culture of larvae until it reaches a weight more than 650 g can be done in 6 month, in addition nirwana has also thick flesh compare with other strain [2]. However, its production is still can be increased by masculinization using natural hormone obtained from bull testicle.

Growth rate of male tilapia is faster than female one, so it is important to develop a technique to produce male monosex of tilapia by administration of male hormone (testosterone). The technique usually use synthetic hormone 17α-*methyltestosteron* to revers fish sex [3, 4]. However the distribution of synthetic hormone is limited, so the price is high. Hormone residue might become an environment pollutant [5]. One alternative to produce male tilapia with masculinization is by using environmental friendly natural testosterone originated from bull testicles. The purpose of this research was to increase the production of Nile tilapia by masculinization with oral and dipping administration of the bull testicle extract.

2. Materials and Methods

2.1. *Materials*

The materials used in this research:

a) Nirwana tilapia originated from The Development Center for Tilapia and Carp Wanayasa, Purakarta, West Java. Male and female broodstocks weight was in the range of 300–400 g and 200–300 g, respectively.
b) The testes of bull was obtained from the slaughter house in Ciroyom Bandung.
c) Ether solution for immersion the bull testicles.
d) ETTS was made from the fresh testicle of bull or local male cow. The testicle tissue was chopped into a size of 1 mm^2. The chopped testicle was dried by freeze dryer at a temperature of -80 $^{\circ}$C to obtain powder.
e) Acetocarmine solution is obtained by dissolving 0.6 g carmine into 100 mL acetic acid 45 % for staining fish gonadal tissue in sexing.
f) 'Artha' feed contained 25 % protein.

2.2. Methods
The research method used in this study was the experimental Completely Randomized Design (CRD) with five treatments and three replications. The treatments were:

A: without immersion and without oral treatment (control)
B: immersion at 600 $\mu g \cdot L^{-1}$ without oral treatment
C: immersion at 600 $\mu g \cdot L^{-1}$ and oral administration at 30 $mg \cdot kg^{-1}$ feed
D: immersion at 600 $\mu g \cdot L^{-1}$ and oral administration at 40 $mg \cdot kg^{-1}$ feed
E: immersion at 600 $\mu g \cdot L^{-1}$ and oral administration at 50 $mg \cdot kg^{-1}$ feed

$$Xij = \mu i + \tau j + \varepsilon ij$$

Xij = result of observation on treatment to -i replication to-j
μi = general average
τj = the effect of treatment to-i
εij = effect of random factors treatment to-i replication to-j

2.3. Procedure
Fifteen aquaria contain 20 L water were prepared, each aquarium was equipped with aeration for oxygen supply from the blower. Another a set of fifteen 60 L-aquaria were prepared to raise the masculinized larvae. When the fish reached 3 cm of total length, they were transferred to the net cage in a pond.

One day old tilapia larvae or at eye spot phase were treated with an extract of bull testicle powder (ETTS). The extract of hormone was poured into the water where the larvae were reared. The larvae were immersion in the extract of hormone for seven days.

Until the yolksack larvae were depleted, the larvae were fed with yolksack of a chicken egg. Then the larvae were fed with natural food, namely Chlorella and Artemia. Two weeks old larvae were then fed with proper commercial feed. After one month old, the treated fish were then placed in floating cages and fed with commercial feed with the amount of five percent/body weight/day, three times a day. Fish sampling to measure growth and to adjust the amount of feed to be given, is done every ten days in three months.

In the third month, survival and growth rates were calculated. In addition, thirty fish from each treatment were sexing. Sexing was conducted by analyzing the gonad stained with *acetocarmine*. The results were grouped into three, namely male, female and intersex.

2.4. Parameters

2.4.1. Survival rate. Survival Rate was observed every day by counting and removing the dead fish. The calculation of survival rate was done by using the formula [6]:

$$SR = \frac{Nt}{No} \times 100\ \%$$

SR = Survival Rate (%)
Nt = the amount of fish of three months age
No = the amount of fish of the initial rearing

2.4.2. Growth. Absolute growth using formula:

$$GA = Wt - Wo$$

GA = absolute growth
Wt = fish weight by three months old
Wo = initial weight

Specific growth rate using the formula:

$$SGR = \frac{(Lnt - LnWo)}{T} \times 100\ \%$$

SGR = Specific Growth Rate (%)
Wt = biomass of fish weight in the end of study (g)
Wo = biomass of fish weight in in the end of study (g)
T = observation periode (day)

2.4.3. Sex ratio: male, female and intersex. At the end of the research, gonad of thirty fish from each treatment were analyzed. The gonad was placed in a petri dish, chopped and mixed with acetocarmine. The gonad was then analyzed under a microscope to determine whether the fish was male, female or intersex. The formula is:

$$\mathrm{M} = Gm\, s^{-1} \times 100\ \%$$
$$\mathrm{F} = Gf\, s^{-1} \times 100\ \%$$
$$\mathrm{I} = Gi\, s^{-1} \times 100\ \%$$

M: Male, F: Female, I: Intersex, s: amount of sample, Gm: male fish, Gf: female fish, GI: Intersex fish

2.5. Data analysis
Survival rate, growth rate and sex ratio were analyzed statistically using F test and Duncan test. Water quality parameters were compared with existing standards.

3. Result and Discussion

3.1. Result

3.1.1. Percentage of male. The result of this research showed that the production of male in the range from 25.55 to 86.67 %. The highest of male percentage of Nirwana was obtained from D treatment (600 μg·L^{-1} + 40 mg·kg^{-1} feeds), that reached 86.67 %. The lowest percentage of Nirwana male was 25.55 % obtained in a treatment (without hormone).

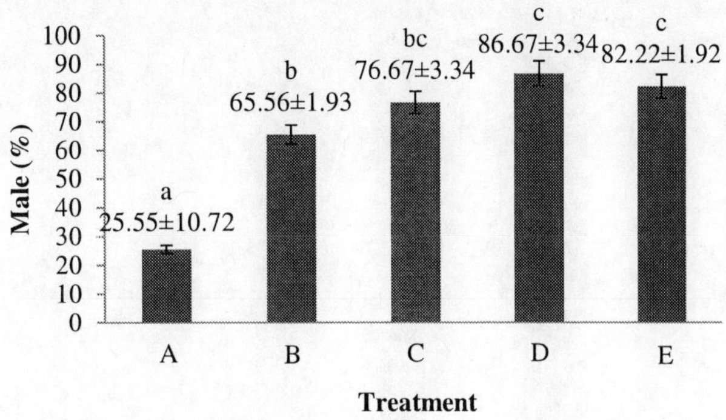

Figure 1. Average Percentage of Male in Tilapia treated with various dosages of bull testicle extracts.

D treatment produced high percentage of male and significantly different from treatment A (without hormone) and B (600 μg·L^{-1} + 0 mg·kg^{-1} feeds) (p, 0.05) but insignificantly different to C treatment (600 μg·L^{-1} + 30 mg·kg^{-1} feeds) and E (600 μg·L^{-1} + 50 mg·kg^{-1} feeds). In normal spawning, tilapia usually produced less than 50 % male. It means that the addition of bull testicle extract into feed and into the media increased the percentage of males.

3.1.2. Percentage of female. Percentage of female obtained from the research ranged from 12.22–74.45 %. The high percentage of the female was produced from a treatment (Without hormone) that reached 74.45 %. In contrast, the low percentage of female was produced from the treatment (600 μg·L^{-1} + 40 mg·kg^{-1} feed) 12.22 % (figure 2).

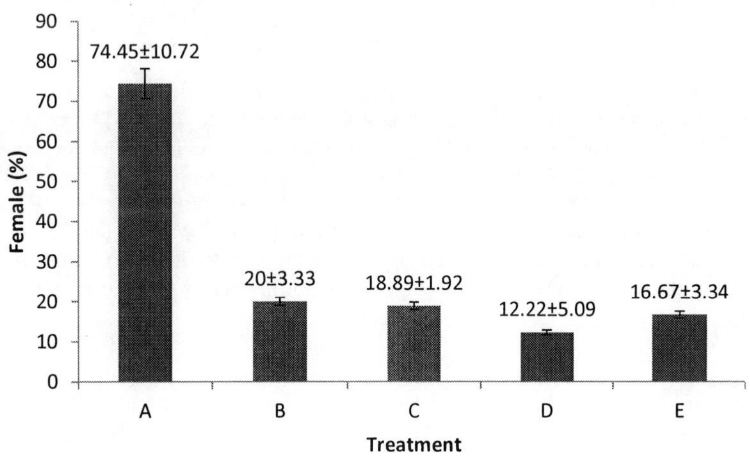

Figure 2. Average percentage of female in Tilapia treated with various dosage of bull testicle extracts.

3.1.3. Percentage of Intersex. Percentage of nirwana intersex obtain from the research ranged from 0 % to 14.44 %. The highest (14.44 %) Nirwana intersex production was obtained from B treatment (600 μg·L^{-1} + 0 mg·kg^{-1} feeds). Meanwhile the lowest one (0 %) was obtained from a treatment (without hormone).

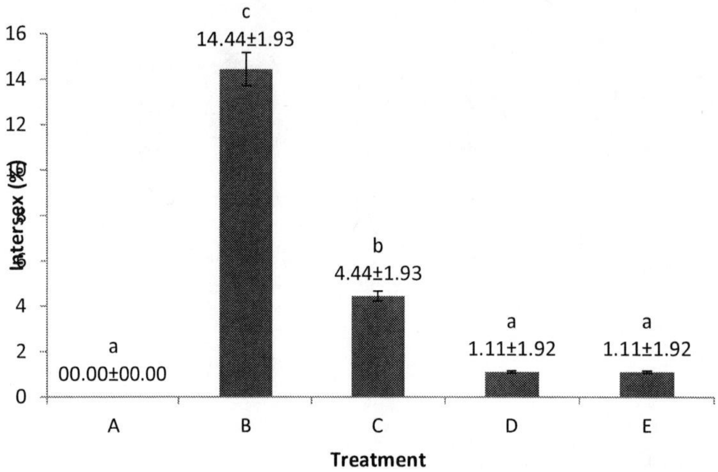

Figure 3. Average Percentage of Intersex in Tilapia treated with various dosage of bull testicle extracts.

3.1.4. Nirwana sex ratio. The data of nirwana sex ratio is presented in table 1.

Table 1. Average percentage of sex rasio in tilapia treated with various dosage of bull testicle extracts.

Treatment	Percentage (%) male	Percents (%) Female	Percents (%) Intersex
A: 0 $\mu g \cdot L^{-1}$ + 0 $mg.kg^{-1}$ of feed	25.55	74.45	0
B: 600 $\mu g \cdot L^{-1}$ + 0 $mg \cdot kg^{-1}$ of feed	65.57	20	14.44
C: 600 $\mu g \cdot L^{-1}$ + 30 $mg \cdot kg^{-1}$ of feed	76.67	18.89	4.44
D: 600 $\mu g \cdot L^{-1}$ + 40 $mg \cdot kg^{-1}$ of feed	86.67	12.22	1.11
E: 600 $\mu g \cdot L^{-1}$ + 50 $mg \cdot kg^{-1}$ of feed	82.22	16.67	1.11

Treatment D at a dose of 600 $\mu g \cdot L^{-1}$ + 40 $mg \cdot kg^{-1}$ of feed produced the highest male fish hatching percentage of 86.67 %. Treatment A without incorporating the hormone (control) produced the highest female fish percentage of 74.45 %. Treatment B at the dose of 600 $\mu g \cdot L^{-1}$ + 0 $mg \cdot kg^{-1}$ feed produced the highest intersex fish percentage in 14.44 % (table 1).

3.1.5. Survival rate. The statistical analysis result showed that there was no significant difference among treatments (figure. 4). The highest survival percentage of 69.50 % were found in treatment D (600 $\mu g \cdot L^{-1}$ + 0 $mg \cdot kg^{-1}$ feeds), while the lowest survival percentage of 46.33 % was found in treatment C (600 $\mu g \cdot L^{-1}$ + 30 $mg \cdot kg^{-1}$ feeds) (figure. 4).

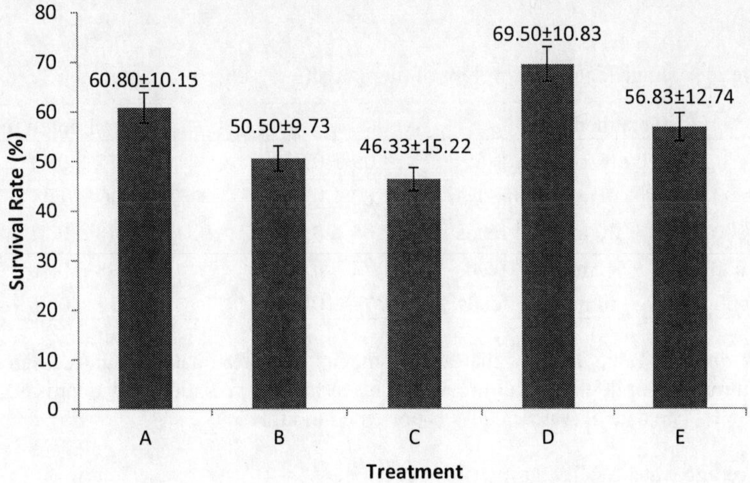

Figure 4. Survival rate of Tilapia treated with various dosage of bull testicle extracts.

3.1.6. Average length and weight of Tilapia fish. Weight and length of tilapia were measured when they got three months old or when the gonads were ready to be observed. This study indicated that treatment a produced fish with average maximum length of 6.18 cm average maximum weight of 3.09. Treatment D produced fish with average minimum length of 4.98 cm and average minimum weight of 1.75 (figure 5).

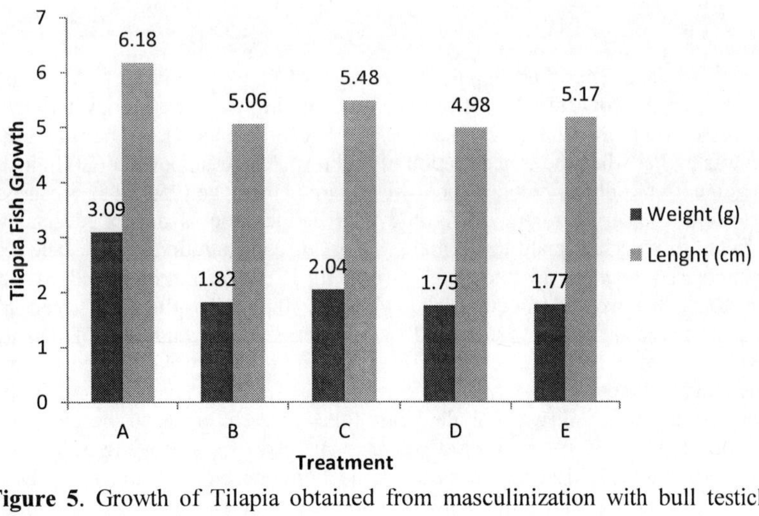

Figure 5. Growth of Tilapia obtained from masculinization with bull testicle extract.

The statistical analysis showed that treatment A (without hormone administration) produced fish with the highest average length and weight, being different significantly from treatment, B, C, D and E (table 2). The statistical analysis showed that treatment A (without hormone administration) produced fish with the highest average length and weight, being different significantly from treatment, B, C, D and E (table 2).

Table 2. Average length and weight of tilapia fish.

Treatment	Average Weight (g)	Average Length (cm)
A: 0 μg·L^{-1}+ 0 mg·kg^{-1} feeds	3.09 ± 0.27^{b}	6.18 ± 0.17^{d}
B: 600 μg·L^{-1} + 0 mg·kg^{-1} feeds	1.82 ± 0.08^{a}	5.06 ± 0.08^{ab}
C: 600 μg·L^{-1} + 30 mg·kg^{-1} feeds	2.04 ± 0.24^{a}	5.48 ± 0.21^{c}
D: 600 μg·L^{-1} + 40 mg·kg^{-1} feeds	1.75 ± 0.09^{a}	4.98 ± 0.10^{a}
E: 600 μg·L^{-1} + 50 mg·kg^{-1} feeds	1.77 ± 0.10^{a}	5.17 ± 0.06^{b}

3.1.7. Water quality. Table 3 shows that all parameters of water quality in this research were still in the optimum range for nile tilapia culture according to Indonesia National Standards/Standar Nasional Indonesia (SNI). The data of water quality is presented in table 3.

Table 3. Average water quality during tilapia rearing.

Treatment	Average Temperature (°C)	Average of pH	Average Dissolved Oxygen (mg·L^{-1})	Average Ammonia (mg·L^{-1})
A: 0 μg·L^{-1} + 0 mg·kg^{-1} feeds	25.7	7.2	6.7	0
B: 600 μg·L^{-1} + 0 mg·kg^{-1} feeds	24.5	7.1	6.1	0.001
C: 600 μg·L^{-1} + 30 mg·kg^{-1} feeds	23.5	7.3	6	0.001
D: 600 μg·L^{-1} + 40 mg·kg^{-1} feeds	23.0	7.5	5.8	0.003
E: 600 μg·L^{-1} + 50 mg·kg^{-1} feeds	25	7.6	5.5	0.003
Indonesia National Standards	25 to 32 °C	6.5 to 8.5	> 3	< 0.002

3.2. Discussion

The treatments of D and E resulted female percentages of 86.67 % and 82.22 %, respectively. This result suggested that the concentration of testosterone entering the larva's body through osmotic and diffusion processes was sufficient to reverse the larva sex to be male. It can be sound that percentage of Nile male was higher when the concentration of bull testicle extract powder (EBT) increased.

The A treatment or control produced the lowest male percentage (25.55 %) because there was no addition testosterone entering to the fish's body, so in the process of sex reversal there was no supporting factor to reverse female to be male. Although concentration bull testicle extract powder (EBT) was increased, such as in treatment E (600 μg·L^{-1} + 50 mg·kg^{-1} feed) resulted the male percentage of 82.22 % lower 4.45 % comparing with D (600 μg·L^{-1} + 40 mg·kg^{-1} feed) treatment. The decrease in male percentage might be caused by an excessive concentration of bull testicle extract powder (EBT). A Treatment at dose of 600 μg·L^{-1} + 40 mg·kg^{-1} feed was assumed as a maximum dose which resulted high male percentage.

The high production of nirwana male from this research occurred because of testosterone containing in the EBT. This hormone is effective to maintain spermatogenesis process and a main mediator in spermiation [7]. Exogenous testosterone hormone entering the fish's body will affect hypothalamus to release LH and FSH. While LH stimulates by Leydig cell to produce testosterone. Fish affect Sertoli cell to stimulate CAMP (Cyclic Adenosine Mono Phospat). Which induces the ABP (Androgen Binding Protein). The ABP has a function to bind testosterone release by Leydig cells. The bind the complex will interact directly in second meiosis, which change secondary spermatocyte to be spermatid.

Extract of bull testes (EBT) contains steroid, ie androgen. This hormone is responsible for male gonadogenesis, testicle. The more EBT added into feed, the more males produce and the concentration of 40 mg·kg^{-1} feed yields the highest males [9]. Beyond 40 mg·kg^{-1} feed, it occurs the phenomenon of paradoxical effect. Previous work using only immersion of 600 μg·L^{-1} produces less than 75 % male.

Statistical analysis indicated that a treatment (without hormone) produced a high percentage of females, significantly different from other treatments. There is no paradoxical affect a course when the use of hormone is too high, then the reproduction of male will not be rich, even reproduction of female will increase. Except exogen hormone causes testosterone synthetic in aromatase inhibition stopped. Again, it shows that in normal spawning tilapia produces much females than males. And the treatment of immersion larvae in EBT combined with oral treatment, feed mixed with EBT, reduced the number of female dramatically until 12 %.

Intersex fish was found because the effect of EBT in masculinization proses was not single effect. It was proved in the B treatment (600 $\mu g \cdot L^{-1}$ + 30 $mg \cdot kg^{-1}$ feed). In this treatment the larvae was only immersed in water mix with 600 $g \cdot L^{-1}$ for twelve hours starting from eye spot phase. Without feeding with hormones-containing food. The use of EBT, also showed the phenomenon of intersex. Intersex fish has ovarium and testicle in one gonad. The dose of 30 mg EBT kg^{-1} feed is not sufficient to alter the gonad into males. The higher oral dose was able to produce less intersex fish.

In terms of time efficiency and treatment as well as the amount of hormone used, immersion is considered to be the most efficient method. However, it has not been fully effective in producing a higher percentage of male fish seeds. Steroid hormone used in the process may leach into the water [8]. In addition, excessive steroid hormone usage and prolonged immersion may cause a paradoxical effect on fish. This indicates that monosex fish, seeds produced by masculinization through immersion is not very effective and, therefore, an additional method is required in order to attain the higher per cent of male fish seeds. The process of masculinization through immersion can be incorporated with oral administration of feed [9].

The survival rate of each treatment did not vary significantly, which suggested that bull testicle extract in each treatment did not considerably affect the survival rate of the fish. The dosage of hormone was given within the safe dose limit, so that it did not affect the survival of the fish. It means that the addition of EBT into the media and in the feed did not give any harmful effect to the fish.

The water temperature fluctuated between 23.0–25.7 °C during the research, which was stills a suitable range of temperature for the growth of tilapia. The optimum water temperature for growing tilapia fish larvae is around 25–32 °C [10]. Optimum temperature in fish breeding will influence the digestive enzymes and metabolism. High level consumption of feed combined with effective digestion and metabolism process will produce optimum energy for growth and reproduction [11].

pH value (Power of Hydrogen) is the hydrogen ion value in water. Water with high H+ ion content will be acidic and water with low H+ ion content will be alkaline. The determining factor in water quality is the acidity (pH). The pH value during the research ranged between 7.2–7.6, which was good for the growth of tilapia fish. Optimum pH range for growing tilapia fish is between 6.5–8.5 [12]. The low temperature-liquid will decrease the activity of the body or cause fish to become weak, easily get infected and usually followed by the high rate of mortality [13].

Dissolved oxygen content during the research fluctuated between 5.5–6.7 $mg \cdot L^{-1}$. Based on DO needed to sustain tilapia fish seeds is > 3 $mg \cdot L^{-1}$ [10]. Oxygen availability is one of determining factors that contribute to fish fed consumption (appetite) since oxygen is required to turn food into energy. A decrease in appetite leads to low consumption of feed, which cause fish to lack the energy needed for growth. Low dissolved oxygen causes a slowing down in fish growth.

The level of amino concentration during the research was 0–0.003 $mg \cdot L^{-1}$. Ammonia (NH_3) is the main substance generated by protein decomposition, which is toxic for fish. Therefore, NH_3 concentration in water should not be more than 1 ppm [14]. The source of ammonia in water is the breakdown of nitrogen (protein and fertilizer) and inorganic nitrogen contained in soil and water [15].

4. Conclusion

Based on the results of research, it can be concluded: Immersion with bull testicle extract of 600 $\mu g \cdot L^{-1}$ combined with oral administration of 40 $mg \cdot kg^{-1}$ feed produces the highest level of male fish percentage (86.67 %) and the highest survival rate (69.50 %).

Suggestions

Based on the result of the research, it can be suggested that in order to attain the maximum amount of male fish population, immersion and oral administration can be incorporated, with the dosage of bull testicle extract of 600 $\mu g{\cdot}L^{-1}$ and 40 $mg{\cdot}kg^{-1}$ of feed.

Acknowledgements

This publication is based on a work supported by Academic Leadership Program Universitas Padjadjaran 2015.

References

[1] Sofi N 2006 *Model Permintaan Filet Ikan Nila Beku Indonesia di Pasar Amerika Serikat* [skripsi]. (Bogor: Institut Pertanian Bogor)

[2] Ghufran M dan K Kordi 2010 Budidaya Ikan Nila di Kolam Terpal (Lily Publisher: Yogyakarta)

[3] Macintosh D J dan Litte D C 1995 *Broodstock Management and Egg and Larval Quality* ed N R Bromage and J B Ronald (USA: Blackwell Science) p 277–330

[4] Phelps R P and Thomas J P 2000 *Tilapia Aquaculture in the Americas The Word Aquaculture Society* ed B A Costa-Pierceand and J E Rakocy (Louisiana: Baton Rounge,) p 34–59

[5] Phelps R P, Sanchez W C, Couturier G M, Abiado M and Dabrowski K 2001 *Studies on Fate of Methyltestosteron and Its Metabolism In Tilapia and on The Use of Phytochemicals as an Alternative Methode to Produce a Monosex Population of Tilapia* Reproduction Control Research 1 (10RCR1/Experiment/Mexico).

[6] Ogunji J, Toor R S, Schulz C and Kloas W *Journal of Fisheries and Aquatic Sciences* **8** 141–7

[7] Goetz F W 1983 Fish *Physiology IX Reproduction* ed W S Hoar, D J Randall and E M Donaldson (Orlando: Academic Press. Inc.) p 117–61

[8] Tan-Fermin I D, Garcia L M B and Castillo A R 1997 *Japanese Journal* of *Ichthology* **40** 413–20

[9] Aragi R A 2013 *Efektivitas Pemberian Tepung Testis Sapi Terhadap Maskulinisasi Larva Ikan Nilem* (*Osteochilus hasselti*). [Skripsi] (Bandung: Fakultas Perikanan dan Ilmu Kelautan Universitas Padjadjaran)

[10] SNI: 7559: 2009. *Produksi Ikan Nila* (*Oreochromis Niloticus* Bleeker) Kelas Pembesaran di Kolam Air tenang

[11] Lucas 2002 *Bioenergetics* of *Aquatic Animals*. (UK: Departemen of Zoology University of Aberdeen Scotland)

[12] Popma T and Michael 1999 *Tilapia Life History* and *Biology*. SRAC Publication No. **283**.

[13] Kordi K and M Ghufran H 2009 *Budidaya Perairan Buku Kedua* (Bandung: PT Citra Aditya Bakti) p 445–964.

[14] Effendie M I 1997 *Biologi Perikanan* (Bogor: Yayasan Pustaka Nusantara)

[15] Effendie 2003 *Metode Biologi Perikanan* (Jakarta: Gramedia Pustaka Utama)

IOP Conf. Series: Earth and Environmental Science **139** (2018) 012009 doi:10.1088/1755-1315/139/1/012009

Impact of Red Water System (RWS) application on water quality of catfish culture using aquaponics

Zahidah[1]*, Y Dhahiyat[1], Y Andriani[1], A Sahidin[1] and I Farizi[1]

[1] Fisheries Study Program Faculty of Fisheries and Marine Sciences, Universitas Padjadjaran, Jl. Raya Jatinangor km. 21 Sumedang. Indonesia
*Email: ibuzah@gmail.com

Abstract. This study aim was to analyze the effect of Red Water System (RWS) probiotics application on water quality in aquaponic system. The research used experimental method using Completely Randomized Design (CRD) with five treatments and three replications. Treatment A: RWS 7.5 $\mu L \cdot L^{-1}$/week without aquaponic probiotic, Treatment B: aquaponic without RWS probiotics, treatment C: RWS probiotic addition in aquaponic media at 7.5 $\mu L \cdot L^{-1}$/week, treatment D: addition of RWS probiotics in aquaponic media at 10 $\mu L \cdot L^{-1}$/week and treatment E: addition of RWS probiotics in in aquaponic media at 12.8 $\mu L \cdot L^{-1}$/week. Parameters measured were pH, temperature, ammonia, nitrate and phosphate. The results showed that water temperature and pH relatively unchanged in all treatments. The addition of RWS probiotics did not improve the concentration of ammonia, nitrate and phosphate. In fact, the catfish culture with only aquaponic resulted lower concentration of ammonia, nitrate and phosphate than other treatment. The lowest value of ammonia, nitrate and phosphates was obtained in the experimental groups of aquaponic with RWS of 10 $\mu L \cdot L^{-1}$/week (Treatment D). Treatment D has the lowest average ammonia of 0.50 ppm, reduced nitrate up to 60.78 % and temperature and pH relatively unchanged.

1. Introduction

Limited land size, limited water source, reduced water quality, and lack of technology inputs have become issues in increasing the production of fresh water fish, especially catfish. The increase in catfish production is only 47.21 %, which is far below the expected target [1]. Therefore, efforts are needed to improve the production of catfish by adding the technology inputs. The technology that can be adopted is the integration of fish culture with plant culture through the aquaponics system [2]. Aquaponics is one of the culture technologies that combine fish and plant culture [3]. The plants in the aquaponics system function as the biofilter that absorbs and uses nitrate (NO_3), ammonia (NH_3) and phosphate (PO_4) as the fertilizer that they reduce the concentration of N and P contaminants and improve water quality [4].

In addition to producing nitrate (NO_3) and phosphate (PO_4), catfish culture also produces ammonia (NH_3) as a result of metabolism and the primary protein element reshuffling that tends to cause physiological disturbance, trigger stress and is also toxic for fish [5]. To deal with the problem, a new innovation that collaborate aquaponics and Red Water System has appeared.

Red Water System (RWS) is a new technology in catfish culture activity that utilizes the result of *Lactobacillus casei* bacteria and *Saccharomyces cerevisiae* fungus fermentation during the culture. RWS is an extension of probiotic application in shrimp farming that is then applied in intensive catfish culture. Adding probiotics in the culture pond leads to decomposition of organic materials in the pond that will produce lower ammonia concentration compared to the control pond [6]. The application of probiotics gives a positive effect on the water quality (dissolved oxygen, ammonia, nitrite and nitrate) compared to control and is able to support the sangkuriang catfish survival [7]. The probiotic microorganisms can also oxidize ammonia, reducing the amount of ammonia in the culture pond [8]. This will directly help the plant to reduce the excessive ammonia in the media that cannot be absorbed by the plants. Hence, this study aim was to analyze the effect of application of RWS probiotics on water quality in aquaponics system fish culture.

2. Materials and Methods

This study was conducted during the period of February to March 2016 at the Laboratory of Fish Culture, Ciparanje, Faculty of Fishery and Marine Science, Universitas Padjadjaran. The method used in this study was the experimental study using Complete Randomized design (CRD) that included five treatments and three replications. The treatments in this study were:

Treatment A : 7.5 $\mu L \cdot L^{-1}$/week RWS solution in fiber without aquaponics (Control A)
Treatment B : aquaponics without RWS solution (Control B)
Treatment C : aquaponics with 7.5 $\mu L \cdot L^{-1}$/week RWS solution
Treatment D : aquaponics with 10 $\mu L \cdot L^{-1}$/week RWS solution
Treatment E : aquaponics with 12.5 $\mu L \cdot L^{-1}$/week RWS solution

In this study, the culture tanks were fibre glass of 70 cm × 70 cm × 70 cm and water depth of 50 cm, with total density of 100 fry per tank. The tanks were equipped with rectangular PVC pipes and styrofoam for plants media holder. Lettuce plants were planted in the netpot with rockwool. The water from the tanks was pumped and distributed into plants media through PVC pipe (figure 1).

Figure 1. Aquaponics design used in this study.

Parameters measured were pH, temperature, ammonia, nitrate and phosphate (table 1). The measurement was performed three times: in the beginning of the study and was repeated every 10 days culture. The RWS solution was applied to the trial pond every 3 days. Data analysis on the physical and chemical parameters was performed descriptively by comparing the collected data to the Indonesian National Standard (*Standar Nasional Indonesia*, SNI) for water quality parameters.

RWS used in this study contained *Lactobacillus casei* and *Saccharomyces cerevisiae* in extract of bean sprouts.

Table 1. Measured parameters, units and its methods.

Parameters	Units	Methods/instrument
pH		Potensiometric/pH meter
Temperature	°C	Potensiometric/thermometer
Ammonia	ppm	Spectrophotometric/Spectrophotometer
Nitrate	ppm	Spectrophotometric/Spectrophotometer
Phosphate	ppm	Spectrophotometric/Spectrophotometer

3. Results and Discussion

The success of fish culture is very much influenced by the quality of water in the culture media. Poor water quality will reduce appetite, delay growth, trigger diseases and may cause deaths among the fish. In fish culture, waste in the form of feces, urine and feed residuals are unavoidable [9]. But not all feed provided is consumed by fish because 15–30 % of the feed goes to the water [10]. In addition, some of the consumed feed will be removed in the form of feces [11, 12]. The unconsumed feed and fish feces are sources of contaminants in fish culture. A new innovation has been made to minimize the waste in fish culture, which includes ammonia, nitrate and phosphate that is referred to as aquaponics system [4] using Red Water System (RWS) method [6, 7]. Water quality analyzed includes the analysis on ammonia, nitrate, phosphate, temperature and acidity (pH).

3.1. Ammonia (NH_3)

This study showed that the concentration of ammonia in catfish culture using aquaponics with RWS was reduced after the second sampling (figure 2).

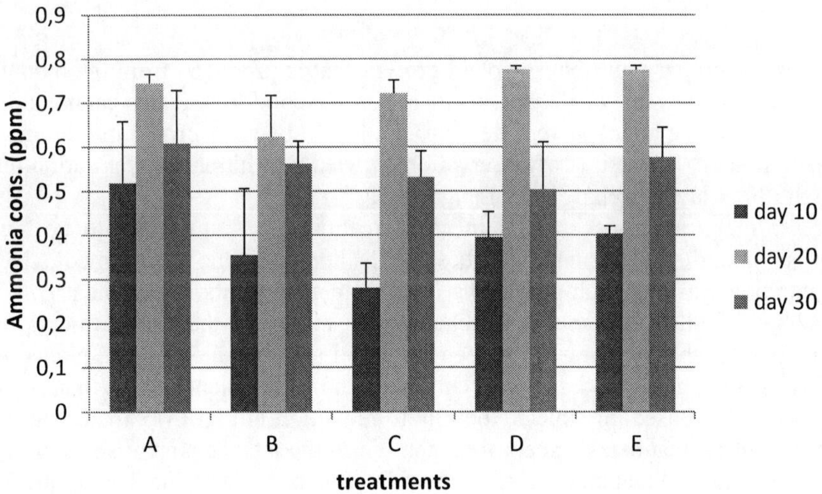

Figure 2. Ammonia concentration (ppm) in catfish tank water added with RWS at various concentrations. A: RWS without aquaponics, B: aquaponics without RWS, C: aquaponics with 7.5 $\mu l \cdot L^{-1}$/week RWS, D: aquaponics with 10 $\mu l \cdot L^{-1}$/week RWS and E: aquaponics with 12.5 $\mu l \cdot L^{-1}$/week RWS.

The measurement of ammonia concentration was performed when the fish was reared for 10 days. Based on (figure 2), ammonia concentration in all treatments increased in the middle of observation (Day 20). The increase of ammonia was caused by the accumulation of residual feed and metabolism waste of fish [9, 10, 12]. Meanwhile, at the end of the study (day 30), from three samplings, decrease in ammonia concentration was found in all treatments with a concentration reduction range of 0.04–

0.27 ppm). The decrease in the ammonia concentration might be caused by the ammonia absorption by the plants, which was getting higher. The reduction in ammonia concentration will reduce the toxicity [13]. The ammonia concentration reduction on the third observation (day 30) was also caused by the starting microbial activities (RWS bacteria) in converting ammonia into nitrate. Microorganisms in probiotics are able to degrade the organic residuals of the feed and feces rapidly that there is no excessive accumulation in fish culture [14]. Table 1 shows the ammonia fluctuations during the study. It was seen that ammonia increased at day 20. Ammonia increasement in treatments C, D and E were higher than controls. Among RWS aquaponics, treatment D showed that the bacteria in RWS increased the decomposition of organic matter in fish culture medium. While on the 30th day, ammonia concentration decrease in RWS treatments were higher than controls. At the end period of culture, highest ammonia decrease was found in treatment D. These results suggested that the bacteria in the RWS solution converted ammonia into nitrate and also was utilized by plants. Inorganic nitrogen used for forming the structure formation of plants, chlorophyll formation, photosynthesis, cell growth and metabolic processes [15].

Table 2. Fluctuation of ammonia concentration (ppm) in catfish tank water added with RWS at various concentrations.

Treatment	D10	D20	D30	D20-D10	D30-D20
A	0.51 ± 0.139	0.73 ± 0.021	0.60 ± 0.119	0.22 ± 0.124	-0.13 ± 0.129
B	0.35 ± 0.151	0.61 ± 0.093	0.57 ± 0.051	0.26 ± 0.243	-0.04 ± 0.137
C	0.28 ± 0.055	0.72 ± 0.031	0.53 ± 0.059	0.44 ± 0.044	-0.19 ± 0.028
D	0.39 ± 0.058	0.77 ± 0.008	0.50 ± 0.108	0.38 ± 0.055	-0.27 ± 0.107
E	0.40 ± 0.018	0.77 ± 0.009	0.57 ± 0.007	0.37 ± 0.008	-0.20 ± 0.059

3.2. Nitrate (NO_3)

Nitrate (NO_3) is the primary nutrient for plant growth that is produced from the oxidation process of ammonia in the nitrogen cycle which happens in an aerobic condition. Plants cannot absorb nitrogen in the form of N^{2+} gas in the atmosphere. The Nitrogen has to be transformed into ammonium (NH_4) or nitrate (NO_3) first [16]. The results of observation showed that nitrate concentration decreased in the beginning to the 30th day in each treatment (figure 3).

The decrease in nitrate concentration in each treatment was suspected to be the effect of the utilization of compound by plants and microbes (RWS) applied in the culture media. Nevertheless, the nitrate concentration was still high due to the low number of plants in treatment B, C, D and E that they were unable to utilize the abundant concentrated nitrate in the culture media caused by feed residuals and feces. Unbalanced ratio between the pond size, media size and plants will lead to the failure of utilizing all concentrated nitrate in culture media by the plants which then accumulates in the fish culture pond [18]. Based on table 2, the nitrate reduction in the RWS treatment was higher than those of in two control treatments, except treatment C. In the later treatment showed better absorption of nitrate by the plant in aquaponics. In treatment B (aquaponic without RWS) nitrate concentration was lower at day 30 than those of in other treatments. In the treatment B there was no addition of bacteria that can accelerate the process of decomposition of organic materials into inorganic materials. The nitrate present in treatment B was derived from natural decomposition so the concentration was lower than that of in RWS treatment. Therefore treatment B gave highest decrease in nitrate concentration.

Table 3. Nitrate reduction (ppm) in catfish tank water added with RWS at various concentrations.

Treatment	D10	D30	D30-D10	% reduction
A	0.789 ± 0.122	0.370 ± 0.033	0.419 ± 0.008	53.147
B	0.658 ± 0.408	0.219 ± 0.035	0.439 ± 0.420	66.737
C	0.634 ± 0.072	0.306 ± 0.056	0.328 ± 0.045	51.739
D	0.802 ± 0.214	0.314 ± 0.076	0.487 ± 0.249	60.780
E	0.824 ± 0.150	0.325 ± 0.064	0.499 ± 0.193	60.547

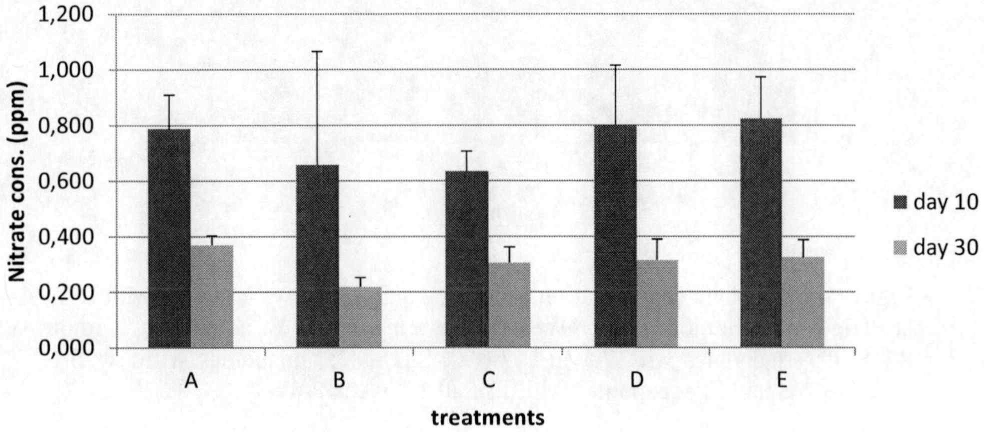

Figure 3. Nitrate concentration (ppm) in catfish tank water added with RWS at various concentrations. A: RWS without aquaponics, B: aquaponics without RWS, C: aquaponics with 7.5 μl·L^{-1}/week RWS, D: aquaponics with 10 μl·L^{-1}/week RWS and E: aquaponics with 12.5 μl·L^{-1}/week RWS.

3.3. Phosphate (PO_4)

Besides ammonia, phosphate is also produced through the decomposition of unconsumed feed [18, 19]. Phosphate concentration in fish feed is 0.96 % [20]. Around 10–15% of the feed provided is not consumed by the fish. The accumulation of phosphate in fish culture media produces a negative impact on the fish survival. On the other hand, phosphate is very beneficial for the life of plants because it is the basic ingredient to strengthen the cell wall, making the plants resistant to disease attacks [21]. The results of the study showed an increase and decrease in phosphate concentration during the study (figure 4).

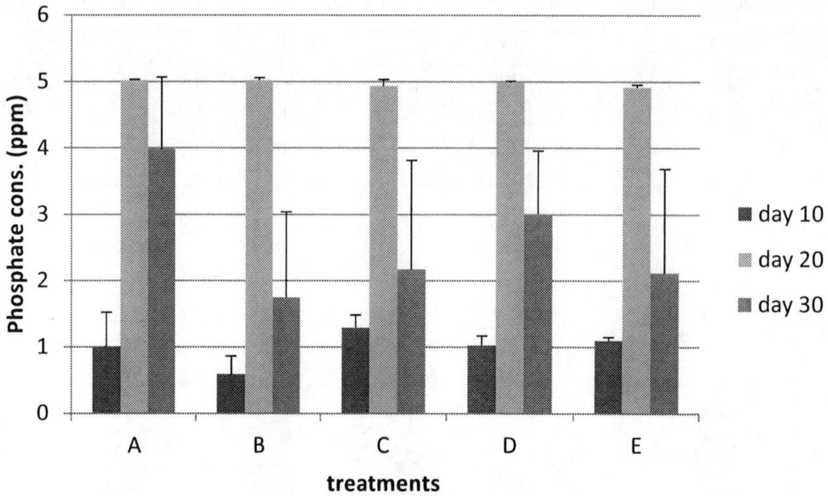

Figure 4. Phosphate concentration (ppm) in catfish tank water added with RWS at various concentrations. A: RWS without aquaponics, B: aquaponics without RWS, C: aquaponics with 7.5 μl·L^{-1}/week RWS, D: aquaponics with 10 μl·L^{-1}/week RWS and E: aquaponics with 12.5 μl·L^{-1}/week RWS.

Phosphate concentration decreased in all treatments during the third sampling (day 30) suggested that aquaponics and RWS were able to reduce phosphate concentration in fish culture media. Good phosphate concentration for fish culture is 0.2–1 ppm [22]. The highest reduction of phosphate was seen in treatment E (3.71 ppm) and B (3.28 ppm), showing the effectiveness of phosphate absorption by the lettuce plants in the aquaponics system. This fact is supported by the result in treatment A with the lowest phosphate reduction because no RWS was applied in the aquaponics system.

3.4. Acidity (pH) and temperature

Acidity (pH) is one of the important parameters that determine the quality of water in the culture media. Most fish is able to adapt well to the environment with a pH of 5–9. Good water pH for catfish culture is between 6.5 and 8 [23]. This study showed that the pH value was still suitable for the life of fish in the culture media (figure 5).

In overall, the results of the 40 days observation revealed that the pH value for each treatment was still in the normal range. The pH value in all treatments with RWS tended to be lower because the microbes growing in the culture media produce acid compound to slow down the growth of pathogenic microbes in the media [14].

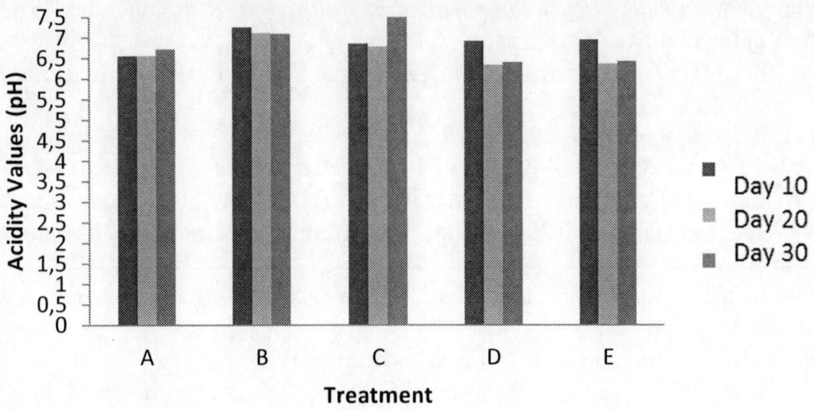

Figure 5. pH values catfish tank water added with RWS at various concentrations. A: RWS without aquaponics, B: aquaponics without RWS, C: aquaponics with 7.5 $\mu l \cdot L^{-1}$/week RWS, D: aquaponics with 10 $\mu l \cdot L^{-1}$/week RWS and E: aquaponics with 12.5 $\mu l \cdot L^{-1}$/week RWS.

The temperature in this study ranged between 24–25°C for each treatment. This value was in the normal range for supporting the life of catfish. The Indonesian National Standard: 01-6484.4-2000, states that 25–30°C is appropriate temperature for growth of catfish.

4. Conclusion

The addition of RWS probiotics increased concentration of nutrient for plant such as ammonia, nitrate and phosphate, while the aquaponics system reduces the concentration of ammonia, nitrate and phospate.

The addition of RWS probiotic increases the decomposition process and availability of nutrient for plant, while the aquaponics system reduces those nutrients. The addition of 10 $\mu L \cdot L^{-1}$/week RWS solution gave best result.

Acknowledgment

This study is funded through the 2015 ALG (Academic Leadership Grant) grant scheme of Universitas Padjadjaran. Authors would like to extend their gratitude to the Rector of Universitas Padjadjaran.

References

[1] Direktorat Jenderal Perikanan Budidaya 2014 *Laporan Tahunan Direktorat Produksi Tahun 2013* (Jakarta) p 45
[2] Diver S 2006 *Aquaponics-integration of hydroponics with aquaculture* National Sustainable (Australia: Agriculture Information Service) p 28
[3] Nelson R L 2008 Aquaponic Equipment: The Biofilter. *http://www.aquaponicsjournal. com/ docs/Aquaponic-Equipment-The-BioFilter.pdf.*
[4] Taufik I 2010 *Uji Multi Lokasi Budidaya Ikan Nila dengan Sistem Akuaponik* (Bogor: Balai Riset Perikanan Budidaya Air Tawar)
[5] Wijaya O, Rahardja B S and Prayogo 2014 *Jurnal Ilmiah Perikanan dan Kelautan* **6** 55–8
[6] Aquarista F, Iskandar and Subhan U 2012 *Jurnal Perikanan dan Kelautan* **3** 133–40
[7] Muartiati K, Simbolon, Wahyuni T and Subadri 2006 *Jurnal Budidaya Air tawar Sukabumi* **3** 1–7
[8] Hartini S, Sasanti A D and Taqwa F H 2013 *Jurnal Akuakultur Rawa Indonesia* **1** 192–202
[9] Rakocy J E, Masser M P and Losordo T M 2006 *Recirculating Aquaculture Tank Production*

Systems L Aquaponics-Integrating Fish and Palnt Culture Souther Regional Aquaculture Center Publication No. 454
[10] Fujaya Y 2004 *Fisiologi Ikan: Dasar Pengembangan Teknik Perikanan* (Jakarta: Rineka Cipta) p 179
[11] Huisman E A 1976 *Aquaculture* **9** 259–73
[12] Utomo N B P, Hasanah P and Mokoginta I 2005 *Jurnal Akuakultur Indonesia* **4** 49–52
[13] Crab R, Avnimelech Y, Defoirdt T, Bossier P and Verstraete W 2007 *Aquaculture* **270** 1–14
[14] Suwoyo H and Mangampa M 2010 *Jurnal Prosiding Forum Inovasi Teknologi Akuakultur* **1** 239– 46
[15] Somerville C 2014 *Small-Scale Aquaphonic Food Production* FAO Fisheries and Aquaculture Technical Paper No. 589
[16] Campbell N A 2003 *Biologi* (Jakarta: Penerbit Erlangga)
[17] Samsundari S and Wirawan G A 2013 *Jurnal Gagasan Ilmiah Masa Depan* **8** 86–97
[18] Garno Y S 2002 *Jurnal Teknologi Lingkungan* **3** 112–120
[19] Susila N 2015 *Jurnal Ilmu Hewani Tropika* **4** 71–4
[20] Brahmana S S, Sumariani Y and Ahmad 2010 *Kualitas Air dan Eutrofikasi Waduk Riam Kanan di Kalimantan Selatan* Prosiding Seminar Nasional Limnologi V
[21] Siregar H R, Sumono S B, Daulay and Susanto E 2013 *Jurnal Rekayasa Pangan dan Pertanian* **3** 60–5
[22] Ebeling J M, Timmons M B and Bisogni J J 2006 *Aquaculture* **257** 346–58
[23] Khairuman and Amri K 2011 *Pembenihan Lele 21 Hari Balik Modal* (Jakarta: Agromedia)

The axenic treatments for *Kappaphycus alvarezii* (Rhodophyta) seedling in laboratory culture

R I Adharini[1*], A R Setyawan[1], A D Jayanti[1], Suadi[1] and E A Suyono[2]

[1] Department of Fisheries, Faculty of Agriculture, Universitas Gadjah Mada, Jl. Flora 1A, Bulaksumur, Sleman, Yogyakarta, Indonesia
[2] Department of Biotechnology, Faculty of Biology, Universitas Gadjah Mada, Jl. Teknika Selatan, Sekip Utara, Sleman, Yogyakarta, Indonesia
[*]Email: ratih.adharini@ugm.ac.id

Abstract. Obtaining an algae axenic culture in the culture medium are challenging. This study aims to determine the effectiveness of the application of several methods to reduce bacterial contamination in culture medium. The study was conducted using a complete randomized design in 3 stages, stage 1 using iodine immersion method with 4 treatments, stage 2 using immersion of antibiotics mixture with 3 different concentrations, stage 3 using mixture of antibiotic and medium with 3 concentration. The results showed that in stage 1, the lowest bacteria density was in treatment 4 (15 % betadine with 60') (675.27 idv·mm^{-2}), stage 2 showed that treatment 1 (50 mg·L^{-1}) had the lowest bacteria density (265.62 idv·mm^{-2}). Stage 3 showed that treatment 3 (10 mg·L^{-1}) had the lowest density of bacteria (24.78 idv·mm^{-2}). Based on the ANOVA test, stage 1 has no significant difference (> 0.05), in stage 2 there was a significant difference with treatment 1 was the best treatment, in stage 3 there was a significant difference and treatment 4 was the best result. Iodine 15 % with 60' immersion time; immersion with 50 mg·L^{-1} antibiotic, and mixture of medium with 10 mg·L^{-1} antibiotic gave the best results in reducing bacterial contamination.

1. Introduction

Seaweed is one of the leading commodities of aquaculture fisheries that have high economic value both in local and export markets. Seaweed is one of the three main commodities in the revitalization of fisheries other than fish and salt. Based on Ministry of Fisheries and Marine Affairs production data of 2015, the production volume of seaweed cultivation is 11.68 million tons or reach 66.87 % of total aquaculture production [1]. This shows that seaweed is the most dominant commodity in the production of aquaculture fishery.

Kappaphycus alvarezii is a seaweed that has a very high economic value. *Kappaphycus alvarezii* seaweed contains carrageenan that has been used in several industries. The use of carrageen in the industry is almost equal to the agar, among others, as a regulator of balance, thickening, gelling, emulsifying and so on. Utilization of carrageenan by industry includes food industry, pharmaceuticals, cosmetics, textiles and organic fertilizer [2]. Kappaphycus species are also highly sought after by coastal communities due to relatively easy cultivation technology, short maintenance time and relatively cheap production costs [3].

Provision of good seed is one of the activities that determine the success of seaweed cultivation. Currently the seeds are used and developed by the community is still obtained from the results of vegetative development by way of setting aside thallus cultivation of his own, so the availability of good quality seeds are sometimes limited. The low technological mastery ranging from seeding, cultivation, harvesting and drying of seaweed is also a crucial problem in increasing the productivity of seaweed farming [4]. Efforts to provide superior seaweed seedlings can be done with the application of biotechnology. Technology which has been developed in seaweed in general is a technique of tissue culture, protoplast culture, embryogenesis techniques and culture of the spore [2].

Seed propagation through *in vitro* as tissue culture and culture is one seeding spores that can be done at any time and does not depend on the season. However, one of the obstacles in the application of biotechnology in the development of seaweed seedling is the difficulty to make an axenic culture. Axenic culture is a culture that obtained from an algal species, without any contamination from other living organisms [5]. The contamination in culture is common, generally by diatoms and bacteria. Development of the axenic culture of seaweed is very important in the preparation of seed stock and mass culture from seeds so that it will determine the success and quality of the seed produced. This study aims to determine the effectiveness of the application of several methods of antibiotic treatments to reduce bacterial contamination in the culture medium.

2. Materials and methods

2.1. Experimental design

The experiments were performed using a complete randomized design divided into 3 stages with one control (without treatment). Stage 1 was the application of iodine immersion with treatments: control (C); T1 (10 % betadine for 30'); T2 (10 % betadine for 60'); T3 (15 % betadine for 30') and T4 (15 % betadine for 60') with three replications. Stage 2 was the application of antibiotic (Penicillin and streptomycin) immersion with 3 different concentrations: control (C); T1 (50 $mg \cdot L^{-1}$); T2 (100 $mg \cdot L^{-1}$); T3 (150 $mg \cdot L^{-1}$) for 24 hours with three replication. Stage 3 was the application of medium and antibiotic mixture (the mixture of medium f/2 with penicillin and streptomycin antibiotics) with 4 different concentrations: control (C); T1 (6 $mg \cdot L^{-1}$); T2 (8 $mg \cdot L^{-1}$); T3 (10 $mg \cdot L^{-1}$). The samples were cultured in the room temperature and light intensity of 4.617 $\mu mol \cdot m^{-2} \cdot s^{-1}$.

2.2. Sample preparation, betadin and antibiotics

The sample was *Kappaphycus alvarezii*. The samples were collected from seaweed farming site in Popayato, Pohuwato district, Province of Gorontalo, Sulawesi, Indonesia 0° 30' 53" North and 121° 28' 24" East. Before the experiment, the samples were cleaned by washing using sterile sea water and the samples were cleaned from the dirt and epiphytes on the thallus surface using a paintbrush and a razor blade. Once clean, then thallus cut into small pieces of 3–5 mm in size. Making betadine by mixing 100 % betadine with sterile seawater with the ratio according to treatment that is equal to 10 % and 15 %. Preparation of antibiotics by mixing Penicillin and streptomycin with sterile sea water by comparison according to treatment i.e. 50 $mg \cdot L^{-1}$, 100 $mg \cdot L^{-1}$ and 150 $mg \cdot L^{-1}$. A mixture of medium and antibiotics was prepared by mixing the f/2 medium and the antibiotics with concentrations of 6 $mg \cdot L^{-1}$, 8 $mg \cdot L^{-1}$ and 10 $mg \cdot L^{-1}$.

2.3. Data collection

In the first stage, iodine immersion was applied, the concentration used was 10 % and 15 % with 30' and 60' immersion period respectively. The samples used in each well were 18 pieces of thallus. Observations were made after 24 hours of immersion. The data taken is the bacterial density of each well from the microplate.

In stage 2 with the application of antibiotic placement, the concentrations used were 50 $mg \cdot L^{-1}$, 100 $mg \cdot L^{-1}$, and 150 $mg \cdot L^{-1}$. The numbers of thallus used in each well were 3 thallus with 5 mm size per well. Observation of bacterial density was done after 24 hours of immersion using microscopic observation with 100x magnification.

In stage 3, the application of a mixture of medium f/2 with antibiotics, the concentration used was without the addition of antibiotics, 6 $mg \cdot L^{-1}$, 8 $mg \cdot L^{-1}$ and 10 $mg \cdot L^{-1}$. The number of thallus used in each well is 3 thallus. Observation of bacterial density was done at 24 hours, 48 hours and 72 hours of immersion.

2.4. Data analysis

The obtained bacterial density data were then descriptively graphed and one way ANOVA test was performed using SPSS program to find out the best treatment.

3. Results and Discussion

3.1. Stage 1 (Iodine immersion)

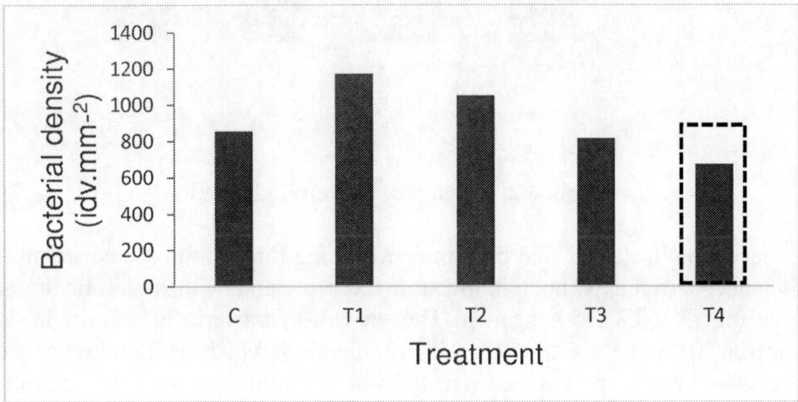

Figure 1. The density of bacteria in stage I.

Based on the observation of bacterial density in stage 1 by immersion on betadine, can be seen the influence between the amount of concentration of betadine and the duration of immersion to bacterial density. The results showed that in 10 % betadine (1 % provine-iodine) immersion (T2 and T3) had the highest bacterial density value and there was no difference between immersion time. The magnitude of bacterial density at T2 and T3 is higher than control, this may be possible with concentrations used still under betadine concentrations to kill bacteria. Figure 1 shows that the best treatment is in T4 which was immersion with 15 % betadine (1.5 % provine-iodine) with 60' immersion time. However, based on ANOVA test, there was no significant difference in each treatment, but T4 treatment showed the best result. Application of higher concentrations and longer periods of time is possible to kill the thallus and spores of seaweed. Betadine contains provine-iodine which is commonly used as a disinfectant. Application of iodine as a disinfectant in the culture of seaweed has been done to some researchers, such as [6] using provine-iodine 2 % to obtain pure cultures, [7] provine-iodine using a mixture of 3 $g \cdot L^{-1}$ and 15 % ethanol, [8] using provine-iodine of 0.5 %. Iodine works by reducing the oxygen needs of aerobic microorganisms. Iodine would disrupt the rate of the respiratory chain of microorganisms by blocking the transport of electrons through electrophilic reactions with respiratory chain enzymes [9].

3.2. *Stage 2 (Antibiotic immersion)*

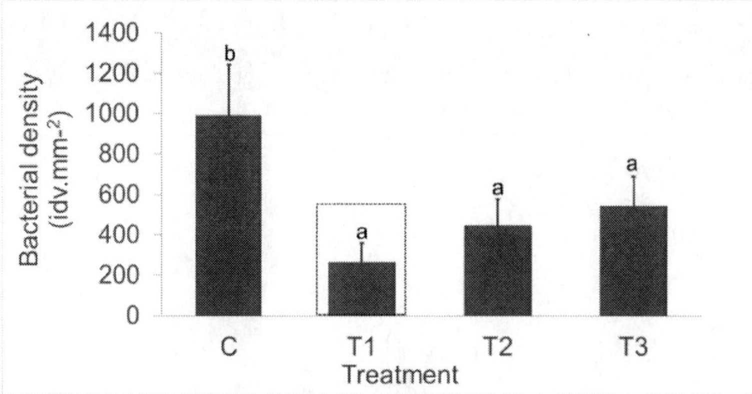

Figure 2. Density of bacterial phase II.

Based on the observation in stage 2 that by immersion using three antibiotic concentration there was a significant difference between without immersion (control) and with antibiotic immersion (T1 (50 $mg{\cdot}L^{-1}$), T2 (100 $mg{\cdot}L^{-1}$), T3 (150 $mg{\cdot}L^{-1}$). The growth of bacteria in control in density was 991 $idv{\cdot}mm^{-2}$, compared with the treatment of T3 immersion which is the higher density between antibiotic immersion treatments having density 544 $idv{\cdot}mm^{-2}$. It is showed that the antibiotic immersion treatment inhibits the bacterial growth effectively. It is also indicated by ANOVA test which has a significance value of 0.04 or below 0.05 which shows that there are different in between treatment and there was an effect of immersion antibiotic with bacterial growth. The best treatment in stage 2 was T1 (50 $mg{\cdot}L^{-1}$) with a density of 266 $ind{\cdot}mm^{-2}$ (figure 2). The best treatment is the smallest concentration because in the larger concentration the bacteria become more resistant to antibiotics therefore ineffective inreducing bacterial growth. The best treatment is the smallest concentration, possible in the larger concentration the bacteria become resistant so that they become more resistant to antibiotics and become ineffective in reducing bacterial growth. Antibiotics have been widely applied to remove or inhibit bacterial growth. Penicillin G., streptomycin sulfate, chloramphenicol and gentamycin are commonly used antibiotics [10]. Antibiotics immersion has been widely applied, as in [7] used a mixture of penicillin and streptomycin of 200 $mg{\cdot}L^{-1}$ to obtain the axenic culture. While, [11] using a rifampicin (300 $mg{\cdot}L^{-1}$), streptomycin/penicillin (25 $mg{\cdot}L^{-1}$) and nystatin (10 $mg{\cdot}L^{-1}$) to obtain axenic culture.

3.3. *Stage 3 (Medium and antibiotic mixture)*

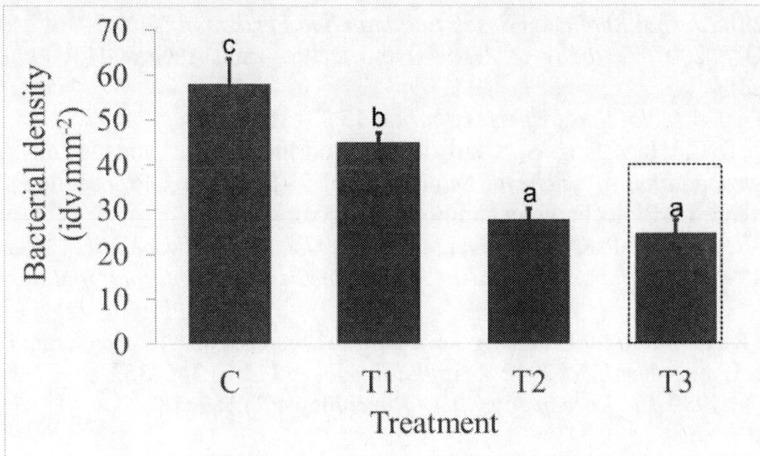

Figure 3. Density of bacterial phase III.

Based on the observation at stage 3 that is, using the mixture of medium f/2 and antibiotic, showed the influence of different concentration with bacterial growth. Figure 3 shows that treatment of C (control) which is a sample without antibiotics had the highest bacterial growth value with the density of 58,11 idv·mm^{-2}. This is different from the treatment of 1 (6 mg·L^{-1}), 2 (8 mg·L^{-1}) and 3 (10 mg·L^{-1}) which is a mixture of medium and antibiotics. The best treatment was found in the treatment of T3 which was the highest concentration of medium and antibiotic (10 mg·L^{-1}) with the lowest bacterial growth of 25 idv·mm^{-2}. Based on the ANOVA test, it shows a significance value of 0.000 or below 0.05 so there is a real difference between treatments, with the best treatment being at T3. Penicillin G. and streptomycin sulfate are included in B-lactam antibiotics, which have a broad spectrum and have an effect on cell wall synthesis in both gram-positive and negative bacteria [11]. This makes both types of antibiotics are effectively used in inhibiting the growth of bacteria in seaweed culture. The application of medium and antibiotic mixtures was shown to be effective in inhibiting bacterial growth, [12] used a mixture of antibiotics (Penicillin G., Streptomycin sulfate, Polymyxin B sulfate and Chloramphenicol) on the medium for ten days to obtain axenic culture and [10] stated an addition of 2–5 drops of antibiotic mixture on 10 mL of medium for three days can inhibit bacterial growth and produce axenic culture.

4. Conclusion

Application of several methods to reduce bacterial growth can be done by using immersion of 15 % betadine for 60 minutes, followed by a mixture of penicillin G. and streptomycin sulfate immersion at concentrations of 50 mg·L^{-1} for 24 hours and using a medium and antibiotic mixture of 10 mg·L^{-1}.

Acknowledgments
This research work was supported by Indonesian Ministry of Research, Technology and Higher Education in project of PUPT Universitas Gadjah Mada 2017.

References

[1] KKP 2016 *Laporan Kinerja Kementerian Kelautan dan Perikanan Tahun 2015* (Jakarta: Kementerian Kelautan dan Perikanan)
[2] Parenrengi A, Rachmansyah dan Suryati E 2012 *Budidaya Rumput Laut: penghasil karaginan (karaginofit)* (Jakarta: BPPKP Kementerian Kelautan dan Perikanan)

[3] Indriani H dan Sumarsih 2003 *Budidaya, Pengolahan dan Pemasaran Rumput Laut* (Jakarta: Penebar Swadaya)
[4] Hikmah 2015 *Jurnal Kebijakan Sosek Kelautan dan Perikanan* **5**(1) 27–36
[5] Sharma O P 2007 *Textbook of Algae* (New Delhi: Tata McGraw-Hill Publishing Company Limited)
[6] Sahidin N 2014 *Malaysian Fisheries Journal* **13** 33–38
[7] Yong W T L, Hoo Ting S, Chin W L, Rodrigues K F and Anton A 2011 *In vitro Micropropagation of Euchema Seaweeds* vol 7 (Singapore: Proceedings 2nd International Conference on Biotechnology and Food Science)
[8] Kumar G R, Reddy C R K and Bhavanath Jha 2007 *J. Appl. Phycol.* **19**(1) 15–25
[9] Maris P 1995 *Scientific and Technical Review of the Office International des Epizooties* **14**(1) 47–55
[10] Andersen R A 2005 *Algal Culturing Techniques* (London: Elsevier Academic Press)
[11] Wilkens S L and Maas E W 2012 *J. Appl. Microbiol.* **112**(2) 346–352
[12] Tatewaki M 1989 *The Korean Journal of Phycology* **4**(2) 183–189

Similarity microalgal epiphyte composition on seagrass of *Enhalus acoroides* and *Thalasia hemprichii* from different waters

R Hartati [1]*, M Zainuri[2], A Ambariyanto [1], Widianingsih [1], A Trianto[1] and R T Mahendrajaya[1]

[1]Marine Science Department, Faculty of Fisheries and Marine Science, Diponegoro University. Jl. Prof. Soedharto, SH. Tembalang, Semarang, Indonesia.
[2]Oceanography Department, Faculty of Fisheries and Marine Science, Diponegoro University
*Email: retnohartati.undip@yahoo.com; retnohartati.undip2@gmail.com

Abstract. The epiphytes are all autotrophic organisms that are permanently attached to rhizomes, roots, and leaves of seagrasses. The epiphyte is an important primary producer for the seagrass ecosystem and contributes significantly to the food chain. This study aims to identify the composition of microepiphyte algae on *Enhalus acoroides* and *Thalassia hemprichii* and their similarity levels of both compositions. The 20 leaves samples of *E. acoroides* and *T. hemprichii* were observed. The epiphytic microalgae which found on the surface of the seagrass leaves were scrapped, collected in a bottle sample, and fixed with 70 % alcohol and identified into genera. The relation of epiphytic microalgal genera to the location and species of seagrass was analyzed using similarity analysis. The Chrysophyta, Cyanophyta, and Chlorophyta epiphytic microalgal were found. There were similarity variations of the microalgal epiphyte in seagrass of *E. acoroides* and *T. hemprichii* and seagrass habitat sites. Morphology and seagrass life affects the abundance and diversity of the epiphytic microalgal attached to the seagrass and it may be associated with the epiphytic lifetime in the seagrass.

1. Introduction

Seagrasses are a major functioning element resulting high productivity of tropical marine regions. They occur in the form of multispecific beds (constructed by more than one species of seagrass) or monospecific (constructed by only one species of seagrass) [1]. The presence of seagrasses provide habitat and consequently, increase the abundance and diversity of marine organisms [2, 3], such as epiphytes that live on leaves and stems of seagrasses. Selection by organisms to seagrass habitat may be influenced by several factors including habitat structures which leads to differences in habitat suitability, food availability and protection provided [4].

Epiphytes are the foremost component of majority seagrass ecosystems. They are also the primary food resource for grazers. Microalgal epiphyte was the most abundant and diverse found in seagrass, especially seagrass leaves [5]. The epiphyte abundance and species composition are strongly influenced by the seagrass species which is related to the variation of leaf turnover of different species of seagrass [6]. Faster leaf turnovers limit the time for colonization of epiphytes compared to the slower ones. The size of seagrasses and the rate of seagrass turnover are important in determining loads of epiphytic organisms attached to them [5]. Therefore, the difference of seagrass bed concerning the number of species (mono or multispecific) will influence the diversity and biomass of epiphyte attached to seagrass leaves in such meadows. The studies on ephypite of the seagrass mostly were macroalgal [7, 8], unspecified epiphytic algae [9] or microalgal ephypite on certain species of seagrass, such as *Halodule wrightii* [10], *Cymodacae serrulata* [11]. Comparing ephypitic microalgae on two different species has only conducted in *Posidonia oceanica* and *Cymodocea nodosa* [12]. Therefore this study aims to identify the composition of microepiphyte algal live on the seagrass of *Enhalus acoroides* and *Thalassia hemprichii* and their similarity levels of both compositions.

2. Materials and Methods

The study was performed in June 2016 at Bandengan (06°33.905'S, 110°39.315'E, and Ujungpiring (06°30'74,0"S, 110°40'18,4" E) Waters of Jepara Regency. *E. acoroides* and *T. hemprichii* are the most dominated seagrass species in those study area [13, 14, 15].

Epiphyte microalgal was scraped from the seagrass leaves which came from two study sites. A total of twenty leaves of *E. acoroides* and *T. hemprichii* were cut at the base of the leaves, stored in a sample bag sealed filled with seawater and placed into cool-box and taken to the laboratory for observation. The leaves were cut and categorized (tip, middle, and base of the leaves). The epiphytic microalgae found on the surface of the leaves portion was carefully scrapped, collected in a bottle sample, fixed with 70% alcohol and identified into species according to [16]. The relation of epiphytic microalgal species to the location and species of seagrass was analyzed using Sørensen similarity analysis [17]. The similarity is high when the value more than 50% and vice versa.

$$\text{Sørensen similarity (S)} = \frac{2a}{2a+b+c} \tag{1}$$

a = the number of species shared by both samples (fraction a)

b and c = numbers of species occurring only in the first and only in the second sample, respectively)

3. Results and Discussion

The seagrass species observed in present work are *E. acoroides* and *T. hemprichii* which live on the intertidal coast with sandy to coarse rubble substrates. A total of 32 genera of ephypite microalgal found on the leaves of two species of seagrasses, i.e., 25 and 21 genera found on the leaves of *E. acoroides* and *T. hemprichii* respectively from Bandengan waters and while the leaves of *E. acoroides* and *T. hemprichii* from Ujungpiring waters provide habitat for 26 and 21 genera of microalga ephypite (table 1, 2). Among the genera found in both location, the highest genera number belong to Chrysophyta Division, followed by Chlorophyta and Cyanophyta.

It is shown in table 3, that almost all comparison have high similarity index (SI > 50 %) with the value in the range of 51,43–80,95 %. There were only 4 value with high dissimilarities, such as TBU (Basal of *T. hemprichii* leaves of Ujungpiring Waters)-TMU (Middle of *T. hemprichii* leaves of Ujungpiring Waters) (50 %), TBB (Basal of *T. hemprichii* leaves of Bandengan Waters) -EMB (Middle of *E. acoroides* leaves of Bandengan Waters) (48,48 %), TTU (Tip of *T. hemprichii* leaves of Ujungpiring Waters)-EMU (Middle of *E. acoroides* leaves of Ujungpiring Waters) (47,37 %) and TBU (Basal of *T. hemprichii* leaves of Ujungpiring Waters)-EMU (Middle of *E. acoroides* leaves of Ujungpiring Waters) (45,16 %). The water quality of both study sites was presented in figure 1 and table 4.

Table 1. "Presence (+)-absence(-)" genera list of epiphytic microalgae identified on the leaves of *E. acoroides* and of *T. hemprichii* in Bandengan waters of Jepara.

No.	Genera	*Enhalus acoroides*			*Thalassia hemprichii*		
		Tip of the leaves	Middle of the leaves	Base of the leaves	Tip of the leaves	Middle of the leaves	Base of the leaves
Division Chrysophyta							
1	Achnanthes	+	+	-	+	+	+
2	Amphora	+	+	-	+	+	+
3	Ardissonea	+	+	-	-	+	+
4	berkeleya	+	+	+	+		+
5	Cocconeis	+	+	+	+	+	+
6	Coscinodiscus	+	+	+	+	+	+
7	Diatomae	+	+	+	+	-	-
8	Diplonesi	+	+	+	-	+	-
9	Gomphonema	-	+	-	-	+	-
10	rammatiophora	+	+	-	-	-	-
11	Gyrosigma	+	+	+	+	+	-
12	Hyalodiscus	+	+	+	+	-	-
13	Hyalosira	+	+	+	-	-	-
14	Leptocylindrus	-	-	+	+	+	+
15	Licomorpha	+	+	+	+	-	-
16	Melosira	+	+	+	+	+	+
17	Navicula	+	-	+	+	+	+
18	Neosynedra	-	+	-	-	-	-
19	Nitzschia	-	-	-	+	+	+
20	Pleurosigma	-	+	+	-	-	-
21	Rhizosolenia	+	+	-	-	-	-
22	Synedra	-	-	-	-	-	-
23	Thalassionema	-	-	-	-	-	+
24	Thalassiothrix	-	-	+	-	-	-
Division Chlorophyta							
1	Atractomorpha	+	+	+	+	+	+
2	Characium	+	-	-	-	-	-
3	Stigeoclonium	+	-	+	-	+	-
Division Cyanophyta							
1	Anabaenopsis	-	+	-	-	-	-
2	Merismopedia	+	+	-	-	+	-
3	Microcystis	+	-	-	+	+	-
4	Planktolyngbia	-	-	-	+	-	-
5	Trichodesmium	-	-	-	-	-	-

Table 2. "Presence (+)-absence (-)" genera list of epiphytic microalgae identified on the leaves of *E. acoroides* and of *T. hemprichii* in Ujungpiring waters of Jepara.

		Enhalus acoroides			*Thalassia hemprichii*		
No.	Genera	Tip of the leaves	Middle of the leaves	Base of the leaves	Tip of the leaves	Middle of the leaves	Base of the leaves
Division Chrysophyta							
1	Achnanthes	+	+	+	+	+	+
2	Amphora	-	-	-	+	+	+
3	Ardissonea	+	+	+	+	+	+
4	Berkeleya	+	+	+	-	+	+
5	Cocconeis	+	+	+	+	+	+
6	Coscinodiscus	+	+	+	+	+	+
7	Diatomae	+	+	+	+	+	-
8	Diplonesi	+	+	+	+	-	-
9	Gomphonema	-	+	-	-	-	-
10	Grammatiophora	+	+	-	-	-	-
11	Gyrosigma	+	+	+	+	+	+
12	Hyalodiscus	-	+	+	+	+	-
13	Hyalosira	+	+	+	-	-	-
14	Leptocylindrus	+	+	+	+	+	+
15	Licomorpha	+	+	+	+	+	-
16	Melosira	+	-	+	+	-	+
17	Navicula	+	-	+	+	-	-
18	Neosynedra	-	-	-	-	-	-
19	Nitzschia	+	-	+	+	+	-
20	Pleurosigma	+	+	+	+	+	-
21	Rhizosolenia	+	-	+	+	-	+
22	Synedra	-	+	-	-	-	-
23	Thalassionema	+	+	-	-	-	-
24	Thalassiothrix	+	+	+	-	-	-
Division Chlorophyta							
1	Atractomorpha	+	+	+	+	+	+
2	Characium	+	+	-	-	-	-
3	Stigeoclonium	+	+	+	-	-	-
Division Cyanophyta							
1	Anabaenopsis	-	+	-	-	-	-
2	Merismopedia	+	-	+	+	+	-
3	Microcystis	+	+	+	-	-	-
4	Planktolyngbia	+	-	-	+	+	-
5	Trichodesmium	+	-	-	-	-	-

Table 3. The result of Sørensen similarity index of ephypite microalga genera among the seagrass species (E : *E. acoroides*; T: *T. hemprichii*), the portion of leaves (T : tip, M: middle, B: basal) middle and location (B: Bandengan waters, U : Ujungpiring Waters).

	ETB	EMB	EBB	TTB	TMB	TBB	ETU	EMU	EBU	TTU	TMU	TBU
ETB	0,00	80,95	70,27	70,27	73,68	54,55	73,91	66,67	71,79	63,16	51,43	58,06
EMB	80,95	0,00	66,67	59,46	63,16	59,46	65,22	66,67	71,79	63,16	62,86	58,06
EBB	70,27	66,67	0,00	71,43	60,61	71,43	68,29	64,86	76,47	66,67	53,33	53,85
TTB	70,27	59,46	71,43	0,00	72,73	71,43	71,43	54,05	64,71	72,73	73,33	61,54
TMB	73,68	63,16	60,61	66,67	0,00	66,67	72,73	52,63	74,29	64,71	58,06	66,67
TBB	54,55	48,48	68,75	71,43	72,73	0,00	63,41	54,05	64,71	72,73	73,33	61,54
ETU	73,91	65,22	68,29	63,41	61,90	63,41	0,00	69,57	79,07	71,43	56,41	51,43
EMU	66,67	66,67	64,86	54,05	52,63	54,05	69,57	0,00	56,41	47,37	51,43	45,16
EBU	71,79	71,79	76,47	64,71	74,29	64,71	79,07	56,41	0,00	80	62,50	64,29
TTU	63,16	63,16	66,67	72,73	64,71	72,73	71,43	47,37	80,00	0,00	77,42	59,26
TMU	51,43	62,86	53,33	73,33	58,06	73,33	56,41	51,43	62,50	77,42	0,00	50,00
TBU	58,06	58,06	53,85	61,54	66,67	61,54	51,43	45,16	64,29	59,26	50,00	0,00

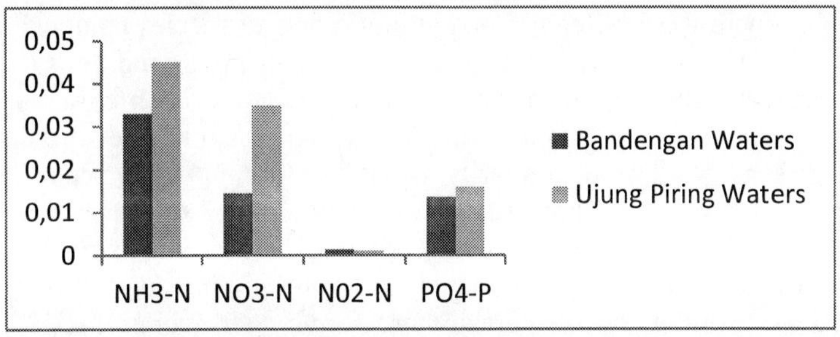

Figure 1. The Nutrient concentration in the water of study sites.

Table 4. The Water quality in Bandengan and Ujungpiring Waters.

Parameter	Bandengan Waters	Ujungpiring Waters
Temperature (°C)	27-33	30-32
Salinity (ppt)	26-31	27-32
pH	8	6
DO (ppm)	3,54-6	2,5-6
Light intensity (cm)	98-110	85-110
Depth (cm)	98-110	85-110

It is known that there were 5 and 3 species of seagrass inhabit in Bandengan and Ujungpiring Waters [14, 15]. *E. acoroides* and *T. hemprichii*, belong to Family of Hydrocharitaceae, were found growing in mixed species meadow. *E. acoroides* is a very distinctive seagrass with very long and ribbon-like (30 – 150 cm long, approx. 1.25 – 1.75 cm wide), with many parallel veins, generally dark green in color, thick and tough leaves. The 'toughness' apparent were caused by the raised margins along the sides of the leaves, which generated by inrolling and thickening of the lateral edges. *T. hemprichii*, exhibits some variation in leaf width and length. However, the basic structure remains the same. The rhizome is thick (up to 5 mm thick), and distinctive, since the nodes, where the old shoots joined the leaf-bearing branch, are plainly visible with a prominent scale at each. Their pale basal leaf sheath is 3–7 cm long and well developed. Leaves are 10–40 cm long, ribbon-like and often slightly curved laterally. Leaf width is generally in the range of 0.4–1.0 cm. There are 10–17 longitudinal leaf veins. The leaves have numerous large tannin cells grouped in short black bars running parallel to the long axis of the leaf. These 'bars' are visible to the naked eye and are one of the diagnostic features of this species. The leaf tip is rounded and sometimes slightly serrated [18].

Compared to other research, there were 27 genera of microalgal ephypite in Teluk Awur [15] and 28 in Karimunjawa Waters [19]. The rapid growth of epiphytes, such as diatoms belongs to Chrysophyta Division which was considered the most important structural elements of the epiphyte on seagrass [19], which usually the result of the high nutrients.

Some epiphytic algae were known as specific and obligate epibionts on certain hosts [20], although most of them were essentially facultative and were not specifically associated with a host species [21]. Algal epiphytes play an important role in coastal benthic communities, as they provide a potential mutualistic interspecific association [22], as food and habitats for animals as well as considered as primary producers in the food chain [23, 24]. The epiphytic algae associated with seagrass are contributed 62, 50 and 44 % of primary production for *Syringodium filiforme*, *T. testudinum* and *Halodule wrightii*, respectively [25], 19–37 % for *T. hemprichii* [26] and 2–9 % for *E. acoroides* [27]. This contribution might be affected by the seagrass species which compose one bed. Other study done by [12] appeared that there were different epiphytic microalgae species composition on the leaves of *P. oceanica* and *C. nodosa*. It may vary due to local environmental conditions (hydrodynamics, light penetration), host characteristics (meadow type, shape forms of leaves, lifespan, and growth rate), and grazing effect which seem to be responsible for those dissimilarities in epiphytic microalgae communities. In present work, the similarity of ephypite microalga seems due to the similar morphology of seagrass leaves, i.e., ribonlike of *E. acoroides* and *T. hemprichii*.

Previous studies by [28, 29, 30] showed that epiphytic community structure was influenced by abiotic factors such as light, temperature, nutrients, and water motion, as well as by biotic factors such as leafage, seasonal cycle of the host, and grazing pressure by herbivores [31, 32, 33, 34]. The similar of water quality of both study sites (figure 1; table 4) also affect the similarity ephypite microalga inhabitant ion seagrasses leaves. The density of the seagrass canopy and shoot size can significantly modify epiphytic biomass, presumably due to effects of light penetration [35, 36], whereas shoot morphology, leaf, and stem ages can influence epiphyte distribution and abundance due to differences in the surface area that is available for epiphyte settlement [37, 38].

4. Conclusion

Most of the ephypitic microalgae comparison among the portion of the leaves, species of seagrasses and study sites have high similarity index (SI > 50 %) with the value in the range of 51,43–80,95 %. Only four dissimilarities happened. It showed the diversity of ephypitic microalgae on the seagrass leaves.

References

[1] Green E E P and Short F T 2003 *World Atlas of Seagrasses* (Berkeley USA: University of California Press) p 298
[2] Bostrom C, Jackson E L and Simenstad C A 2006 *Estuarine Coastal Shelf Sci.* **68** 383–403
[3] Lee S Y, Fong C W and Wu R S S 2001 *J Exp. Mar. Biol. Ecol.* **259** 23–50
[4] Ambo-Rappe R 2016 *J. Environ. Sci. Technol.*, **9** 246–256
[5] Borowitzka M A, Lavery P S and van Keulen M 2006 *Epiphytes of Seagrasses*. Ed Larkum, A W D, Orth R J and Duarte C M. *Seagrasses: Biology, Ecology and Conservation*

(Netherlands: Springer) chapter 19 pp 441–461
[6] Chung M H and Lee K S 2008 *Algae* **23** 7581
[7] Leliaert F , Vanreusel W, De Clerck O and Coppejans E 2001 *Belg. Journ. Bot.* **134** 13–20
[8] Mounir B B, Moufida A, Wafa F, Mabrouka M and Asma H 2016 *J. of Coast. Life Med.* **4**(3) 211–216
[9] Aho K and Beck E 2011 *DUJS* 43–44
[10] Moncreiff C A, Sullivan M J, Daehnick A E 1992 *Mar. Ecol. Prog. Ser.* **87** 161–171
[11] Govindasamy C and Anantharaj K 2013 *Bot. Res. Internat.* **6**(3) 67–70
[12] Mabrouk L M B, Brahim A, Hamza M, Mahfoudhi and Bradai M N 2014 *J.Mar. Biol.* **2014** 1–10
[13] Riniatsih I, Widianingsih, Rejeki S, Endrawati H dan Agus E L 2013 *ILMU KELAUTAN* **18**(2) 84–90
[14] Hartati R, Trianto A and Widianingsih 2016 *Sea Ranching of Indonesian Sea Cucumber* Research and Community Services Institution (Semarang: Diponegoro University) p 104
[15] Hartati R, Trianto A and Widianingsih 2017 *Habitat characteristic of two selected locations for sea cucumber ranching purposes*. IOP Conf. Series: Earth and Environmental Science **55**
[16] Tomas C 1997 *Identifying Marine Phytoplankton* 1st Edition Academic Press p 858
[17] Kwak T J and Peterson J T 2007 *Community Indices, Parameters, and Comparisons* Ed Guy C S and Brown M L *Analysis and Interpretation of Freshwater Fisheries Data* The American Fisheries Society p 961
[18] Lanyon J 1986 *Seagrasses in the Great Barrier Reef Region* (Townsville, Queensland: GBRMPA) p 51
[19] Hartati R, Widianingsih, Trianto A, Zainuri M and Ambariyanto 2017b *BIODIVERSITAS* **18**(3) 947–953
[20] Pearson G A and Evans L V 1990 *J. Phycol.* **26** 597–603
[21] Wahl M and Mark O 1999 *Marine Ecol. Progress Series* **187** 59–66
[22] Stachowicz J J and Whitlatch R B 2005 *Ecolog.* **86** 2418–2427
[23] Danilov R A and Ekelund N G A 2000 *Sci. Environ.* **248** 63–70
[24] Fong C W, Lee S Y and Wu R S 2000 *Aquatic Bot.* **67** 251–261
[25] Wear D J, Sullivan M J, Moore A D and Millie D F 1999 *Mar. Ecol. Progr. Ser.* **179** 201–213
[26] Heijs F M L 1984 *Aquat. Bot.* **20** 195–218
[27] Brouns J J W M and Heijs F M L 1986 *Aquat. Bot.* **25** 21–45
[28] Lavery P S, Reid T, Hyndes G A and Van Elven B R 2007 *Mar. Ecol. Progress Series* **338** 97–106
[29] Lee K S, Park S R and Kim Y K 2007 *J. of Exp. Mar. Biol. and Ecol.* **350**(1–2) 144–175
[30] Mabrouk L, Hamza A, Mahfoudi M and Bradai M N 2012 *Cah. Biol.Mar.* **53**(4) 419–427
[31] Gambi M C, Lorenti M, Russo G F, Scipione M B and Zupo V 1992 *Mar. Ecol.* **13**(1) 17–39
[32] Mabrouk L, Hamza A, Brahim M B and Bradai M N *Tunisia Mar. Ecol.* **32**(2) 148–161
[33] Mazzella L, Buia M C and Spinoccia L 1994 *Biodiversity of epiphytic diatom community on leaves of Posidonia oceanica* Ed Marino D and Montresor M *Proceedings of the 13th Diatom Symposium* (Bristol UK: Biopress)
[34] Prado P, Alcoverro T, Mart´ınez-Crego B, Verg´es A, P´erez M and Romero J 2007 *J. Exp. Mar. Biol. Ecol.* **350**(1–2) 130–143
[35] Carruthers T J B 1994 *Leaf Production, Canopy Structure and Light Climate in a Density Manipulated Amphibolis griffithii Meadow* Thesis (Australia: University of Western Australia)
[36] Castej´on-Silvo and Terrados J 2012 *Mar. Ecol.* **33**(2) 165–175
[37] Borowitzka M A, Lethbridge R C and Charlton L 1990 *Mar. Ecol. Progress Series* **64** 281–291
[38] Bulthuis D A and Woelkerling W J 1983 *Aquatic Bot.* **16**(2) 137–148

IOP Conf. Series: Earth and Environmental Science **139** (2018) 012012 doi:10.1088/1755-1315/139/1/012012

Zooplankton communities in Cenderawasih Bay National Park, West Papua: can their composition be used to predict whale shark Rhincodon typus Smith, 1828 appearance frequencies?

S N Marliana[1*], M Bataona[1] and E N Ihsan[2]

[1]Department of Ecology and Conservation, Faculty of Biology, Universitas Gadjah Mada, Jalan Teknika Selatan, Sekip Utara, Yogyakarta 55281, Indonesia
[2] WWF Indonesia Site Cenderawasih Bay National Park, Jalan Huntap Iriati 2 Lorong 1 Sebelah Kiri, Teluk Wondama 98362, West Papua, Indonesia
*Email: sn.marliana@ugm.ac.id

Abstract. The use of lift net fishing vessels in Cenderawasih Bay National Park (CBNP) along with the increased popularity of CBNP as an ecotourism area is suspected to have an impact on the behavior and population of its whale sharks Rhincodon typus Smith, 1828. The differing frequency of whale shark appearances along the waters of CBNP has been alleged to be related to the distribution of the whale sharks' food sources, one of which is zooplankton. This preliminary research aimed to investigate the composition of the zooplankton community in CBNP based on distance from the coast and difference in locations, and to use the pattern of zooplankton compositional variation as a basis for indication of the frequency of whale shark appearances. There were clear differences in the composition and diversity of zooplankton communities among sampling stations, but these differences were not strong enough to infer the cause of the different whale shark appearance frequencies in different locations. Nevertheless, the waters of CBNP had an equal availability of zooplankton for whale sharks. With the increasing popularity of whale shark tourism, understanding the species' feeding habits is critical to the sustainability of both the industry and the enigmatic species on which it depends.

1. Introduction

Located in Indonesia's West Papua Province, Cenderawasih Bay National Park (CBNP) has been identified as one of the aggregation sites of the whale shark Rhincodon typus Smith, 1828 in the Indo- Pacific region. Previous long-term monitoring of whale sharks conducted in CBNP waters has provided an overview of the sex composition and size of the population inhabiting this area [1, 2]. Nevertheless, knowledge of this population is still very limited compared with that in other areas [3], such as in Australia [4, 5, 6, 7], Mexico [8, 9], Belize [10], or Seychelles [11, 12]. In recent years, whale sharks have been reported to often roam near CBNP's coasts, allegedly as a response to the booming of bagan (stationary fishing vessels with lift nets) along the coast. This has resulted in CBNP gaining popularity as a tourist destination, the results of which has been both economically rewarding and potentially damaging to the whale shark population itself. An increasing number of tourists visit the bay for a glimpse of the whale sharks around the lift net vessels. These sharks are lured to the surface by small fish being used as bait, a method that may cause a direct disturbance to whale sharks by altering their natural feeding behavior, along with accustoming them to interactions with humans, which can make them an easier target for poachers. The whale sharks may also be indirectly disturbed due to the vulnerability of their habitat to increased human activity in the water and along the mainland, as pollutants entering the ocean affect the survival of the marine organisms on which the sharks feed.

The impact of both the use of lift net vessels for fishing and the increasing popularity of CBNP as a whale shark ecotourism site on the behavior and population of its whale sharks has not been studied in depth [3]. Previous studies indicated a correlation between the number of vessels and frequency of whale shark sightings [1, 2]. They also reported differences in the frequency of whale shark sightings in different parts of CBNP, specifically along the waters of the villages of Kwatisore and Napan Yaur. Between 2013 and 2016, there were reportedly no sightings of whale sharks in Napan Yaur, in contrast with Kwatisore, where sightings were frequent. The prevailing speculation that this disparity is related to the abundance and distribution of one of the whale shark's food sources, anchovies, known locally as puri is supported by one study whose catch data revealed that anchovies composed a very small fraction of Napan Yaur's annual catch, whereas in Kwatisore, they were the major fish caught by fishermen [2].

Baseline data on whale sharks' preferred food sources and their availability in CBNP is essential for monitoring the national park's water quality and the impact of human activities on its whale shark population, as well as providing a basis for estimating the carrying capacity of its ecosystem. This preliminary research aimed to investigate the background composition of zooplankton in CBNP, along with the pattern of variations in composition, which could be used as an indicator of whale shark sighting frequencies.

2. Methods

Research was carried out in Cenderawasih Bay National Park, in the waters off the coast of Nabire District, West Papua Province, Indonesia (approximately 2°53'04.5"S 134°50'15.4"E to 3°13'47.2"S 134°57'43.5"E), in August 2016. Zooplankton sampling was performed following the procedures described in Motta et al. and Taylor, with modifications [5, 10]. Because of the darkness of the water, a search for naturally feeding whale sharks to directly collect the zooplankton samples on which they were feeding was not possible. We therefore opted for an indirect approach. Zooplankton samples were collected from eight stations located between the villages of Kwatisore and Napan Yaur (Figure 1). In each station, a transect was laid out perpendicular to the shoreline, and zooplankton samples were taken at four sampling points at distances of 0.5, 1.0, 2.0 and 4.0 nautical miles from the coastline. Sampling was carried out in the morning between 8:00 and 11:00 a.m., except for the 0.5 nautical mile sampling point on the first transect, where sampling was carried out at 3:00 p.m. because of bad weather. Samples were collected by horizontally towing a 125 μm mesh conical plankton net with a 450 mm opening just under the water surface behind the boat for 2 minutes at a constant speed of 2 knots. Each zooplankton sample was then homogenized in the cod end, concentrated to 10 ml, then fixed with 1 cc of 4 % formaldehyde and 2 to 3 drops of Lugol's iodine solution. In total, 32 zooplankton samples were collected from all sampling sites. Zooplankton identification was conducted according to Shirota [13]. Using the program SPADE [14], Morisita's index of similarity was used to measure the similarity of zooplankton community composition between sampling sites, while diversity estimates were performed using Shannon's index and the inverse Simpson's index.

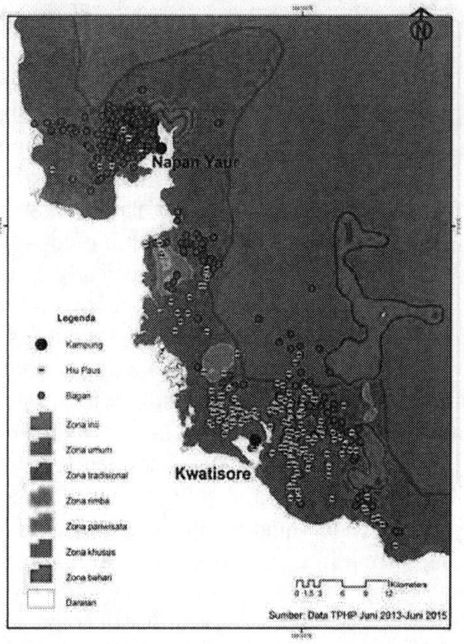

Figure 1. Approximate placement of plankton sampling stations and transects. Yellow lines represent transects, numbers indicate station number. (Source of map: Tania 2015, with modifications)

3. Results and Discussion

3.1. Result

3.1.1. Zooplankton communities composition in Cenderawasih Bay National Park

Across distances (0.5, 1.0, 2.0, and 4.0 nautical miles), the zooplankton communities were found to have almost the same taxonomic group composition. Crustaceans dominated CBNP at all distances, with a proportion of 78–88 %, 59–65 % of which were copepods (Figure 2). This result is in accordance with Tait [15], who reported that at least 70 % of marine zooplankton are crustacean, with Copepoda as the predominant group. The dominance of crustaceans and similarity in taxonomic composition was also found in the majority of sampling stations (Figure 3). Crustaceans dominated the composition of seven of the eight stations, ranging from 76 to 97 %, whereas rotifers had the lowest proportion, ranging from 2 to 20 %. The only station in which this was not the case, Station 4, was composed of 51 % crustaceans and 34 % rotifers, respectively the lowest and highest proportions among the stations. The zooplankton community in Stations 7 and 8, located in Napan Yaur, was composed almost entirely of crustaceans 91 % and 97 %, respectively. Sergestids were not found in Stations 4 and 5, while barnacles were not found in Stations 2, 4 and 5 (Figure 4). Copepods and branchiopods were ubiquitous in all sampling points, except for a couple in Station 8 where both groups were absent.

In general, all of the sampling sites were highly similar in zooplankton community composition. Morisita's similarity index for multiple communities estimated a pronounced similarity level of 0.97 among sampling points based on distance, while the similarity of communities among stations was comparatively lower with 0.75 (Table 1). The conspicuously high similarity across distances was consistent with the pairwise similarities between the sampling points, whose values ranged from 0.94

to 0.98 (Table 2). Pairwise similarity values between sampling stations showed greater variations, ranging from 0.43 to 0.93, with the lowest values found between Station 8 and the other stations (Table 3).

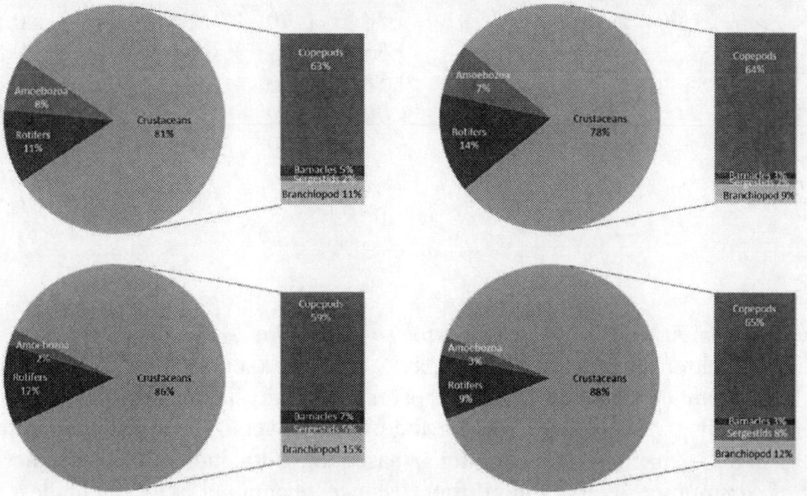

Figure 2. Composition of zooplankton taxonomic groups in Cenderawasih Bay National Park based on distance from the coast: a) 0.5 nautical miles; b) 1.0 nautical miles; c) 2.0 nautical miles; d) 4.0 nautical miles.

Table 1. Morisita's similarity index for multiple zooplankton communities in Cenderawasih Bay National Park. SE, standard error.

Zooplankton communities similarity	Estimate (SE)	95% Confidence interval	Zooplankton communities similarity
Among stations	0.75 (0.03)	(0.70; 0.81)	Among stations
Among distances	0.97 (0.02)	(0.94; 1.00)	Among distances

Table 2. Pairwise Morisita's similarity indices for zooplankton communities found in various distances in Cenderawasih Bay National Park. NM, nautical miles.

Sampling distance	0.5 NM	1.0 NM	2.0 NM	4.0 NM
0.5 NM	1.00	0.98	0.94	0.98
1.0 NM		1.00	0.94	0.98
2.0 NM			1.00	0.95
4.0 NM				1.00

Table 3. Pairwise Morisita's similarity indices for zooplankton communities found in various sampling stations in Cenderawasih Bay National Park.

Station	1	2	3	4	5	6	7	8
1	1.00	0.91	0.78	0.78	0.90	0.86	0.87	0.58
2		1.00	0.84	0.85	0.94	0.93	0.87	0.54
3			1.00	0.58	0.66	0.83	0.88	0.84
4				1.00	0.89	0.82	0.58	0.43
5					1.00	0.81	0.77	0.43
6						1.00	0.85	0.67
7							1.00	0.76
8								1.00

3.1.2. Zooplankton communities diversity in Cenderawasih Bay National Park
A total of 11 zooplankton genera from three phyla and six taxonomic groups were identified from the samples. Both Shannon's and Inversed Simpson's diversity indices resulted in similar values in six of the eight stations, with Station 6 having the highest diversity (Table 4). Stations 5 and 8 were noticeably less diverse than the other stations, based on both indices. Both indices also showed similar levels of community diversity at all four distances examined, with the highest diversity being found at a distance of 2.0 nautical miles (Table 5).

3.2. Discussion

3.2.1. Potential factors influencing the sighting frequency of whale sharks
Whale shark monitoring in the CBNP area revealed that whale sharks were often seen near lift net vessels stationed around the Kwatisore area, but not in Napan Yaur. People associated this behavioral pattern with the low quantity of anchovies caught in Napan Yaur compared with Kwatisore, where they made up the largest proportion of fish caught by lift net vessels. This resulted in a public perception that anchovies were the main food source of whale sharks in CBNP. In order to prove this assumption, information on the natural diet of both the anchovies and whale sharks in this region is required; however, to date no such data have been made available.

In general, anchovies are known to prey on small zooplankton, especially crustaceans [16]. Based on 20 specimens of anchovies taken during the study, all were identified to belong to one species, *Encrasicholina punctifer* Fowler, 1938 (buccaneer anchovy). This anchovy species is commonly found near coasts in the Indian Ocean, and spawns continuously throughout the year [17]. Buccaneer anchovies are the most important food source for large pelagic fish [18]. Thus, it is possible that, in the CBNP area, buccaneer anchovies have become whale sharks' main source of food. On the other hand, it is also possible that the emergence of whale sharks at sites where the anchovies form the majority of the catch is merely due to both species sharing the same food preferences. Both possibilities require further research if the location and frequency of whale shark sightings is to be more clearly understood.

3.2.2. Composition and diversity of zooplankton as an indicator of whale shark appearances
Kwatisore and Napan Yaur's different whale shark sighting frequencies has previously been attributed to either the presence of anchovies or the type of zooplankton on which the whale sharks feed. Our results showed that the zooplankton communities' similarities in composition and diversity at different distances were very high. This suggests that, up to 4.0 nautical miles from the coast, physicochemical factors in CBNP did not play a deterministic role in zooplankton

distribution, as well as that CBNP's waters ensured the availability of zooplankton as a food source for whale sharks with a relatively uniform abundance and diversity. Thus, within that range, zooplankton composition and diversity were not a reliable indicator or predictor of the appearance of whale sharks in CBNP.

The similarities in zooplankton composition between Station 8 in Napan Yaur and the other stations were low. If this were assumed to be an indicator of the lack of whale shark sightings around this station, then the main constituents of the whale sharks' diet in CBNP and in this case, also the anchovies' diet could not possibly be crustaceans, especially copepods, as they were the group that dominated the zooplankton communities in the Napan Yaur region. More data are needed to substantiate this hypothesis, as studies on whale shark diets in various locations around the world have generated highly variable results. In Mexico, it was reported that nearly 85 % of whale sharks' diet consisted of copepods [8]. In Australia, whale sharks were found to feed primarily on krill (Pseudeuphausia latifrons), copepods, and schools of small fish [5], while in Tanzania, sergestid shrimp Lucifer hanseni composed over 50 % the whale sharks' diet [19]. There is an indication that whale sharks have no preference for certain types of prey, but rather show preference for the quantity of biomass of their prey. Generalizations therefore cannot be made based on our results.

Given that this preliminary study aimed to get a general idea of whale sharks' natural diet in CBNP, further research should be conducted on the differences in zooplankton biomass at various locations in CBNP in relation to the frequency of whale shark appearances. It is also important that zooplankton samples are taken on the spot where whale sharks are feeding, so that the composition and biomass of prey can be identified and estimated more accurately. There is no reliable information on the daily distribution of plankton and CBNP's whale sharks' natural feeding behavior, so the results of this study leave a gap of information in terms of the actual availability of whale shark food sources. Research on the temporal abundance of anchovies in CBNP's waters should also be carried out, since there is a possibility that the frequency of occurrence of whale sharks will be positively correlated with the abundance of anchovies at any given time.

3.2.3. Implications for Whale Shark Habitat Management in CBNP

The wide gap of information on the ecology of whale sharks in CBNP could have a negative impact on the sustainability of the population. Efforts to safeguard whale shark populations in CBNP need to be based on knowledge of their natural diet and how to maintain it. Ignorance of their natural diet has also driven negative and conservational damaging stereotypes of the whale sharks among some fisherman in CBNP, who hold that the sharks harm their livelihood by preying on their catches. Any solution to this problem requires collaboration between stakeholders so as to align the economic and ecological interests of the parties involved, education of the public on whale shark ecology and its importance to the local economy, and of course more extensive research on the whale shark itself, all of which will ensure a sustainable, environmentally friendly, and mutually beneficial whale shark-based tourism industry.

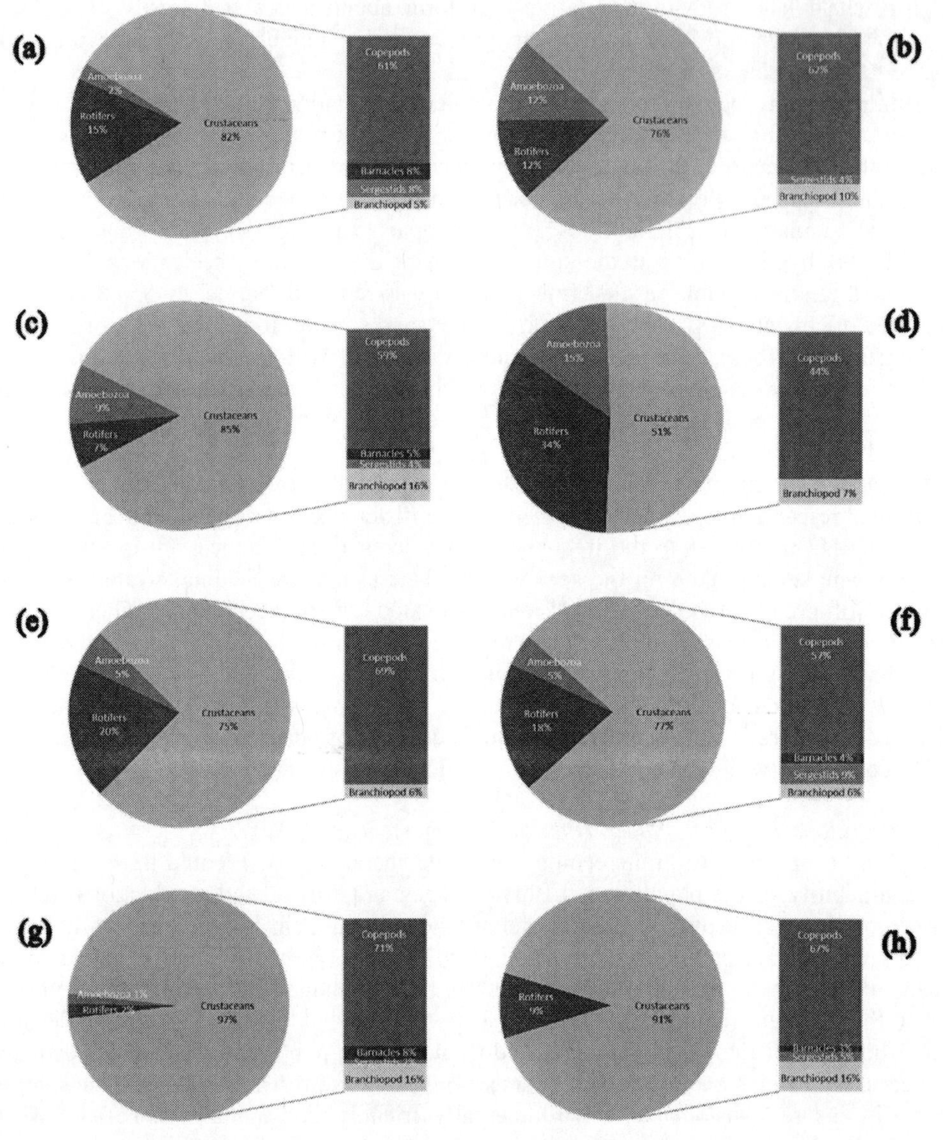

Figure 3. Composition of zooplankton taxonomic groups in Cenderawasih Bay National Park in Nabire, from the east (Kwatisore) to the west (Napan Yaur): (a) Station 1; (b) Station 2; (c) Station 3; (d) Station 4; (e) Station 5; (f) Station 6; (g) Station 7; (h) Station 8.

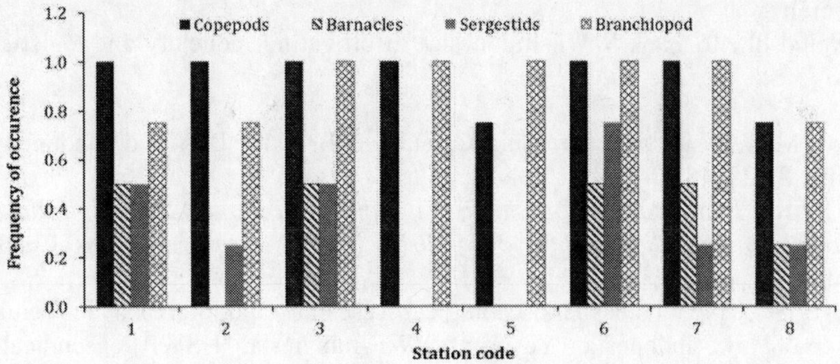

Figure 4. Distribution of crustaceans, the dominant zooplankton group in Cenderawasih Bay National Park, in various sampling stations based on their occurrence in four sampling points of each transect.

Table 4. Zooplankton community diversity of various sampling stations in Cenderawasih Bay National Park. SE, standard error.

Location	Diversity estimator	
	Shannon Index (SE)	Inversed Simpson Index (SE)
Station 1	2.15	7.64 (0.11)
Station 2	2.22	8.30 (0.11)
Station 3	2.29	8.74 (0.08)
Station 4	2.02	6.41 (0.12)
Station 5	1.78	5.37 (0.10)
Station 6	2.34	10.04 (0.09)
Station 7	2.11	7.16 (0.13)
Station 8	1.88	5.01 (0.14)

Table 5. Zooplankton community diversity at various distances in Cenderawasih Bay National Park. NM, nautical mile; SE, standard error.

Sampling distance	Diversity estimator	
	Shannon Index (SE)	Inversed Simpson Index (SE)
0.5 NM	2.26 (0.03)	8.27 (0.11)
1.0 NM	2.26 (0.03)	8.51 (0.12)
2.0 NM	2.32 (0.03)	9.21 (0.09)
4.0 NM	2.30 (0.03)	8.94 (0.09)

4. Conclusions

From Kwatisore to Napan Yaur, Cenderawasih Bay National Park was found to have uniform potential to provide zooplankton as a whale shark food source. Although there was a clear distinction in the composition and diversity of the zooplankton community at one station in Napan Yaur, this study could not provide conclusive results on what causes differences in the frequency of whale shark sightings at various locations in CBNP. With an increasing number of people visiting CBNP to see its whale sharks, understanding their feeding habits is critical to sustaining both the industry and the enigmatic species on which it depends. The manipulation of whale shark appearances for the sake of tourism, such as using bait to lure the sharks to the surface, should be also done in accordance with their natural feeding habits, so as not to potentially alter the behavior of the whale sharks and disrupt their natural habits and habitat.

Acknowledgments
The authors would like to thank WWF Indonesia for providing monetary and logistical support.

References

[1] Himawan M R, Wijonarno A, Tania C, Subhan B, Noor B A and Maduppa H 2015 AACL Bioflux **8** 123–133
[2] Tania C 2015 *Pemantauan dan Studi Hiu Paus di Taman Nasional Teluk Cenderawasih: Laporan Pemantauan Tahun 2014–2015* (Taman Nasional Teluk Cenderawasih/WWF Papua)
[3] Stewart B S 2014 Whale shark ecological research and outreach in Teluk Cenderawasih National Park, Indonesia (Yayasan WWF-Indonesia: HSWRI Technical Report 2013–382)
[4] Norman B M and Stevens J D 2007 *Fish. Res.* 84 81–86
[5] Taylor J G 2007 *Fish. Res.* 84 65–70
[6] Bradshaw C J A, Fitzpatrick B M, Steinberg C C, Brook B W and Meekan M G 2008. Biol. Conserv. 141 1894–1905
[7] Sleeman J C, Meekan M G, Fitzpatrick B J, Steinberg C R, Ancel R and Bradshaw C J A 2010 *J. Exp. Mar. Biol. Ecol.* 382 77–81
[8] Nelson J D and Eckert S A 2007 *Fish. Res.* 84 47–64
[9] Motta P J, Maslanka M, Hueter R E, Davis R L, de la Parra R, Mulvany S L, Habegger M L, Strother J A, Mara K R, Gardiner J M, Tyminski J P dan Zeigler LD 2010 *Zool.* 113 199–212
[10] Graham R T and Roberts C M 2007 *Fish. Res.* 84 71–80
[11] Rowat D and Gore M 2007 *Fish. Res.* 84 32–40
[12] Rowat D, Gore M, Meekan M G, Lawler I R and Bradshaw C J A 2009 *J. Exp. Mar. Biol. Ecol.* 368 1–8
[13] Shirota A 1966 *The Plankton of South Vietnam: Freshwater and Marine Plankton* (Japan: Overseas Technical Cooperation Agency)
[14] Chao A and Shen T J 2010 Program SPADE: Species Prediction and Diversity Estimation. Program and user's guide (Taiwan: CARE, Hsin-Chu)
[15] Tait R V 1980 *Elements of Marine Ecology: An Introductory Course* (London: Butterworths) p 33
[16] Carpenter K E and Niem V H 1999 FAO species identification guide for fishery purposes: The living marine resources of the Western Central Pacific Volume 3. Batoid fishes, chimaeras and bony fishes, part 1 (Elopidae to Linophrynidae) (Rome: FAO) pp 1397–2068 (accessed via http://www.fao.org/3/a-x2401e on 10 Dec 2016)
[17] Maack G and George M R 1999 *Fish. Res.* 44 113–120
[18] Froese R and Pauly D 2013 FishBase (accessed via http://www.fishbase.org/Country/CountrySpeciesSummary.php?c_code=356&id=558 on 10 Dec 2016)
[19] Rohner C A, Armstrong A J, Pierce S J, Prebble C E M, Cagua E F, Cochran J E M, Berumen M L and Richardson A J 2015 *J. Plankton Res.* 37 1–11

IOP Conf. Series: Earth and Environmental Science **139** (2018) 012013 doi:10.1088/1755-1315/139/1/012013

Biodiversity of gastropod in the Sombu Beach, Wakatobi, Indonesia

F Rahmayanti[1]*, F K Nazira[1], A K Dewi[1], D F Oktaviani[1], I N K Millaty[1], T A Prasetya[1], H B Sasmita[1], M Nashrurrokhman[1], D A Roshitafandi[1], D Febiansi[1], H W Sartika[1], W G Zulfikar[1], R N Kurnia[1], D A Islami[1] and Y A Tranggono[1]

[1] Marine Study Club, Faculty of Biology, Universitas Gadjah Mada, Jl. Teknika Selatan, Sekip Utara, Bulaksumur, Depok, Sleman, D.I. Yogyakarta, Indonesia
*Email: ferintarahma@gmail.com

Abstract. The Sombu Beach is located in Wangi-wangi Island, Southeast Sulawesi, Indonesia. The beach is still natural and the substrates are dominated by fine sand and coral reef flat which is suitable for the habitats of gastropods. The aim of this research was to understand the diversity of gastropod in the intertidal zone of the Sombu Beach, Wangi-wangi Island. The research was conducted in January 2017 in the Sombu Beach, Wangi-wangi Island, Indonesia. The collection of the samples was conducted using quadrant plot method. Ten transects along the beach were made with four plots on every transect. So that, there were 40 plots data which were analyzed using Shannon-Wiener index of diversity (H'), evenness index (E), and dominancy index (D). The result of this research shows 13 species of classes Gastropods. The most abundant species in this study were Mitra sp. (86 individuals), Columbella sp. (40 individuals) and Conus sp. (35 individuals). The diversity index (H') was 1.8729, the evenness index was 0.7302, and the dominance index was 0.2071.

1. Introduction

Indonesia is an archipelagic country and known as a mega biodiversity area. Wakatobi is one of the islands in Indonesia, located in the Southeast Sulawesi. The island consists of four clusters island, they are Wangi-wangi, Kaledupa, Tomia and Binongko, with marine waters covering area is 18,377.31 km^2 [1]. Wakatobi is a National Park area, the occupation is one of the prime locations of marine conservation in Indonesia [2]. These islands are located in the center of Coral Triangle Initiative (CTI), known as the region of coral reef biodiversity [3]. Coral reefs are habitat for many marine species such as molluscs [4].

Gastropod is a soft-bodied animal and runs using the abdominal muscles. In general, the body of mollusc consists of head, body and mantle [5]. Mollusc is a phylum with the highest species diversity after arthropods. Gastropods have extensive habitats such as rivers, lakes, swamps, ponds, beaches and even underground aquifers [6]. The structure of the gastropod community was influenced by many factors such as the way of life, competition, food availability, substrate type, water temperature, and salinity [7]. However, substrate conditions are a major factor affecting the distribution of gastropods [8].

This study was conducted in the intertidal zone of the Sombu Beach within the fine sand and coral reef substrates. Based on the previous research [9] the beach with a substrate of fine sand and coral reefs, gastropods most often found comes from family Cypraeidae. This study aimed to know the biodiversity of gastropods in intertidal zones of the Sombu Beach which located in the Coral Triangle Initiative and National Park areas. Gastropods are benthos organisms that are used a bio-indicator of healthy water ecosystems.

2. Materials and Methods

The research was conducted on January 4th, 2017 at the intertidal zone of the Sombu Beach, Wangi-wangi Island, Indonesia (05°16'12.2"S 123°31'08.8"E). The sampling was done by using quadratic plot (1 m x 1 m) which was able to determine gastropod diversity. The instruments used in this research were rolled meter, thermometer, refractometer, GPS, tweezers, aquades, 70 % alcohol, a camera for documentation and identification book.

Gastropods were taken by quadratic plot on the ten transect lines, transects has 20 m in length, vertically with the shoreline. In each transect, four plots were placed. There were 40 plots in total. The pictures of the obtained samples taken, then they were preserved by using aquades (for relaxation), then the visceral mass was separate from the shell. For gastropod that has no shell (e.g. Nudibranch), alcohol 70 % was used for preservation. Gastropod samples identification was conducted by using literature such as The Living Marine Resources of The Western Central Pacific Volume 1 [10].

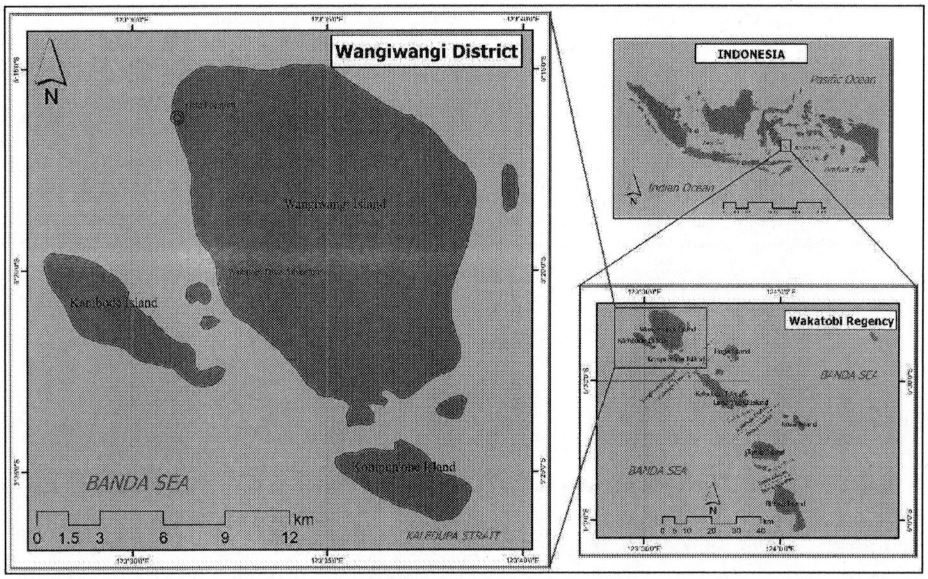

Figure 1. Location of the Sombu Beach

2.1. *Data Analysis*

Several characteristics of the community are the diversity of Gastropod species, which were able to be determined by using Shannon-Wiener index [11], of diversity (H'), evenness index (E), and dominance index (D) as follows:

$$H' = -\sum (ni/N) \ln (ni/N) \quad (1)$$

$$E = \frac{H'}{Hmax} \quad (2)$$

$$D = \sum_{i-1}^{s} (Pi)^2 \quad (3)$$

An notational:

S = Total number of species; N = Total number of observed individual; ni = The number of I individual; $Pi = n.N^{-1}$; $Hmax = \log^2 S$

The criteria for Shannon-Wiener diversity shown:

H' ≤ 3.32	: diversity is low
3.32 < H' < 9.97	: diversity is moderate
H' ≥ 9.97	: diversity is high

3. Result
The abundance and diversity of gastropods in the Sombu Beach, Wangi-wangi Island, Wakatobi, Indonesia as follows:

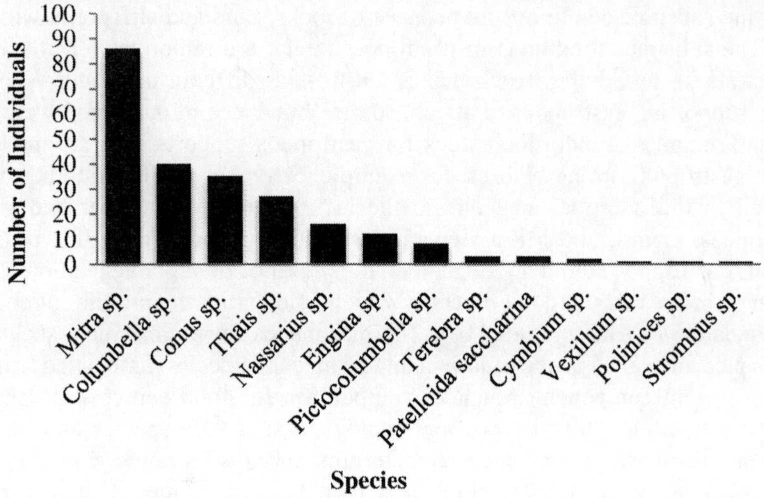

Figure 2. The abundance of Gastropods in the Wangi-wangi Island, Wakatobi, Indonesia

Figure 1 shows there were 13 species of gastropods in the Sombu Beach with the largest number of individuals was *Mitra* sp. with total 86 individuals. The index diversity index (H') obtained was 1.873 (low diversity), the evenness index (E) was 0.730229 (high), and dominance index was 0.2071 (low).

Table 1. Shimpson's diversity index, evenness index and dominance index of gastropods.

Index	Index Number	Category
Diversity Index (H')	1.8729	Low
Evenness Index (E)	0.7302	High
Dominance (C)	0.2071	Low

4. Discussion
The characteristic of gastropods which found in the intertidal zone of the Sombu Beach were small which less than 5 cm and had a distinctive color. Most of the gastropod samples found had a rather non-distinctive color pattern with variations between black and white or yellow. Engina had a yellow-black pattern and Cymbium had a distinctive shiny shell shape with a wide inner lip side. Patteloida had a distinctive shell shape like a symmetric cone but still in the Gastropods class [12]. Intertidal

zone of Krakal beach was dominated by carnivorous gastropods. In the same substrates, gastropods were mostly found from the family Cypraeidae as an omnivore organism [9, 10].

Based on sampling conducted in the Sombu Beach, obtained 13 genera of gastropod. Figure 1 shows that the highest abundance of Gastropods is from genus Mitra consists of 86 individual, following by Columbella with 40 individual and Conus with 35 individual. Table 1 shows the evenness index, diversity index, and dominance of Gastropods in Sombu Beach. Index of evenness measures the relative abundance of the different species making up the richness of an area. Based on the result of evenness calculation, high evenness index value was 0.7302. Otherwise, the diversity of gastropods in the Sombu Beach shows a low level of diversity with a diversity index was only 1.8729. This caused by the substrate conditions that consist of rocks, sand and cliffs only without the variation of vegetation. The substrate conditions in gastropod habitat are important for attaching and grazing. The sandy substrate in the Sombu Beach makes Gastropods difficult to attach. Whereas on the coral cliffs area were supporting gastropods to attach, so the abundance of gastropod were high. Vegetation such as algae and seagrass as main food stock for gastropods supports the gastropods variety because certain types of gastropods are herbivores, for example, *Strombus* sp. The number of gastropods was also found small in the intertidal area due to the lack of seagrass number and variety. Vexillum, Cymbium, Polinices, Conus, and Mitra Genera were found abundance on this beach because those were carnivorous gastropod and did not depend on the presence of slight vegetation [10,13,14,15].

The dominance index measured the diversity which calculating the present number of species, and the relative abundance of each species [16]. The dominance means the most striking and abundant species. Dominance also means the relative importance of species associated with the degree of influence on ecosystem components, such as competition for food sources. Dominance is used for community characterization, habitat types and ecological sites of a species and helps in identifying climate and disease responses. Dominance can determine the species function in the group [17].

The results obtained were 0.2071 which less than 0.5 which means there are no species that dominate other species. Gastropods are benthic organisms whose existence depends on substrate conditions and environmental conditions such as waves. The substrate conditions on the Sombu coast were sandy with several coral spots. Gastropods depend on substrate conditions which could be functioned as a shelter or hindrance from the waves. Also, the substrate provides food sources such as seagrass and organism remnants. The benthic organism has a strong relation to the substrate conditions. Herbivore type gastropods would be difficult to be found in the sandy beach which covered only by few seagrasses but would be easy to find both carnivorous and scavenger types [17].

From 40 plots, the most abundance gastropods found were in the 1 to 10 plot with a coral substrate. Substrates may affect the presence of gastropods and coral substrate is an ideal substrate for most Gastropods because Gastropods need to be attached to life. *Mitra* sp. known as carnivores and active predators [10]. Besides *Conus* sp. also known as a predator whose prey is other Molluscs, Polychaeta or small fish [18]. Also, Conus has conotoxin that serves to paralyze the prey [19]. Therefore these two genera can be found abundantly in the coral substrate although algae are not available because other living organisms can be prey to them.

5. Conclusion

From this research, we concluded that the diversity of gastropods was moderate due to the value of diversity index (1.8729), the evenness index (0.7302) and the dominance index (0.2071). There were only three genera that found in the Sombu Beach because the variety of the substrate is low, that will impact the food resources and habitats for gastropods.

Acknowledgments

This document was jointly prepared with Division of Mollusca KSK Biogama. We deliver our gratitude to KSK Biogama for the support and facility.

IOP Conf. Series: Earth and Environmental Science **139** (2018) 012013 doi:10.1088/1755-1315/139/1/012013

References

[1] Badan Pusat Statistik (BPS) 2010 *Kabupaten Wakatobi dalam Angka Tahun 2010* (Wanci: Badan Pusat Statistik) p 1–15

[2] Ayiful R A 2004 Strategi Pengembangan Kegiatan Pariwisata di Taman Nasional Kepulauan Wakatobi Sulawesi Tenggara [Tugas Akhir] (Semarang: Universitas Dipenogoro)

[3] Reef Check Indonesia 2007 *Satu Dekade Pemantauan Reef Check: Kondisi dan Kecenderungan pada Terumbu Karang Indonesia* www.reefcheck.or.id

[4] Glynn P W, Manzello D P and Enochs I C 2017 *Coral Reefs of the Eastern Tropical Pasific: Peristence and Lost Dynamyic Environment* (Miami: University of Miami, Springer)

[5] Notosoedarmono S, Martosupomo M, and Tuheteru M 2014 *Distribusi Gastropods Di Ekosistem Mangrove* Proceeding National Seminar Raja Ampat

[6] Bouchet P, Gargominy O, Ponder W F and Strong E E 2008 Hydrobiologia **595** 149–166.

[7] Mujiono N 2008 J. Biologi

[8] Bolam S G, Fernandez T F and Huxham M 2002 *Diversity, biomass, and ecosystem process in the marine benthos* Ecological Monograph.

[9] Dibyowati L 2009 *Keanekaragaman Moluska (Bivalvia dan Gastropods) di Sepanjang Pantai Carita Pandeglang, Banten* [Skripsi] (Bogor : Institut Pertanian Bogor).

[10] Carpenter K E and Niem V H 1998 *The Living Marine Resource of the Western Central Pacific* Vol. 1 Seaweeds, Corals, Bivalves and Gastropods (Rome: Food and Agriculture Organization of the United Nations) p 363-381

[11] Stevenes M H H 2005 *A Primer of Ecology with R* (London: Springer) p 293

[12] Nakano T, and Ozawa T 2005 J. Mollusc. Studies **71** 357– 370

[13] Duda T F, Khon A J, and Palumbi S R 2001 Biol. J. Linn. Soc. **73** 391– 409

[14] Waki T, Noseworthy R G, and Choi K S 2016 Korea J. of Asia-Pacific Biodiversity **9** 100 – 103.

[15] Wolff W J and Montserrat F J. Mar. Biol. Assoc. UK **85** 949–953

[16] Danovaro R 2009 Methods for The Study of Deep-Sea Sediments, Their Functioning and Diversity (Boca Raton: CRC Press) p 200

[17] Dirzo R and Mendoza E 2008 Encyclopedia of Ecology vol 1 (Amsterdam: Elsevier) p 146

[18] Islami M M 2009 Oseana **34** (2) 33– 40

[19] Junior V H, Neto J B P and Cobo V J 2006 Revisto da Sociedade Brasiliera de Med. Trop. **39** (5) 498–500

Analysis of habitat characteristics of small pelagic fish based on generalized additive models in Kepulauan Seribu Waters

A A Rivai[1*], V P Siregar[2], S B Agus[2] and H Yasuma[3]

[1] Department of Marine Technology, Graduate School of Bogor Agricultural University, Indonesia
[2] Faculty of Fisheries and Marine Science, Department of Marine Science and Technology, Bogor Agricultural University, Indonesia
[3] Faculty of Fisheries Sciences, Hokkaido University, Japan
*Email: andi.alamsyah.03@gmail.com

Abstract. One of the required information for sustainable fisheries management is about the habitat characteristics of a fish species. This information can be used to map the distribution of fish and map the potential fishing ground. This study aimed to analyze the habitat characteristics of small pelagic fishes (anchovy, squid, sardine and scads) which were mainly caught by lift net in Kepulauan Seribu waters. Research on habitat characteristics had been widely done, but the use of total suspended solid (TSS) parameters in this analysis is still lacking. TSS parameter which was extracted from Landsat 8 along with five other oceanographic parameters, CPUE data and location of fishing ground data from lift net fisheries in Kepulauan Seribu were included in this analysis. This analysis used Generalized Additive Models (GAMs) to evaluate the relationship between CPUE and oceanographic parameters. The results of the analysis showed that each fish species had different habitat characteristics. TSS and sea surface height had a great influence on the value of CPUE from each species. All the oceanographic parameters affected the CPUE of each species. This study demonstrated the effective use of GAMs to identify the essential habitat of a fish species.

1. Introduction

The distribution of fisheries resources varies according to their species and their preferred habitat. The distribution of fisheries resources have a specific pattern, such as small pelagic fishes are converged on specific chlorophyll concentrations and temperature [1]. Fish tend to distribute in areas that are good for spawning, feeding and growing during their life cycle. Also, the distribution of resources is also affected by other factors, such as the fish behavior and the dynamics of the population [2]. Identification of important areas that have high fish distribution is one way to support the ecosystem approach to fisheries management (EAFM) and then will also support the sustainability of a fishery resource. Information of the distribution of fisheries resources can be obtained by analyzing the relationship between the fish distribution and the oceanographic characteristics of an area. This is because there is a significant correlation between the distributions of fishes with their environmental [3].

Analysis of habitat characteristics can provide information about habitat characteristics which were favored by fish. This kind of information can be used to obtain important areas to fish, fish distribution and areas with high abundance of fish resource. Analysis of fish habitat characteristics had been widely done before. Some previous studies used a statistical approach to determine the abundance of certain fish species [1, 4] and to map the hotspots of certain fish species [5, 6]. These studies analyzed of habitat characteristics by evaluating the relationship between the presences of fishes with oceanographic parameters in an area. Various oceanographic parameters had been analyzed in these studies, such as sea surface temperature and chlorophyll-*a* concentration. Total suspended solid is one of the oceanographic parameters that affect the distribution of fish, especially in coastal waters and estuaries. [7] reported that the changes in total suspended solid concentration may affect the growth and development of fish. However, TSS parameter is still not widely used as one of the oceanographic parameters in habitat characteristic analysis so that the information of TSS on fish distribution is lacking. Since TSS were assumed to affect the fish distribution, TSS along with the other oceanographic parameters were included in the analysis of habitat characteristics in this study. The addition of TSS in this analysis was one of the novelty in this research.

Analysis of habitat characteristics of fish can be done by using a statistical approach. There are several statistical approaches that can be used and one of them is generalized additives models (GAMs). Some previous studies had used GAMs and reported that this method could effectively analyze the relationship between oceanographic parameters and fish abundance in a waters [6, 8, 9]. GAMs utilize a non-linear and non-parametric technique to analyze the relationship between the variable response and the explanatory variables, which are highly non-linear and non-monotonic [10, 12].

Data of oceanographic parameters have been widely available from various sources by utilizing a variety of technologies, such as satellite remote sensing technology. Utilization of this technology can provide an effective way for monitoring and managing an area. This technology can provide information within a large area in a periodical time. In using data from satellite remote sensing technology, some constraints can be encountered, such as the accuracy need to be improved and the processing of data which require advanced skills. However, with the availability of such information, allowed us to identify the oceanographic characteristics of preferred habitat by a specific fish. With the availability of fisheries information, such as catch data, fishing locations data and oceanographic data,combined with statistical approaches, better information about the distribution of fishery resources can be obtained.

Kepulauan Seribu is an archipelago that is located in the northern part of Jakarta, the capital of Indonesia. This area has complex geomorphology, consisting of various ecosystems, such as coral reefs, mangroves, seagrass beds and supplies a wide range of fisheries products to the greater Jakarta area. Kepulauan Seribu has high fisheries resources, with total catch landing from marine fishing activity in 2013 was 1555 ton [13]. These fisheries resources require a good fisheries management so that it can be used sustainably. Information about the characteristics of fish habitat in Kepulauan Seribu waters can support the implementation of sustainable fisheries management.

The aim of this study was to analyze the habitat characteristic of four fish species in the Kepulauan Seribu waters by evaluating the relationship between CPUE and various oceanographic parameters that influence it. The results of this study were expected to provide information about the distribution and the habitat of these fish. Such information can be used to map the potential fishing ground, to increase fishing benefit and to support better fisheries management.

2. Methods

2.1. Study area

The study area encompassed the fishing ground of lift net fisheries which was located in the waters of Kepulauan Seribu with Lancang Island as their fishing base. Geographically, the area was bounded by 106°19'E to 106°51'E and 5°33'S to 6°5'S, including some small islands in this region (figure 1).

2.2. Data Acquisition

The data that was used in this analysis were obtained from the lift net fisheries in Lancang Island. These fisheries were artisanal fisheries and did a one-day fishing trip that operates at night and took the advantage of light to attract and to concentrate fishes. The data used in this analysis were fishing location data, catches landing data and some oceanographic parameter data.

The fishing location data was derived from GPS recorded data. The data were collected from 11 fishing vessels in Lancang Island from 2014 to 2015. The catch landing data was derived from four dominant species that were caught by lift net, namely anchovies (*Stolephorus* sp.), squid (*Loligo* sp.), scads (*Selaroides leptolepis*) and sardines (*Sardinella fimbriata*). The catch landing data during 2014–2015 was collected from an officer who was appointed by Ministry of Marine and Fisheries of Indonesia and was in the form of monthly catch landing by lift net fisheries in Lancang Island.

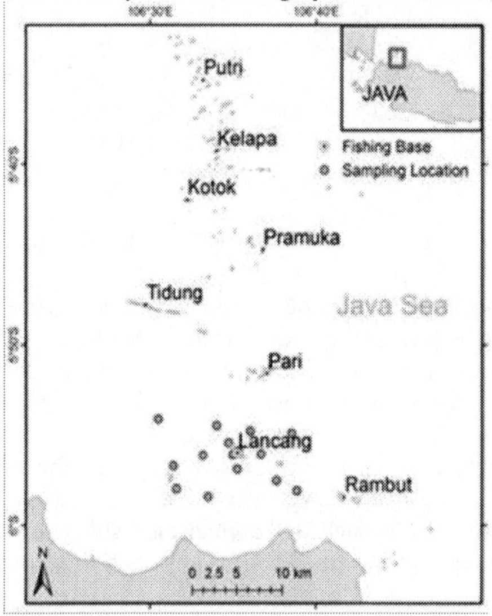

Figure 1. Map of the study area. The sampling site distributed around Lancang Island, which was the fishing base of lift net fisheries in Kepulauan Seribu Waters.

The oceanographic parameter data that were used in this study were sea surface temperature (SST), chlorophyll-a concentration, sea surface height (SSH), currents, bathymetry and total suspended solids (TSS). These data were used as an explanatory variable in the analysis of habitat characteristic. These oceanography parameters were satellite-derived data and downloaded from 2014 to 2015 (table 1). To validate these oceanographic data, the in situ data were collected in June 2015 and taken from several locations around the waters of Lancang Island (figure 1). The in situ data, namely SST and TSS, were collected using multiparameter water quality checker.

Table 1. Oceanographic parameter data which were used in analysis of habitat characteristics.

Variable	Unit	Source	Resolution
Sea Surface Temperature (SST)	Celcius (^{o}C)	INDESO	0.02 degree
Chlorophyll-*a* concentration	$mg{\cdot}m^{-3}$	INDESO	0.02 degree
Sea Surface Height (SSH)	m	INDESO	1/12 degree
Currents	$m{\cdot}s^{-1}$	INDESO	1/12 degree
Bathymetry	m	ETOPO1	1 Arc minute
Total Suspended Solid	$mg{\cdot}L^{-1}$	Landsat 8	30 meters

2.3. Data Analysis

2.3.1. Catch per Unit Effort (CPUE). CPUE data was calculated from monthly catch landing data which was divided by the monthly fishing effort. Fishing effort data was present in the monthly fishing trip form and was obtained from direct interviews with several captains of lift net fisheries in Lancang Island.

2.3.2. Estimation of Total Suspended Solid (TSS). TSS data was extracted from Landsat 8 images that were downloaded from the ESPA website (http://espa.cr.usgs.gov/). The downloaded Landsat 8 image was 24 images in total and in the form of Surface Reflectance (SR) which had been corrected atmospherically. Correction of Landsat 8 image to SR was done using Landsat Surface Reflectance Code (LaSRC) algorithm [14, 15]. Landsat 8 SR data was then calibrated by dividing all digital number values by 10,000. The calibrated SR data was then masked using CFMask algorithm to remove clouds and land in an image.

TSS in $mg{\cdot}L^{-1}$ was derived from Landsat 8 surface reflectance data using algorithm in [16],

$$TSS = A*\exp^{(S*R(0)\text{ red band})}$$

where A = 8.1429, S = 23.704 and R(0) is surface reflectance.

2.3.3. Accuracy Assessment. The accuracy assessment was used to evaluate the accuracy of estimated TSS data from Landsat 8 image and downloaded SST data from INDESO. This analysis evaluated the accuracy of these data based on the coefficient of determination (R^2) value, Mean Absolute Percentage Error (MAPE) and Root Mean Squared Error (RMSE). This analysis used the in situ data as the comparison data.

2.3.4. Analysis of Habitat Characteristic using Generalized Additive Models (GAMs). This analysis used fishing locations data, CPUE and some oceanographic parameters data. This analysis was done to understand the preferred habitat of four fish species which were investigated in this study. CPUE and some oceanography parameters were transformed in term of logarithm natural to normalize the distribution of the data. Generalized Additive Models (GAMs) with Gaussian distribution and identity link function was used to evaluate the relationship between CPUE with some oceanography parameters. GAMs approach non-linear and non-parametric technique for regression modeling and its main advantage compared to the traditional regression methods is its ability to analyze the relationship between the variable response and the explanatory variables, which are highly non-linear and non-monotonic [10, 12]. GAMs was constructed using the gam function within the mgcv R package (R x64 3.3.2).

The forward stepwise and shrinkage approaches were used in the final selection model. This approach allowed us to avoid collinearity among the explanatory variables. The model selection was based on the lowest value of Akaike Information Creation (AIC), the level of Deviance Explained (DE) and inspection of residual plots (i.e., QQ plots). The degree of smoothing for each explanatory variables was chosen based on the Restricted Maximum Likelihood (REML). The likelihood methods are less prone to local minimal than other smoothness selection criteria (i.e., Generalized Cross Validation, Un-biased Risk Estimator, etc.), allowed us to avoid over-fitting and to simplify the interpretation result [17]. Followed the method in [9], the thin plate spline smoother regression was applied to the models, with the maximum degree of smoothing freedom (k) for the main effects was limited to 5. The explanatory variables with significant levels of < 0.05 were kept for the final models.

3. Results

Figure 2 showed the increasing of anchovy CPUE from year to year, in contrast with sardine, scad and squid CPUE which were decreased in 2014. Seasonally, sardine and squid had high CPUE in west monsoon (December–February), while anchovy and scad were relatively stable. The catch of the lift net fisheries in Kepulauan Seribu waters was dominated by anchovies, followed by sardines and scad and squid. IT showed that the Kepulauan Seribus waters were the preferred habitat of anchovy.

3.1. Accuracy Assessment

The results of accuracy assessment in figure 3 showed that in situ SST and downloaded SST from INDESO had a good relationship ($R^2 = 0.9028$). The difference in the value of in situ SST and downloaded SST from INDESO can be categorized as small (RMSE = 0.86 °C; MAPE = 2.93 %). For TSS parameters, the results of this analysis showed that in situ TSS and estimated TSS from Landsat 8 image have a low relation ($R^2 = 0.3384$). The error of TSS data can be categorized as small (RMSE = 6.1 $mg{\cdot}L^{-1}$; MAPE = 19.65 %). The estimated TSS from Landsat 8 image had a lower value than in situ TSS. However, the results of the comparison showed that *in situ* TSS and estimated TSS from Landsat 8 had the same pattern. From this result, it could be concluded that the INDESO SST data and the estimated TSS data from Landsat 8 could be used in this study.

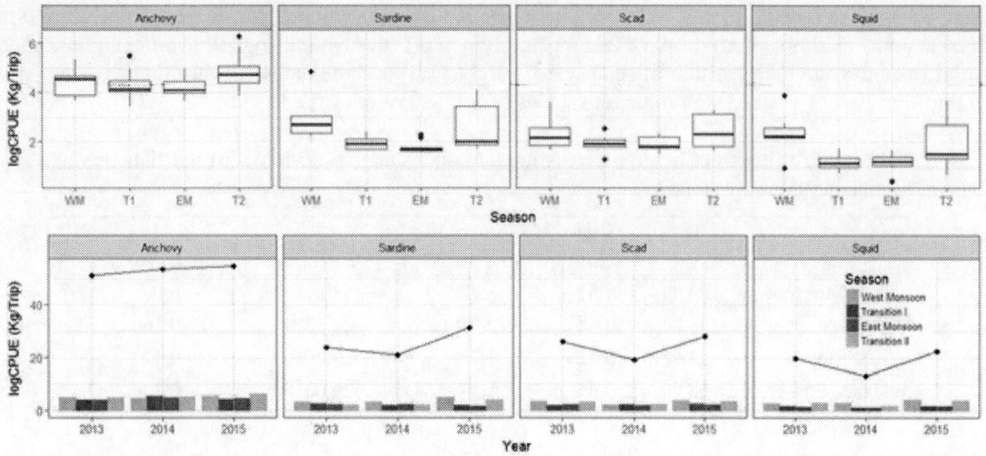

Figure 2. Profile of CPUE data of lift net fisheries in Kepulauan Seribus waters. WM: West Monsoon (Dec–Feb); T1: Transition 1 (Mar–May); EM: East Monsoon (Jun–Aug); T2: Transtition 2 (Sep–Nov).

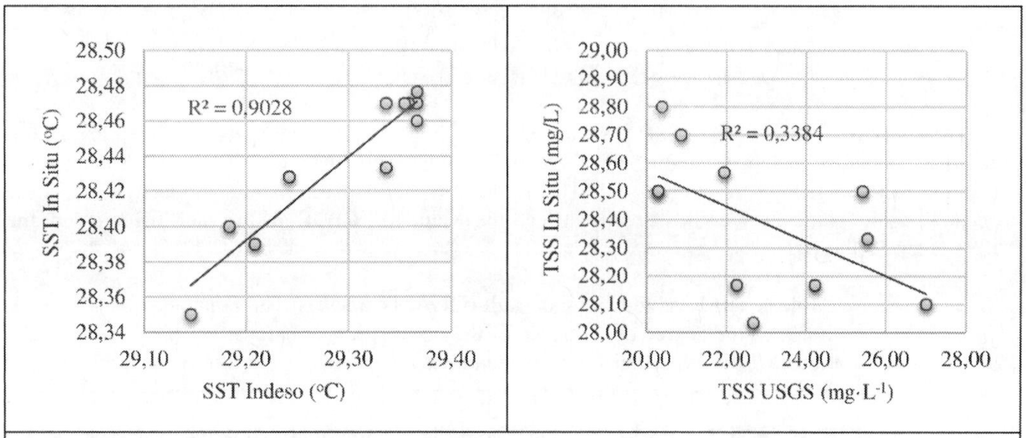

Figure 3. The linear relationship between sea surface temperature (SST) data from INDESO and total suspended solid (TSS) data from Landsat 8 with in-situ data in Kepulauan Seribu waters.

3.2. Analysis of Habitat Characteristic

Table 2 showed the seasonal average of oceanographic parameters on fishing locations in the Kepulauan Seribu waters. Oceanographic conditions on fishing locations were changed every season. The highest average of total suspended solids (TSS) occurred in the transitional II. Sea surface temperature (SST) relatively had a stable condition. The average of current speed and sea surface height (SSH) in each season were fluctuated and tended to had the same pattern, where when the current speed increased, the SSH also increased.

The result of habitat characteristic analysis with GAMs showed that the habitat characteristic was varied by the species (figure 4–7). Analysis of habitat characteristics was done after considering all the statistical measurements and the diagnostic plot results. Deviance explained in the final model varied between 77.3 % (anchovy) to 92.9 % (sardine). The natural logarithm of sea surface temperature (sst), the natural logarithm of chlorophyll-*a* concentration (chl), the natural logarithm of total suspended solid (tss), the natural logarithm of sea surface height (ssh) and the natural logarithm of current were found significant ($p < 0.05$) for all target species of lift net fisheries (table 3).

Table 2. Profile of oceanographic parameters on fishing location of lift net fisheries in Kepulauan Seribu waters.

Season	TSS ($mg \cdot L^{-1}$)	SSH (m)	Chlorophyll-a ($mg \cdot m^{-3}$)	SST (^{0}C)	Current ($m \cdot s^{-1}$)
West Monsoon	20.99	0.61	3.22	28.87	0.06
Transition I	19.06	0.68	2.05	29.17	0.16
East Monsoon	23.65	0.72	1.76	29.27	0.17
Transition II	58.00	0.65	5.47	29.08	0.14
Average	32.56	0.68	2.98	29.16	0.15

Table 3. Final GAMs model from each target species of lift net fisheries in Kepulauan Seribu waters. CPUE and all explanatory variable were transformed in terms of the natural logarithm. DE: Deviance Explained.

CPUE	Final Model	DE
Anchovy	s(ssh) + s(chl) + s(sst) + s(current) + s(tss)	77.3%
Squid	s(ssh) + s(sst) + s(current) + s(tss)	87.7%
Sardine	s(ssh) + s(chl) + s(sst) + s(current) + s(tss)	92.9%
Scad	s(ssh) + s(chl) + s(sst) + s(current) + s(tss)	87.6%

Table 4 showed that TSS and SSH parameters had high F value on each species. From this result, it could be concluded that these parameters gave a high influence on the value of CPUE. TSS parameters could affect the fish distribution and SSH parameters could affect the fishing activities. These oceanographic parameters could directly affect the value of CPUE of lift net fisheries in the Kepulauan Seribu waters.

Table 4. The F value from the result of GAMs analysis. The F value indicated the degree of influence of the oceanographic parameter on the CPUE of fish species in Kepulauan Seribu waters.

Species	F Value				
	s(ssh)	s(chl)	s(sst)	s(current)	s(tss)
Anchovy	36.097	20.669	21.437	9.372	75.29
Squid	53.287	7.288	11.399	18.578	137.306
Sardine	78.741	0.955	39.148	8.5	102.289
Scad	75.772	5.141	6.041	8.654	44.509

The plot of the best smoothing showed a higher probability of finding high CPUE of anchovy and squid at SST from 28 to 29 °C and sardine and scad from 29 to 30°C (figure 4–7). In chlorophyll-a concentration parameter, the plot of the best fitting smooth showed high CPUE of sardine and scad could be found at 0.61–1.1 $mg \cdot m^{-3}$ and at 0.41–1 $mg \cdot m^{-3}$ for anchovy. Their preferred habitat corresponded to Eutrophic and upper Mesotrophic waters (Oligotrophic waters: < 0.084 $mg \cdot m^{-3}$; Lower Mesotrophic waters: 0.084–0.359 $mg \cdot m^{-3}$; Upper Mesotrophic waters: 0.36–0.793 $mg \cdot m^{-3}$; Eutrophic waters: > 0.793 $mg \cdot m^{-3}$) [18].

The probability of finding higher CPUE, according to GAMs plot in sea surface height (ssh) parameter, for anchovy and squid were at 0.74–0.8 m and for sardine and scad were at 0.6–0.65 m. Total suspended solid is an important parameter for environmental monitoring in the coastal area. The plot of the best fitting smooth showed high CPUE of all target species found at TSS concentration of 54–67 $mg \cdot L^{-1}$ and at current velocity of 0.12–0.22 $m \cdot s^{-1}$ (figure 4–7).

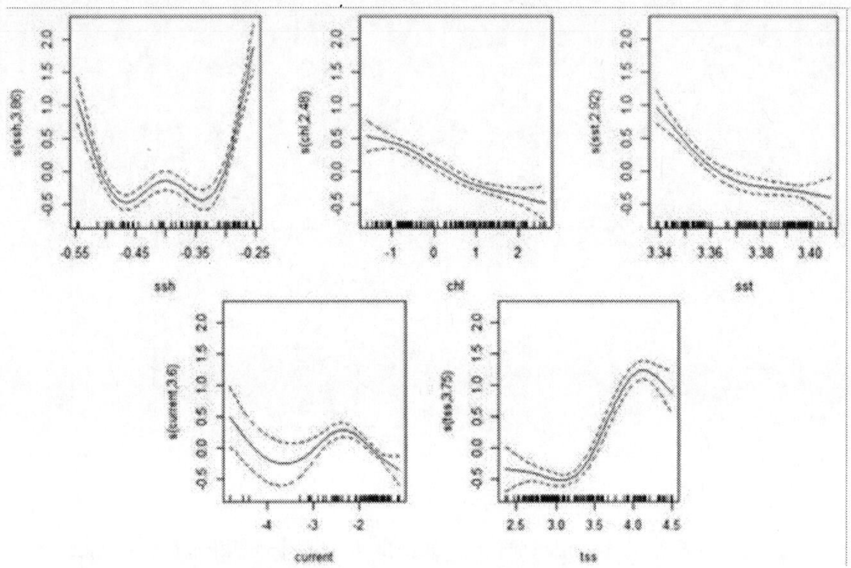

Figure 4. Plots of the best fitting smooth in GAMs analysis for anchovy.

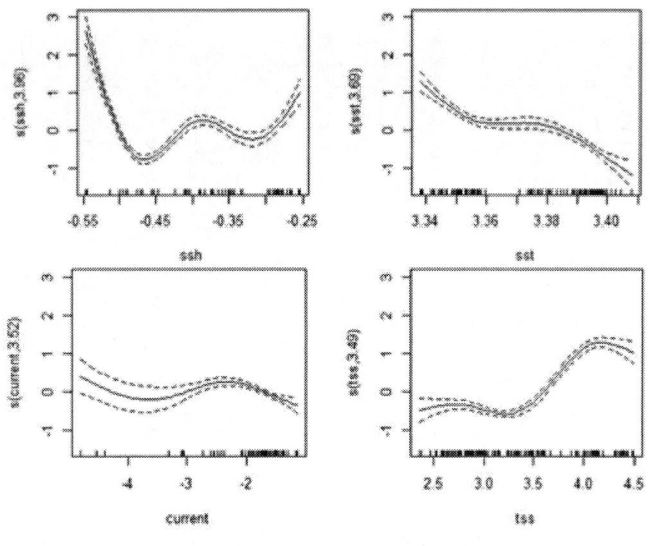

Figure 5. Plots of the best fitting smooth in GAMs analysis for squid.

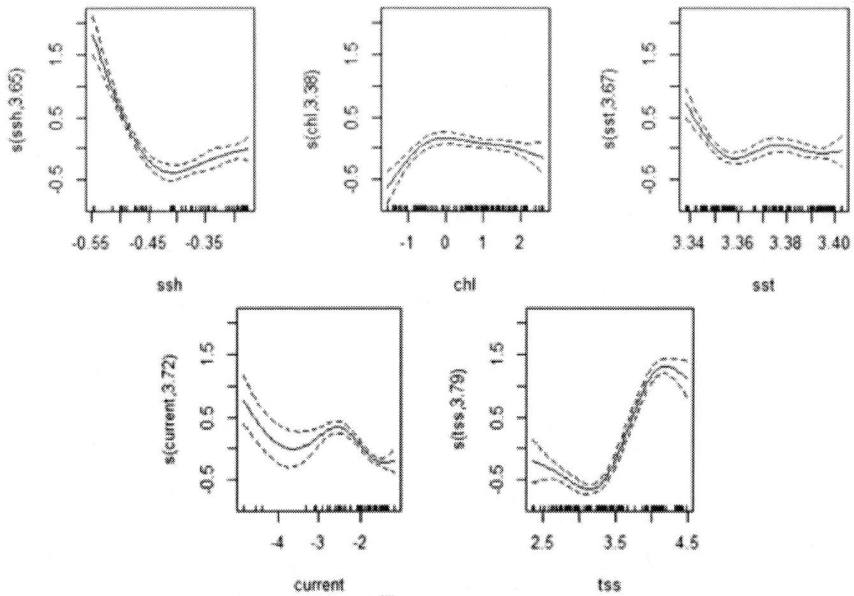

Figure 6. Plots of the best fitting smooth in GAMs analysis for sardine.

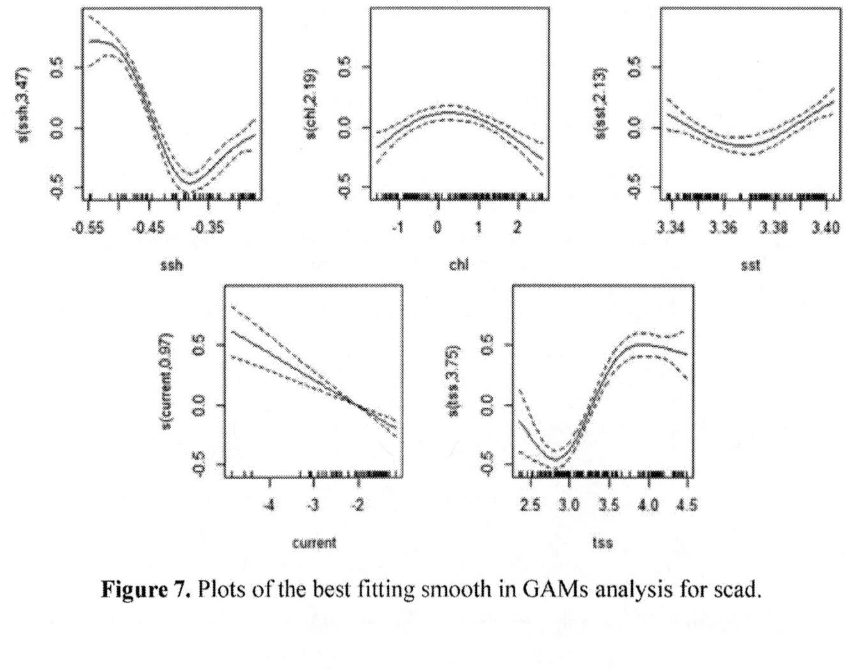

Figure 7. Plots of the best fitting smooth in GAMs analysis for scad.

4. Discussion

Using Generalized Additive Models (GAMS) for habitat characteristic analysis, CPUE data, fishing location data and some oceanographic parameters data were integrated. The result of this analysis showed that each target species had different preferred oceanographic characteristics and distributed on varying range of oceanographic parameters. Sea surface height (SSH) and total suspended solids (TSS) have a high influence on the value of CPUE. SSH could affect fishing activities and TSS could affect the distribution of fish.

In nature, organisms distributed uniformly or randomly but converged in an area with particular characteristics [19]. On fisheries, the spread of efforts and catches had some relationships with the distribution of targeted species that were affected by their preferred habitat characteristic (i.e. the availability of coral reefs and seagrass) and oceanographic parameters (i.e. temperature, chlorophyll concentration and currents) [20]. Based on the results of habitat characteristics analysis, several oceanographic parameters influence the CPUE. These oceanographic parameters were sea surface temperature, chlorophyll-*a* concentration, sea surface height, current and total suspended solids.

Temperature is one of the factors that significantly affect the distribution of fish [21]. Changes in the temperature of the waters can affect the metabolism of the fish, the growth of gonads and the predation and can indirectly affect the level of reproduction, survival and the catches of fish [22]. The concentration of chlorophyll-*a* in the waters can be used as an indicator of the productivity of a marine environment and may affect the distribution of certain fish. The concentration of chlorophyll-*a* widely had been used to identify the level of primary productivity [23] and at a level above 0.2 $mg \cdot m^{-3}$, the concentration of chlorophyll-*a* can support some fishing activities in an area [24].

Sea surface height (SSH) and currents can be analyzed to identify the mass movement of water, heat and nutrient flow, which can affect the productivity of marine environment [25]. Also, SSH can also be used to predict the other oceanography features, such as fronts, eddies and convergences [26].

Sediment, an essential element in aquatic ecosystems cycle, is in the form of suspended particles, transported and deposited by water [27]. The increase of sediment concentrations often caused by anthropogenic activities, such as mining, agriculture and urban development. The increasing of precipitation, the increasing of temperatures and the melting of ice which were associated with climate change also indirectly affects the increased of sediment concentration [28]. The increasing of total suspended solids (TSS) concentration in the waters can affect spawning, migration, movement pattern, development and growth of the aquatic organisms [7]. In fish, the effect of TSS was depended on several factors, such as the species, the temperature at exposure time the type of sediment [29], the frequency and the concentration of TSS [28].

Bathymetry parameter was not found significant to CPUE value. This was probably due to the CPUE data which was obtained from small pelagic species, where their habitat was located on the sea surface. However, bathymetry data which showed the depth profile of an area is considered as one of the factors that may affect the distribution of certain fish species [9]. Depth has a significant relationship with various important environmental parameters, such as temperature, oxygen and nutrient concentration, light intensity and primary productivity of a waters.

By utilizing the CPUE relationship with some oceanographic parameters as input in spatial analysis assessments, allowed us to generate information of the potential fishing ground and the distribution of fisheries resources. With that information, stakeholders could define a better fisheries management policy to ensure the sustainability of a fishery resource.

5. Conclusion

Each observed fish species had different habitat characteristics. Sea surface temperature, chlorophyll-*a* concentration, total suspended solids, sea surface height, and current were found to significantly affect the CPUE of lift net fisheries in Kepulauan Seribu waters. Total suspended solid and sea surface height gave a high influence on CPUE value. Total suspended solid affected the distribution of fish and sea surface height affected the fishing activities. All species were found to be highly distributed in the TSS range of 49.40–67 $mg \cdot L^{-1}$. The catch of anchovy and squid were high when the SSH value was in the range of 0.74–0.78 m, while the catch of scad and sardine were high when SSH value was in the range 0.60–0.64 m. High CPUE values occurred in certain water conditions.

The use of GPS data and catch data at the fishing locations needs to be improved so that the results of this analysis can be more reliable. For further research, by using the result of this study, mapping the potential fishing ground area can be done. Improving the fishery management in Kepulauan Seribu can be done by enforcing the logbook policies to be more optimal and better information on fisheries activity can be obtained.

Acknowledgments
This research was supported by Indonesia Endowment Fund for Education (LPDP) and the Japan Student Services Organization (JASSO). Thanks to Laboratory of Mapping and Spatial Modeling, IPB and Fisheries and Marine Technology Laboratory, Hokkaido University which had provided facilities and infrastructure during the research process.

References

[1] Valavanis V D, Georgakarakos S, Kapantagakis A, Palialexis A and Katara I2004 *Ecol. Model.* **178**(3–4) 417–427
[2] Fulton E A, Smith A D M and Punt A E 2005 *ICES J. Mar. Sci.* **62**(3) 540–551
[3] Li G, Chen X, Lei L and Guan W 2014 *Int. J. Remote Sens.* **35**(11–12) 4399–4421
[4] Hattab T, Lasram F B R, Albouy C, Sammari C, Romdhane M S, Cury P, Leprieur F and Le Loc'h F 2013 *PLoS One* **8**(10)
[5] Zainuddin M, Kiyofuji H, Saitoh K and Saitoh S I 2006 *Deep-Sea Research II* **53** 419–431
[6] Coen L D 1995 *A review of the potential impacts of mechanical harvesting on subtidal andintertidal shellfish resources* (South Carolina: South Carolina Division of Natural Resourcesand Marine Resources Research Institute)
[7] Zwolinski J P, Emmett R L and Demer D A 2011 *ICES J. Mar. Sci.* **68**(5) 867–879
[8] Maina I, Kavadas S, Katsanevakis S, Somarakis S, Tserpes G and Georgakarakos S 2016. *Fisheries Research* **183** 326–339
[9] Hastie T J and Tibshirani R J 1990 *Generalized Additive Models* (London: Chapman & Hall/CRC)
[10] Guisan A, Edwards T C and Hastie T 2002 *Ecol. Model.* **157** 89–100
[11] Wood S N 2006 *Generalized Additive Models: An Introduction with R* (New York: Chapman & Hall/CRC)
[12] BPS Kepulauan Seribu 2014 *Kepulauan Seribu in Figures* (Kepulauan Seribu: BPS Kepulauan Seribu)
[13] USGS 2017 *Landsat 8 Surface Reflectance Code (LaSRC) Product* (South Dakota: U.S. Geological Survey)
[14] Vermote E, Justice C, Claverie M and Franch B 2016 *Remote Sensing of Environment* **185** 46–56
[15] Budiman S 2004 *Thesis* Enschede: Geo-Information Science and Earth Observation ITC
[16] Wood S N 2011 *J. R. Stat. Soc.***73**(1) 3–36
[17] Karydis M and Tsirtsis G 1996 *Sci. Total Environ.* **186**(3) 209–219
[18] Downing J A 1986 *Nature* **323** 255–257
[19] Syah A F, Saitoh S I, Alabia I D and Hirawake T 2016 *Fishery Bulletin* **114**(3) 330–342
[20] Tseng C T, Su N J, Sun C L, Punt A E, Yeh S Z, Liu D C and Su WC 2013 *ICES J. Mar. Sci.* **70** 991–999
[21] Cheung W W L, Brodeur R D, Okey T A and Pauly D 2015 *Prog. Oceanogr.* **130** 19–31
[22] Garcia S M, Staples D J and Chesson J 2000 *Ocean Coast. Manage.* **43** 537–556
[23] FAO 2003 *The Application of Remote Sensing Technology to Marine Fisheries: an Introductory Manual* (Section 7) (Rome: Food and Agriculture Organization)
[24] Ayers J M and Lozier M S 2010*J. Geo-phys. Res.* **115** C05001

[25] Polovina J J and Howell E A 2005 *ICES J. Mar. Sci.* **62** 319–327
[26] Beussink Z S 2007 *Thesis* Missouri State University
[27] Kjelland M E, Woodley C M, Swannack T M and Smith D L 2015 *Environment Systems and Decisions* **35**(3) 334–350
[28] Muck J 2010 *Biological effects of sediment on bull trout and their habitat guidance for evaluating effects* (Washington: Fish and Wildlife Service) p 57

IOP Conf. Series: Earth and Environmental Science **139** (2018) 012015 doi:10.1088/1755-1315/139/1/012015

Pattern of relative growth in cockle *Anadara antiquata* in Ihamahu coastal waters, Central Maluku

L Siahainenia, S F Tuhumury, P A Uneputty*, and N C Tuhumury

Fishery and Marine Science Faculty, Pattimura University. Jl. Mr. Chr. Soplanit. Kampus Unpatti Poka, Ambon
*e-mail: y_louhen@yahoo.com

Abstract. *Anadara antiquata* is economically important species for fisheries and belong to phylum mollusc which also found in Maluku. However, the density of population begins to decrease recently due to high exploitation by people. The objective of the present study was to analyze relative growth pattern of *A. antiquata* including size distribution and growth pattern based on shell dimensions. The study was conducted from December 2016 to February 2017 in Ihamahu coastal waters. Data were collected by using purposive random sampling. All the individuals of *A. antiquata* found were measured the length, width and height by using a vernier caliper to the nearest 0.01mm. Data were analyzed by using SPSS 20 and Microsoft Excel software. The results indicated that the A. antiquata mostly found in seagrass bed with a muddy substrate. The size distribution of shell dimensions was different during sampling. Overall, the length ranged from 15.87 mm to 57.5 mm, the width from 15.50 mm to 48.60 mm and the height was from 9.36 mm to 35.9 mm. The population of *A. antiquata* consisted of juvenile and mature size. The mature size (> 30 mm) was more dominant in the population. The *A. antiquata* showed allometric relative growth pattern based on shell dimensions.

1. Introduction

Maluku province is known as Thousand Island archipelago which having high biodiversity of marine and fisheries resources such as fish, echinoderms, molluscs, algae. Mollusc is one of the marine and fisheries resources which has been utilized by coastal communities in Maluku due to easily collected during low tide. Some species are usually consumed or sold by local people namely the top shell (*Trochus niloticus*), abalone (*Haliotis* spp.), blood cockle (*Anadara granosa*), and ark cockle (*Anadara antiquata*).

Ihamahu is a coastal village and has three keystone ecosystems, mangrove, seagrass and coral reef. Beside of the three ecosystems, there are at least nine coastal lagoon ecosystems with differences in size and they distribute along the coast [1]. Therefore, this village has a special and unique environmental condition where high diversity and potency of marine species. Some previous studies have shown that the lagoon ecosystems in Ihamahu village had high potency and diversity of marine resources inhabiting lagoon ecosystems [2, 3, 4].

There has been coastal environmental degradation in Ihamahu village mostly due to anthropogenic activities such as collecting marine resources during low tide (day and night), domestic waste, bridge building, fishing and uphill development [5]. One of the species collected during low tide is the arc cockle, *A. antiquata* by the local community in Ihamahu village for their daily needs due to its delicious meat. The continual utilization of this shellfish by local communities will decline the number of this species in the future. Recently, it is difficult to collect this species in large size. Therefore, the in-depth study is needed to sustain this commercial species of marine mollusc.

Growth is one of the important events in the life history of an organism besides reproduction. Growth in mollusc is usually determined by measuring the shell linear increment which is then converted to somatic growth. This conversion was done by using the relationship between body (tissue) weight and shell size [6]. Knowledge of growth of organisms is crucial in understanding their life history. Information on growth is useful for management of the species. By understanding the growth of the species and its relationship to reproduction, further knowledge will be gained on the status of the species, aiding decisions on whether they could be exploited or have already been overexploited.

Information available on the growth of *A. antiquata* is limited. Some previous studies have shown different growth of *A. antiquata* [7, 8, 9]. Shell size determined the growth rate of *A. antiquata*. The shell size 30 mm grow faster than the 40 and 50 cm shell [7]. Also, the growth coefficient for this species was 0.23/year and lifespan was 12.04 year [8]. Whereas, water quality and food availability also affected the growth rate of this species [9]. The objective of the present study was to analyze relative growth pattern of *A. antiquata* including size distribution and growth pattern based on shell dimensions.

2. Materials and Methods

2.1. Study sites

Geographically, Ihamahu village located at 03°30'15"S–03°31'42"S and 128°41'13"E–128°41'25"E (figure 1). It has various substrates such as coarse sand, medium sand, and pebbles. There were seven species of seagrasses found namely *Enhalus acoroides*, *Thalassia hemprichii*, *Cymodocea rotundata*, *Cymodocea serrulata*, *Halophila minor*, *Halodule uninervis* and *Syringodium isoetifolium*.

2.2. Sampling and handling procedure

Fieldwork was carried out from December 2016 to February 2017. Samples were taken using purposive random sampling in the intertidal area zone without replacement [10]. A quadrate 1 x 1 m was used to collect the *A. antiquata*. All individuals of *A. antiquata* within a quadrate were counted and collected. Approximately 30–50 individual were collected at each sampling time depending on their abundance. When samples were collected, the shell length, width and height were measured in situ (figure 2) using a vernier caliper to the nearest 0.1 mm.

2.3. Data analysis

The relationship between shell length (SL) and width (SW) as well as height (SH) was analyzed by fitting simple linear regressions [12]. The patterns of growth were analyzed by using a regression equation which was $\log y = \alpha \log x \pm \log b$ where y is the dependent variable, x is independent variable; b is constant, and α is the regression coefficient (constant differential growth ratio). This equation was used to assess the allometric status [13]. A slope (b) less than 1 indicates negative allometry, while (b) greater than 1 indicates positive allometry. If a slope is not significantly different from 1, there is isometry [14]. The slope of two-regression lines was analyzed using student t-test [15] SPSS 20 and Microsoft Excel software.

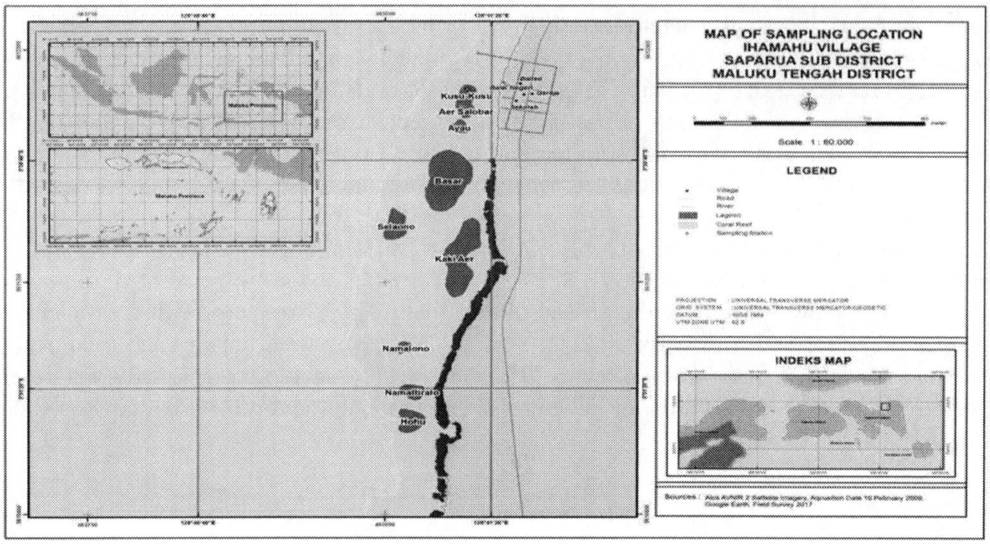

Figure 1. Study sites.

Figure 2. Measurement of shell dimension of *A. antiquata. A.* Shell length (SL); B. Shell width (SW); C. Shell height (SH). Modified after [11].

3. Results and Discussion

3.1. Size distribution of Anadara antiquate

Overall, the shell dimensions including length, width and height were quite different monthly (table 1). The shell length (SL) ranged from 15.87–57.5 mm, the shell width was from 15.51–48.6 mm and the shell height was from 9.36–35.9 mm. As the shell length is the main parameter of growth, the results indicated that the population of *A. antiquata* could be divided into two groups which were juvenile and mature.

The mature size was more dominant (91.85 %) than the juvenile size (8.15 %). Mostly the shell length of juvenile size for *A. antiquata* was less than 30 mm in Tanzania and the mature size was more than 31.33 mm [16]. The shell length was higher in Ihamahu coastal waters than that found in Lada Bay, Sunda Strait [17]. Similarly, the length of *A. antiquata* was slightly high in Ihamahu coastal waters compared to *A. antiquata* in Banten waters with 54.35 mm [18]. The maximum shell length for this species was 100.5 mm and common 70 mm [11]. The maximum shell length was different because of the different environmental condition, difference sampling time and fishing pressure activity in the area.

Table 1. Formatting sections, subsections and subsubsections.

Month	Shell length (SL)	Shell width (SW)	Shell height (SH)
December	15.87-49.7 (37.10±1.03)	15.51-35.56 (26.43±0.69)	9.36-33.12 (22.60±0.71)
January	26.1-52.79 (40.92±1.27)	18.9-40.52 (30.69±0.94)	15.7-31.81 (25.3±0.80)
February	15.87-57.5 (42.47±1.03)	35.5-48.6 (31.71±0.83)	9.36-35.9 (26.58±0.70)
Pooled	15.87-57.5 (40.08±0.66)	15.51-48.6 (29.48±0.51)	9.36-35.9 (24.78±0.45)

The size distribution (SL) of this edible cockles *A. antiquata* was shown in figure 3. It could be seen from figure 3 that during the sampling time the population of *A. antiquata* was mostly composed of mature size. Based on the class interval, the dominant size classes were 35–40 mm and 40–45 mm in shell length (SL). The smallest and the largest size class were found in small number. The shell length was more than 50mm found only in small proportion 5.92 % respectively. It was probably caused by exploitation. Local women and children are always gathering this species during low tide especially for large size.

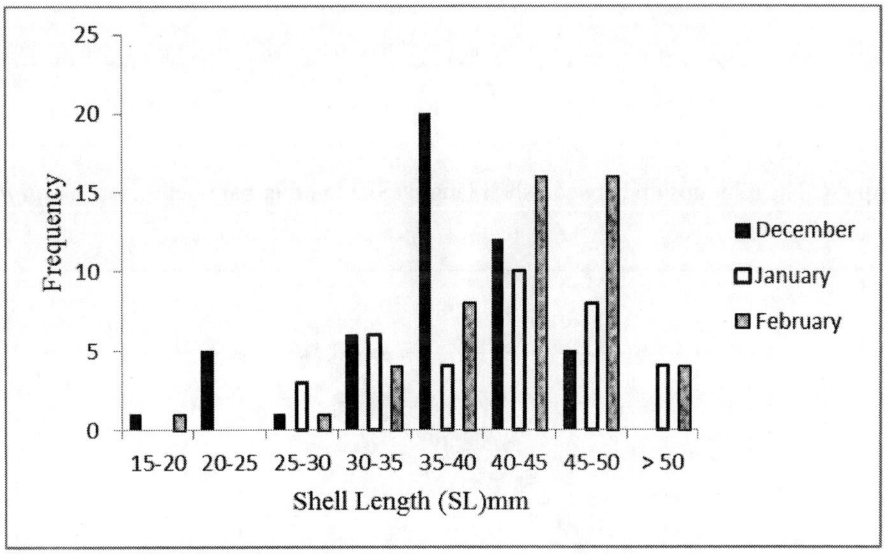

Figure 3. Size frequency distribution (SL) of *A. antiquata* found in Ihamahu coastal waters.

3.2. Growth pattern of Anadara antiquata

A total of 135 individuals of *A. antiquata* were used in this study collected from December 2016 to February 2017. The first reference dimension was shell length (SL) and dependent variables were shell width (SW) and shell height (SH). The result indicated all dimensions of shell increased as increasing in size.

IOP Conf. Series: Earth and Environmental Science **139** (2018) 012015 doi:10.1088/1755-1315/139/1/012015

A. antiquata showed a straight line which is evidence that the shell length (SL) is highly associated with shell width (SW) and shell height (SH) (figure 4). In other words, an increase in shell length resulted in increasing shell width and shell height.

There was a relationship between shell lengths (SL) against other parameters. This species showed a strong correlation between SL against SW ($r = 0.837$, $t = 17.6$, $p < 0.01$), and SL against SH indicated very strong correlation ($r = 0.941$, $t = 32.2$, $p < 0.01$). This suggests that the increase in shell length (SL) contributed more to the increase in shell height (SH).

Similarly, the shell width (SW) also associated with shell height (SH) (figure 5). The relationship between SW and SH was also strong ($r = 0.857$, $t = 19.21$, $p < 0.01$). Increasing in shell width (SW) as a second reference dimension resulted in increasing in shell height (SH) as a dependent variable (figure 5).

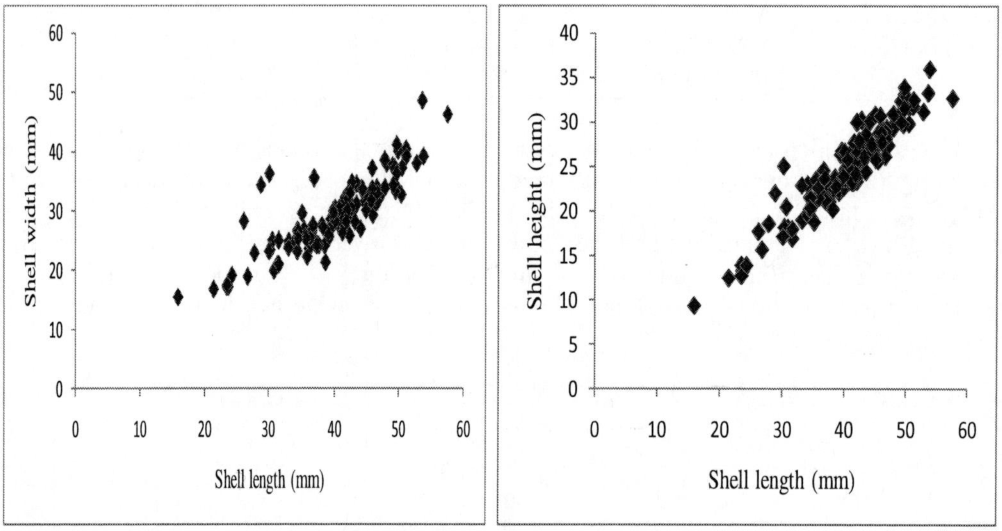

Figure 4. The relationship between shell length (SL) to other parameters for *A. antiquata.*

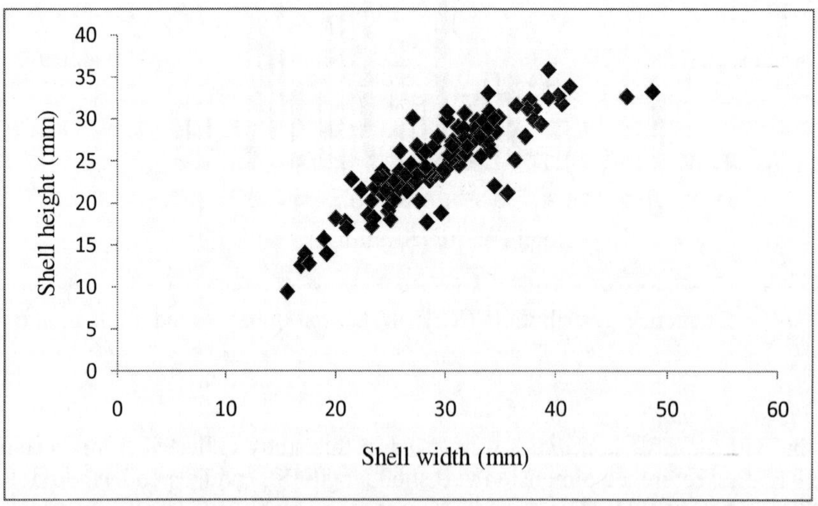

Figure 5. The relationship between shell width (SW) to shell height (SH) for *A. antiquata.*

The simple linear regression of shell dimensions was given in table 2. This species exhibited allometric growth. The value of the slope (b) ranged from 0.637 to 0.751, indicating that this species showed negative allometric. If the slope (b) less than 1 showing negative allometric. The concept has been applied to understand the relative growth pattern of the organisms such as macroinvertebrates [14].

Table 2. The linear regression equation of shell parameters dimension for *A. antiquate* (* = p <0.01; - = negative allometric).

Shell parameters (mm)	Equation	Correlation Coefficient (r)	SE	t value	*P*	Allometric status
Shell length with						
SW	y = 0.646x +3.581	0.837	0.37	17.6	*	-
SH	y = 0.637x -0.749	0.941	0.20	32.2	*	-
Shell width with						
SH	y = 0.751x + 2.625	0.857	0.51	19.21	*	-

Some previous studies of *A. antiquata* on length-weight relationship also showed allometric growth [3, 9, 16]. Even though no studies on the growth of *A. antiquata* based on shell geometry were available for comparisons, the use of shell dimension in the present study to determine growth pattern is still useful. The shell length, shell width or shell height was commonly used to estimate the growth in mollussc [13, 19, 20].

4. Conclusion

The maximum length recorded of *A. antiquata* was 57.5 mm. Whereas the maximum width was 48.60 mm and the height was 35.9 mm. The size distribution of shell dimensions was different during the time of sampling. The population of *A. antiquata* was dominant in size 35–40 mm and 40–45 mm showing mature individuals. However, the largest size was found in small numbers. The *A.antiquata* showed negative allometric relative growth pattern based on shell dimensions. To sustain the species for the future, some recommendations are needed namely some period and area to be closed for fishing activity (closed season) to allow this species reached maximum size and recovered after spawning.

References

[1] Sahetapy D 2006 *Diversity ecosystem and coastal resources in Ihamahu village, Saparua sub-regency, Central Maluku regency*, Indonesia (Unpublished)

[2] Tubalawony S 1990 *The abundance of fish in seagrass bed Ihamahu coastal waters* BSc Honour Thesis (Maluku: Fishery and Marine Science Faculty Pattimura University)

[3] Selanno D A J, Loppies C Z and Tuhumury S F 2006 *J. Ichthyos* **5** 3742

[4] Selanno D A J, Tuhumury S F, Uneputty P A, and Matruty D 2014 *Empowerment and strengthening of economic of sea cucumber fishers in Central Maluku and South east Maluku by environmental systems approach* Research report MP3EI (Maluku: Pattimura University)

[5] Selanno D A J, Natan Y, Uneputty Pr A and Lewerissa Y A 2013. *Empowerment and strengthening of economic of sea cucumber fishers in Central Maluku and South east*

Maluku by environmental systems approach Research report MP3EI (Maluku: Pattimura University)

[6] Hughes R N 1986 *A functional biology of marine gastropods* (Great Britain: Croom Helm Ltd.) p 245

[7] Nurdin J, Marusin N, Izmiarti, Asmara A, Deswandy R and Marzuki Z 2016 *Makara Sains* **10** 96101

[8] Pattikawa J A 2007 *J. Ilmu Kelautan* **12** 18196

[9] Setiawan A, Bahtiar and Wa N 2016 *J. Manajemen Sumberdaya Perairan* **1** 13-27

[10] Khouw A S 2009 *Methods and quantitative analysis in marine bioecology* (Jakarta: P4L.DKP) p346

[11] Poutiers J M 1998 *The Living Marine Resources of the Western Central Pacific* 1 Seaweeds, coral, bivalvia and gastropods, ed K.E. Carpenter and V.H. Niem (Rome: FAO Species Identification Guide For Fisheries Purposes)

[12] Fowler J and Cohen L 1990. *Practical statistic for field biology*. vol vii (Wiltshire : John Willey and Sons Redwood Books) p227

[13] Mieskiewicz A G 1980 *A comparative study of the biology of Nerita spesies from the Townsville region, North Queensland*, Honour thesis, JCUNQ p 96

[14] Gayon, J 2000. History of the concept of allometry. *American zoologist*. **40** 748-758

[15] Zar J H 1999. *Biostatistical analysis.* Fourth edition, (New Jersey: Prentice Hall International, Inc.) p 663

Characteristic sediment and water column chlorophyll-a in the sea cucumber's *Paracaudina* sp. habitat on the Kenjeran Water, Surabaya

W Widianingsih[1]*, M Zaenuri[2], S Anggoro[2], H P Kusumaningrum[2] and R Hartati[3]

[1]Student of Programme Doctorate Coastal Resources Management, Diponegoro University
[2]Faculty of Fisheries and Marine Sciences, Diponegoro University
[3]Faculty of Sciences and Mathemathic, Diponegoro University Jln. Prof Soedarto, SH, Kampus UNDIP Tembalang Semarang, Indonesia
*Email: widia2506@yahoo.com

Abstract. The study of characteristic sediment and water column chlorophyll-a has an important role in the sea cucumber habitat. Sediment chlorophyll-a represents a productivity primer for the benthic community. This research has a purpose to investigate characteristic sediment and water column chlorophyll-a on the Kenjeran water, Surabaya. Sediment samples were collected by the ekman grab for analysis, grain size and nutrient. The sample for sediment chlorophyll-a was taken by core sampler. The water samples were taken with Nansen Bottles. According to the research result, the values of sediment chlorophyll-a at station 10, 11 and 12 were higher than the other stations. In contrast, the value of chlorophyll-a in the column water had almost the same value for each station. The sediment chlorophyll-a value on clay and silt sediment type was higher than the fine sand and coarse sediment type. The suitable habitat characteristic for *Paracaudina* sp. was clay and silt sediment with sediment chlorophyll concentration ranging from 347.82 mg·m^{-2} to 1135.52 mg·m^{-2}.

1. Introduction

Kenjeran waters are semi-enclosed waters located in the Madura Strait, Surabaya. There are many marine resources such as sea cucumber, bivalve, gastropod and others marine benthic organisms. In these area, there are seven species of sea cucumber such as *Phyllophorus* sp., *Paracaudina australis, Colochirus quadrangularis, Holothuria* sp., *Holothuria sanctori*, *H. forskali* and *H. turriscelsa* [1]. Sea cucumber *Phyllophorus* sp. and *Paracaudina* sp. were dominant species in the Kenjeran Water [1]. The abundance of these two species is largely determined by habitat and environmental conditions.

Most of the sea cucumbers are sediment feeder which passes large amounts of sediment through their gut system to assimilate a fraction of the low content of organic matter, mainly benthic diatom (microphytobenthos), bacteria and detritus [2]. Sea cucumbers play an important role for recyclers of inorganic nutrient and thus a part of the close cycling of materials. According to the research, high sea cucumber densities in experiments may reduce microphytobenthos production and biomass [2].

IOP Conf. Series: Earth and Environmental Science **139** (2018) 012016 doi:10.1088/1755-1315/139/1/012016

In the shallow water, the benthic microalgae play an important role as primary producers. The primary production in the benthic ecosystem is the main food resources for most all benthic organisms such as micro-zoobenthic, meiobenthic, macrozoobenthos and macrobenthos organisms. Most all of benthic microalgae such as benthic diatom can excrete high extra-polymeric substance (EPS) that contribute to supply food resources for many benthic organisms [3]. Furthermore, benthic microalgae can significantly reduce the release of nutrients when the benthic microalgae diffuse across the surface and indirectly by oxygenating surface sediment [4, 5].

The value of microalgae benthic is represented by sediment chlorophyll-a which depend on the nutrients content not only in the water column but also in the benthic ecosystem. There is a relationship between the dynamic of nutrient in the water column and the benthic ecosystems. Recently, most of the researches in the Kenjeran Water, were about sea cucumber biology, nutrition content, the population of marine biota such as bivalve, crustacea and fish [1, 6]. However, some studies of the physical and chemical characteristic of sea cucumber habitat on Kenjeran Water Surabaya have not explored yet. This research aimed to determine characteristic sediment and water column chlorophyll-a in the habitat of sea cucumber *Paracaudina* sp, the Kenjeran Water, Surabaya.

2. Materials and Methods

2.1. Sampling

This research was conducted in the Kenjeran Waters, Surabaya. Sediment samples and water samples were taken from August to October 2016. Samples were taken each per month. Sediment and water samples were taken from 12 stations (figure 1).

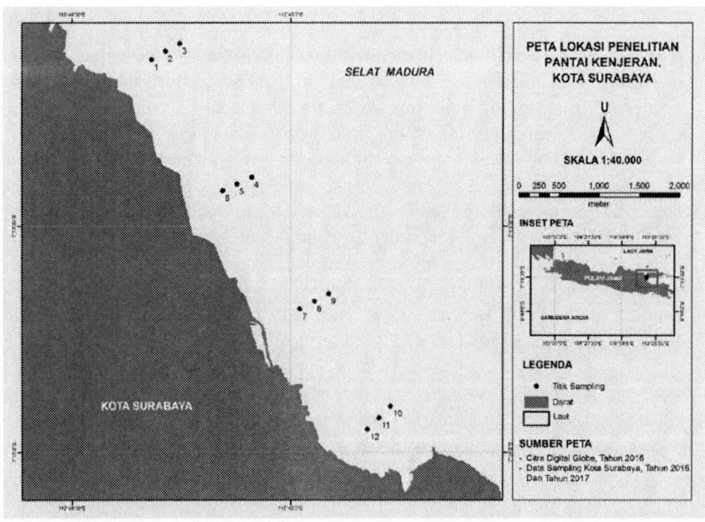

Figure 1. Sampling Station at the Kenjeran Waters, Surabaya.

Sediment samples were taken by sediment grab (Eikman Grab) and core samples. The uppermost layer of sediment was sliced in 0–1 cm. Water samples for measuring Chlorophyll-a were taken by Nansen Bottle. Salinity in the water column was measured directly in situ by refractometer (S·mill^{-1}, Atago, Japan).

2.2. Analysis

Lorenzen Method has been used for analysis of sediment chlorophyll-a content [7]. Grain size was analyzed by ASTM (American Society of Testing Material) method. A spectrophotometric method was used to analyzed orthophosphate, nitrate, and water chlorophyll-a content [8].

Sediments grain size were determined by sequential sediment-sieving, using analytical sieve shaker, to evaluate the relative abundance (% dry weight) of > 1000, 1000–500, 500–250, 250–125, 125–63 and < 63 um size fraction. Sediments were classified into sediment type according to the method from Folk 1954 [3]. Organic matter was measured as ash-free dry weight (AFDW) loss by ignition after 5 hours at 550 °C.

SPPS 16 was used to analyze the normality variable chlorophyll-a in the sediment and the column water. Comparing averages between variable were analyzed using one-way ANOVA and non-parametric Kruskal-Wallis [9].

3. Result and Discussion

3.1. Result

3.1.1. Chlorophyll-a in Sediment and Water Column. According to the research, from station 1 to 9 chlorophyll-a in the sediment fluctuated not only in August and September but also on October 2016. The highest value of chlorophyll-a in sediment was found at station 10 in September as amount 1135.52 mg·m^{-2}, and then followed by August 2016 as amount 1104.83 mg·m^{-2} (figure 2). The magnitude of the chlorophyll-a content in the sediment at stations 10, 11 and 12 compared to stations 1 to 9 were also followed by the amount of total organic matter, percentage C–organic, phosphate and nitrate value (figure 4 and 5).

On the graph shows that the value of chlorophyll-a concentration in the water column on August showed greater than on September and October 2016 (figure 2). The value of concentration chlorophyll-a tended to decrease from August to October 2016. The effect of La Nina throughout 2016 can cause heavy rainfall in some parts of Indonesia. Of course, this can affect the primary productivity of waters that indicated by the decreasing chlorophyll-a concentration in the water column in September and October compared to August (figure 2).

Furthermore, according to Kolmogorov-Smirnov test, the chlorophyll a content distribution in sediment at 12 sampling sites on 3 months (August, September and October 2016), showed that the chlorophyll-a normal distribution was not found in sediment (Asymp sig (2–tailed) = 0.014 < 0.05). The test non-parametric was continued with Kruskal-Wallis analysis. According to the result, showed that the Chi-square value was 1,195 df 3 (7.81) at P = 0.05 was smaller than Chi-square table (df = 3, P = 0.05). It concludes that there were no significant differences between the variables chlorophyll-a content in the sediment for sampling time in August, September and October 2016.

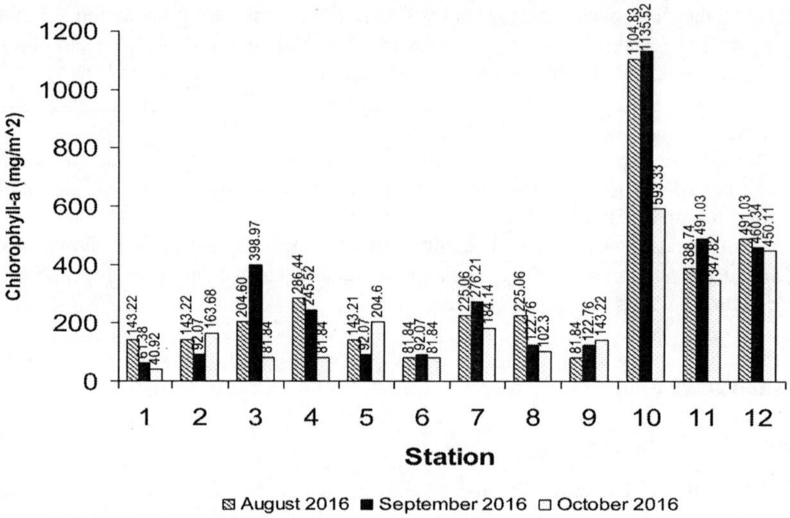

Figure 2. Chlorophyll-a Content in Sediment from Kenjeran Water, Surabaya.

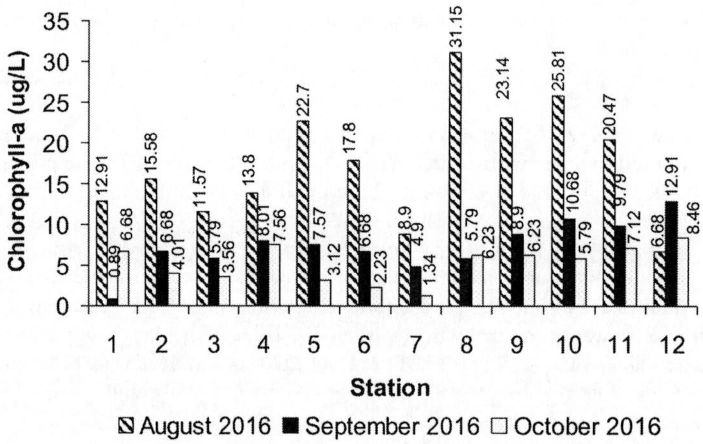

Figure 3. Chlorophyll-a Content in Water Column from Kenjeran Water, Surabaya.

According to one sample Kolmogorov-Smirnov test (SPPS 16 test), the distribution of chlorophyll a in the water column for 3 months (August, September and October 2016) was normal (value of asymp sig. (2-tailed) = 0.114 was higher than 0.05). After that, one-way ANOVA test was continued. The result showed that there was significant different content chlorophyll-a in the water for August, September and October 2016. Difference between months (August, September and October 2016) influence the chlorophyll-a content in the water column (One–Way ANOVA; F3, 44 = 18,217, $P < 0.01$).

3.1.2. Nutrients in the Sediment and Water Column. The concentrations of ortho-phosphate in the sediment station 10–12 were higher than other stations (station 1 to 9). Moreover, at the station 11 and

12, on August, the concentration of nitrate was higher than others stations. The concentration value of nitrate in the sediment on October 2016 at station 8, 9 and 12 were higher than the others stations (figure 5). Contrastly, at the station 10. the concentration of nitrate was lower than the other stations.

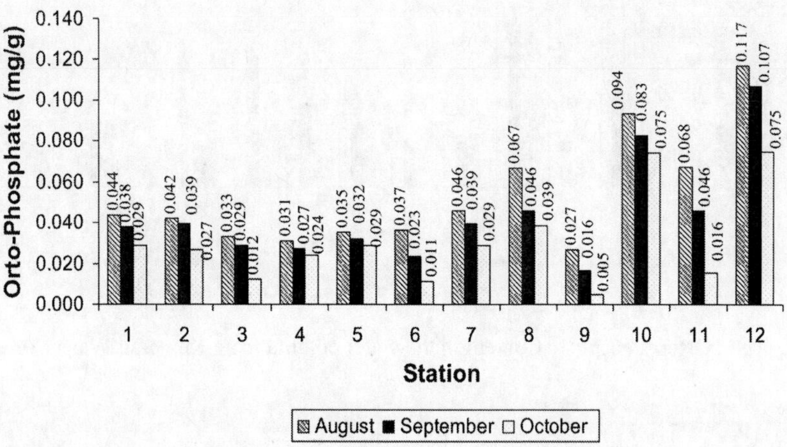

Figure 4. Orto-Phosphate content in the sediment from Kenjeran Water, Surabaya.

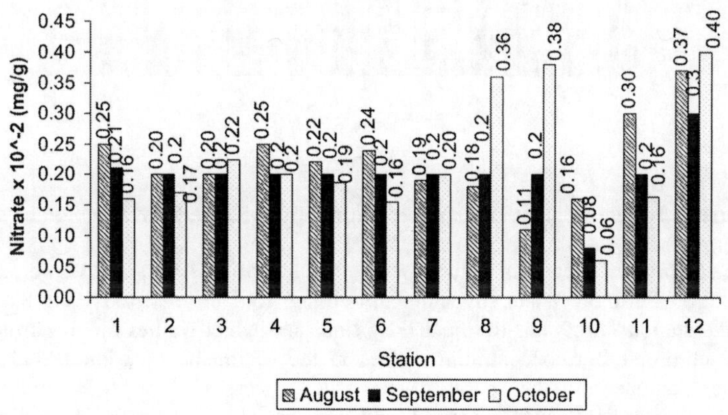

Figure 5. Nitrate content in the sediment from Kenjeran Water, Surabaya.

According to the variance homogeneity test (Levene statistic sig. = 0.024 < 0.05), showed that the nitrate content variant in the sediment was not homogenous. Whereas, the variant data distribution was normal (One-sample Kolmogorov-Smirnov test, Asymp. Sig. (2–tailed) = 0.063 > 0.05). Therefore, time for sampling (August, September and October) had no influence on Nitrate content in the sediment (One Way Anova; $F2, 33 = 0.360, P > 0.05$).

The variance homogenity test showed that concentration of phosphate in the water column was homogenous. However, according to statistic analysis, there is no significant differences concentration phosphate within August, September and October 206. Besides, Nitrate content in the water column showed a normal distribution (One-sample Kolmogorov-Smirnov test, Asymp. sig. (2–tailed) = 0.201

> 0.05). The variation in the nitrate content in the water column of the months (August, September, October) was significantly different (One-way Anova; F2, 33 = 5,156, P < 0.05).

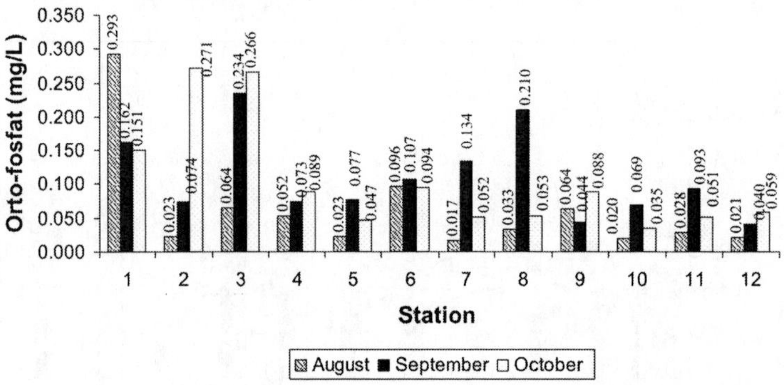

Figure 6. Ortho-Phosphate Content in the water column from Kenjeran Water, Surabaya.

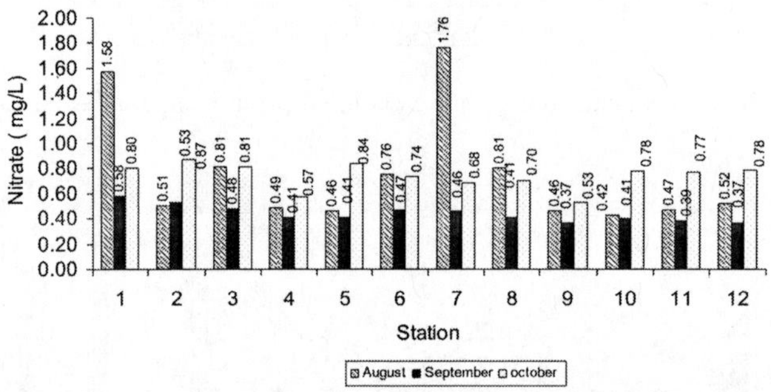

Figure 7. Nitrate Content in the water column from Kenjeran Water, Surabaya.

3.1.3. Grain Size Sediment. According to the research result, station 10, 11 and 12 were dominated by clay sediment ranging 35 to 58 % and silt sediment ranging from 41 to 61 % (figure 8, 9,10). On the other hand, on the station 1 to 9 was dominated by fine sand which values ranging from 35 to 87 % (figure 10). The chlorophyll-a concentration values in the sediment at station 10, 11 and 12 were higher than other stations. In the shallow water, the processes of physical conditions such as hydrodynamic (bottom current, surface current, and tidal process) were affecting the particles grain size [10].

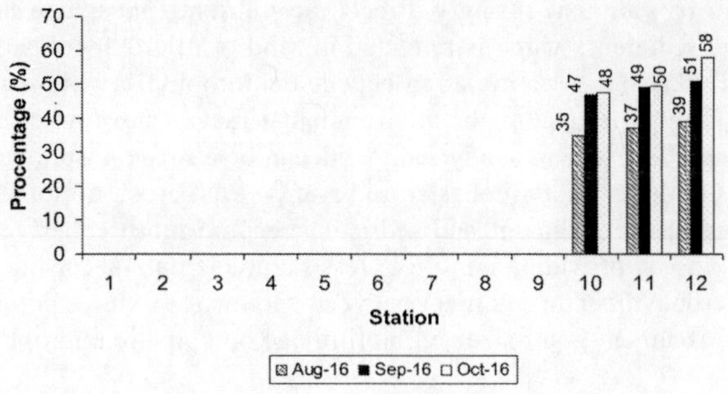

Figure 8. Percentage (%) clay for each station on the Kenjeran Water, Surabaya.

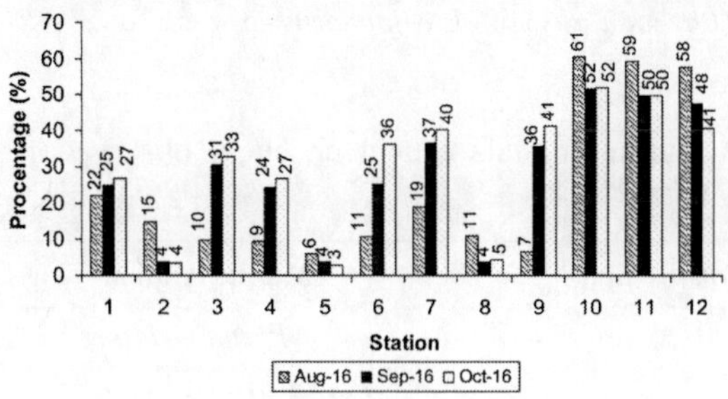

Figure 9. Percentage (%) silt for each station on the Kenjeran Water, Surabaya.

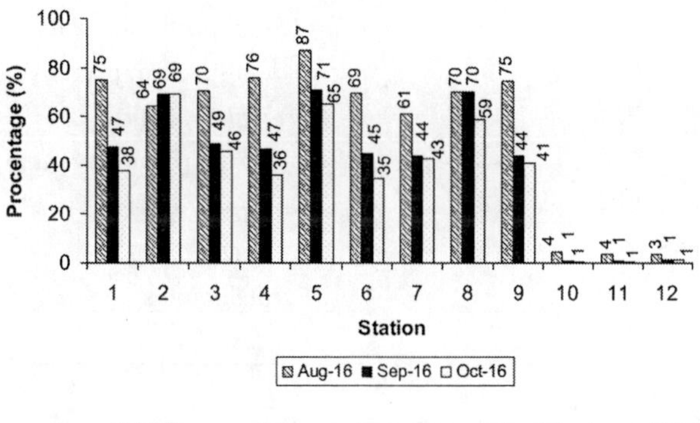

Figure 10. Percentage (%) fine sand for each station on the Kenjeran Water, Surabaya.

3.2. Discussion

The phytoplankton in the water column, as measured by chlorophyll-a concentration, was a function of the concentration of total nitrogen and phosphate [5]. Moreover, phytoplankton and micro-phytoplankton (benthic microalgae) affect organic matter turnover and nutrient cycling in the shallow water [11].

On August 2016, the station 1 showed the highest phosphate concentration 0.293 $mg \cdot L^{-1}$ compared to other stations. Additionally, the highest concentration of nitrate value was found at station 7 (1.76 $mg \cdot L^{-1}$). As for the other stations had almost the same value (figure 5). The value of chlorophyll-a in sediment could be caused by the physical mixing of the sediment, bioturbation and nutrient content in the sediment [5].

The characteristic of sediment strongly affects the nutrient (phosphate and nitrate) and physical conditions within the sediment, which is reflected in kind of microphytobenthos [12]. Some research revealed that there was a stronger correlation between chlorophyll-a sediment and characteristic type of sediment. Physical factor is one of the environmental factors have an effect on sediment stability against biological ones. The type of sandy sediments can be easily transported to the water column by the bottom current. On the contrary, cohesive sediments resist erosion. Some microphytobenthos can secret mucus on the surface sediment and sediment become more cohesive. Sediment composition impacts biofilm shaping by providing a more or less favorable habitat for microphytobenthos, leading to differences in microphytobenthos content between sediment mixtures composition [10]. There is a relationship between sediment grain-size and biofilm age on epipelic microphytobenthos resuspension [10].

Base on this research result, sea cucumber *Paracaudina* sp. was widely caught at stations 10, 11 and 12. Meanwhile, for station 1-9 were found a lot of marine organisms such as *Pinna* sp., *Sascrostrea* sp., shrimp, *Perna viridis*, shells, *Corbicula javanica* (Kupang), fish, crab, and another kind of sea cucumber (*Phyllophorus* sp.). *Phyllophorus* sp. were found at the station 3, 4, 5, 6, 8 and 9 (table 1).

Table 1. Marine animals caught on every observation station in the Kenjeran Water.

Location/ Station	Marine Animal
1	*Perna viridis*
2	*Perna viridis*, *Pinna* sp.
3	*Phyllophorus* sp.
4	*Phyllophorus* sp.
5	*Phyllophorus* sp, shrimp, crab
6	*Phyllophorus* sp, shrimp, crab
7	*Corbicula javanica*
8	crab, *Phyllophorus* sp, *lorjuk kerang*
9	*Phyllophorus* sp.
10	*Paracaudina* sp.
11	*Paracaudina* sp.
12	*Paracaudina* sp.

Even though, although there is no effect of sampling time on chlorophyll a content (Kruskal-Wallis test, Chi-square = 1.195 df 3 (7.81) at P = 0.05. was smaller than Chi-square table (df = 3, P = 0.05) but at the stations 10, 11 and 12 found many Holothurid. *Paracaudina* sp. was found on sediment which was higher in chlorophyll-a content (station 10, 11 and 12). Those station had value of chlorophyll-a concentration within range 347,82–1135,52 mg·m^{-2}. Chlorophyll-a content in the sediment is high because there are a lot of communities of benthic microalgae that inhabit the intertidal flats of water, from dense and highly productive diatom-dominated biofilms at the surface the fine sand grain sediment [13]. This fit with this research, the high chlorophyll-a content in sediment affected the abundant of Paracaudina. These species prefer to live in the clay and silt sediment type. In the meantime, Phyllophorus preferred habitat with fine sand sediment type and less chlorophyll-a concentrations compared to the habitat of *Paracaudina* sp.

4. Conclussion

In the waters of Kenjeran, Paracaudina sp prefers habitats with clay and silt sediment types and high concentration of chlorophyll-a in the sediment, and high value of the nutrient. The high value of nutrient (phosphate and Nitrate) is a need for growing microphytobenthos (benthic microalgae) as food resources for marine benthic organisms such as Paracaudina sp. The high and low chlorophyll content in the water column does not affect the presence of *Paracaudina* sp.

References

[1] Winarni D, Affandi M, Masithah E D and Damayanti H 2012 *Proceeding of The International Fisheries Symposium IFS* 2012, Can Tho City-Viet Nam, 6–8 December 2012
[2] Uthicke S and R. Karez 1999 J. Exp. Mar. Biol. Ecol. **236** 69–87
[3] Jesus B, Brotas V, Ribeiro L, Mendes C R, Cartaxana P and Paterson D M 2009 *Continen. Shelf Res.* **29** 1624–1634
[4] Sundback K., Miles A and E. Goransson. 2000. *An annual study Mar Ecol Prog Ser* **200** 59-76.
[5] Sin Y, Ryu SO and E.Song 2009 *Algae* **24**(3) 149–161
[6] Widianingsih W, Zaenuri M, Anggoro S and Kusumaningrum H P S 2016 2nd International Symposium on Aquatic Products Processing and Health (ISAPPROSH 2015). *Aquatic Procedia* **7** 271–276
[7] Lorenzen C J and Jeffrey S W 1980 *UNESCO Tech. Paper in Mar. Sci.* **35** 1–27
[8] Strickland J D H and Parson T R 1972 *Fisheries Res.Board of Canada.* Bulletin **167**
[9] Fowler J, Cohen L and P. Jarvis 1998 *Practical Statistic for Field Biology* 2nd Ed (USA: *John Wiley & Sons Inc)* **p 259**
[10] Ubertini M, Lefebvre S, Rakatomalala C and Orvian F 2015 *J. Exp. Mar. Biol. Ecol.* **467** 52-64
[11] Middleburg J J, Barrangquet C, Boschker H T S, Herman P M J, Moens T and Heip C H R 2000 *Limnol. Oceanogr.* **45** 1224–1234
[12] Jesus B, Mendes C R, Brotas V and Peterson D M 2006 *J. Exp. Mar. Biol. Ecol.* **332** 60-74
[13] Serodio J, Vieira S and Crus S 2008 *Cont. Shelf Res.* **28** 1363–1375

IOP Conf. Series: Earth and Environmental Science **139** (2018) 012017 doi:10.1088/1755-1315/139/1/012017

DNA barcode of *Acropora hyacinthus* of Karimunjawa Archipelago

D P Wijayanti[1*], E Indrayanti[1], H Nuryadi[2,3], S A Rintiantono[1] and A Sabdono[1]

[1] Marine Science Department, Diponegoro University, Jl. Prof. Soedarto SH, Tembalang, Semarang

[2] Tropical Marine Biotechnology, Central Laboratory of Research and Services, Diponegoro University, Jl. Prof. H. Soedarto, SH, UNDIP, Tembalang, Semarang, Indonesia

[3] Department of Biology, Chemistry and Marine Science, University of the Ryukyus, 1-Senbaru, Nishihara-cho, Okinawa

*Email: diah_permata@mail.com

Abstract. Karimunjawa is one of the earliest marine parks in Indonesia. Karimunjawa National Park (KNP) was designated as a marine conservation area to conserve marine resources from destructive fishing activities. Scleractinian corals in the genus Acropora are among the most dominant distributed in the KNPs, including the species of *Acropora hyacinthus*. Here, we present a comprehensive analysis of intra- and interspecific COI variabilities in *A. hyacinthus* to analyze genetic diversity and to describe the kinship relationship of the coral between 5 localities of the reefs. Genetic marker Cytochrome Oxidase I of the mitochondrial genome DNA (mtDNA) was used to analyze genetic diversity. Reconstruction of phylogenetic tree and genetic diversity were made by using software MEGA 5.05 (Molecular Evolutionary Genetics Analysis). The results indicate corals *A. hyacinthus* from five localities of Karimunjawa Archipelago are in the high category of genetic diversity. However, the five populations showed a close genetic relationship of kinship. This is likely due to the small size of the population and few numbers of samples that may not represent the population. The results may aid managers of the park in the selection of appropriate propagules sources which can help to restore important data for conservation and sustain coral reef resources.

1. Introduction

Karimunjawa waters are one of the earliest marine parks in Indonesia. The park lies in the Java Sea, approximately ± 45 nautical miles from Jepara, on the north coast of Central Java, Indonesia. The park was designated as a Strict Natural Reserve on 9 April 1986 by Minister of Forestry under decree (PHKA No. 123/Kpts-II/1986) as an effort to conserve marine resources from destructive fishing activities. In 2001, all marine waters of Karimunjawa National Park (KNP) were designated as a marine conservation area by the Ministry of Forestry Decree No. 74/Kpts-II/2001 [1]. The park consists of 27 big and small islands, seven of which are inhabited. Coral communities are found vigorously to a depth about 15 m with Pocillopora and Acroporidae being the dominant family. According to survey conducted by Wildlife Conservation Society (WCS), there are 63 genera of 15 families of scleractinian corals. Recent trends in the average of coral cover were moderate (around 40–60 %) [2] with similarity index between 0.43 to 0.91 [3].

KNP represents well the population-related reef management issues that are common in developing countries [4], particularly in the Indo-Pacific where efforts to mitigate local impacts that reduce reef resilience are currently scarce [5]. With most populated island in Indonesia is nearby (Java Island, > 60 % of total nation population), KNP pose a significant challenge for reef managers to address issues of balancing demands for reef-generated resources with conservation efforts [6]. Though it was reported that co-management frameworks as adaptive measures to gain trust, collaboration and a sense of ownership by reef users, improved some ecological condition of the reefs [1], however, extensive human activity related to tourism [7], coupled with emergence of coral disease [8, 9, 10], coral bleaching [11, 12] and increasing local demands of coral reefs resources by a growing local population [13] can contribute to reefs's decline condition.

For coral reefs, potential adaptation ability to re-establish after disturbances is important. There are various factors that may facilitate local adaptation including genetic factors. Understanding coral diversity is essential in managing coral reefs ecosystem especially when many of the reefs are in degraded condition [14]. When MPA network is designated, it is important to understand genetic connectivity among reefs and which areas are particularly rich in species to ensure that the designated protected areas are self-sustaining and possibly export propagules to neighboring reefs [15]. Most corals release gametes in a great event namely mass corals spawning [16] which were then dispersed by ocean currents into various distances from the natal reefs. Knowledge of genetic diversity is a need to understand the coral distribution and the source of coral larvae which is important to replenish coral reefs ecosystem. However, genetic diversity has rarely been corporate in designing a Marine Protected Area [14].

DNA barcoding (DBC) is an alternative to traditional taxonomic methods that could become a useful tool for coral reef conservation [17]. Cytochrome c oxidase subunit 1 (COI) has been proposed as the main barcoding gene for metazoans [18]. DNA barcoding has been extensively used to discriminate between closely related species, to identify new, cryptic or invasive species, and to assess biodiversity across many animal phyla [19]. The effectiveness of this approach clearly depends on the availability of extensive databases of COI-5 barcode sequence standards [17]. Based on the results of DNA barcode of 90 species of corals from 44 genera belonging to 14 families, suggested that the criterion was not being appropriate for most scleractinian corals (or anthozoans in general) because this region evolves very slowly in these organisms and consequently both inter- and intra-specific variation are extremely low [20]. However, taxonomy of *Stylophora pistillata* was successfully demonstrated comprises of four deeply divergent clades based on COI divergence and phylogeny of the coral [21].

Genus Acropora, are by far the most diverse living reef-building coral genus, with 114 species recognized worldwide [22] and others possibly unnamed. The genus distributes on most tropical reefs throughout the world, except the eastern Pacific Reefs [23]. Acropora is distinguished from all other extant coral genera by a unique form of polyp structural. Corallites are small, divide into two types axial and radial. Corallites have septa in two cycles or less and columellae are seldom developed [24]. Acropora plays a dominant role in the species composition and abundance of many Indonesian reefs [25]. Acroporid is a good example of studying reefs connectivity since they have the most genetic information available for any coral genus worldwide. Acroporids also knew to have largest geographical ranges compare to other corals family [26].

Here, we present a comprehensive analysis of intra- and interspecific COI variabilities in *Acropora hyacinthus* to determine connectivity patterns and genetic diversity of the coral on Karimunjawa Archipelago reefs. The results may aid managers of the park in the selection of appropriate propagules sources which can help to restore important data for future coral reef conservation and sustain coral reef resources.

2. Materials and Methods

2.1. Sample collections

Sampling was carried out at Karimunjawa Archipelago (KNP) at five study sites namely Seruni, Genting, Sambangan, Menjangan Kecil and Cilik Island. One site each were established at three islands, in approximately 2–7 m depth. Sampling trip was carried out in May 2014 and June 2015. Snorkeling or scuba was used to sample > 20 individuals of *A. hyacinthus* per reef, a sample size adequate for genetic diversity tests [27].

Each coral colony as a sample was documented for identification. The samples from branches were cut for 1–2 cm of random colony using cutting pliers. At least 1 m distanced from one colony to another colony to avoid collected clone colony. As branches colony, *A. hyacinthus* can reproduce through fragmentation [28]. Small pieces were then put into a 2 mL labeled cryovial. To avoid re-sampled, each colony that was already cut was tagged using the small plastic number and cable ties. Table 1 has specific site information along with the number of corals collected at each site. Samples were preserved in 95 % ethanol until used. In total, 104 coral branches were obtained from KNP.

Table 1. Sample locations and number of collected colonies.

Location	Ordinate of sample locations	Number of colonies
Genting Is.	E 110°35’11.6” S 05°50’38.8”	24
Seruni Is.	E 110°26’24.0” S 05°53’02.7”	20
Sambangan Is.	E 110°34’57.8” S 05°51’15.0”	20
Menjangan Kecil Is.	E 110° 24’ 22,3” S 05° 53’ 23,3”	20
Cilik Is.	E 110° 30’ 24,7” S 05° 49’ 18,6”	20
	Total	104

2.2. *DNA Extraction*

A quick and simple DNA extraction method with the resin, Chelex 100 (BioRad) was done. A 0.2 mm section of tissues were scrapped using sterile spatula and place into 1.5 mL tubes filled with 300 uL 10 % Chelex. The samples were then vortexed in Chelex slurry for 10–15 seconds briefly at high speed in a microcentrifuge, incubated for 20 minutes at 95 °C and vortexed again for 10–15 seconds to ensure that all contents are in the bottom of the microcentrifuge tube. The supernatant was directly used for PCR amplification. DNA concentration was quantified using Nanodrop with spectral length on $260{\cdot}280\ \mu m^{-1}$ and $260{\cdot}230\ \mu m^{-1}$.

2.3. *PCR Amplification*

Amplifications were performed in 25 µl volume master mix PCR solution. Universal primer for COI gene, consisted of forward primer LCOI1490: 5'-CAAATCATAAAGATATTGG-3' and reverse primer HCOI2198: 5'-TAAACTTCAGGGTGACCAAAAAATCA-3' [29] were used for amplification. Master mix PCR solution containing 1 µl of each primer, 1 µl of DNA Template (< 250 $ng{\cdot}\mu l^{-1}$), 12.5 µl of GoTaq ® Green Master Mix (Promega, USA) and 9.5 µl of Nuclease-Free Water (Promega, USA). Thermal cycle condition was following protocol by [20]. PCR products were analyzed using 1 % agarose gel. PCR products that have DNA band were directly sequenced at 1st Base Laboratory using both of primers. Sequencing results were analyzed using MEGA 6.0 for alignment process and Phylogenetic tree was constructed using Phylogenetic Analysis Using Parsimony (PAUP) software [30].

3. Results and Discussion

In total, there were 32 sequences resulted from Menjangan Kecil Is., Cilik Is., Genting Is., Seruni Is., and Sambangan Is., Karimunjawa Archipelago, Java Sea. The resulting Oxidase I (CO I) of the mitochondrial genome DNA (mtDNA) sequences were corresponding to the genotype that was analyzed for homologies of sequences in the database using BLAST searches. BLAST analysis showed that all 32 sequences were closely related to *Acropora hyacinthus* under 99 % similarity. All nucleotide sequences have been deposited in the GeneBank database under accession number LC189268–LC189316.

3.1. Phylogenetic Tree

Reconstruction of the phylogenetic tree was done using PAUP* 4.0 (Phylogenetic Analysis Using Parsimony) [30] with parsimony method and 1,000 times boots trap. Kinship relationship was observed using TreeView program.

Phylogenetic construction based on COI sequences found 6 distinct clusters of *A. hyacinthus*. Outgroup using DNA data base of *Montipora* sp. Phylogenetic tree of *A. hyacinthus* was shown in figure 1.

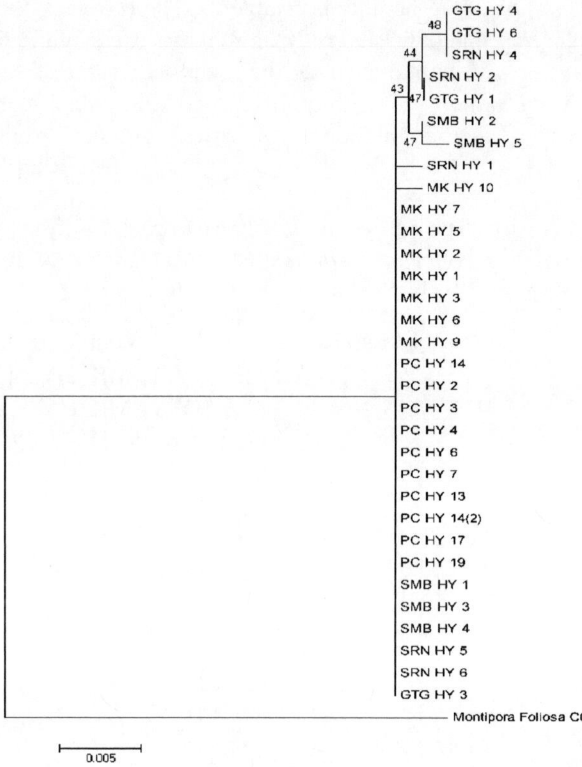

Figure 1. Phylogenetic tree construction of *Acropora hyacinthus* collected from Menjangan Kecil Is. (MK), Cilik Is. (PC), Genting Is. (GTG), Seruni Is. (SRN) and Sambangan Is. (SMB), Karimunjawa Archipelago, Java Sea. There 6 different clades observed from the tree. *Montipora foliosa* was used as outgroup.

Clade 1 was comprised of three samples of Genting and Seruni Isl. (GTG HY4, GTG HY 6 and SRN HY 4), clade 2 consisted each colony of SRN HY 2 and GTG HY 1, clade 3 consisted only two samples from Sambangan Isl. (SMB HY 2 and SMB HY 5), while clade 4 (SRN HY 1) and 5 (MK HY 10) consist only one sample each. Clade 6 is the biggest clade that comprised of mix samples collected from five different reefs (Menjangan Kecil, Cilik, Genting, Seruni and Sambangan island) (MK HY 7, MK HY 5, MK HY 2, MK HY 1, MK HY 3, MK HY 6, MK HY 9; PC HY 14, PC HY 2, PC HY3, PC HY 4, PC HY 6, PC HY 7, PC HY 13, PC HY 14(2), PC HY 17, PC HY 19; SMB HY 1, SMB HY 3, SMB HY 4, SRN HY 5, SRN HY 6 and GTG HY 3) (figure 2).

When comparison was made with *A. hyacinthus* population from Raja Ampat, a similar result was obtained. *A. hyacinthus* of Raja Ampat has grouped into 6 different clades [31]. Samples of *A. hyacinthus* from Kofiau District was collected from Deer Island and Boo Island that was separated by more than 100 km, while samples from Karimunjawa Archipelago was collected from two different

IOP Conf. Series: Earth and Environmental Science **139** (2018) 012017 doi:10.1088/1755-1315/139/1/012017

cluster islands namely Seruni, Sambangan and Genting cluster and Cilik and Menjangan Kecil cluster islands that separate by most far 60 km only. However, the Menjangan Kecil Island and Cilik Island were set behind the main Karimunjawa Island that faces the open sea while Seruni, Sambangan and Genting were set inside lagoonal area of Karimunjawa main island.

3.2. Analysis of Genetic Diversity

Genetic diversity of *A. hyacinthus* was observed using DNAsp [32]. Based on statistic analysis of genetic diversity of 32 samples showed 15 haplotypes (table 2). There were 11 haplotypes (73.3 % of total samples) are unique haplotypes that consisted only one sample each. While four haplotypes (26.7 %) are more common haplotypes with haplotype 1 as the highest member of samples (13 samples). Estimation of haplotype diversity value of *A. hyacinthus* was 0.8286 which categorized has a high genetic diversity [33]. It is likely that population of *A. hyacinthus* at Karimunjawa Archipelago showed high genetic diversity.

Table 2. Haplotype diversity distribution of *Acropora hyacinthus* from Karimunjawa Archipelago, Java Sea; HY, *Acropora hyacinthus*; MK, Menjangan Kecil Is.; PC, Cilik Is.; GTG, Genting Is.; SRN, Seruni Is.; SMB, Sambangan Is.

No	Haplotype	Number of samples	Samples code
1.	Haplotype 1	13	MK HY 1, MK HY 5, MK HY 6, MK HY 7, MK HY 9, PC HY 14, PC HY 6, PC HY 7, PC HY 13, PC HY 14(2), PC HY 17, PC HY 19, GTG HY 3
2.	Haplotype 2	3	MK HY 2, SRN HY 5, SRN HY 6
3.	Haplotype 3	1	MK HY 3
4.	Haplotype 4	1	MK HY 10
5.	Haplotype 5	3	PC HY 2, PC HY 3, PC HY 4
6.	Haplotype 6	1	SMB HY 1
7.	Haplotype 7	1	SMB HY 2
8.	Haplotype 8	1	SMB HY 3
9.	Haplotype 9	1	SMB HY 4
10.	Haplotype 10	1	SMB HY 5
11.	Haplotype 11	1	SRN HY 1
12.	Haplotype 12	2	SRN HY 2, GTG HY 1
13.	Haplotype 13	1	SRN HY 4
14.	Haplotype 14	1	GTG HY 4
15.	Haplotype 15	1	GTG HY 6
	Total	32	

Genetic distance within island showed that samples collected from Seruni Isl. has farthest genetic distance (0.003 ± 0.003) followed by samples from Sambangan (0.002 ± 0.001) and Genting Isl. (0.002 ± 0.003). While samples from Menjangan Kecil and Cilik Isl. Showed nearest genetic distance. When comparison was made between a distance of different reefs, the closest genetic distance was shown by samples from Menjangan Kecil and Cilik Isl., (0.000) (table 3).

Table 3. Genetic distance within distance of *Acropora hyacinthus* from Karimunjawa Archipelago; d, distance; S.E., standard error.

Island name	D	S.E.
Menjangan Kecil	0.000	0.000
Cilik	0.000	0.000
Sambangan	0.002	0.001
Seruni	0.003	0.003
Genting	0.002	0.003

Distributions of species of the coral genus Acropora in the Indonesian archipelago show a duality reminiscent of the Wallace's Line patterns seen in terrestrial animals and plants, rather than the concentric pattern predicted by the "center of origin" model [34]. This duality is due to an overlap of Indian Ocean species distributions diminishing eastwards and Pacific Ocean species distributions diminishing westwards within the archipelago [34]. Additionally, a large number of species with broad Indo-Pacific distribution, as well as some regional endemics, occur within the archipelago. Thus although Indonesia has a high species number overall, this is due to the presence of a composite fauna with strong regional differences.

Using multi-locus nucleotide sequence data, [26] recently described two, previously unsuspected, cryptic species complexes within the *Acropora syngameon* (i.e., groups of species connected through a genetic exchange). Acropora is categorized as the most diverse and abundant genera of reef-building corals [22, 24]. The genus is notorious for high levels of shared polymorphism between species, which is due, in part at least, to widespread genetic exchange through introgression [35, 36, 37]. Despite the absence of any fixed sequence differences, utilized allele frequency variation at 10 loci to demonstrate that two common coral species in Australia, *A. cytherea* and *A. hyacinthus*, actually represent species complexes minimum two and four cryptic species. These cryptic species were discovered in complete sympatry and form part of an extended syngameon with a complex network of gene flow among species. Cryptic species were identified by the 'genotypic cluster' species definition, which defines species as distinguishable genotypic groups.

The results of this research indicate that corals *A. hyacinthus* from Karimunjawa Island is categorized as high of genetic diversity, though they showed a relatively close genetic relationship of kinship. We need run a second gene such as microsatellite or intron sequencing to confirm possibility that population of *A. hyacinthus* at Karimunjawa Archipelago may have a cryptic variation as suggested by [26]. However, the possibility that population of *A. hyacinthus* at Karimunjawa Archipelago has high diversity could not be excluded. Therefore, managers of the KNP need more appropriate approach to resources include the genetic diversity.

Acknowledgments
Authors acknowledge Diponegoro University for provided the research grant to DPW through Fundamental Research project under decree Ministry of Cultural and Education through Penerimaan Negara Bukan Pajak (PNBP) Diponegoro University, the fiscal year 2014 and PT PURA Baruna Jaya, Sambangan Island for logistic and accommodation during sampling collection. Samples collection permit was obtained from Karimunjawa National Park Authority under number 889/BTNKJ-1.6/SIMAKSI 2015. Further, authors are indebted to reviewers for their contribution to improve the manuscript.

References

[1] Campbell S J, Kartawijaya T, Yulianto I, Prasetia R and Clifton J 2013 *Mar. Policy.* **41** 72–79
[2] Ardiwijaya R L, Baird A H, Kartawijaya T and Campbell S J 2008 *Changes in reef fish biomass in Karimunjawa national Park: a test of the effectiveness of government gazetted marine parks in Indonesia* (Florida: Proc. of the 11 Intl. Coral Reef Symp., Ft. Lauderdale, 164–168)
[3] Balai Taman Nasional Karimunjawa 2008 *Statistik Balai Taman Nasional Karimunjawa.* (Semarang: Balai Taman Nasional Karimunjawa)
[4] Bell J D, Ratner B D, Stobutzki I and Oliver J 2006 *Ocean and Coastal management* **49** 976–985
[5] Hughes T P, Baird A H, Bellwood D R, Card M, Connolly S R, Folke C, Grosberg R, Hoegh-Guldberg, Jackson J B C, Kleypass J, Lough J M, Marshall P, Nystrom M, Palumbi S R, Pandolfi J M, Rosen B and Roughgarden J 2004 *Sci.* **301** 929–933
[6] Edinger E N, Kolasa J and Risk M J 2000 *Diversity and Distributions* **6** 113–127
[7] Hehalatu A 2015 *Thesis* (unpublished) Universitas Sriwijaya Palembang Indonesia
[8] Sabdono A and Radjasa O K 2006 *Jurnal Imu Kelautan Indonesia* **11**(3) 158–162
[9] Hakim H I, Radjasa O K, Khoeri M M, Pratama G A, Nasima D, Ambariyanto, Sarjito and Sudoyo H 2012 *Ecologia* **2** 52–59
[10] Sabdono A, Wijayanti D P and Sardjito 2017 *Res J. Microbiol.* **12** 154–160
[11] Wilkinson C 2000 *Status of coral reefs of the world: 2000* (Australia: Australian Inst. of Marine Science)
[12] Habibi A, Setiasih N and Sartin J 2007 *A decade of reef check monitoring: Indonesia coral reefs, condition and trends* (The Indonesian Reef Check Network)
[13] Taruc S A K 2011 *Thesis* The University of Queensland Australia
[14] Baums I B 2008 *Mol. Ecol.* **17** 2796–2811
[15] Palumbi S R 2003 *Ecological Appl.* **13** S146–S158
[16] Baird A H, Guest J R and Willis B L 2009 *Annu. Rev. Ecol. Evol. Syst. 2009* **40** 551–71
[17] Neigel J, Domingo A and Stake J 2007*Coral Reefs* **26** 487
[18] Hebert P D N, Cywinska A, Ball S L and DeWaard J R 2003 *Biological identifications through DNA Barcodes* (London: Proc of the Royal Soc of London Series B 270, 313–321)
[19] Valentini A, Pompanon F and Taberlet P 2008 *Trendsin Ecology and Evolution* **24** 110–117
[20] Shearer T L and Coffroth M A 2006 *Mol. Ecol. Res. (2008)* **8** 247–255
[21] Keshavmurthy S, Yang S Y, Alamaru A, Chuang Y Y, David Obura M, Fontana S, Palmas S D, Stefani F, Benzoni F, Donald A M, Noreen A M E, Chen C, Wallace C C, Pillay R M, Denis V, Amri A Y, Reimer J D, Mezaki T, Sheppard C, Loya Y, Abelson A, Mohammed M S, Baker A C, Mostafavi P G, Suharsono B A and Chen C A 2013 *Sci. Rep.* **3** 1520
[22] Wallace C C 1999 *Staghor corals of the world: A revision of Genus Acropora* (Australia: SCIRO)
[23] Wallace C C 2001 *Wallace's line and marine organisms: the distribution of staghorn corals (Acropora) in Indonesia* (Honduras: Tethyana Expedition Report) pp 1–9
[24] Veron J E N 2000 *Corals of the world* (Australia: Australian Institute of Marine Science)
[25] Suharsono 1992 *3rd ASEAN Science and Technology Week Conference Proceedings, Vol. 6, Marine Science: Living Coastal Resources* (Singapore: National University of Singapore) p 41–54
[26] Ladner J L and Palumbi S R 2012 *Mol. Ecol.* **21**(9) 2224–2238
[27] Wallace C C 1999 *Staghorn corals of the world: a revision of the coral genus Acropora (Scleractinia; Astrocoeniina; Acroporidae) worldwide, with emphasis onmorphology, phylogeny and biogeography CSIRO* (Collingwood Victoria)
[28] Harrison P and Wallace C C 1990 *Reproduction, dispersal, and recruitment of scleractinian corals. In "Ecosystems of the world Vol. 25: Coral Reefs" Ed by Z Dubinsky* (Amsterdam: Elsevier Press) pp 133–207
[29] Folmer O, Black M, Hoeh W, Lutz R and Vrijenhoek R 1994 *Molecular Marine Biology and*

Biotechnology **3** 294–299
[30] Swofford D L 1998 *PAUP*: Phylogenetic Analysis Using Parsimony (*and Other Methods). Version 4.0* (USA: Sinauer Associates)
[31] Wijayanti D P and Sabdono A 2016 *Patho-Biotechnology Strategies for Conservation of Indonesian Coral Reefs against The Emergence of Outbreak of Coral Diseases* (Research report)
[32] Rozas J, Sánchez-DelBarrio J C, Messeguer X and Rozas R *Bioinformatics* **19** 2496–2497
[33] Nei M 1987*Molecular Genetics* Columbia University New York Press
[34] Wallace C C 1997 *Proceedings of the 8th International Coral Reef Symposium* (Panama) **1** 365–370
[35] van Oppen M, McDonald B, Willis B and Miller D 2001 *Mol. Biol. Evol.* **18** 1315–1329
[36] van Oppen M, Willis B, van Rheede T and Miller D 2002 *Mol. Ecol.* **11** 1363–1376
[37] Vollmer S V and Palumbi S R 2002 *Sci.* **296** 2023–2025

IOP Conf. Series: Earth and Environmental Science 139 (2018) 012018 doi:10.1088/1755-1315/139/1/012018

Food preference of red devil (*Amphilophus labiatus*) in the Sermo Reservoir, Kulon Progo Regency

A Ariasari[1,2]*, S Helmiati[1], E Setyobudi[1]

[1] Department of Fisheries, Universitas Gadjah Mada
[2] Faculty of Geography, Universitas Gadjah Mada
*Email: anaariasari@gmail.com

Abstract. Food preference is one of the important information that can be used to know the food chain in order to manage fisheries resources. This study aims to determine the food habits and preference of red devil (*Amphilophus labiatus*) in the Sermo Reservoir, Kulon Progo Regency. Samples were collected randomly each month from September 2013 to February 2014. Each sample collected was measured its total length, body weight, and determined sex, then dissected to measure the gut length and to observe gut contents. Results showed that red devil is omnivorous (relative gut length = 3.83) with food composition consisted of fish, crustaceans, detritus, phytoplankton, zooplankton, plants, insects, insect's larvae, *Chironomus* sp., and annelids. A change occurred in the food preference of red devil, i.e. the young fish prefers to feed *Chironomus* sp. larvae (86.02 %) whereas the adult fish prefers fish/fish chunk (81.82 %). Trophic level status of red devil showed as carnivorous and niche overlapping between male and female of the adult.

1. Introduction

The Sermo Reservoir is an artificial reservoir that located in the Hargowilis Village, Kokap District, Kulon Progo Regency, Yogyakarta. It has some endemic fishes that consists of wader pari (*Rasbora latestriata*) while some introduction fishes are tilapia (*Oreochromis* sp.), tambaqui (*Cyclossoma macropomum*), common carp (*Cyprinus carpio*) and gouramy (*Osphronemus goramy*). Since 2012, red devil (*Amphilophus labiatus*) has become a dominant catch by fisherman (~75 %). The composition of total catch in 2012 consisted of *Mozambique tilapia* (7 %), tilapia (15 %), red devil (73 %), spotted barb (2 %), and silver rasbora (3 %) [1]. The dominance of red devil has depressed other species. Although the population was increased, the price of red devil reduced than tilapia or other species. As a consequence, it decreased the income of local fisherman [2].

The uncontrolled growth of red devil lowers other fish population for long-term period and affects food chain. Red devil is a part of reservoir community and controlling its population should be based on the ecological balance. A good management of resources can be carried out by monitoring and collecting data frequently through sampling of total catch that records species composition, length and weight composition, reproduction aspect, otolith, and fish gut content [3, 4]. Previous studies have investigated the relationship of length and weight as well as otolith of red devil in the Sermo Reservoir. However, the food habits and preference of red devil have not been studied until today [5, 6].

Food preference consists of the relative length of fish gut and food preference level. Food preference shows information about food composition, food habits, and preference level of red devil on certain food type. Information about food types and habits are important in the fish resources management in order to restore the balance of the reservoir resources, maintain the fish resources, and increase catch productivity. Aim of this research was to investigate the food composition, habits, and preference of red devil (*Amphilophus labiatus*) in the Sermo Reservoir, Kulon Progo Regency.

2. Materials and Methods

Samples were collected biweekly for six months, from September 2013 to February 2014. Samples were taken in four stations: (a) outlet, (b) reservoir body, (c) reservoir basin, and (d) inlet (figure 1). Samples were collected from the fisherman total catch that used gillnet. 50 fishes were randomly collected every month. Fish morphology was observed to assure the research sample. Red devil has thick lips, steep forehead elongation and red-black devil has 5–8 vertical lines on the body [7]. All of the samples were analyzed at the Department of Fisheries of Universitas Gadjah Mada.

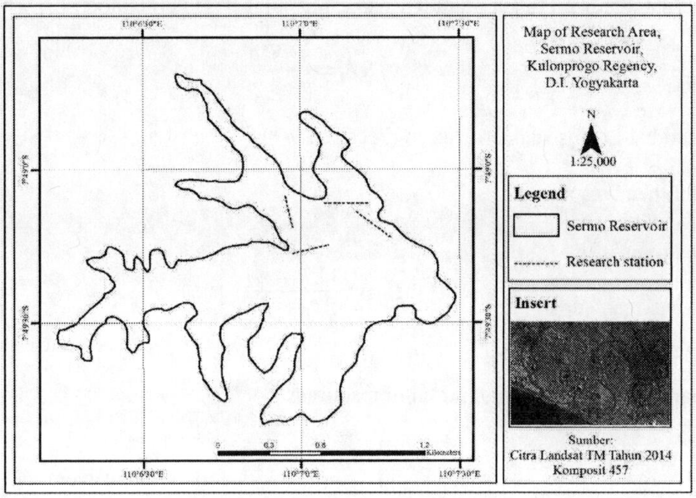

Figure 1. Map of research area in the Sermo Reservoir.

The total length of red devil was measured by ruler and total weight was measured by scales. Samples were then dissected for the gonad and digestive tract observation. Digestive tract preserved with formalin 4 % was measured. Volume of each food type was measured by modified Sedgwick rafter. Each food type was observed by quantitative method, i.e. numerical and volumetric method. Numerical method counted food types component while gravimetrical method counted volume of food types [8]. Each food type was observed using microscope with 4x to 10x magnification and identified by following [9]. Data were analyzed using the following formulas:

Index of Preponderance [10]

$$IP = \frac{V_i O_i}{\sum_{i=1}^{n} V_i x O_i} \quad (1)$$

where:
IP = index of preponderance
Vi = percentage of food volume
Oi = percentage of occurrence

Trophic level [11]

$$T_t = 1 + \sum \left\{ \frac{T_{tp} x I_p}{100} \right\} \quad (2)$$

where:
Tp = trophic level

Ttp = trophic level at food level
Ii = index of preponderance

Niche breadth [12]

$$B_{ij} = \frac{1}{\sum_{i=1}^{n} \sum_{j=1}^{m} P_{ij}^2} \tag{3}$$

where:
Bij = niche breadth of fish group to food resources
Pij = proportion of fish group related to food resources
N = number of fish group (i=1,2, 3…, n)
M = number of food resources (j=1,2, 3…, m)
Niche breadth is standardized to get point 0-1 using formulas based on Hulbert [3] which is:

$$BA = \frac{B-1}{N-1} \tag{4}$$

where:
BA = niche breadth's Levins standardization (between 0-1)
B = nihce breadth's levins
N = number of utilized resources

Niche overlapping [12]

$$C_h = \frac{2\sum P_{ij}P_{ik}}{\sum P_{ij}^2 + P_{ik}^2} \tag{5}$$

where:
Ch = Simplified Morisita Index
Pij, Pik = proportion of resources from total of resources

3. Results

The number of samples were 425 species and samples with food content were 211 species. Length distribution of red devils was about 7.20–15.5 cm and most of them were male (56.87 %). Weight distribution of red devil was about 6.2–74.6 g. Food habits were measured by food composition that fed by fish. Food of red devils consisted of fish (34 %), crustacea (2 %), detritus (10 %), phytoplankton (13 %), zooplankton (17 %), macrophyte (3 %), insects (4 %), larva insects (1 %), *Chironomus* sp. (12 %), and annelids (4 %). The food composition of red devils in the Sermo Reservoir is shown in figure 2.

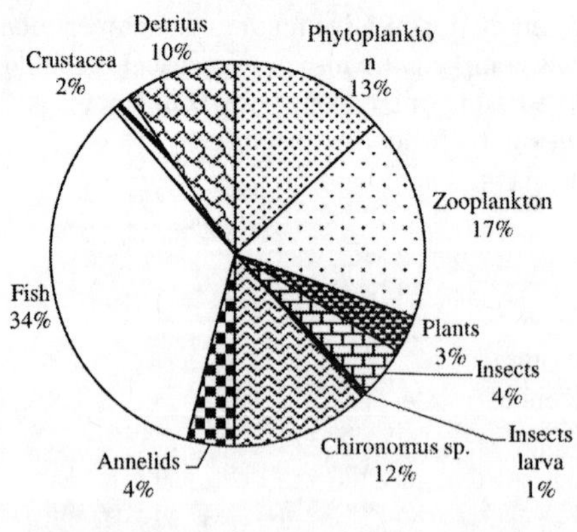

Figure 2. Food composition of red devil in the Sermo Reservoir.

Food habits can be known based on the food composition of red devil. Food composition of red devil consists of fish (34), crustacea (2 %), detritus (10 %), phytoplankton (13 %), zooplankton (17 %), plants (3 %), insects (4 %), insects' larvae (1 %), *Chironomus* sp. (12 %), and annelids (4 %). The dominant food composition of red devil is fish. Various food types of food composition of red devil show that red devil is omnivorous.

Table 1. Index of preponderance of red devil in the Sermo Reservoir.

Food type	Percentage (%)	Group
Phytoplankton	2.98	Additional food
Zooplankton	1.97	Additional food
Macrophyte	0.13	Additional food
Insects	0.51	Additional food
Insects larvae	0.01	Additional food
Chironomus sp.	6.92	Complementary food
Annelids	0.36	Additional food
Fish	82.79	Main food
Crustacean	0.06	Additional food
Detritus	3.91	Additional food
Unidentified	0.36	Additional food

Chironomus sp. (86.2 %) has dominated the main food of red devil which has length group of < 9 cm, whereas piece of fish is the main food for the 9–< 14 cm of length group. *Chironomus* sp., detritus, and *Trebonema,* sp. is the complementary food. Other food is an additional food.

Table 2. Index of preponderance based on the length size.

Food Type	Index of Preponderance (%)			
	< 9 cm	9–11	12–14	> 14
Phytoplankton				
-*Tribonema* sp.	6.99	1.41	4.68	1.23
- *Ephitemia* sp.	0.00	0.00	0.01	0.00
- *Nitzschia* sp.	0.00	1.41	0.03	0.00
- *Spyrogira* sp.	0.00	0.02	0.00	0.00
Zooplankton				
-*Aapidiaca* sp.	0.00	0.01	0.02	0.00
- *Daphnia* sp.	0.00	0.01	0.00	0.00
- *Cyclops* sp.	0.00	0.10	0.00	0.00
- *Diaphtomus* sp.	0.41	1.43	0.32	0.11
- *Euglypha* sp.	1.82	0.68	1.97	0.47
Makrofita	2.33	0.05	0.10	0.20
Insects	0.00	0.69	0.32	0.78
Insects larvae	0.00	0.02	0.00	0.00
Chironomus sp.	86.02	5.30	6.68	5.78
Annelids	0.00	0.80	0.00	0.03
Fish	2.43	84.44	80.45	76.41
Crustacean	0.00	0.12	0.00	0.00
Detritus	0.00	2.91	4.99	14.99
Unidentified	0.00	0.54	0.24	0.00

The Index of Preponderance shows that food changing is based on the length group of red devil. The younger red devil preferred to feed *Chironomus* sp. than other food whereas the adult red devil fed fish as the main food. In addition, the complementary food of adult red devil is detritus.

Table 3. Trophic level of red devil in the Sermo Reservoir.

Length Class (cm)	N (species)	Trophic Level	Status
< 9	4	3.36	Carnivore
9–11	132	3.82	Carnivore
12–14	16	3.74	Carnivore
> 14	59	3.63	Carnivore

Trophic level shows the fish level of food chain in the ecosystem. Red devil is a carnivore in the Sermo Reservoir (3.36–3.82). Both younger and adult red devil are carnivore. All of the red devil consumed the carnivore food type in the Sermo Reservoir.

Table 4. Niche breadth of red devil.

Length Class (cm)	N (species)	Male		Female	
		Niche Breadth	Standardization	Niche Breadth	Standardization
< 9	4	1.868	0.072	1.061	0.005
9–11	132	1.241	0.020	1.736	0.061
12–14	16	2.714	0.143	1.404	0.034
> 14	59	1.916	0.076	1705	0.059

Niche breadth shows the various type of foods which are consumed by red devil and it is an indicator of food habit. Male red devil has wider niche breadth than female. Niche breadth of male red devil ranges from 1.241 to 2.714 with Levins standardization ranges from 0.020 to 0.143. Male red devil consumed higher various food type on the 12–14 cm of length group than others. Niche breadth of female red devil ranges from 1.061 to 1.736 with Levins standardization ranges from 0.005 to 0.061. Female red devil consumed higher various food type on the > 9 cm of length group than others.

Table 5. Niche overlapping of male red devil.

Length Class (cm)	< 9	9–11	12–14	>14
< 9	1	0.133	0.101	0.108
9–11		1	0.868	0.942
12–14			1	0.960
> 14				1

Niche overlap shows the similarity of food type which is consumed between the male and female fishes or some of the length group. The niche overlap of male red devil is about 0.089–1. Male red devil on the > 9 cm of length group consumed the same food type of > 12 cm that cause competition. The younger red devil has no competition on the < 9 cm of length group.

Table 6. Niche overlapping of female red devil.

Length Class (cm)	< 9	9–11	12–14	>14
< 9	1*	0.112	0.069	0.006
9–11		1*	0.992	0.973
12–14			1*	0.963
> 14				1*

The niche overlap of female red devil is about 0.006–1. Adult female red devil consumed the same food type between > 9 cm and > 12 cm of length group. The younger female red devil has no competition, however there is a competition between > 9 cm and > 12 cm of length group. Both adult male and female red devil consumed the similar food type causing higher competition.

4. Discussion

The Index of Preponderance shows the variation of food habits and preference [10]. Fish and/or other part of fishes are the main food of red devil in the Sermo Reservoir (IP > 40 %; 82.79 %). Whereas *Chironomus* sp. is the complementary food (4% ≤ IP ≤ 40%; 6.92%). The additional food included phytoplankton, detritus, zooplankton, macrophyte, insects, insect's larvae, annelids, crustacea, and unidentified (IP < 4 %; 0.01–3.91 %).

As comparison, some closed related species have different food habit. *Amphilohus ocellatus* prefers to feed rotifer, cladocera, copepod, and fish, whereas *Amphilophus citrinellus* tends to feed cyanophyceae, bacillariophyceae, rotifer, cladocera, and fish [13]. Every species has relatively food preference that can be affected by several factors [10]. The abundance of natural food in waters is not necessarily used by fish. Factors that affected food consumption of fish are the spreading of food, food viability, fish preference, and physical factors by time and place. The number of food and food type are affected by age, time, and place.

The food preference of red devil shows a change, with the main food of younger red devil (< 9 cm) is *Chironomus* sp. and of adult red devil is fish. The changing food preference has been demonstrated by *Amphilophus citrinellus* in the Djuanda Reservoir. It exhibited a change in food preference along with length [4]. This change is affected by food viability and abundance in waters. The pattern of food habit can be affected by some factors like age, stage, time, and environmental factor that affect food natural viability [14]. Viability of food in waters is also affected by biotic and abiotic factors like temperature, light, space, and surface area [10].

Male red devil has wider niche breadth than female so that male red devil eats various food types than female. Niche breadth standardization of male and female red devil is relative narrow. Most of red devils consumed piece of fish than other food types so that influence the narrow niche breadth although the food types are diverse. An organism consumes various foods and when the consumed food is relatively same, it would affect the high niche breadth [15].

Red devil has different niche breadth based on the species and length size. Organism consumes a lot of food resources will increase the niche breadth although the viability of food resources is low [16]. Food niche breadth will change in every fish group, with small fish used narrow niche breadth [10].

Trophic level is the sequence of food utilization level or material and energy like food web [19]. Trophic level of red devil in the Sermo Reservoir is about 3.36–3.82. It shows that red devil is a carnivore. Both younger and adult red devil are carnivore in the trophic level status. Younger red devil mainly consumed *Chironomus* sp. and adult red devil consumed fish. Red devil can prey herbivore fish in the food web.

Niche breadth of red devil varies at each length size class. Fish abundance and capability of food utilized in waters isn't determined based on the fish length growth. Niche breadth of red devil in the Sermo Reservoir is generally about 1.061–2.714. Niche breadth of *Amphilophus citrinellus* in the Djuanda Reservoir is about 7.5472 [3]. Niche breadth difference can be affected by food viability, food abundance, and fish habitat. The number and food types consumed by a species usually based on age, habitat, and time [10].

Niche overlap is several food resources that utilized by some organisms [17]. Male red devil is overlap on the > 9 group cm and > 12 group cm. Male red devil in the length group that utilized similar food can cause prey or competition. Niche overlap between length group shows that similar food utilized by different group. Utilization of the same food will cause food predation. The greater niche overlap between length group causes greater predation between group and a uniform food type [18]. Niche overlap in the Sermo Reservoir generally shows a great chance on the food predation between class group. Small red devil (< 9 cm) has the lowest prey than adult (> 9 cm).

Niche overlap of red devil is about 0.0006–0.973. *Amphilophus citrinellus* in the Djuanda Reservoir has niche overlap of about 0.8825. *Amphilophus amphilopus* and *Amphilophus citrinellus* use the same food type. Same food type utilization will cause food prey between species or group. High value of niche overlap will cause a prey and low value of niche overlap will decrease prey between group [13].

Red devil in the Sermo Reservoir prefers to feed fish body than other foods. This condition makes certain fish population decreased. Fish composition in the Sermo Reservoir is dominated by red devil [1]. Sustainable handling needs to be done for balancing the fish population in the Sermo Reservoir. One of the stages is fish introduction that can suppress the population of red devil and achieve the balance of food chain in the Sermo Reservoir.

5. Conclusion

Food type of red devil in the Sermo Reservoir consists of fish, crustacea, detritus, phytoplankton, macrophyte, insects, insect's larva, *Chironomus* sp., and annelids. Small fish is the highest preference of adult red devil (> 9 cm) whereas *Chironomus* sp. is the highest preference of small red devil (< 9 cm). Red devil in the Sermo Reservoir is an euryphage that consumes several food types and is a carnivore. Red devil has changed the food type by increasing length group. High predation in a grabbing food occurs in the adult red devil.

Acknowledgement

We would like to thank Lembaga Penelitian dan Pengabdian Masyarakat (LPPM) Universitas Gadjah Mada for providing us a financial support and the Department of Fisheries Universitas Gadjah Mada for supporting this research.

References

[1] Djasmani S S, Rahmawati N and Sihwardoyo R W 2013 *Symp. on Komposisi Jenis Ikan dan Perkembangan Gonad Hasil Tangkapan Jaring Insang Ukuran Mata Jaring Berbeda di Waduk Sermo Kulon Progo* (Yogyakarta: Universitas Gadjah Mada)

[2] Setyobudi E 2013 *Changing Biodiversity Case of Red Devil* (Zheijang: University Hangzhou)

[3] Nikolsky G V 1980 *Theory of Fish Population Dynamics as The Biological Background for Rational Exploitation and Management of Fishery Resources* (New Delhi: Jayyed Press)

[4] Tampubolon P A R P 2013 *Karakter Ekobiologi Ikan Oskar (Amphilophus citrinellus Gunther 1864) Sebagai Dasar Pengendalian Ikan Asing di Waduk IR H Djuanda Jawa Barat* (Bogor: Sekolah Pascasarjana Institut Pertanian Bogor)

[5] Habibie S A and Djumanto 2014 *Penggunaan Otolith untuk Penentuan Umur dan Laju Pertumbuhan Yuwana Ikan Red Devil (Amphilophus labiatus) di Waduk Sermo* (Yogyakarta: Universitas Gadjah Mada)

[6] Wijaya A B and Djumanto 2014 *Hubungan Panjang-Berat dan Faktor Kondisi Ikan Red Devil (Amphilophus labiatus) di Waduk Sermo Kabupaten Kulon Progo* (Yogyakarta: Universitas Gadjah Mada)

[7] Klingenberg C P, Barluenga M and Meyer A 2003 *Journal of The Linnean Society* **80** pp 397–408

[8] Zacharia P U 2013 *Trophodynamics and Review of Methods for Stomach Content Analysis of*

Fishes (Kochi: Central Marine Fisheries Research Institute Kochi) pp 1–12
[9] Shirota A 1996 *The Plankton of South Viet-Nam* (Japan: Oyerseas Technical Cooperation Agency Japan) p 416
[10] Effendie M I 1979 *Metode Biologi Perikanan* (Bogor: Yayasan Dewi Sri) p 112
[11] Caddy J F and Sharp G D 1986 *An Ecological Framework for Marine Fishery Investigations* (USA: FAO Fish Tech. Pap) p 283
[12] Collwell R K and Douglas J F 1971 *Ecology* **52**(4) pp 567–576
[13] Nurnaningsih M F, Rahardjo and Sukimin S 2005 *Jurnal Iktiologi Indonesia* **4**(2) pp 61–65
[14] Pradini S, Rahardjo M F and Kaswadji R 2001 *Jurnal Iktiologi Indonesia* **1**(1) pp 41–45
[15] Krebs C J 1999 *Ecological Methodology* (New York: Harper Collins Publ. Inc.) p 654
[16] Anakotta A R F 2002 *Studi Kebiasaan Makanan Ikan-Ikan yang Tertangkap di Sekitar Ekosistem Mangrove Pantai Oesapa dan Oebelo Teluk Kupang Nusa Tenggara Timur* (Bogor: Institut Pertanian Bogor)
[17] Sinaga T P 1995 *Bioekologi Komunitas Ikan di Sungai Banjaran Kabupaten Banyumas Jawa Tengah.* Tesis (Bogor: Institut Pertanian Bogor)
[18] Sjafei D S, Affandi R and Fauziah R 2004 *Jurnal Iktiologi Indonesia* **4**(1) pp 15–23

Performance of rumpon-based tuna fishery in the Fishing Port of Sendangbiru, Malang, Indonesia

D G R Wiadnya[1]*, A Damora[2], M M Tamanyira[2], D Nugroho[3] and A Darmawan[4]

[1] Department of Fisheries, Faculty of Fisheries and Marine Science, University of Brawijaya, Jl. Veteran 65145 Malang, Indonesia
[2] Coral Triangle Program Directorate, WWF-Indonesia, Graha Simatupang, Tower 2 Unit C 7th Floor, Jl. Letjen T.B. Simatupang Kav 38 Jakarta 12540, Indonesia
[3] Research Center of Fisheries, Ministry of Marine Affairs and Fisheries, Balitbang KP II Building, 2nd Floor Jl. Pasir Putih II, Ancol Timur, Jakarta 14430, Indonesia
[4] Department of Aquatic Resource Management, Faculty of Fisheries and Marine Science, University of Brawijaya Malang, Indonesia
*Email : gwiadnya@gmail.com

Abstract. Catch records on FADs-based tuna fishery (hand-line) in the southern of East Java was conducted in fishing base of Sendangbiru, from February 2016 to May 2017. Total 45 rumpon (FADs) were found within space area of 64,081 square nautical miles, with average between FADs distant of 25 nautical miles. All FADs were located outside Fisheries Management Areas, except one FAD of code 31. Hand-line fishery was designed for a maximum fishing trip of 14 days. Total catch biomass mainly determined fishing trip. When catch exceeded 1,000 kg, fishermen tend to turn back although with < 10 days fishing trip. Yellowfin tuna and skipjacks were two main species that formed total catch with average catch biomass varied amongst 425–1,360 kg trip^{-1}. Apart from baby-tuna, yellowfin tuna and albacore within catch were at size class higher than its first maturity stage. However, length-class of skipjack indicated that it still in immature stage. Despite of FAD license, Ministry of Marine Affairs and Fisheries neither informed nor aware the position of these FADs.

1. Introduction

The development of fishing gears and methods may not always based on the new concept of science [1]. It is rather due to the ingenuity of fishermen to cope with different fishing conditions and to meet their socio-economic requirements. It was long time ago before, fishermen have realized the drifting materials as a potentially good material for trolling [2]. They anchored the floating materials that lately known in science as Fish Aggregating Device, FADs [3, 4]. In Indonesia, traditional bamboo rafts anchored close to the coast have been practiced since early 1900s to attract pelagic fishes [5]. The device was locally named as 'rumpon' or 'rumpong'. As fishing technology developed, more rumpon were deployed in deeper area using metal pontoon as floating device and with submerged coconut leaf frond. This technology was first introduced in the eastern part of Indonesia [6, 7]. Since early 1980s, many fishermen migrated to the southern coast of East Java, and settled in the East Java Province, and brought with them the deep-sea FAD's technology.

Floating materials such as FAD can attract both juveniles and adult fishes in variety of number and diversity [8]. This ecological trap may result in negative impact to fishery. FADs associated fish stocks will quickly overfished as they more vulnerable to fishing pressure [9, 10]. Also, FAD acts as a catalyst for cannibalism between large (> 90 cm length-frequncy (FL)) and small (< 30 cm FL) size of yellowfin tuna as both groups aggregated at FADs [3]. Currently, the number of rumpon in Indonesia has grown rapidly. As a precautionary approach in fisheries management, the Ministry of Marine Affairs and Fisheries of the Repubilc of Indonesia has regulated the placement of FADs (through Decree No.30/2004 and No.26/2014) within the Indonesia's marine jurisdiction and ZEE. The regulation stated that all of the FADs must be licensed and the distance between FAD should be minimum ten nautical mile (nmi), and do not result in hedge effect, such as zigzag placement. This study aims to assess the performance of FAD-based tuna fishery in the southern coast of East Java, operated by mainly migrant fishers from the South Sulawesi.

2. Materials and Methods

Catch species and biomass of FAD-based tuna fishery were monitored from February 2016 to May 2017. All the catches were landed at the fishing port of Sendangbiru, East Java (figure 1) and auctioned. So, the price and total revenue per species biomass were also recorded. Catch species was identified based on the general guideline mainly scombrid [11]. Weighing of the catch followed the auction system. Identified yellowfin (*Thunnus albacares*) and bigeye (*Thunnus obesus*) were measured individually, and albacore (*Thunnus alalunga*) per three individuals. The rest of the catches were weighed per group. On the willingness of fishermen to cooperate, position of each rumpon was recorded using GPS and all positions were digitized to calculate the total area of FADs within network (figure 1).

When a fishing boat went out for fishing, boat name and date were recorded. The date of the same boat came back to auction was also recorded. Total fishing trip for each fishing boat was calculated as total time (days) allocated by a boat from fishing auction to fishing ground, plus actual fishing time, and time to go back to fishing port of origin (Sendangbiru). The fishing boat dimension (GT, HP, and crews) was estimated based on sampling of 80 out of total 400 fishing boats. Analysis was conducted for total area of FADs within network, trip, catch species per month, and total revenue from catch auction. Length-frequency analysis was done for four dominant species in the catch, i.e. yellowfin, albacore, baby tuna, and skipjack.

3. Results and Discussion

3.1. Total area of FADs network

All 45 rumpon positions were mapped and formed a FADs network as shown in figure 1. Estimated total area within FAD network is 219.792 km^2. The closest distance of each FADs varied between 4 and 100 km, with an average distance of 46.8 ± 6.4 km or equal to 25.2 ± 4.4 nmi. Only three FADs are placed less than ten nmi one to the other. Each FAD, on average, occurred at around 1,800 nmi^2. All FADs are located outside the 12 nmi except one FAD (FAD 31) (figure 1). The license for these FADs should be given under central government in Jakarta. However, the Ministry of Marine Affairs and Fisheries was neither informed, nor aware the placement of these FADs (rumpon) in the southern coast of Java.

3.2. Fishing trip and catch species

The catch composed of at least eight different families, i.e. Scombridae, Istiophoridae, Xiphiidae, Coryphaenidae, Gempylidae, Lampridae, Carangidae, and Mobulidae. The most dominant species in the catch belong to family Scombridae, and auctioned together with family Istiophoridae and Xiphiidae. The first tuna group within scombrid consists of three species: yellowfin (*Thunnus albacares*), albacore (*Thunnus alalunga*), and big eye (*Thunnus obesus*). The smaller size of these species (< 50 cm FL) could not be separated properly and was categorized by buyers as baby-tuna. The second tuna group is swordfishes (Xiphiidae) and billfishes (Istiophoridae). The third group composed of one species called skipjack (*Katsuwonus pelamis*). The last group is little tuna, composed of mixed species, but mainly kawa-kawa (*Euthynnus affinis*), frigate tuna (*Auxis thazard thazard*), bullet tuna (*Auxis rochei rochei*), striped bonito (*Sarda orientalis*), and leaping bonito (*Cybiosarda elegans*). To some extents, the Indo-

Pacific mackerel, that composed of *Scomberomorus commerson*, *S. guttatus*, *S. lineatus*, and wahoo (*Acanthocybium solandri*) is identified in the catch. Occasionally, fishermen caught species of Carangidae, specifically rainbow runner (*Elagatis bipinnulata*) and *Lepidocybium* sp. (Gempylidae) and the species of family Mobulidae (Rays). All of these species are sold locally.

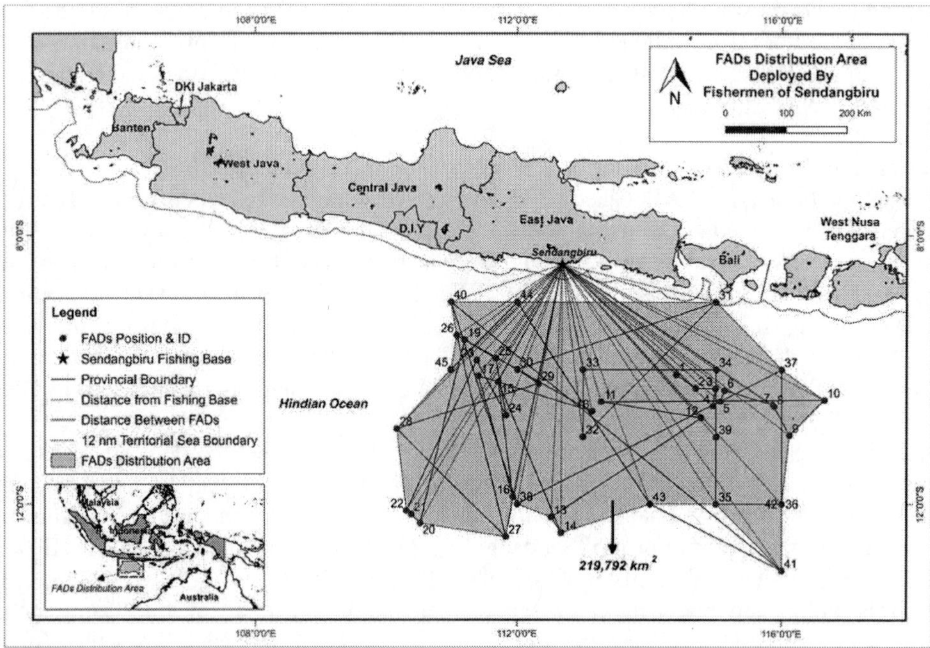

Figure 1. Location of 45 FADs deployed by fishermen from which polygon was created to calculate the total area (km^2) covered by rumpon.

Fishermen designed fishing trip for a maximum of 14 days. In reality, average fishing trip varied amongst fishermen and changed times (figure 2a). The lowest fishing trip is found in March and this definitely influenced the total catch-per-unit-of-effort (figure 2b). Fishermen tend to stop fishing when they reach a total catch biomass of > 1,000 kg (1 ton). On the other hand, when fishing trip is nearly 14 days, they would return home whatever the catch. Fishing capacity and limited time in fishing (trip) seemed to be the two main factors of the rumpon-based tuna fishery in Sendangbiru, East Java.

3.3. CPUE per species

Yellowfin tuna and skipjack are two main species that formed the total catch biomass (CPUE). Yellowfin is known as the strongest species that associated with floating materials such as FAD [3, 4, 9]. Bigeye tuna is caught occasionally, with N = 6 (present in the catch only six times out of 1,083 landings). The swimming depth of this species is likely deeper than fishermen hand-line could reach. Little tuna and Albacore are also found very rare in the catch (table 1). Little tuna species was more ecologically found in the coastal areas and formed a strong schooling compared to yellowfin and skipjack [11].

During February–April, catches are initially dominated by skipjack, little tuna, and baby-tuna respectively. Gradually, the yellowfin tuna replaced these smaller size fishes within catch (not shown in the table). Yellowfin tuna is known as a cannibal in the FAD area, apart from its normal behavior to follow floating materials [3]. The more the skipjacks and baby-tuna found in the FADs, it will give a strong sign of higher catch after the school of these smaller fish size.

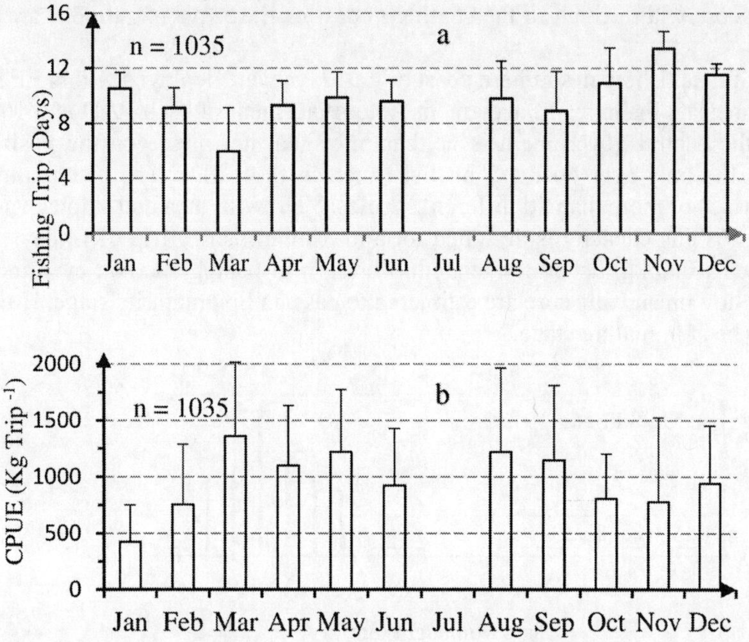

Figure 2. Avarage fishing trip, in days (a) and Catch-per-unit-of-effort, in kg·trip^{-1} (b) of rumpon-based tuna fishery (hand-line) in the southern coast of East Java.

Table 1. Catch-per-Unit-of-Effort, CPUE (kg·trip^{-1}) of rumpon-based tuna fishery in the Sendangbiru Fishing Port

Species	N	CPUE (kg·trip^{-1})		
		Means + SE	MIN	MAX
Baby tuna	894	221 ± 6^{b}	23	1.511
Skipjack tuna	831	277 ± 9^{a}	22	2.020
Yellowfin tuna	802	291 ± 11^{a}	13	1.857
Albacore	102	225 ± 21^{ab}	19	1.025
Bigeye tuna	6	117 ± 11^{b}	63	130
Little tuna	17	240 ± 69^{ab}	34	1.162

3.4. Length-class distribution

Length-class distribution from four species and group within catches is shown in figure 3. Yellowfin tuna (locally madidihang) is mainly caught at size of 140 cm FL, although it is found at lower size (figure 3). Female yellowfin tuna mainly reproduced at minimal size of 94.8 cm FL [12, 13, 14]. So, apart from 8 % of juveniles found in the catch (figure 3a), most of the adult yellowfin tuna in the catch are assumed to pass its first maturity and spawning cycle. Baby-tuna may consist of yellowfin species at length-class between 25 and 50 cm FL (figure 3c). All tuna species at this size category are immature and cannot be avoided from the catch.

Skipjacks found in the catch varied between 25 and 50 cm FL (figure 3d) with average length of 38.4 cm FL. Studies conducted around Philippines [14, 15] found that this female skipjack (*Katsuwonus pelamis*) usually matured at length size of 40–50 cm. So, most of the skipjacks within catch in the southern coast of Java would be immature considering the spawning size above. It is difficult to suggest the mature size of Albacore (*Thunnus alalunga*) around the area. Studies conducted around the Pacific up to South Africa showed that albacore usually matured at size between 85 and 95 cm FL [11, 15].

Length-classes of albacore within catch in the southern coast of East Java (figure 3b) are higher than first mature size.

Rumpon or FADs of tuna fishery in southern coast of East Java were deployed within EEZ, or outside the 12 nmi of the Indonesia's Fisheries Management Area. Placement of each FAD is > 10 nmi one to the other. However, the central government is neither informed, nor aware on the position of each rumpon. Hand-line is the only gear operated by fishermen in the FADs. The catch composed of 8 families, totally comprised of more than 20 different species. Yellowfin tuna and skipjack are two main species in the catches. Fishing capacity is designed for a maximum fishing trip of 14 days. When total catch biomass exceeded 1,000 kg, fishermen tended to bring it to fishing auction, even the trip is less than ten days. Most yellowfin and albacore are estimated to pass its first maturity stage. However, most skipjacks are found to be at immature stage.

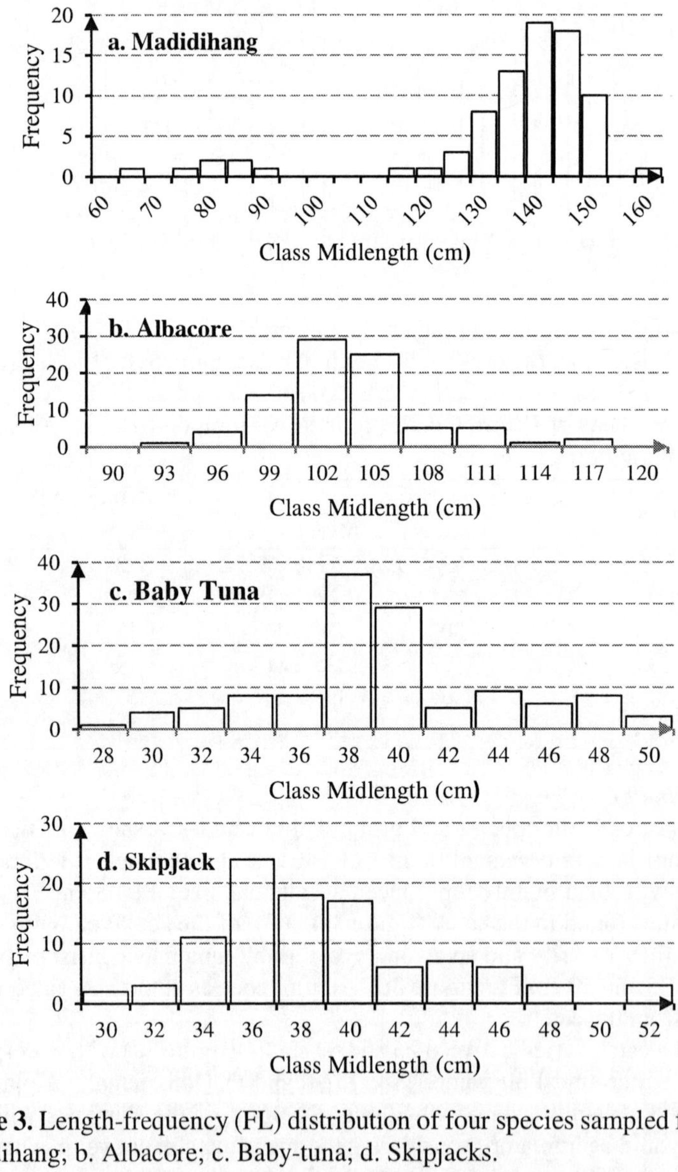

Figure 3. Length-frequency (FL) distribution of four species sampled from catches: a. Madidihang; b. Albacore; c. Baby-tuna; d. Skipjacks.

Acknowledgement
Authors appreciate and wish to thank to WWF-Indonesia who financially support this study. Also, we wish to thank to field enumerators: Ofan, Arief and Anam, for the data collection

References
[1] Kawamura G, Monintja D R and Mangunsukarto K 1983 *Indonesia Occasional Papers* **2** 9–40
[2] Gooding R M and Magnuson J J 1967 *Pacific Science* **21** 486–497
[3] Buckley T W and Miller B S 1994 *Bulletin of Marine Science* **55**(2–3) 445–459
[4] Girard C, Benhamou S and Dagorn L 2004 *Animal Behaviour* **67**(2) 319–326
[5] Bergstrom M 1983 *Review of experiences with and present knowledge about fish aggregating devices* (India: Development of Small-Scale Fisheries in Bay of Bengal Programme FAO Report) p 57
[6] Nasution C H, Merta G S and Arifuddin R 1986 *J. Mar. Fish. Res*. **37** 31–58
[7] Monintja D R 1993 *Mar. Technol. Bull.* **3**(3) 11–37
[8] Dempster T and Taquet M 2004 *Reviews in Fish Biology and Fisheries* **14**(1) 21–42
[9] Dagorn L, Holland K N and Itano D G 2007 *Marine Biology* **151**(2) 595–606
[10] Robert M, Dagorn L, Deneubourg J L, Itano D and Holland K 2012 *Mar. Biol.* **159** 907–914
[11] Carpenter K E and Niem V H 2001 *The living marine resources of the Western Central Pacific: FAO species identification guide for fishery purposes, v.6: Bony fishes part 4: (Labridae to Latimeriidae), estuarine crocodiles, sea turtles, sea snakes and marine mammals* (Rome: FAO) pp 3686–3758
[12] Wild A 1994 *FAO Fisheries Technical Paper* **336**(2) 439
[13] Mardlijah S and Patria M P 2012 *BAWAL*. **4**(1) 27–34
[14] Tandog-Edralin D D, Cortez-Zaragoza E C, Dalzell P, and Pauly D 1990 *Asian Mar. Biol.* **7** 15–29
[15] Froese R and Pauly D 2017 FishBase World Wide Web electronic publication www.fishbase.org (downloaded on July 4th, 2017)

The impact of ENSO on regional chlorophyll-a anomaly in the Arafura Sea

D M P R Dewi[1]*, D Fatmasari[1], A Kurniawan[1] and M A Munandar[2]

[1] Indonesian State College of Meteorology Climatology and Geophysics. Perhubungan Street I No. 5 Meteorology Complex, Pondok Betung, Bintaro 15221, Indonesia
[2] Indonesia Agency of Meteorology, Climatology and Geophysics. Angkasa I Street No.2, Kemayoran, Capital Jakarta, 10620 Indonesia

*Email: peraithadewi@gmail.com

Abstract. The El Niño-Southern Oscillation (ENSO) is a naturally occurring phenomenon that involves fluctuating ocean temperature in the equatorial Pacific. ENSO influences ocean climate variability in Indonesia including the Arafura Sea. The relationship between oceanic chlorophyll-a and ENSO has been the focus of study over the past decade. Here we examine the impact of ENSO on regional chlorophyll-a anomaly in the Papua waters using 14 years of chlorophyll-a and sea surface temperature (SST) data from AQUA MODIS and sea level anomaly data from AVISO. It is found that when El Niño events occur the negative SST anomaly in the Papua waters as well as the enhanced upwelling cause the increase of chlorophyll-a concentration. The highest chlorophyll-a concentration (> 1 mg·cm^{-3}) occured during El Niño and observed around the Aru archipelago. In contrast during La Niña event, the positive SST anomaly in Papua waters and the suppressed upwelling cause the decrease of chlorophyll-a concentration. Our results suggest that during El Niño (La Niña), the enhanced (suppressed) upwelling related to the significant decreasing (increasing) of sea level anomaly.

1. Introduction

Indonesia is one of the largest archipelagic countries in the world. Besides being in the tropics, it's also located between two oceans and two continents that makes Indonesia as a center of global atmospheric and ocean circulations [1]. The ENSO is a naturally occurring phenomenon that involves fluctuation of ocean temperature in the central and eastern equatorial pacific, coupled with changes in the atmosphere [2]. ENSO has a major influence on climate pattern in Indonesia and many studies have shown the social-economic impacts of ENSO on human. Marine phytoplankton plays a vital role in the global carbon cycle via photosynthetic carbon fixation and produces nearly half of the world's oxygen via photosynthesis [3]. As the base of marine food chain, marine phytoplankton is also widely used as an indicator for fishing ground. Estimating the amount of marine phytoplankton, the biology variable derived from satellite sensor, chlorophyll-a is used so the term of phytoplankton and chlorophyll-a is used interchangeably [4]. Since September 1997, the ocean color sensor Sea Viewing Wide Field of View Sensor (SeaWiFS) has provided an outstanding data set for a wide range of studies, involving large-scale oceanic biological productivity [4]. And the released of the Moderate Resolution Imaging Spectroradiometer (MODIS) satellite as the new generation of SeaWIFS in 2002 has given a better understanding of physical processes in the ocean. Studies on the relationship between oceanic chlorophyll-a and ENSO has received a lot of attention in Indonesia. Previous study has found that during E1 Niño episodes, the Java-Sumatra upwelling extends spatially and temporally [5]. However, the effect of ENSO on oceanic chlorophyll-a in the Arafura Sea remains lacking until today [6]. In the present study, we investigate the interannual variability of oceanic chlorophyll-a in the Arafura Sea and its relation to ENSO.

2. Materials and Methods

The study area is shown in figure 1. The area of black box (5 °S–8 °S; 136 °E–140 °E) is examined for spectral analysis of chlorophyll-a.

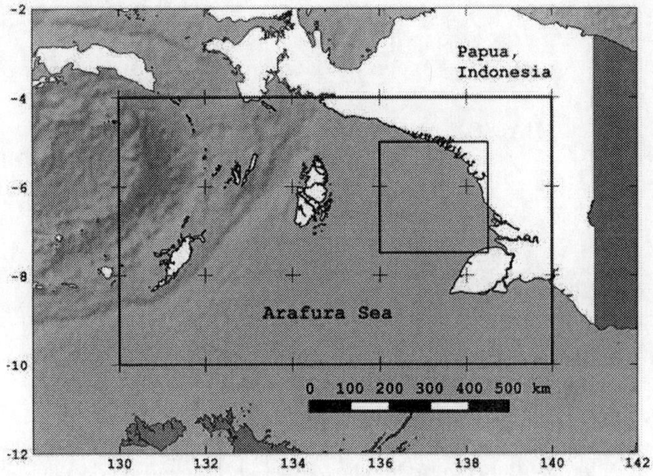

Figure 1. Map of the study area.

Monthly mean chlorophyll-a data used in this research is the Modis level 3 with spatial resolution of 4 x 4 km^2 for period of July 2002–December 2016. Chlorophyll-a data were estimated by applying the Ocean Chlorophyll MODIS algorithm (OC3M) and were obtain from The National Aeronautics and Space Administration (NASA) Ocean Color website (https://oceancolor.gsfc.nasa.gov/). Monthly mean SST data were extracted from the NOAA satellite imagery of with spatial resolution of 4x4 km^2. SST data were analyzed for period of July 2002–December 2016 and were obtain from The National Aeronautics and Space Administration (NASA) website (http://www.podaac.jpl.nasa.gov).

Sea Level Anomaly data used in this research is satellite altimetry data from AVISO. *(ftp://ftp.aviso.altimetry.fr/global/delayed-time/grids/climatology/monthlymean/)* The data were analyzed for period of July 2002–December 2016. Sea level anomaly was obtained from the following formula:

$$\mathrm{SLA} = \mathrm{SSH} - \mathrm{MSSH} \quad (1)$$

where,
SLA = Sea level anomaly
SSH = Sea surface height
MSSH = Mean of sea surface height

ONI indices used of the three months mean SST anomaly value for the Niño 3.4 region (5° N–5° S, 120°–170° W). Events are defined as five consecutive overlapping three-month periods at or above the +0.5° anomaly for warm (El Niño) events and at or below the -0.5 anomalies for cold (La Niña) events. ONI indices were obtained from the National Weather Service Climate Prediction Center *(http://www.cpc.ncep.noaa.gov/products/analysis_monitoring/ensostuff/)*. Based on this category, we grouped the months of ENSO each year (table 1). DJF refers to December, January and February, MAM refers to March, April, May, JJA refers to June, July, August, and SON refers to September, October and November.

Table 1. List of ENSO months used in this study.

Year	El Niño	La Niña
2002–2003	JJA, SON, DJF	-
2003–2004	-	-
2004–2005	JJA, SON, DJF, MAM	-
2005–2006	-	-
2006–2007	SON, DJF	-
2007–2008	-	SON, DJF, MAM
2008–2009	-	-
2009–2010	JJA, SON, DJF, MAM	-
2010–2011	-	JJA, SON, DJF, MAM
2011–2012	-	SON, DJF
2012–2013	-	-
2013–2014	-	-
2014–2015	MAM	-
2015–2016	JJA, SON, DJF, MAM	-

3. Results and Discussion

3.1. Spatial and temporal variability of chlorophyll-a in the Arafura Sea

In the Arafura Sea, the dominant factor that affects ocean condition is Australian-Indonesian Monsoon. Spectral analysis of chlorophyll-a (figure 2a) shows that chlorophyll-a concentration has a dominant pattern of annual oscillation. This also observed in the wavelength analysis (figure 2b) that temporally the annual oscillation looks dominant in the 12–month period. This indicates that monsoon has dominant impact on the distribution of chlorophyll-a.

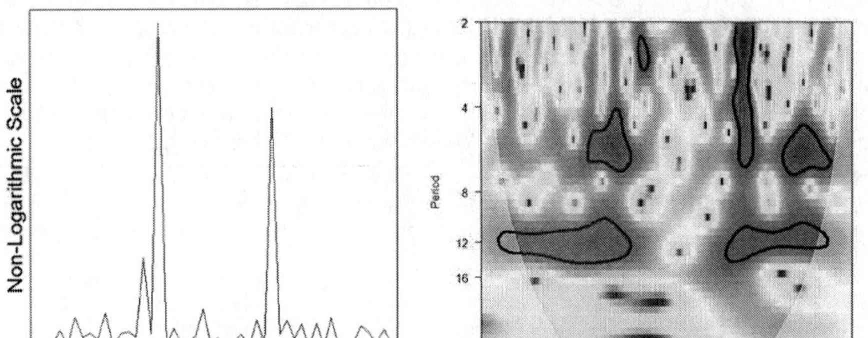

Figure 2. (a) Spectral analysis of chlorophyll-a. (b) wave lenght tranfomation of chlorophylla-a from July 2002 to December 2016.

During the northwest monsoon, northwesterly wind from Asia blows over the Arafura Sea with relatively low speeds ranging from five to eight knots [6]. The SST over the Arafura Sea ranges from

29–32 °C (figure not shown). This observation suggests that there is no occurrence of upwelling in this season. This condition is supported by chlorophyll-a distribution (figure 3a) that shows relatively low concentration with an exception in the coastal regions. The high concentration of chlorophyll-a in coastal region might be due to the process of surface runoff from the land as a natural source of nutrients [7].

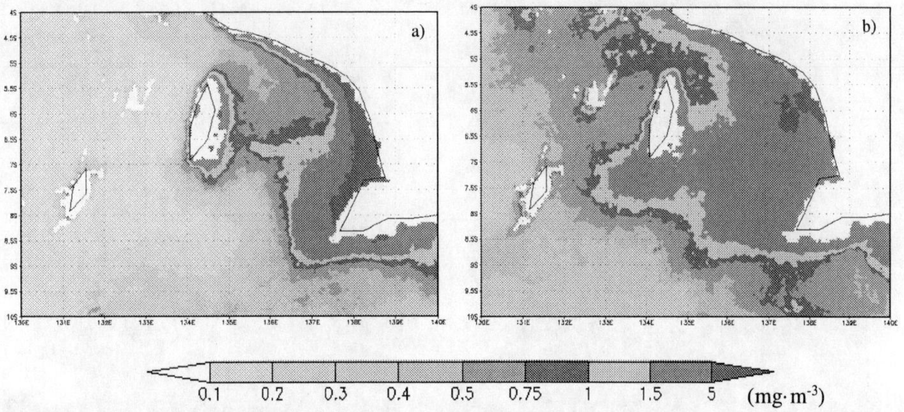

Figure 3. a) Average chlorophyll-a concentration during the northwest monsoon (DJF) and (b) during the southeast monsoon (JJA).

While during the southeast monsoon, southeasterly wind from Australia [6] generates upwelling in the Arafura Sea, Banda Sea, and Timor Sea. The wind speed during this period ranges from 13 to 20 knots. Upwelling brings cooler water from deeper layer to the surface and increases chlorophyll-a concentration in the region (figure 3b).

3.2. Chlorophyll-a anomaly during ENSO

ENSO not only affects climate variability and weather in Indonesia but also the variability of marine productivity [8]. In 2015, a strong El Niño occurred and lasted from June to November (figure 3). Our results show that this strong El Niño affected the chlorophyll-a concentration in the Arafura Sea. During JJA 2015 anomalous negative SST (figure 4c) appeared in the Arafura Sea, followed by positive anomaly of chlorophyll-a concentration (figure 4a). The positive anomaly of chlorophyll-a reached > 1 mg cm^{-3} and the negative anomaly of SST ranged from 0.1 to 0.6 °C. Whereas during SON 2015 the negative anomaly of SST became more significant with values < -0.8°C (figure 4d), indicating the mature phase of El Niño. During this season upwelling around the Aru archipelago demonstrates a decrease.

IOP Conf. Series: Earth and Environmental Science **139** (2018) 012020 doi:10.1088/1755-1315/139/1/012020

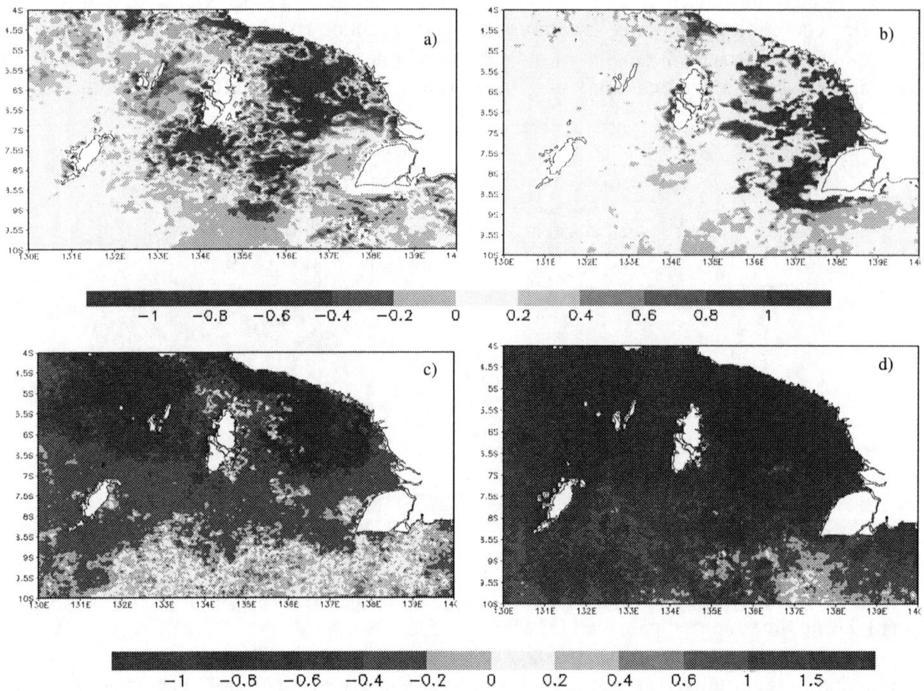

Figure 4. Chlorophyll-a anomaly during (a) JJA and (b) SON 2015. SST anomaly during (c) JJA and (d) SON 2015.

In contrast to El Niño condition, in particular when moderate La Niña occurred in 2010, positive anomaly of SST appeared and followed by negative anomaly of chlorophyll-a concentration in the northern part of the Arafura Sea. During JJA, the anomalous positive anomaly of SST (figure 5c) ranged from 0.2 to 1.5 °C. During this period the negative anomaly of chlorophyll-a concentration reached >1 mg cm^{-3} (figure 5a) This condition continued during SON with more significant positive SST in the Arafura Sea (figure 5d). The southern part of the Arafura Sea shows a positive anomaly of chlorophyll-a concentration. This might be generated by dynamics of the Arafura Sea like current and tide and it needs further analysis. Overall, the warmer SST during La Niña and low chlorophyll-a concentration indicated a suppressed upwelling in the Arafura Sea.

During El Niño, air pressure in the eastern Pacific is weaker compared to the western Pacific. As a consequence, wind and warmer water flows from the western Pacific to the eastern Pacific. This causes a decrease in SST and sea level in the western Pacific [9]. The decreased temperature in the western Pacific induces a SST decrease in the Indonesia Seas. In addition, the declined sea level in the western Pacific during El Niño might affect the intensity of upwelling in the Indonesian Seas. We hypothesize that this is the reason why the intensity of upwelling during El Niño is stronger than that during La Niña and neutral year [10].

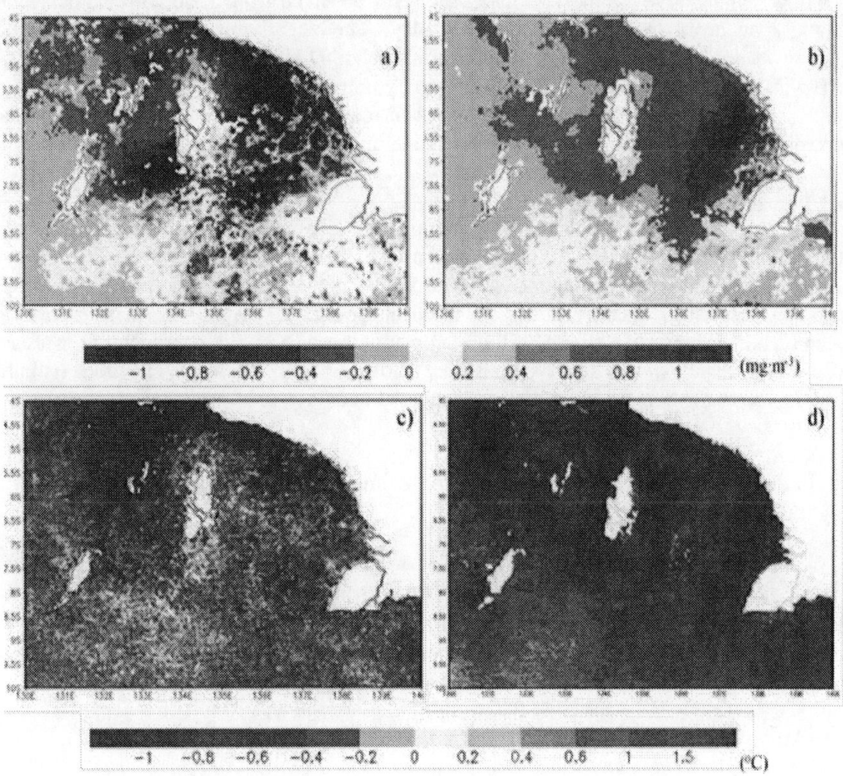

Figure 5. Chlorophyll-a anomaly during (a) JJA and (b) SON 2010. SST anomaly during (c) JJA and (d) SON 2010.

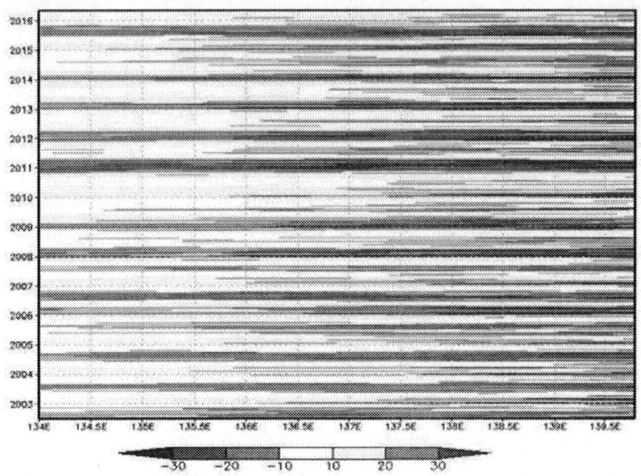

Figure 6. Hovmoler diagram of sea level anomaly in the Arafura Sea.

In the Arafura Sea negative anomaly of mean sea level occurred from June until October, whereas positive anomaly occurred from November until May. The negative sea surface height anomaly suggested the upwelling event and the positive sea surface height anomaly suggested the downwelling event [11]. The influence of El Niño on sea level anomaly in the Arafura Sea is obvious. During strong El Niño 2015, significant decreased of sea level occurred in the Arafura Sea (figure 6), indicating enhanced upwelling. This condition reversed during La Niña.

4. Conclusion

According to the results of this study, it can be concluded that the occurrence of chlorophyll-a anomaly in the Arafura Sea is strongly coupled to ENSO. When El Niño occurred, negative SST anomaly in the Papua waters is associated with the enhanced chlorophyll-a concentration. The highest chlorophyll-a concentration (> 1 $mg \cdot cm^{-3}$) occured during El Niño and observed around the Aru archipelago. In contrast, when La Niña occurred, positive SST anomaly and low chlorophyll-a concentration are evident in the Papua waters. Results of this study suggest that during El Niño (La Niña) enhanced (suppressed) upwelling is coupled to the decreased (increased) sea level anomaly.

Acknowledgment

We thank to Indonesia State Collage of Meteorology, Climatology, and Geophysics for supporting us with funding for this research.

References

[1] Aldrian E 2008 *Meteorologi Laut Indonesia* (BMKG: Jakarta)
[2] WMO 2014 *El Niño–Southern Oscillation* (Geneva: World Meteorology Organization) pp 2–8.
[3] Michael J B, Robert T O, David A S, Charles R M, Jorge L S, Gene C F, Allen J M, Paul G F, Ricardo M L and Emmanuel S B 2006 *Nature* **444** 75
[4] Martin S, Christine P and Alberto R P *J. Geophysical Research* **110** (C11016) 1
[5] Dwi S, Arnold L G and Quanan Z 2001 *J. Geophysical Research Letter* **20** 1599
[6] Dewa G A P 2016 *Pengaruh Angin, Arus Laut dan Suhu Permukaan Laut Terhadap Kosentrasi Krorofil-a di Perairan Indonesia Timur* [Indonesia] (Jakarta: Sekolah Tinggi Meteorologi Klimatologi dan Geofisika)
[7] Hildan I N B 2011 *Validasi Dan Pengembangan Algoritma Suhu Permukaan Laut Pathfinder Satelit Noaa–Avhrr Di Perairan Utara Papua* [Indonesia] (Bogor: Institut Pertanian Bogor)
[8] Ahmad F, Denny N S, Kumarso and Muhammand Z 2016 *J. Earth and Environmental Science* **55** 012021
[9] Bayong T 2003 *Geosain* (Bandung: Institut Teknologi Bandung) p 133
[10] Kunarso, Nining S N and Agus S Karakteristik 2005 *J. Ilmu Kelautan* **10** 17–13
[11] Umaroh, Sutrisno Anggoro, and Muslim 2017 *IOP Publishing* **55** (01204) 1–8

The growth and reproduction of Seren (*Diplocheilichtys pleurotaenia*) in the Jatigede Reservoir Sumedang Regency Province of West Java

T Herawati[1], A Yustiati[1], S Y Diliana[1] and Adhardiansyah[1]

[1] Faculty of Fisheries and Marine Science Universitas Padjadjaran
Jl. Raya Bandung-Sumedang km 21, Jatinangor, Jawa Barat, Indonesia
*Email: herawati.h19@gmail.com

Abstract. The research was conducted to find out the growth and productive pattern of Seren fish in Jatigede Sumedang Reservoir of West Java during the period of November 2016 to December 2016. The research used survey method with quantitative descriptive data analysis. Seren fish samples used in the study was 30 individuals collected in November 2016 and 41 individuals collected December 2016. The parameters observed namely were sex, gonads maturity level and index and fecundity. The results show that the average length of seren fish was 147 mm and the largest one was 273 mm. The fish growth patterns were allometrically negative. The condition factor ranged from 0.973 to 1.105. The ratio of nutritional fish seren was balanced between male and female fish. Gonad Maturity Index of male seren fish and female seren were relatively similar. The GMI male fish ranged between 0.285 to 11.055 % while the female GMI fish ranged from 1.23 to 11.76 %. The Seren fish of 225 mm has average fecundity of 10,032 grains, fish fitting 260 mm average 23,471 grains, there was a relationship between the addition of length size and fecundity.

1. Introduction

The Jatigede Reservoir has high fishery resources potential including indigenous fish Cimanuk River, introduction fish and restocking fish. The Sergeant Fish [1, 2, 3, 4, 5] is one of the fish found in the Cimanuk river basin and in Jatigede Reservoir. Seren is a fish consumption and it has a tasty meat and relatively less bones. So that the fish has high potential for aquaculture. It has also high potential for export commodity due to high abundant resource and big market opportunity.

The absence of management of fishery resources in the Jatigede Reservoir, especially on fishing activities conducted by communities would affect to fish community. Capture fishery activities conducted by the communities pay less attention to the sustainability of fish populations, so it is feared can cause to extinction of the indigenous fish in the Cimanuk River that has adapted to Jatigede Reservoir.

Fishery biological studies cover aspects of growth and reproduction as basic factor in the determination of fishery resources management policy both in capture fishery activities and conservation efforts of fish resources. This research aims to determine the growth and reproduction aspect of Seren fish in Jatigede Reservoir.

IOP Conf. Series: Earth and Environmental Science **139** (2018) 012021 doi:10.1088/1755-1315/139/1/012021

2. Materials and Methods

2.1. Materials

Fish samples were obtained from fishermen that catch the fish in Jatigede Reservoir around Sukamenak Village area, Darmaraja Sub-district from November to December 2016. The fish species (*Diplocheilichtys pleurotaenia*) was identified using some references book [1, 2, 3, 4, 5, 6, 7, 8, 9] and was to identified at LIPPI Cibinong, West Java Province. There were 30 individuals fish collected in November 2016 and 41 fish in December 2016. The fish was measured their total length and weight individually using millimeter block and electric balance, respectively, then egg diameter using micrometric microscope (1 μm accuracy). The fish was measured that Gonad Maturity index, fecundity using dissecting set, petri dish, dropper and measuring cup. The fish samples were obtained from Jatigede Reservoir in Sukamenak water area, Darmaraja sub District in November and Desember 2016. Research on growth and reproductive aspect analysis was conducted at Aquaculture laboratory Faculty of Fishery and Marine Science Universitas Padjadjaran Jatinangor. Fish identification at laboratory Indonesian Institute of Science, Cibinong Bogor Regency, Province of West Java.

2.2. Data analysis

2.2.1. Long-weight relations. The determination of fish growth pattern was done by calculating the relation of the weight length as described [9] as follows:

$$W = aL^b \quad (1)$$

where:
W = fish weight (g)
L = fish length (mm)
a, b = constants

The relationship between the length and weight was analysed using regression equation (R^2) to know the influence of each variable, and analysed correlation value (r) to know the relationship between variables.

2.2.2. Conditional factor. Analysis of condition factors using the metric system [9] with the formula:

$$K = 10^5 WL^{-3} \quad (2)$$

where:
K = condition factor
W = weight of fish (g)
L = length of fish (mm)

2.2.3. Sex ratio. According to [10], the sex ratio is calculated by comparing the number of male and female fish obtained as follows:

$$X = J : B \quad (3)$$

where:
X = sex ratio
J = number of male fish
B = number of female fish

2.2.4. Gonad maturity index (GMI). Calculation of Gonad Maturity Index/Gonado Somatic Index refers to method of fishery biology [9] with the following formula:

$$GMI = \frac{B_g}{B_t} \times 100\,\% \tag{4}$$

where:
GMI = gonad maturity index (%)
Bg = gonad weight (g)
Bt = Body weight of (g)

2.2.5. Fecundity. Fecundity of fish is determined by using gravimetric method by formula [9]:

$$F = \frac{B_g}{B_s} \times F_s \tag{5}$$

where:
F = total number of eggs (grains)
Fs = number of eggs on some gonads (grains)
Bg = weight of all gonads (g)
Bs = partial weight of gonads (g)

3. Results and Discussion

3.1. Weight distribution
Seren fish caught in the Jatigede Reservoir was vary in size, smallest size was 147 mm and weight was 64.90 g, while the largest size was 273 mm and weight was 255.32 g. The total length size catch in November 2016 was found mostly at interval between 147 and 171 mm (43 %) (figure 1a). In December 2016 the size majority was found at interval between 232 and 245 mm (37 %) (figure 1b).

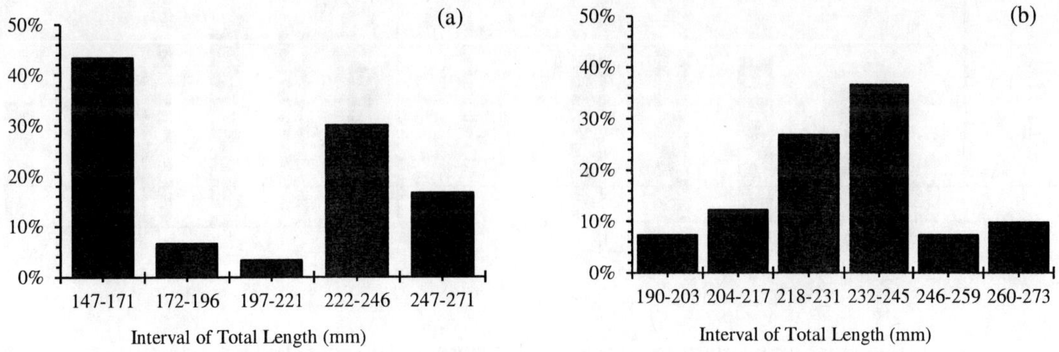

Figure 1. Weight distribution. (a) November 2016 and (b) December 2016.

Most Seren that caught in November are relatively smaller than Seren that caught in December. This indicates that Seren overall grow most in November and in December the growth are focused in reproduction organ.

3.2. Weight-length relation
The growth pattern of the Seren fish was allometric negative with the regression equation of length weight relationship in November 2016 i.e. log W = 1.6166 log $L^{-1.6481}$ (figure 2a), and in December that was log W = 2.7906 log $L^{-4.4422}$ (figure 2b). The correlation value on the long relationship of fish

Seren weight ranged from 0.9308 to 0.9547 which showed very strong relationship [11]. The regression value of the relationship of weight lengths ranged from 0.8663 to 0.9115, meaning 86.63 to 91.15 % growth length affects weight growth.

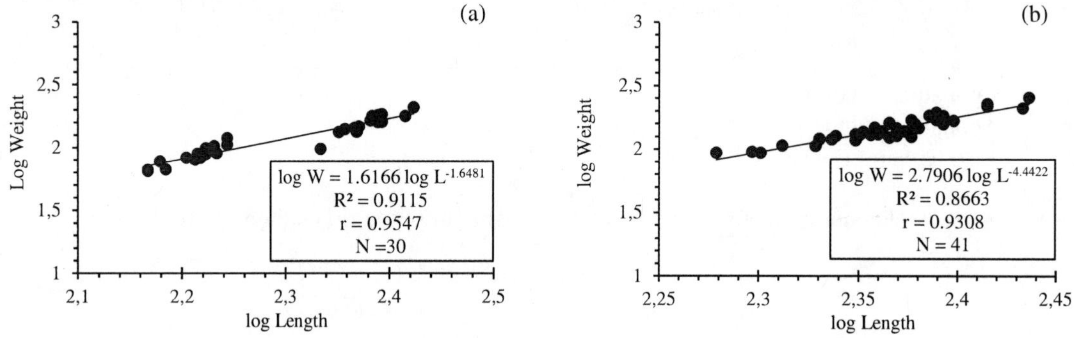

Figure 2. Regression of weight-length relation. (a) November 2016 and (b) December 2016.

3.3. *Condition factor*

Seren fish condition factor in November 2016 ranged from 0.973 to 1.088 (figure 3a), while in December 2016 ranged from 1.029 to 1.105 (figure 3b), the low level of condition factor in November was due to the small dominant fish caught in which the growth of seren fish in size between 147 to 171 mm. In November the maximum condition factor was reached at the size interval between 222 to 246 mm, (figure 3a) meanwhile in December was reached at interval size between 218 to 231 mm (figure 3b). Achieving the maximum condition factor at that size, because the fish's gonad is in mature condition (GML IV). The process of formation of reproductive cells peaked at GML IV [12], so that the condition factor will be high and will decrease after the completion fish spawn or the GML V.

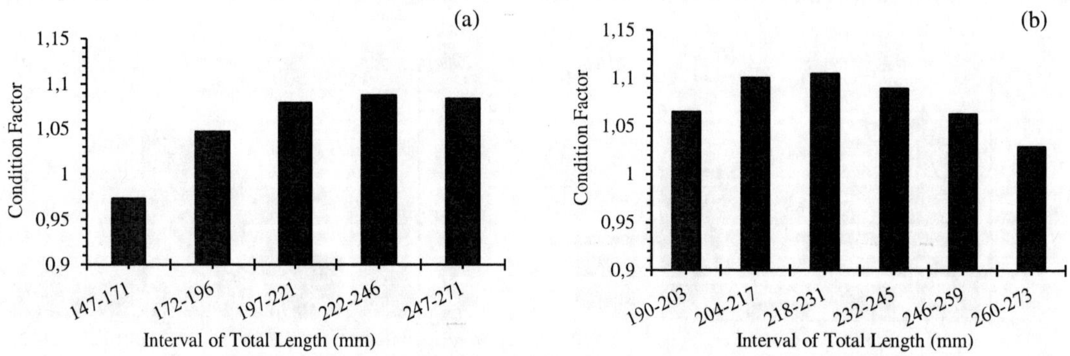

Figure 3. Seren condition factor. (a) November 2016 and (b) December 2016.

3.4. *Sex ratio*

The sex ratio of fish in November 2016 showed the male and female ratio was 47:53 % or 1:1, whereas in December 2016 the male and female ratio was 59:41 % or 3:2. The chi square test of sex ratio in November showed that χ^2 count = 0.36 < χ^2 table = 3.84 means the ratio between male fish and female fish was balanced (1:1) (figure 4a). The result of chi square test of sex ratio of fish in December showed that χ^2 count = 3.24 < χ^2 table = 3.84 means the ratio between male fish and female fish is balanced (1:1) (figure 4b).

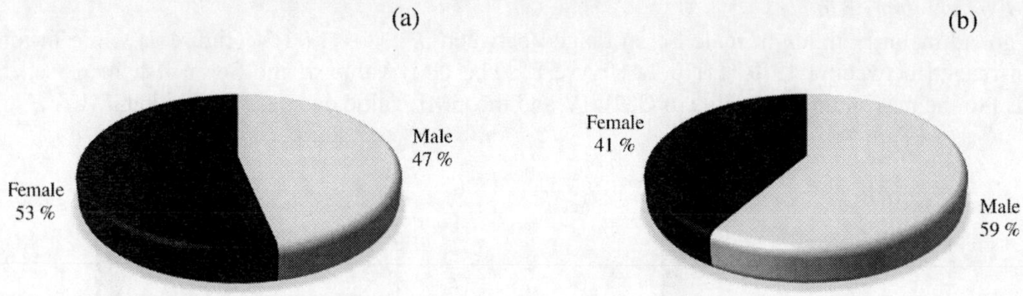

Figure 4. Seren sex ratio. (a) November 2016 and (b) December 2016.

3.5. *Gonad maturity level*

Male gonad maturity level of male Seren was generally at stage I and II, while in female Seren was at stage IV. In November 2016, male Seren was mostly in the condition of GML II of 8 fish, whereas in female Seren mostly at the condition of GML IV counted 10 fish (figure 5). In December 2016, male Seren is mostly in GML I condition as much as 18 fish, whereas in female seren most at GML IV condition are 13 fish (figure 6). It shows that female Seren fish have a chance to spawn every month.

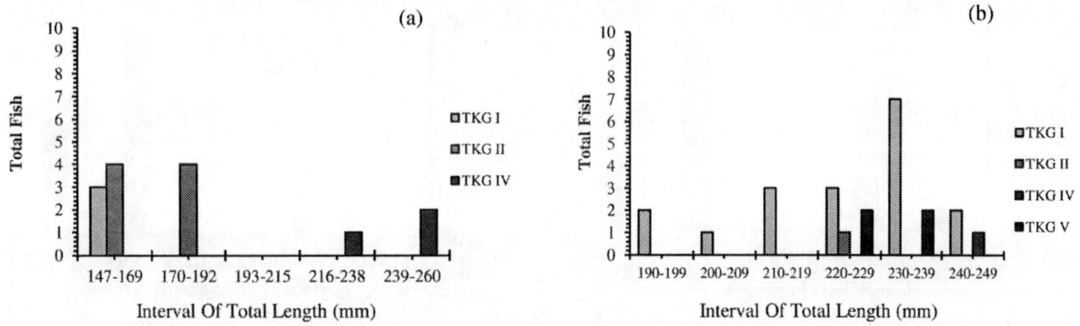

Figure 5. Male fishes gonad maturity level. (a) November 2016 and (b) December 2016.

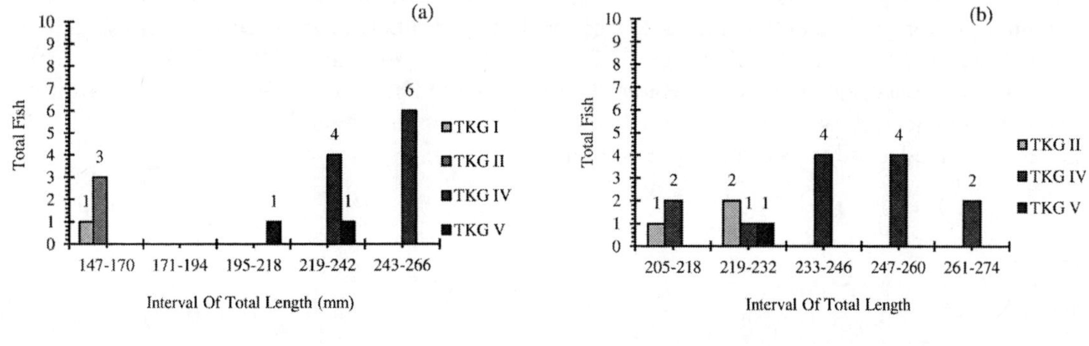

Figure 6. Female fishes gonad maturity level. (a) November 2016 and (b) December 2016.

3.6. Gonad maturity index

The gonad maturity index in male Seren ranged between 0.29 to 11.06 % (figure 7) while in female seren ranged between 1.19 to 11.76 % (figure 8). The GMI value of the Seren fish increased from GML I to the maximum IKG value in GML IV and the GML value decreased again in GML V.

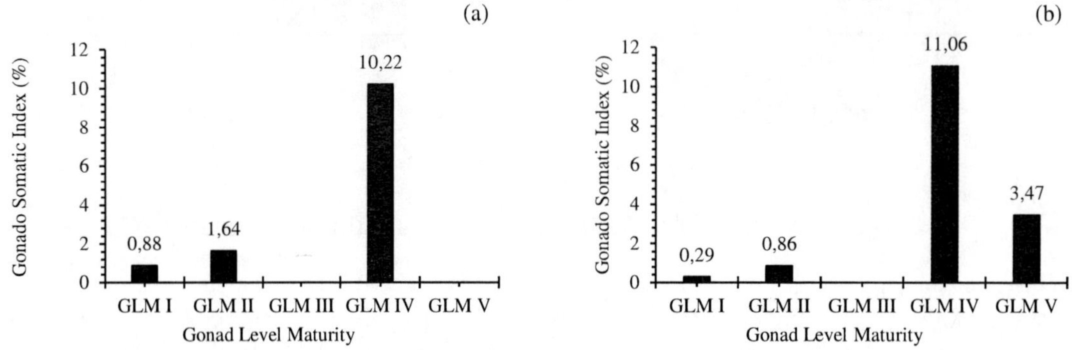

Figure 7. Male fishes gonad maturity index. (a) November 2016 and (b) December 2016.

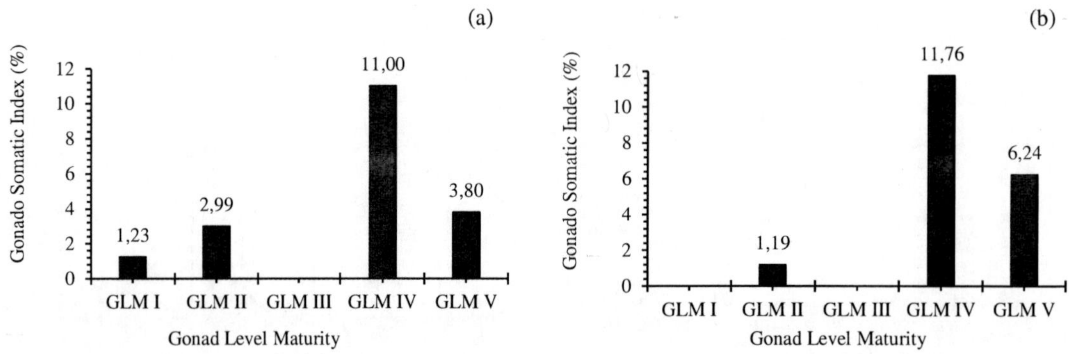

Figure 8. Female fishes gonad maturity index. (a) November 2016 and (b) December 2016.

3.7. Length and fecundity value relation

Seren fish with length 225 mm has an average fecundity value of 10,032 grains, Seren fish with length 260 mm has average fecundity value 23,471 grains. Length relationship with fecundity in seren fish in November 2016 following equation log F = 7,903 log $L^{-14.63}$ with value $R^2 = 0.5291$ means 52.91 % the length influence fecundity value (figure 9 (a)). In December 2016, the length relationship with fecundity equation follows the log F = 0.6858 log $L^{+2.6598}$ with the value $R^2 = 0.0197$ means 1.97 % of the length of the fecundity as can be seen in figure 9.

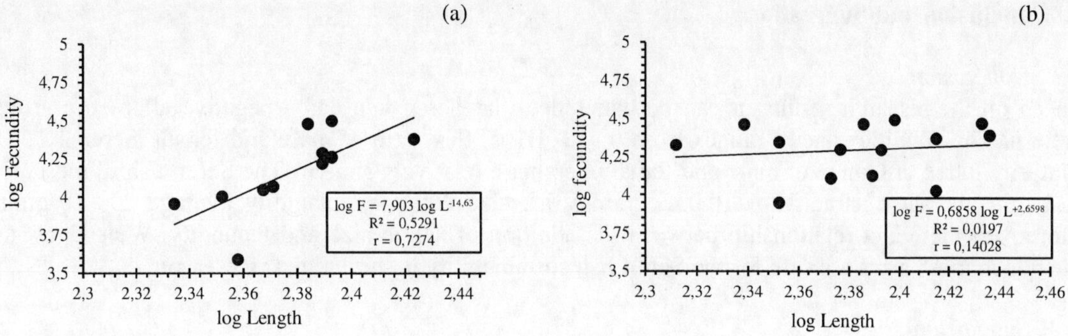

Figure 9. Length and fecundity value relation. (a) November 2016 and (b) December 2016.

3.8. *The first size of mature gonad*

Seren that caught in November and December are in GLM I to GLM V, there are no fish that length 147 to 203 mm that the gonad is already ripe. Seren fish that have been ripe gonad first and have spawn at least 1 times is on the size of 204 to 222 mm (figure 10).

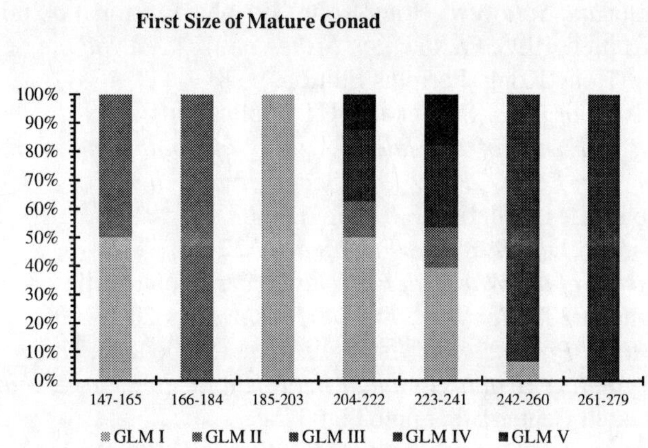

Figure 10. The first size of mature gonad.

Based on the size distribution, Seren that caught in November to December 2016 in Jatigede Reservoir experienced growth and development. In November the smallest fish caught a total length of 147 mm and was the most caught fish, whereas in December the smallest fish caught a total length of 190 mm and the most caught fish measured the total length between 232 to 245 mm as much 37 %.

The growth pattern is allometrically negative, meaning that the length increase is relatively faster than the weight gain, but there is a close correlation between the length increase and the weight, meaning that if there is a long increase followed by weight gain.

The fish condition factor in December was relatively better than in November. In December, the condition factor of the total length between 218 to 231 mm was 1.105 while in November the condition factor of the total length between 222 to 246 mm is 1.088. The occurrence of an increase in the condition of the condition closely related to the development of reproductive organs is the level of gonad maturity. In December there are many female fish that are ready to be mijah female fish with maturity level of gonad IV, these fish added weight is bigger than the length increase, morphology of fish look plump and mathematically can be seen from regression equation as follows, at month November ie $\log W = 1.6166 \log L^{-1.6481}$ and in December that is $\log W = 2.7906 \log L^{-4.4422}$.

IOP Conf. Series: Earth and Environmental Science **139** (2018) 012021 doi:10.1088/1755-1315/139/1/012021

4. Conclusion and Suggestion

4.1. Conclusion

Based on the research results, it can be concluded that Seren fish has a negative allometric growth pattern. The condition factor ranges from 0.973–1.105. Sex ratio of male and female Seren 1:1. The Maturity Index of Gonad of male and female Seren are relatively similar. The Seren fish of measuring 225 mm average fecundity of 10,032 grains, fish size 260 mm fecundity average 23,471 grains. Moreover, there is a relationship between the addition of length size and fecundity. We suggest that the 5 inch mesh sizes are safe for the Seren fish sustainability in the Jatigede Reservoir.

4.2. Suggestion

It is necessary to conduct long-term research in order to know the exact growth pattern and peak spawning.

References

[1] Karnasuta J 1993 *Systematic Revision of Southeastern Asiatic Cyprinid Fish Genus Osteochilus with Description of Two New Species and a New Subspecies* (Thailand: Kasetsart University Fishery Research Bulletin) p 105

[2] Kottelat M, A J Whitten, S N Kartikasari and S Wirjoatmodjo 1993 *Freshwater Fishes of Western* Indonesia *and Sulawesi* (Hong Kong: Periplus Editions) p 260

[3] Kottelat M, A J Whitten 1996 *Freshwater Fishes of Western Indonesia & Sulawesi: Additions and Corrections* (Hong Kong: Periplus Editions) p 8

[4] Kottelat M 2001 *Fishes of Laos* (Srilanka: WHT Publications) p 198

[5] Kottelat M 2013 *The Fishes of the Inland Waters of Southeast Asia: a Catalogue and Core Bibiography of the Fishes Know to Occur in Freshwaters, Mangroves and Estuaries* (Singapore: The Raffles Bulletin of Zoology) p 663

[6] Hadiaty R K and Siebert D J 1998 *Rev. Fr. Aquariol* **22** 1–4

[7] Nelson J S 2006 *Fishes of the Word 4th Ed.* [New Jersey: John Willey & Sons, Inc.] p 601

[8] Hui T H and M Kottelat 2009 *Ichthyol. Explor. Freshwaters* **20** 13-69

[9] Herawati T 2017 *Modul Praktikum Biologi Perikanan* (Sumedang: Unpad Press)

[10] Haryani G S 1998 *Analisa Histologi Gonad Ikan-ikan di Perairan Danau Semayang* (Cibinong: Limnology Research Center LIPI) pp 632–637

[11] Sugiyono 2012 *Metode Penelitian Kuantitatif Kualitatif dan R&D* (Bandung: Alfabeta) p 380

[12] Mote N 2014 *Biologi Reproduksi Ikan Brek (Barbonymus balleroides Cuvier & Val. 1842) di Sungai Serayu Kabupaten Banjarnegara Provinsi Jawa Tengah* Tesis (Bogor: Bogor Agricultural Institute)

IOP Conf. Series: Earth and Environmental Science **139** (2018) 012022 doi:10.1088/1755-1315/139/1/012022

Correlation analysis of Standardized Precipitation Index (SPI) for the water debit and level of the Cisadane River during El Niño and La Niña years

A N Khoir*, M Rohmah and Nuryadi

Indonesian Collage State of Meteorology Climatology and Geophysics
*Email: khoirneko@gmail.com

Abstract. Hydrometeorological factor causes most disaster in Indonesia, and two of them are drought and flood. This study aims to correlate Standardized Precipitation Index (SPI) 3-monthly to water debit and water level in the Cisadane River. The monthly rainfall data from Serpong and Pasar Baru rain station from 2009 to 2011 when moderate El Niño and moderate La Niña happened. The correlation analysis between debit and water level to SPI 3-monthly used rain post of Serpong to represent the condition of the upstream area and rain post of Pasar Baru to represent the condition of the downstream area. The results showed that during La Niña year, the rainfall on the upstream area of the Cisadane River influenced the increase and the decrease in water debit and water level. Meanwhile, the rainfall on the downstream area of the river has an opposite effect on the increase and the decrease of debit and water level of the Pasar Baru. On the upstream area, the correlation between rainfall and water debit is 0.8, and the correlation between rainfall and water level is also 0.8. During El Niño year, the correlation was less than 0.5.

1. Introduction

The public's need for food and other activities is highly dependent on the availability of water. One of the water sources is rain. Rainfall is water grain or ice crystal that falls out from clouds or cloud groups. If the water can reach the earth surface, it is called rain [1]. If the water does not fall to the earth surface, it is called virga [2]. In most cases, the amount of rainfall decreases due to the long dry season that affects the supply of clean water and it will make an imbalance between the needs and availability of clean water. El Niño and La Niña are one of the global climate phenomena that have been proven in several studies can give influence of the rise and fall of the number and intensity of rainfall.

The El Niño and La Niña events occurred when the Pacific Ocean and atmosphere above it changed from their usual condition for some seasons. The central and eastern tropical Pacific warming is associated with the El Niño events. While La Niña events are the opposite, associated with a sustained cooling of the same areas. El Niño which is called hot episode and La Niña which is called cold episode is located between Nino 3 (5 °N–5 °S, 150 °E–90 °W) and Nino 4 (5 °N–5 °S, 160 °E–150 °W) and Nino 3.4 (5 °N–5 °S, 180 °E–120 °W) [3]. The impact of El Niño and La Niña to Indonesia is different in each region [4]. In the case of East Java, seasonal variability of rainfall was dominantly influenced by El Niño in 1997 [5].

Meteorological drought is an early sign of drought. The method that can be used in predicting and evaluating the meteorological drought is the Standardized Precipitation Index (SPI), developed by McKee et al. (1993) [6]. The SPI is the index used to determine the deviation of rainfall to its normal in a long period. The SPI values are calculated using the probabilistic statistical method of the gamma distribution. El Niño in 2009/2010 and La Niña in 2010/2011 are categorized as moderate scale due to the Oceanic Niño Index (ONI) was already 0.5 in July 2009 and 0.8 in July 2010. This study aims to determine the SPI values on the upstream and downstream of the Cisadane River during the 2009/2010 El Niño and 2010/2011 La Niña. Also, this study also aims to determine its effect on the water debit and water level of the Cisadane River

2. Materials and Methods

2.1. Data and location

The data used in this study were obtained from two data sources, namely:

- Monthly rainfall data of the Pasar Baru and Serpong rain station from July 2009 to April 2011.
- Data of debit and water level of the Cisadane River from July 2009 to April 2011.

The location of this study is in the Tangerang district (106° 20'–106° 43' E and 6° 00'–6° 20' S). Debit data of Pasar Baru and Serpong from 2009 to 2011 were obtained from the Cisadane Major River Basin Organization. Data of rainfall and debit used the same time series to provide results that can represent the characteristics of the region.

Figure 1. Map of Cisadane River flow [9].

2.2. Analysis method

2.2.1. 3-Monthly SPI. In this research, three-monthly SPI is used to give the comparison of rainfall for three months period with total rainfall from three months of the same period for all year of historical data (database). Three-monthly SPI can be used to identify short and medium drought. The three-monthly SPI is used to investigate how significant the influence of rain that occurred two months before the drought. So the influence of rainfall that occurs still have an impact to identify hydrological drought. The SPI index value is obtained from the following equation:

$$G(x) = \int_0^x g(x)dx = \frac{1}{\hat{\beta}^{\hat{\alpha}}\Gamma(\hat{\alpha})}\int_0^x x^{\hat{\alpha}-1}e^{-x/\hat{\beta}}dx \quad (1)$$

where, $\alpha > 0$ is form parameter, $\beta > 0$ is scale parameter, $x > 0$ is total precipitation, and $\Gamma(\alpha)$ is Gamma function. An estimation of the coefficient values of α and β for each rain is defined by the following equation.

$$\hat{\alpha} = \frac{1}{4A}\left(1 + \sqrt{1 + \frac{4A}{3}}\right) \tag{2}$$

$$\hat{\beta} = \frac{\hat{x}}{\hat{\alpha}} \tag{3}$$

where,

$$A = \ln(\hat{x}) - \frac{\sum \ln(x)}{n} \tag{4}$$

The gamma function is not defined as x = 0, and since the rainfall distribution can be zero, the value of g(x) becomes:

$$H(x) = q + (1 - q)G(x) \tag{5}$$

where, q is no-precipitation probability (number of zero rain events) and H(x) is cumulative probability of precipitation. The gamma distribution [G(x)], then turns to normal standard with zero average and standard deviation of 1, so we get the SPI index using the formula below:

$$Z = SPI = -t - \left(\frac{C_0+C_1t+C_2t^2}{1+d_1t+d_2t^2+d_3t^2}\right), \quad \text{for } 0 < H(x) \le 0.5 \tag{6}$$

$$Z = SPI = +t - \left(\frac{C_0+C_1t+C_2t^2}{1+d_1t+d_2t^2+d_3t^2}\right), \quad \text{for } 0.5 < H(x) < 1.0 \tag{7}$$

where,

$$t = \sqrt{\ln\left(\frac{1}{(H(x))^2}\right)} \quad , for\ 0 < H(x) \le 0.5 \tag{8}$$

$$t = \sqrt{\ln\left(\frac{1}{(1.0-H(x))^2}\right)} \quad For\ 0.5 < H(x) < 1.0 \tag{9}$$

where x is rainfall (mm) and H(x) is the cumulative probability of observed rainfall, and $C_0, C_1, C_2, d_0, d_1, d_2$ are constants with the following values:

$C_0 = 2{,}515517$, $C_1 = 0.802853$, $C_2 = 0.010328$
$d_0 = 1.432788$, $d_1 = 0.189269$, $d_2 = 0.001308$

2.2.2. Pearson correlation method. Pearson correlation is used to find out the correlation between two variables and to determine the strength of linear relationship between two or more variables [8]. If a variable has the same pattern with another variable, then the two variables have a high correlation. For example, a variable has increased value and on other variables also occur the same thing than the two variables have a positive correlation. However, if there is an increase in the value of a variable but other variables have decreased the value of both variables have a negative correlation. If there is no change in a variable, although other variables change, then the two variables, have no relation (uncorrelated). The correlation equation is as follows:

$$r = \frac{n\sum XY - \sum X \sum Y}{\sqrt{n\sum X^2 - (\sum X)^2}\sqrt{n\sum Y^2 (\sum Y)^2}} \quad (10)$$

where, r is a correlation between variables X and Y, n is numbers of data, X is the independent variable, Y is the dependent variable. The dimension used to measure the degree of linear correlation (correlation) is called the coefficient correlation which is expressed by the notation "r". The correlation value ranges from -1 to +1. The value of correlation close to 1 means that the relationship between the two variables is greater and on the contrary, closer to the number 0 the correlation is smaller.

3. Results and Discussion

3.1. Correlation between SPI and debit and water level during El Niño

Figure 2 shows that the debit of water at Pasar Baru initially around 60 m^3/s. It then decreased until September 2009 and gradually rose from October 2009. Similarly, the initial SPI value is around -0.4 and then it rose until reached the peak with a value of 1.6 November 2009, and dropped significantly in December. Figure 2 showed that the pattern of SPI and debit does not always in line. The correlation between SPI of Pasar Baru and debit is 0.4. It indicates that SPI and water debit have a weak correlation. If the index shows dry during El Niño, the water debit is high.

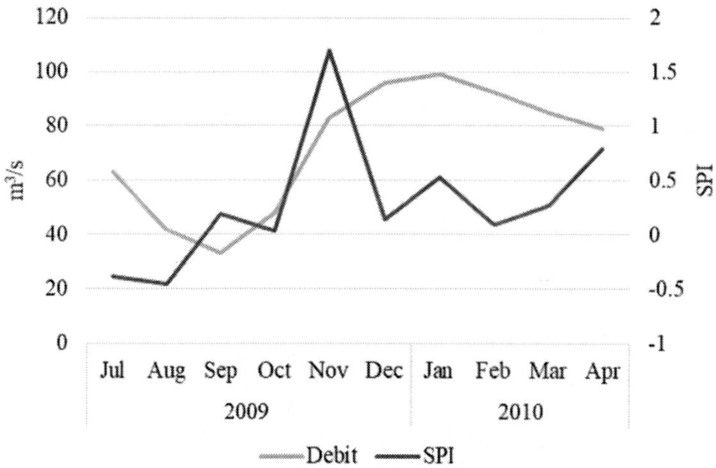

Figure 2. Correlation between the SPI Pasar Baru and three-monthly average debit during El Niño. The black line shows debit value (m^3/s) and grey line denotes the SPI value.

Figure 3 shows that during El Niño the initial debit is 60 m^3/s. It then dropped until September 2009 and gradually rose until reached the peak in January 2009. The variability of SPI is in line with the debit. The initial value of the SPI is around 1 then decreased until September 2009 and gradually rose until December 2009. It then dropped significantly until the end of the El Niño period. The correlation value between SPI of Serpong and water debit is 0.5.

Figure 4 shows that the SPI of Pasar Baru is -0.5. It increased gradually, reached 1.7m in November 2009 and then dropped significantly. The water level was around 1.1m, and it steadily increased, reached 1.5m in February 2010. It then decreased until the end of El Niño period. Both figures initially have the same pattern, but starting in November 2009 they do not show a good agreement. The correlation value between SPI of Pasar Baru and the water level was 0.3. It shows that the SPI in Pasar Baru does not represent water level condition during El Niño year.

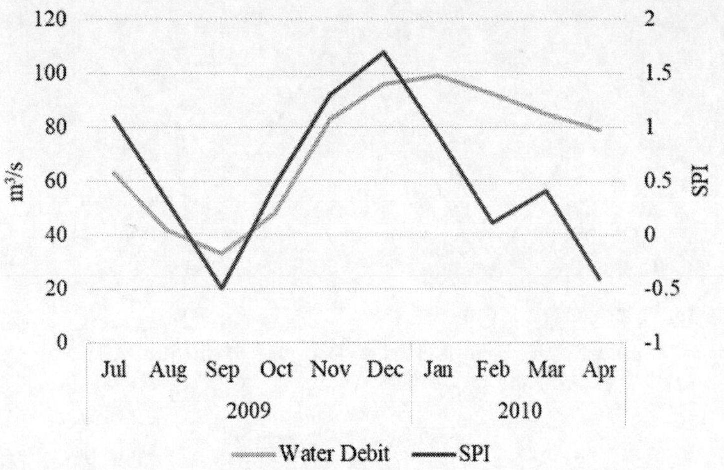

Figure 3. Correlation between SPI Serpong and three monthly average debit during El Niño. The black line shows debit value (m^3/s) and grey line shows SPI value.

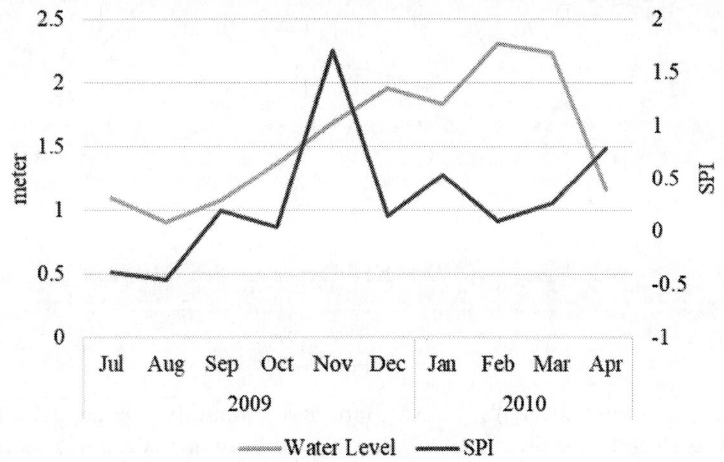

Figure 4. Correlation between SPI Pasar Baru and water level during El Niño. The black line shows SPI value and the grey line shows water level (m).

Figure 5 shows that the value of SPI Serpong was 1.2 and it decreased until September 2009 then rose until December 2009 and dropped constantly. The water level was around 1 m, and it decreased until August 2009 then rose until March 2010 and decreased again. Figure 4 shows that when SPI was dry, the water level was not always following this condition. The correlation value between SPI Serpong and the water level is 0.3.

3.2. Correlation between SPI and debit and water level during La Niña

Figure 6 shows that the water debit at Pasar Baru had the initial value of 60 m3/s. It increased until November 2010 and gradually dropped until the end of the La Niña period. In contrast, SPI which started at around 0.3 decreased until September 2010 and rose from October 2010 until April 2011. It was likely that the temporal variability of water debit in Pasar Baru does not follow the SPI pattern. The correlation value between SPI of Pasar Baru and water debit during La Ni ña was -0.5.

IOP Conf. Series: Earth and Environmental Science **139** (2018) 012022 doi:10.1088/1755-1315/139/1/012022

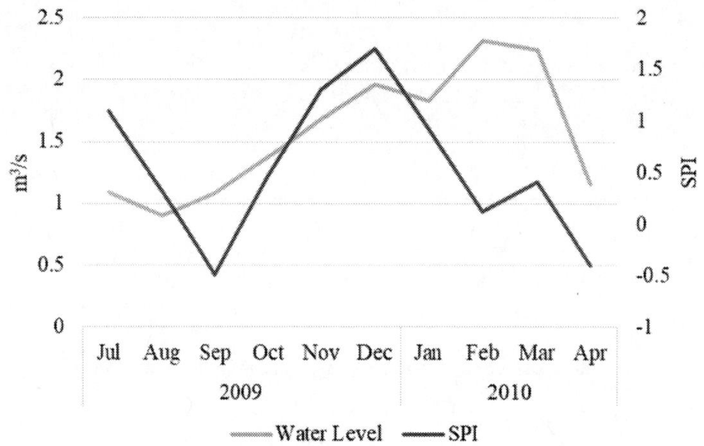

Figure 5. Correlation between SPI Serpong and water level during El Niño. The black line shows SPI value and the grey line shows water level (m).

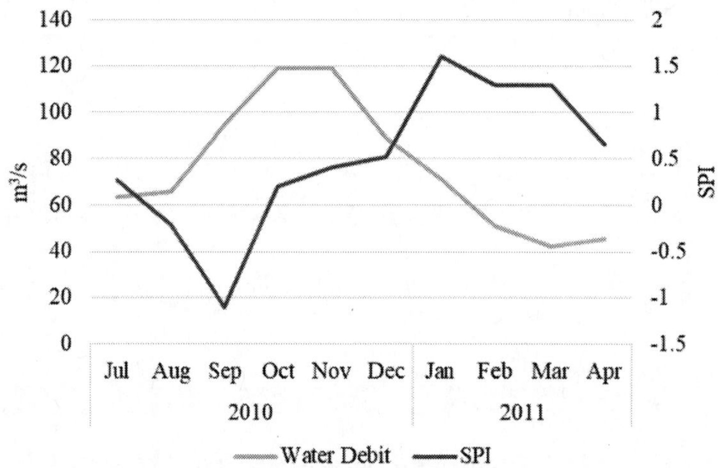

Figure 6. Correlation between SPI Pasar Baru and 3 monthly average debit (m^3/s) during La Niña. The black line shows the SPI value and the gray line is water debit (m^3/s).

Figure 7 shows that during the 2010/2011 La Niña the initial water debit was 60 m^3/s. It then increased until November 2010 and dropped gradually until the end of La Niña. The SPI had a similar pattern for water debit. The initial value of SPI was around 2. It dropped until the end of La Niña. Started in November 2010, the SPI dropped significantly. A good agreement between SPI and water debit suggested that when the SPI value increased, the debit would increased as well. When the SPI value decreased, the water debit dropped. The correlation value of both variables was 0.8.

Figure 8 shows that the SPI of Pasar Baru, had the initial value of 0.3. It decreased until September 2010 and then increased gradually until the end of the La Niña period and reached the peak at 1.5 in January 2011. As for the water level which had initial value of 1.5 meters increased until September 2010 then dropped gradually until the end of the La Niña period. Both graphs had a converse pattern. The correlation value of both variables was -0.9.

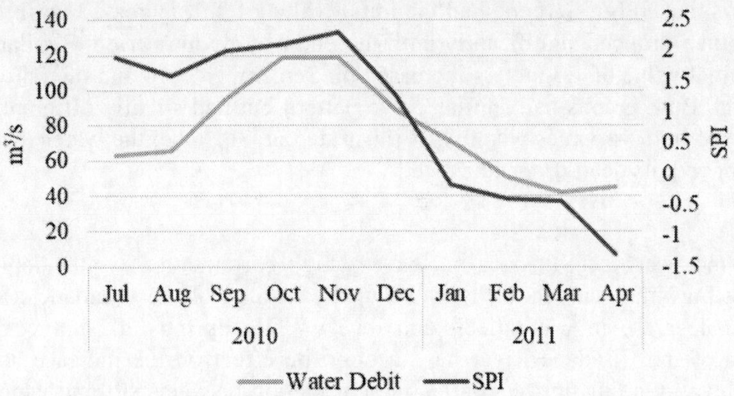

Figure 7. Correlation between SPI Serpong and 3 monthly average debit (m^3/s) during La Niña. The black line shows the SPI value and the gray line is water debit (m^3/s).

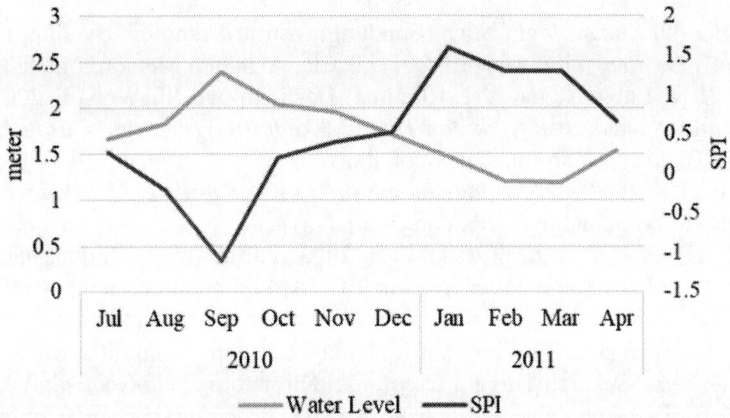

Figure 8. Correlation of SPI Pasar Baru and water level at the time of La Niña. The black line shows SPI value and the grey line shows water level in meter.

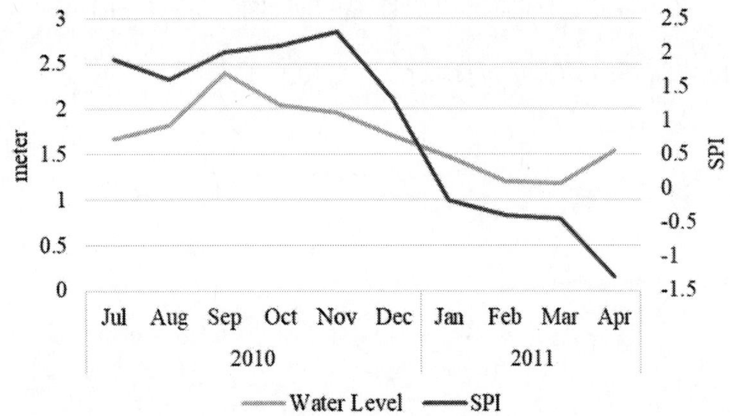

Figure 9. Correlation of SPI Serpong and water level at the time of La Ni ña . The black line shows SPI value and the grey line shows water level in meter.

Figure 9 shows that SPI of Serpong had the initial value of 2. It increased until reached the peak in November 2010 then dropped significantly until the end of La Niña period. Similarly, for water level which had the initial value of 1.7 meters increased on September 2010 and then dropped until the end of La Niña period. Both graphs had similar curve pattern but had slightly different of steepness. The correlation value of both variables was 0.8. It illustrated as SPI rose, the water level rose as well, so did when SPI dropped, its debit dropped as well.

4. Conclusion

During the El Niño year, the SPI does not have a high correlation to water debit and water level. However, during La Niña year, the SPI on the upstream area of the Cisadane River influences the increase and the decrease in water debit and level. While during La Niña year, the SPI on the downstream area of the Cisadane River has an opposite effect on the increase and the decrease of water debit and level. On this upstream area, the correlation between SPI and water debit is 0.8, and the correlation between SPI and water level is 0.8.

References

[1] Tjasyono B 1999 *Klimatologi Umum* (Bandung: Institut Teknologi Bandung)
[2] Soepangkat 1994 *Pengantar Meteorologi* (Jakarta: Akademi Meteorologi dan Geofisika Jakarta)
[3] Bayong T H K, Lubis A, Juaeni I, Ruminta, Harijono and Sri Woro B 2008 *Dampak Variasi Temperatur Samudera Pasifik dan Hindia Ekuatorial Terhadap Curah Hujan di Indonesia* (Jakarta: Badan Meteorologi dan Geofisika)
[4] Tjasyono B H K 2009 *Meteorologi Indonesia 1: Karakteristik & Sirkulasi Atmosfer* (Jakarta: Badan Meteorologi Klimatologi dan Geofisika)
[5] Aldrian E, Gates L D and Widodo F H 2003 Variability of Indonesian Rainfall and the Influence of ENSO and Resolution in ECHAM4 imulations and in the Reanalyses, MPI Report 346, MPI Project (Hamburg: Max-Pack Institute of Meteorology)
[6] Fatony A 2015 Pengaruh El Niño dan La Niña Terhadap Variabilitas Musim di Wilayah Jawa Timur (Jakarta: Sekolah Tinggi Meteorologi Klimatologi dan Geofisika)
[7] McKee T B, Doesken N J and Kleist J 1993 *Proc. on the 8th Conference on Applied Climatology* (California: The Relationship of Drought Frequency and Duration to Time Scales)
[8] Kim M W, *et al.* 2011 *Geophysical Research Letter* **38** (L15809)
[9] Prastiwi A D 2015 Analisis Standardized Precipitation Index (SPI) dan Pendugaan Debit Sungai Cisadane Kabupaten Tangerang (Jakarta: Sekolah Tinggi Meteorologi Klimatologi dan Geofisika)
[10] Sumanto 2014 *Statistika Terapan* (Jakarta: CAPS)

IOP Conf. Series: Earth and Environmental Science **139** (2018) 012023 doi:10.1088/1755-1315/139/1/012023

Fish fauna in the Krueng Geumpang River, Indonesia

M Nasir[1]*, M Munira[2], Z A Muchlisin[3]

[1] Departement of Biology, Univeersitas Syiah Kuala, Banda Aceh 23111, Indonesia
[2] Departement of Pharmacy, Health Polytechnic Aceh, Banda Aceh 23352, Indonesia
[3] Faculty of Marine and Fisheries, Universitas Syiah Kuala, Banda Aceh 23111, Indonesia
*Email : muhd.nasir@fmipa.unsyiah.ac.id

Abstract. The objective of the present study was to examine the diversity and distribution of fishes in the Krueng Geumpang River.The survey was conducted from 1st to 22nd August 2015 with six sampling locations. Fish samples were caught using gillnets and fish traps. Data analyses performed in this study were the frequency of incidence (FOI), diveristy index (H'), and dominance index (C). A total of 88 individual fishes belong to 12 species and six familia. *Tor soro* is the dominant species in this river. There are two species of fish that widely distributed i.e. *Tor soro* (FOI = 66.7 %) and *Neolissochilus thienemanni* (FOI = 50.0 %). The Shannon-Wienner diversity index ranged from 0.00 to 2.05. The low value of the diversity index (H') can be caused by factors such as river morphology, poisoning, mining, and overfishing.

1. Introduction

The Geumpang River is located in the Pidie Jaya District, Aceh Province, Indonesia. The watershed of the river is situated in the rain forest of Ulu Masen Ecosystem, Sumatra. Currently, the river condition is threatened by anthropogenic activities like illegal logging, traditional gold mining, and overfishing.

Riverine biota has undergone changes as a consequence of drastic land conversion. The rapid urban development threats river biota and water quality. The use of detergents in households and fertilizers in crop fields enabled the flow of nutrients into surface waters. Moreover, many adverse changes in river ecosystems occurred due to a conversion of riparian land [1]. This condition is susceptible for aquatic organisms like fish. The pollution will disrupt the ecological balance of aquatic biota and produce negative impact for human being [2]. The waste discharge into waters has the potential to contaminate aquatic biota and disrupt food chain in the aquatic systems [3].

Fishing in the Aceh Province is mostly done in the sea, especially along the coast and continental shelf. Fishing also exists in lakes and rivers. The problems that arise in capture fisheries are overfishing and water pollution. A number of species experiences significant population decline and are under threat of extinction. This results the decline of in number of fish catched in the wild. The objective of present study is to examine the diversity and distribution of fish in the Geumpang River.

2. Materials and Methods

2.1. Time and location

Survey was conducted from August 1–22, 2015 at six sampling sites along the Geumpang River (figure 1). The sampling sites were determined purposively based on the characteristics of river (table 1). The fish was caught using gillnets (mesh size of 0.5 to 1.5 inches), fish traps, and hooks. Sampling activities were conducted for two daysat each sampling site. The sampled fishes were counted and photographed prior to preservation with 10 % formalin and then transported to the Basic Biology Laboratory of the Department of Biology, Faculty of Mathematics and Natural Sciences, University of Syiah Kuala, Banda Aceh Darussalam for further taxonomic identification by following [4].

Table 1. Locations of research, GPS coordinates, and characteristics of each sampling location.

Locations	Coordinates	Characteristics
Lhok Tarok	4°49'40.93"N; 96° 8'10.01"E	Bordering to the protected forest around the settlements and paddy fields additional
Lhok Kuala	4°49'52.66"N; 96° 8'3.95"E	Water from a small river
Kuala Keunee	4°51'29.97"N; 96° 6'46.27"E	Assembly between Kr. Geumpang and Kr. Keunee
Lhok Tunggok	4°52'51.43"N; 96° 5'43.34"E	Many activities taking Quarry
Alue Landong	4°53'43.19"N; 96° 3'38.98"E	Bordered by natural forest and Alue Landong River
Tanjakan	4°55'12.66"N; 96° 1'18.89"E	Deeper river water and larger river

2.2. Data analysis

The frequency of incidence (FOI) was calculated to estimate the local distribution of fish by using the following formula:

$$FOI = Ni\ .\ St/\ N \cdot St \times 100\% \quad (1)$$

where Ni St=total number of locations where the species i is found, N·St = total number of sampling locations.

The Shannon-Wienner diversity index was employed to evaluate the diversity of fish fauna in the Geumpang River.

$$H' = -\sum pi.\ln.pi \quad (2)$$

where H '= Shannon-Wiener Diversity Index, ni = number of individuals of all species i, and N = number of individuals of all species. The diversity index expresses the species richness in a community and shows the balance in individual proportion of every species. The Shannon Winner diversity index is classified into three levels: low ($H < 2$), moderate ($2 < H < 4$), and high ($H > 4$) [5].

The Simpson dominance index was calculated to assess the dominance of the fish species in the community. The formula is as follows:

$$C = \sum (ni/N)^2 \quad (3)$$

where, C = Dominance Index, ni = number of individuals of the species to-i, and N = a total number of individuals. Type Dominance Index (C) ranges from 0 to 1, where value tends to 0 is an indication no predominant species. Value tends to 1 indicates the presence of predominant species in the community [6].

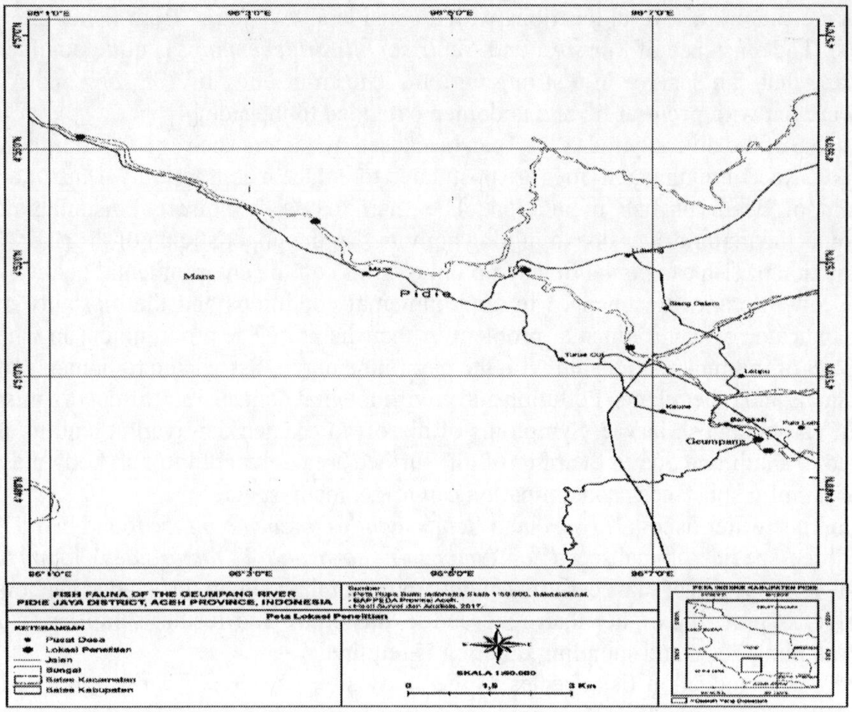

Figure 1. Map of sampling location in the Geumpang River.

3. Results and Discussion

A total of 88 individuals of fish belongs to 12 species, 11generas, and six families were recorded during the survey. The largest number of species was recorded in Lhok Tarok site (table 2). Cyprinidae was a predominant family of fish in the Geumpang River (figure 2). There are two species of fish that distributed widely, i.e. *Tor soro* (FOI = 66.7 %) and *Neolissochilus thienemanni* (FOI = 50.0 %) (table 3). Fish diversity index (H') ranges from 0 to 2.05 indicating low degree of diversity. The dominance index (C) ranges from 0.15 to 1.

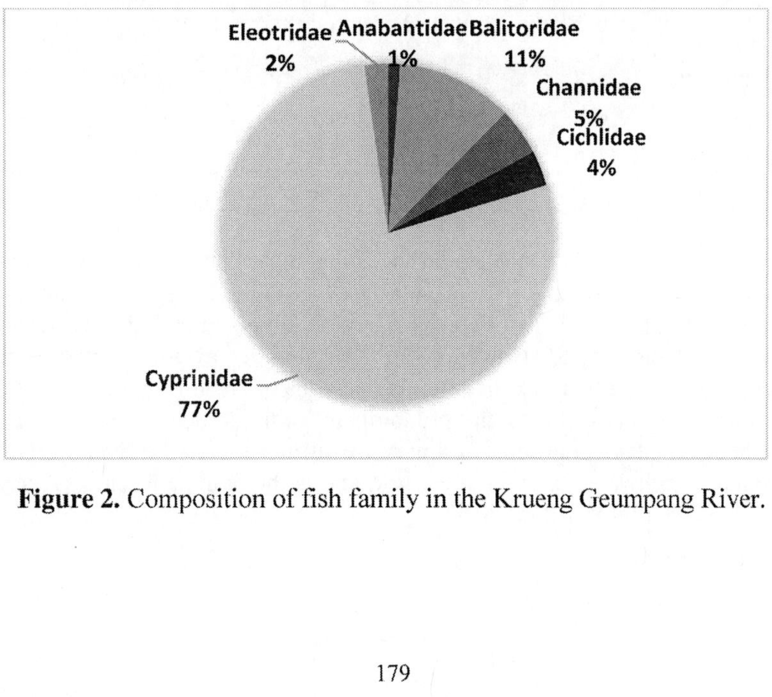

Figure 2. Composition of fish family in the Krueng Geumpang River.

Tor soro and *Neolissochilus thienemanni* are two dominant species in the Krueng Geumpang River. Both species are caught in several locations with a considerable amount. Both fishes are in the family of Cyprinidae. The existence of *Tor soro* and *Neolissochilus thienemanni* is quite dominant because it has body shape that can survive in a strong current. The front body of *Tor soro* and *Neolissochilus thienemanni* are flat with pectoral fin and abdomen extended to the side [4].

The low value of the diversity index (H') can be caused by factors such as river morphology, mining, and overfishing. Recent mining activities are postulated to induce mass mortality. Mass mortality of fish is a symptom of death of fish population. The main cause is poison. Poisoning due to water contamination is the major driver due to mining activities in the upper stream of the river.

The destruction of fish can be the first symptom of stress on the environmental quality. Various fish species have low tolerance for changes in environmental conditions and the mortality of fish in the region is an indicator of environmental problem in their habitat. The environment in which they live also has a direct or indirect relationship with the natural resources associated to human and is linked to the life of animals and other plants. Pollution and environmental damage may affect organisms at certain age levels, in particular fish larvae. Symptoms of dissolved oxygen deprivation tend to affect the big fish compared to small one due to the ratio of gill surface area compared to fish body mass. While the symptoms of harmful substances contamination can affect many species.

Among the freshwater fishes, *Tor soro* and *Neolissochilus thienemanni* are found in most sites. There are at least three species of mahseer (*Tor tambra, T. soro and T. tambroides*) found in Aceh [6]. However, only one species of mahseer was caught from sampling site in this study. The total number of fish in the river is probably higher than recorded in this study and further sampling using a greater diversity of sampling gears and sampling duration is required.

We found one introduction fish species in the study site, i.e. *Oreochromis niloticus* or known as tilapia. This fish presents in the Indonesia waters for long-term period. *Oreochromis niloticus* was introduced in 1969 by the Center of Research and Development for Freshwater Aquaculture, Ministry of Marine and Fisheries Affairs of the Republic of Indonesia. It is estimated that this species became established in the Aceh Province since 1990s [7].

Table 2. Total genus, species, and biological indices of fish collection in the Krueng Geumpang River.

Location	Genus	Species	Number of individuals	Dominance Index (C)	Diversity Index (H')
Lhok Tarok	10	10	42	0.15	2.05
Lhok Kuala	1	1	1	1.00	0.00
Kuala Keunee	1	1	12	1.00	0.00
Lhok Tunggok	1	1	7	1.00	0.00
Alue Landong	4	5	15	0.37	1.23
Tanyakan	1	1	11	1.00	0.00
Total	11	12	88	0.19	1.97

The negative impact of alien fish species on the indigenous fish community includes predation on local fishes, competition for food and habitat, interference with mate selection, and unexpected cross breeding with local fish. The results in a depressed genetic diversity and distinctiveness of local fishes and the transmission of disease [8]. However, the total number of alien fish species in the Krueng Geumpang River is lower than that recorded in other regions of Aceh Province.

Cyprinidae fish is widely distributed and predominant in the Krueng Geumpang River. Fish of the order Cypriniformes is the largest group of primary freshwater fish in the river system [9]. Cyprinidae is the largest group of freshwater fishes in the world and is the dominant family recorded in the Tripa peat swamp, Aceh province. Moreover, Cyprinidae is dominant in many other freshwater habitats in Asia [9].

Table 3. List of species of fish and total individual, and distribution in the Geumpang Krueng River.

Family	Species	Local name	Locations						Total	FOI (%)
			A	B	C	D	E	F		
Anabantidae	*Anabas testudineus*	Krup	1						1	16.7
Balitoridae	*Protomyzon griswoldi*	Ilee	10						10	16.7
Channidae	*Channa striata*	Bace	4						4	16.7
Cichlidae	*Oreochromis mossambichus*	Mujair					1		1	16.7
	Oreochromis niloticus	Nila	1				1		2	33.3
Cyprinidae	*Anematichthys apogon*	Ceperas	9				1		10	33.3
	Hampala macrolepidota	Palung	2						2	16.7
	*Neolissochilus thienemanni**	Kerling		1			4	11	16	50.0
	Puntius brevis	Groe	3						3	16.7
	Rasbora sp.	Depik	6						6	16.7
	Tor soro Valenciennes	Naleh	4		12	7	8		31	66.7
Eleotridae	*Ophiocarasp*	Ntok	2						2	16.7
Number of Individual			42	1	12	7	15	11	88	
Number of species			10	1	1	1	5	1	12	

Note: * IUCN red list
A= Lhok Tarok,
B= Lhok Kuala
C= Kuala Keunee
D= Lhok Tunggok
E= Alue Landong
F=Tanyakan
FOI= Frequency of incidence

4. Conclusion

Cyprinidae fish is widely distributed and predominant in the Krueng Geumpang River. Among the freshwater fishes, *Tor soro* and *Neolissochilus thienemanni* are commonly found in most sites. *Tor soro* and *Neolissochilus thienemanni* are the two dominant species in the Krueng Geumpang River. The low value of the diversity index (H') can be caused by factors such as river morphology, mining, and overfishing.

Acknowledgements

The authors acknowledge the Syiah Kuala University for providing financial support for this project. The appreciation is also dedicated to the Department of Biology and DMAS Programme who have given permission and support for the implementation of this study and as well as to those who participate and help in the implementation of this research.

References

[1] Kruk A, Cieplucha M, Zieba G, Tybulczuk S, Tszydel M, Marszal L, Blonska D, Galicka W and Przybylski M 2016 *Ecological Informatics*. **33** 109–118
[2] Lyons K, Carlisle A, Preti A, Mull C, Blasius M, O'Sullivan J and Lowe C G 2013 *Marine Environmental Research* **90** 27–38
[3] Karrari P, Mehrpour O and Abdollahi M 2012 *Daru: Journal of Faculty of Pharmacy, Tehran University of Medical Sciences* **20**(1) 2

[4] Kottelat M, Whitten S N, Kartikasari and Wirjoatmodjo S 1993 *Freshwater Fishes of Western Indonesia and Sulawesi* (Singapore: Periplus Edition)
[5] Muchlisin Z A and Siti-Azizah M N 2009 *International Journal of Zoological Research* **5**(2) 62–79
[6] Odum E P and Barret G W 2004 *Fundamental of ecology* (Belmont: homson Brooks Cole Publishing) p 624
[7] Chen C H, Hsieh C H, Hwang D F 2012 *Food Control* **28** 240–245
[8] Muchlisin Z A 2012 *Archieves of Polish Fisheries.* **20** 129–135
[9] Muchlisin Z, Akyun Q, Riska S, Fadli N, Sugianto S and Halim A 2015 *Check List the Journal of Biodiversity Data.* **11**(2) 1–8

IOP Conf. Series: Earth and Environmental Science **139** (2018) 012024 doi:10.1088/1755-1315/139/1/012024

Dominancy of *Trichodesmium* sp. in the Biawak Island

D J Prihadi

Faculty of Fishery and Marine Science, Padjadjaran University
*Email: donny.juliandri.prihadi@unpad.ac.id

Abstract. The Biawak Island is one of the small islands in West Java Province with an abundance of marine biological resources. This research was conducted to collect the primary producer zooplankton and water quality parameters. Direct observation is done by field surveys and measurement in situ for plankton and environmental parameters such as temperature, water transparency, water current, salinity, dissolved oxygen, and pH. *Trichodesmium* sp. was found dominance in where some other types of zooplankton were found in the area, like *Scenedesmus* sp., *Sagitta* sp., *Acartia* sp. also occurred. Further, the most abundance of *Trichodesmium* sp. was found in southern of Biawak Island where mangroves, coral and seagrass ecosystem provide nutrients which indirectly support the abundance of planktons. Trichodesmium sp. is plankton that can survive in water with minimum nutrient.

1. Introduction

Indonesia is the largest archipelago in the world with a vast sea and land territorials of about 7.7 million km^2. The nation comprises of 17,504 islands with the shoreline length of 95,181 km^2 stretching from Sabang to Merauke. The difference in values for the length of coastlines in some islands occurs because of climate change that has a great affect (the existence of the islands is submerged, coastal erosion of seawater) and there's also a difference in the calculation methods. About 75% of the area consists of coastal and ocean waters, including the territorial sea and 3.1 km^2 islands, as well as 2.7 million km^2 of the exclusive economic zone (EEZ). Indonesia is one of the countries which has a large potential in the sector of marine fisheries [1]. As an island nation with the longest coastline in the world i.e., 81,000 km^2, Indonesia has a potential of 10.28 million tons of fish production every year [2]. The number of islands that have been registered to the United Nations is about 13,466 islands, but now already 16,065 islands have been ratification to the United Nations.

The study of marine resources in the Biawak Island can be done by looking at the aspects of hydro-oceanography (currents, tides, waves, temperature, etc.), its ecology (ecosystems, mangroves, coral reefs, seagrass and associated biota) and bio-geochemistry (aquatic chemistry, biochemical substrate, etc.). However, this research is only a small part of the study resources and this research was conducted to find out the existing plankton communities in the waters of the Biawak Island. Planktons are the smallest organisms in nature that made life in the diverse marine waters possible. Myriads of marine organisms depend on the presence of planktons (phytoplankton and zooplankton) that exist in the Biawak Island for survival. The cluster of the Biawak Island has been designated as a conservation area sea area (KKLD) according to Decree of Regent Indramayu No. 556/Kep Diskanla 528/2004 dated April 7, 2004 [3].

The purpose of this study is to (i) identify the types of zooplankton that live in and surrounding waters of the Biawak Island (ii) conduct assessments of the water condition in and around the Biawak Island. The benefits of this research are to increase the availability of data and update information about zooplankton communities that live in the waters of the Biawak Island. This research will be published in national and international levels to be a useful information resource of diversity and biodiversity in Indonesia.

Local governments have established this area as one of the marine conservation area regions (KKLD), including the Biawak Island, Gosong Island and Candikian Island. Unfortunately, data related to the natural resources in the area is still very low and have not been integrated with one another. It results in the inconsideration of the marine resource's status and condition in the Biawak island (including its surrounding waters) as a whole.

Much of the island is about 26 miles (± 50 km) from the mainland of Indramayu, Indramayu Regency, and has climate type D (temperate) with an average annual rainfall of 1,428 mm. The daily temperature ranges from 26–27 °C the highest and 18–30 °C the lowest and air humidity ranges between 70–80 %. Western and Eastern wind alternates more or less every six months. The Western wind blows from December to April and the Eastern wind from May to October. Ocean waves in the Biawak Island and its surrounding areas are affected by seasonal waves, namely East and West as well as the season of passage altitude reaches 0.5 to 0.8 m.

2. Methods

The study was conducted in the Biawak Island, Province of West Java and the surrounding waters from July to November 2013. The map of the location of the research can be seen in figure 1.

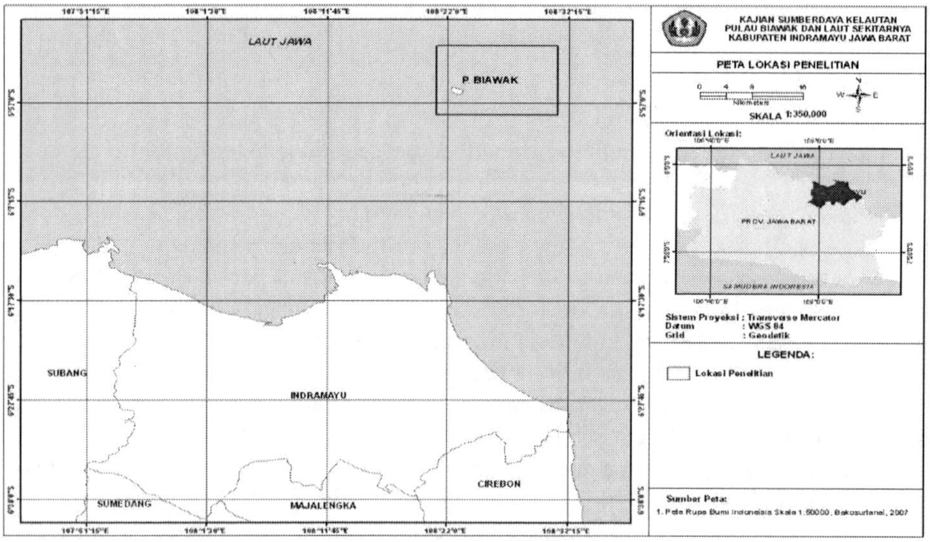

Figure 1. Research location in Biawak Island.

Direct observation was done by field surveys and measurement in situ for plankton and environmental parameters such as temperature, water transparency, water current, salinity, dissolved oxygen, and pH. The stages of this research consist of collected data that begins with the study of the literature related to the general condition of the location. The literature study of the biological resources and research on planktons and further data collection was done in the field (surveys) to get primary data. After the data and information were collected, the next step was data processing and analyzing to get a picture of a community resource and biodiversity of marine planktons that existed in the Biawak Island.

Table 1. The coordinate point of observation station ecosystem.

Observation /Sampling Station	Coordinates Ecosystem Observation	Ecosystem Characteristics
1	05°52'07.0" LS 108°24'18.8" BT	Coral Reefs
2	05°52'07.0" LS 108°24'19.3" BT	Coral Reefs
3	05°56'05.6" LS 108°22'53.0" BT	Coral Reefs
4	05°48'12.6" LS 108°25'29.2" BT	Coral Reefs
5	05°56'05.5" LS 108°23'05.1" BT	Seagrass
6	05°55'59.3" LS 108°23'11.0" BT	Seagrass

For the data collection of plankton, sampling was performed on 6 locations spread over the Biawak Island, where station 1–4 were located in the coral reef ecosystem around Biawak Island and the 5–6 station located in seagrass ecosystem around Candikian Island. The method used was random sampling/retrieval especially using plankton net, with a depth of 0.50–1 m below sea level. In this research, samples of phytoplankton were filtered with a 20 μm plankton net.

3. Results and Discussion

The condition of environmental parameters for temperature, salinity, DO, pH and water current had been calculated and obtained from different sample retrieval time (morning, afternoon and evening) (table 2). The sampling was undertaken in the Biawak Island water column which depth ranged from 19–54 m below sea level. The results of measurements of parameters such as chemical physic SST (Sea Surface Temperature) ranged from 29.80–2.20 °C. Temperatures ranged from 27.6–32.8 °C, salinity ranged from 26–33 ‰, DO range from 4–5 mg·L^{-1} and a pH range from 7.47–7.86 as for the brightness when sampling on-site research i.e. 1.8–7 m.

Table 2. Physical and chemical parameters in the Biawak Island including the temperature of the transparency of the flow of salinity, DO and pH during daylight (9am–6pm).

Temperature (°C)	Transparency (m)	Current (m·s^{-1})	Salinity (‰)	DO (ppm)	pH
30.50	6.00	0.28	30.00	5.10	7.61
31.10	5.30	0.34	30.00	5.30	7.86
29.80	5.00	0.30	30.00	5.90	7.50
30.30	4.70	0.32	31.00	5.70	7.70
30.70	4.70	0.31	27.00	5.70	7.65
31.40	5.50	0.25	27.00	5.80	7.75
30.10	5.00	0.20	26.00	5.00	7.47
31.60	5.40	0.21	27.00	4.70	7.55
32.20	5.10	0.37	33.00	4.00	7.50

The condition of the waters of the island of the Biawak Island can be explained as follows:

a. According to [4], the order *Aspidochirotida* larvae is a characteristic that lives in tropical waters 4.70–6 m of transparency is good for marine plankton lived under seawater.
b. The Biawak Island waters of temperatures ranging between 29.80–31.60 °C, the water temperature has a role in regulating the life of aquatic biota, especially in the process of metabolism. The temperature increase led to the increase in oxygen consumption, therefore,

resulted in the decline of the solubility of oxygen in water [5]. According to [6], larval sea cucumber is zooplankton as well has optimum temperature ranges between 28–29 °C.

c. Salinity value range has been obtained on the waters of 26–33 ‰. In time measurements were done differently at each station but in character location at each station are likely to be the same. The salinity is one of the factors that affect the existence and life of biota in the waters, including marine plankton [6].
d. The pH of the Biawak Island waters ranged on 7.50–7.86. It was classified as alkaline pH, so still supporting the biota life. Coastal waters had more stable pH and had a narrow range, which usually ranges between 7.7–8.4 [8]. The pH level is essential considering that most of the aquatic biota and also marine plankton are sensitive to changes in pH and pH value of around 7–8.5 [5].
e. Dissolved oxygen (DO) has a significant role in the aquatic life biota at once become the limiting factor for life in sea life. The Biawak Island waters had the average value DO 5.3–6 mg·L^{-1}, which classified as the normal range. The dissolved oxygen levels which are suitable for marine organisms is 5–7 mg·L^{-1} [9]. Per increment DO may be followed by the decrease in the number of plankton [5]. DO effect on plankton respiration and against the availability of water.

The presence of planktons in the water plays an important role in regulating marine life. Plankton serves as a natural food source for biota [9]. The base of the marine being the presence of phytoplankton in seawater affects the existence of small reef fish and larvae of free-living marine life that need phytoplankton as a food source. Furthermore, small fish and other marine life would be preyed by other fish as a necessity of survival in an aquatic biological resources sustainability management for coastal and sea areas to be maintained and preserved. In the absence of plankton, interferences will occur at the level as plankton is very important organisms in the aquatic food web. The presence of sunlight can increase the productivity of phytoplankton.

Table 3. Distribution of plankton in tehe Biawak Island

No.	Species	Sampling Stations					
		1	2	3	4	5	6
1	*Closterium* sp.	2	-	-	7	-	-
2	*Thallasiotrix* sp.	-	-	3	-	-	-
3	*Trichodesmium* sp.	8	-	3	114	5	5
4	*Coscinodiscus* sp.	15	-	1	3	-	-
5	*Scenedesmus* sp.	21	-	1	-	9	9
6	*Fragillaria* sp.	-	5	-	-	6	6
7	*Merismopedia* sp.	-	-	-	-	4	4
8	*Melosira* sp.	4	-	-	15	-	-
9	*Thallasionema* sp.	-	-	-	7	-	-
10	*Ceratium* sp.	-	-	-	15	-	-
11	*Rhizosolenia* sp.	-	-	-	6	-	-
12	*Sagitta* sp.	-	30	-	-	-	-
13	*Nauplius* sp.	3	7	-	-	-	-
14	*Proates* sp.	-	1	-	-	-	-
15	*Acartia* sp.	-	12	4	7	-	-
16	*Glaucus*sp.	1	-	-	-	-	-
Number of individual (in 100 L)		54	55	12	174	24	24
Number of Spesies (S)		7	5	5	8	4	4
Abudance (ind. 10 L^{-1})		5	6	1	17	2	2

According to table 3, station 4 was located on the southern the Biawak Island and it has the highest abundance of plankton with diverse species compared to other stations. From this observation, we suggest that station 4 is an excellent habitat for plankton, followed by station 2 and 1. The smallest number of plankton was found in station 3. There are several factors that cause a large population of plankton in the Biawak Island. The factors were the existence of mangrove ecosystems and seagrass ecosystems. In the Biawak Island, the coral reef covers were categorized as a medium condition at station 4, 2 and 1. However, note that the *Trichodesmium* sp. are abundant in station 4. Trichodesmium spp. is a type of Cyanobacteria plankton that could live in the waters with low nutrient levels. In the condition where there are a high nutrient availability and other supportive conditions, blowout (blooming) could occur. The blooming of *Trichodesmium* spp. will initiate the onset of anoxic conditions and deaths as the ones in shellfish shells in India [10]. This phenomenon can cause negative effects and disadvantageous for marine biotas such as fish and some other marine invertebrates. It causes a decrease in the amount dissolved oxygen due to the process of decomposition by bacteria.

One possible factor is the existence of mangrove, coral reef, and seagrass ecosystems in the Biawak Island. We postulate that these ecosystems play a profound role in providing nutrient for the plankton growth. Hence, the population of plankton is abundant in this region. The smallest abundance observed in station 3 may relate to the absence of mangrove and bad condition of the coral reef ecosystem. Less plankton abundance is also observed in station 5. This might be due to the region were covers of the sandbar and lacks marine ecosystem.

The importance of mangrove ecosystems for the growth and survival of planktons is derived from the presence of nutrients from mangrove leaves that have dried up and fell to seawater, as well as mangrove roots that pose multitudes of nutrients in the area that can be used by planktons. Seagrass ecosystems can affect the presence of planktons. Many types of planktons that live in seagrass leaves are known as periphyton. A variety of planktons can utilize the presence of dead seagrass leaves as a source of food (nutrient), as well as shelter.

It is important to note that the *Trichodesmium* sp. is abundant in station 4. Trichodesmium sp. is a type of plankton from the group Cyanobacteria that can live at low nutrient levels and can experience a blowout (blooming) when there's high nutrient availability. The location of station 4 is nearby the ship dock. The abundance of Trichodesmium spp. in this station might be due to leaked oil from the dock which carried by the ocean current. Further research is needed to test this hypothesis.

The blooming of *Trichodesmium* sp. will initiate the onset of anoxic conditions and deaths as the ones in shellfish shells in India [11]. This condition can also bring adverse effects, and it is not advantageous for marine biotas such as fish and some other marine invertebrates. It causes a decrease in the amount dissolved oxygen due to the process of decomposition by bacteria. *Trichodesmium* sp. can only have a high population in certain months, the further study is necessary to investigate the ability of this species to tie up nitrogen in the sea. Seawater, where many of this species lives, looks bright blue and different from the average sea color. However, at the end of the research, the color difference is not visible. Monitoring aquatic environments in the Biawak Island should start from local government as the manager and participation of the people in Indramayu Regency.

The previous study on the presence of phytoplankton and zooplankton in the Biawak Island shows that Bacillariophyceae, Cyanophyceae, Dynophyceae, and some genus of phytoplankton Cyanophyceae were found as Spirulina, Gloeotrichia, and Trichodesmium. For the Bacillariophyceae and Bacteriastrum were found from Dynophyceae such as Ceratium and Peridinium [10].

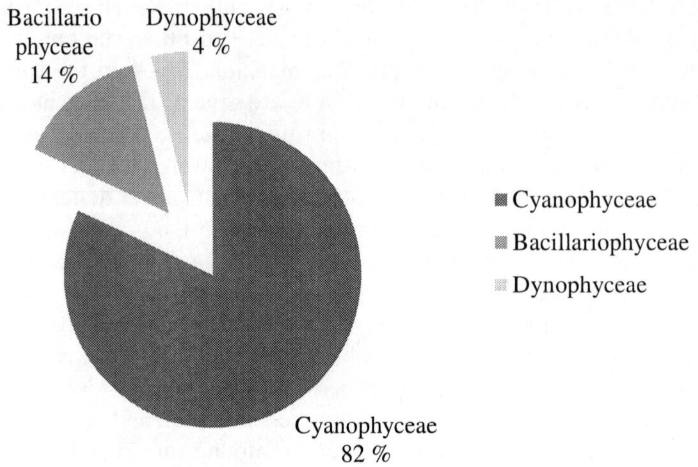

Figure 2. Phytoplankton distribution in the Biawak Island.

Meanwhile, zooplanktons found in the Biawak Island comprises Calanus and Diaphanosoma (Maxillophoda), Synchaeta and Brachionus (Rotifers). Specifically, *Calanus* sp. was found all around the island. *Calanus* sp. is abundant in the morning rather than in noon and evening [10].

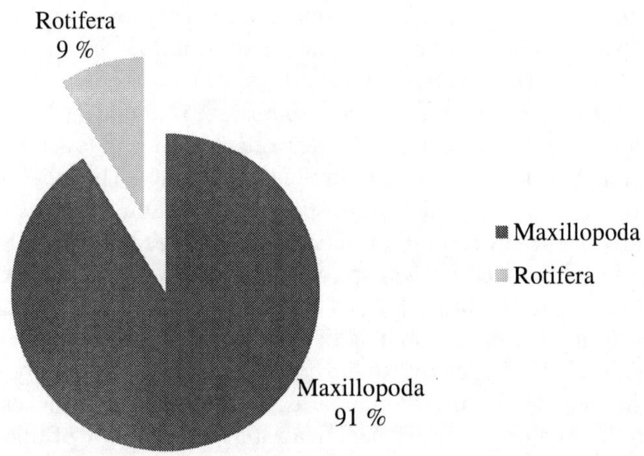

Figure 3. Distribution of zooplankton in the Biawak Island.

Manta rays gather in places where there are lots of zooplanktons like *Acartia* sp. Manta point in Bali Nusa Penida is one of the water territories that are often used as a gathering place for manta rays (*Manta birostris*) [12]. In the coastal area, the region has a lot of *Acartia* sp. and it is favored by sea organisms like whales, manta rays and whale sharks.

4. Conclusion and Suggestion

4.1. Conclusion

The present study shows that the existence of mangrove, seagrass and coral reef ecosystems play a fundamental role in determining plankton population obtained in the Biawak Island. Planktons obtained on the island is diverse with the abundance of plankton are quite diverse is dominated by *Trichodesmium* sp. Zooplankton that exists in this region may have strengthened marine food chain and fishes that live in the coral reef ecosystem. Further research concerning the abundance, diversity, and distribution of plankton community in the waters of Biawak Island, and Candikian Island is required to investigate the connection between fish and other marine biotas' in the areas.

4.2. Suggestion
Research suggests the need for further research concerning the abundance, diversity, and distribution of existing plankton communities in the waters of the Biawak Island and Candikian Island. Moreover, other research to learn more about the connections between fish and sea life in the presence of planktons as the most important producers in the food chains in the sea and coastal areas.

References

[1] Dahuri R 2004 *Sumber Daya Wilayah Persisir dan Laut* (Jakarta: PT. Pradnya Paramita)
[2] Dahuri R 2016 *Pengelolaan Sumberdaya Pesisir dan Laut di Indonesia General* Seminar (Bandung: Padjajaran University)
[3] Purba N P, Taofiqurohman A and Yusuf M A 2012 *Laporan Akhir* (Bandung: Universitas Padjadjaran Jatinangor)
[4] Bakus G J 1973 *The Biology and Ecology of Holothurians* (New York: Academic Press)
[5] Effendi H 2003 *Telaah Kualitas Air bagi Pengelolaan Sumberdaya dan Lingkungan Perairan* (Yogyakarta: Penerbit Kanisius)
[6] Hyman L 1955 *The Invertebrates Echinodermata* Vol IV (New York: Mc Graw-Hill Book Company)
[7] Nybakken J W 1988 *Biologi Laut: Suatu Pendekatan Ekologis* (Jakarta: PT. Gramedia)
[8] Gufran M H, Kordi K and Tanjung A B 2007 *Pengelolaan Kualitas Air Dalam Budidaya Perairan* (Jakarta: Penerbit Rineka Cipta)
[9] Nontji A 2006 *Plankton* (Jakarta: Pusat Penelitian Oseanografi)
[10] Thillai R, M Rajkumar, J Sun, Prabu V A and Perumal P 2009 Seasonal Variations of Phytoplankton Diversity in The Coleroon Coastal Waters, Southeast Coast of India (Sinica: Acta Oceanologica)
[11] Cuncun H 2012 *Struktur Komunitas Fitoplankton Kaitannya dengan Distribusi Nutrien di Perairan Pulau Biawak Kabupaten Indramayu* (Bandung: Padjajaran University Press)
[12] Prihadi D J 2011 *Analysis of Plankton Structure Related to Ecotourism in Manta Point Nusa Lembongan Bali Indonesia* (Pekalongan: Proceeding of International Seminar)

Ocean wave characteristic in the Sunda Strait using Wave Spectrum Model

R Rachmayani*, N S Ningsih, S R Adiprabowo and S Nurfitri
Department of Oceanography, Faculty of Earth Sciences and Technology, Institut Teknologi Bandung, Indonesia
*E-mail: rrachmayani@oceanography.itb.ac.id

Abstract. The wave characteristics including significant wave height and direction, seas and swell in the Sunda Strait are analyzed seasonally to provide marine weather information. This is crucial for establishing secured marine activities between islands of Sumatera and Java. Ocean wave characteristics in the Sunda Strait are simulated for one year (July 1996–June 1977) by using SWAN numerical model. The ocean wave characteristics in the Sunda Strait are divided into three areas of interest; southern, centre and northern part of the Sunda Strait. Despite a weaker local wind, the maximum significant wave height is captured at the southern part with its height of 2.6 m in November compared to other seasonally months. This is associated with the dominated swell from the Indian Ocean contributes on wave energy toward the Sunda Strait. The 2D spectrum analysis exhibits the monthly wave characteristic at southern part that is dominated by seas along the year and swell propagating from the Indian Ocean to the Sunda Strait during December to February (northwest monsoon), May, and November. Seas and swell at northern part of the Sunda Strait are apprehended weaker compared to other parts of the Sunda Strait due to its location is farther from the Indian Ocean.

1. Introduction

Information on weather and sea condition changes is essential to activities on seas, either for sea transportation activities, fisheries, marine resources exploration, or development in marine sector. Natural factor is responsible for two hundred sixty ship accidents [1]. And among various natural factors, one that greatly in sequences marine activities is wave, and therefore, in the marine meteorological services, in addition to wind information, wave information is the most important that should exist in marine weather information [1]. Since waves are generated by the wind, the incidence of waves is strongly influenced by the presence of winds on the sea surface [1]. Wave variations in the Indonesian waters are mostly induced by the northwesterly and southeasterly monsoon winds. Some wave variations caused by the monsoon winds are wave height, wave period, wave direction, seas and swell in the waters. In addition, wave characteristics can also be affected by storm events. In spite of minor events of storm in the Indonesian waters, waves formed in areas passed by storms can spread to the Indonesian waters due to its long travel distance before finally breaking [2, 3].

Recent studies of wave characteristics using spectrum models have been conducted [4, 5]. A global study of wave characteristics is discussed by using low resolution of WAM Model; this indicates that the presence of swells spread from high latitudes to equator and the dominance of swell in tropical waters, especially in Indonesia might happen throughout the year [4]. A regional study with domains covering the Indian Ocean, Java Sea, Karimata Strait, and South China Sea found variation of annual and semi-annual monsoons that dominates the wave variability in the South China Sea and Indian Ocean [5]. The goal of this present work is to further examine the characteristics of seas and swell in the Sunda

Strait associated with its interaction to dominated monsoon. Moreover, this present work utilizes a higher resolution of spectrum model to capture wave characteristics in more details. This paper is organized as follows: data and experiment are introduced in section 2, section 3 presents the results and discussion of the study, and the conclusions are given in section 4.

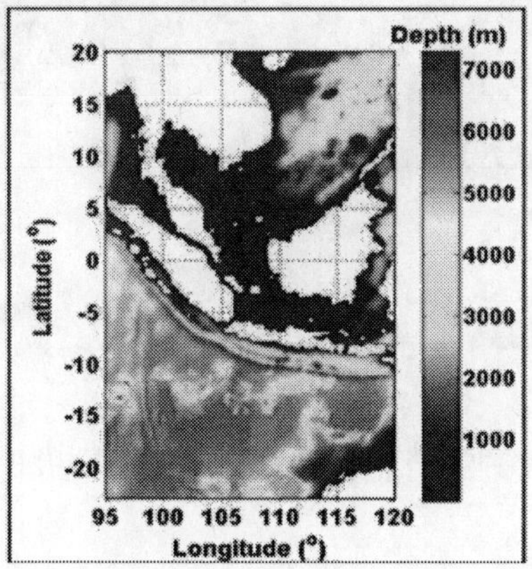

Figure 1. Domain model and bathymetry profile.

2. Materials and Methods

2.1. Data

This study focusses on the Sunda Strait that shown in figure 1 located at 103°–108° E and 5°–8° S. The domains of research simulation include the Java Sea, Sunda Strait, and Indian Ocean. The wide gap in the southwestern part of the Sunda Strait allows the wave characteristics of the Sunda Strait to experience a great influence from the Indian Ocean at depths of about 1000–2000 m. The narrow gap in the northeastern part between the Sunda Strait and Java Sea has a minor effect to the wave variation in the Sunda Strait. In this study, wind and bathymetry data are applied as model input. The wind field is taken from 10 m above sea level called CCMP (Cross Calibrated Multi Purposed) with a gradient resolution of 0.25° × 0.25° and intervals every six hours. Meanwhile, bathymetry data from GEBCO (Global Bathymetric Chart of Oceans) produced with 1' × 1' resolution [6]. The wave characteristics including wave height, wave direction, seas and swell in the Sunda Strait are studied from simulation data using wave spectrum model SWAN (Simulating Waves Nearshore) for one year (July 1996–June 1997).

2.2. Model experiment

SWAN is a third-generation wave model for obtaining realistic estimates of wave parameters in coastal areas, lakes, and estuaries from given wind, bottom, and current conditions. However, SWAN can be used on any scale relevant for wind-generated surface gravity waves. The model is based on the wave action balance equation with sources and sinks [7, 8]. In this simulation, the used input is wind and bathymetry (water depth), while the effect of current is negligible. The wave propagation process that can be simulated by the SWAN model is propagation of waves, refraction, and shoaling related to the spatial variation of the bottom of the water, and diffraction and reflection by a barrier. Meanwhile, the generation and wave dissipation processes contained in the SWAN are generated by wind, white capping-induced dissipation, and dissipation when the wave breaks due to depth factor, dissipation by basic friction, and nonlinear interactions between waves. The SWAN model has three computational spatial grid options, namely box space grid, curved-linear spatial grid, and unstructured space grid. The computational spectral grid is determined with the minimum frequency and frequency of the rated wave as well as the desired frequency resolution. SWAN uses an implicit numerical scheme, so that the accuracy of the model results depends on the time interval.

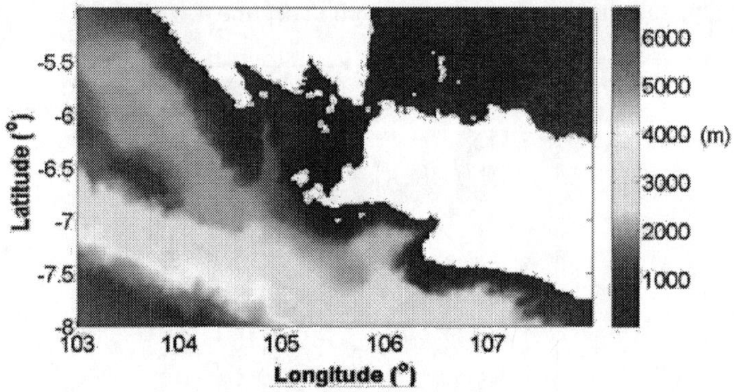

Figure 2. Regional domain model and bathymetry profile.

2.3. Design experiment

The simulation of the study area in the Sunda Strait, Indian Ocean and Java Sea is a nested simulation of the regional model. As shown in figure 2, the regional model domains include the South China Sea, Karimata Strait, Java Sea, and Indian Ocean. Regional domain simulations were performed using Emphirical Orthogonal Function (EOF) model [5]. The wave simulation in the study area uses boundary conditions obtained from regional model results. The scenario used in this wave simulation is a nonstationary hindcasting simulation. The initial spectrum used in the model is the initial spectrum of Joint North Sea Wave Project (JONSWAP). The physical processes that occur in the simulation are the development of waves, quadruplet nonlinear interactions, and dissipation. The discrete wave propagation in the geographic space uses the BSBT (backward in space and backward in time) scheme.

3. Results and Discussion

3.1. Pattern of wave variability

The ocean wave characteristics in the Sunda Strait are divided into three areas of interest, southern part of the Sunda Strait toward the Indian Ocean (area 1), centre of the Sunda Strait (area 2), and northern part of the Sunda Strait toward the Java Sea (area 3) as illustrated in figure 3. The model results show a linier relationship between the higher of wind speed and the higher of significant wave height (Hs). The direction of wave propagation is strongly influenced by the wind direction. Waves generally move in the direction of the wind, however, in certain part of the Sunda Strait, the significant wave height does not correspond to the wind. This is associated with the narrow shape of the Sunda Strait; limiting the occurrence of the development of wave and the wind-distracted swell that dominate the direction of the wave.

Figure 4 presents spatial patterns of wind variability and significant wave height in the Sunda Strait during northwest monsoon months of December-January-February (DJF). The direction of waves in the Indian Ocean is east-northeast and the significant wave heights in the Indian Ocean reach 0.75 to 1.25 m. In Java Sea, the waves move east-northeast in corresponding with the wind direction. The wave characteristics in area 1 are influenced by the open water of the Indian Ocean which lead the waves propagate to the east (area 1). The propagated waves from the Indian Ocean reach area 2 in February and area 3. The significant wave height is significantly reduced (increased) due to the weakened (strengthened) wind in January (February).

IOP Conf. Series: Earth and Environmental Science **139** (2018) 012025 doi:10.1088/1755-1315/139/1/012025

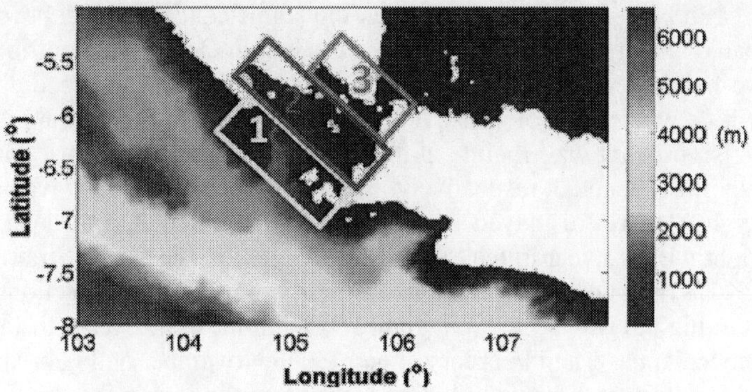

Figure 3. Divided areas in the Sunda Strait.

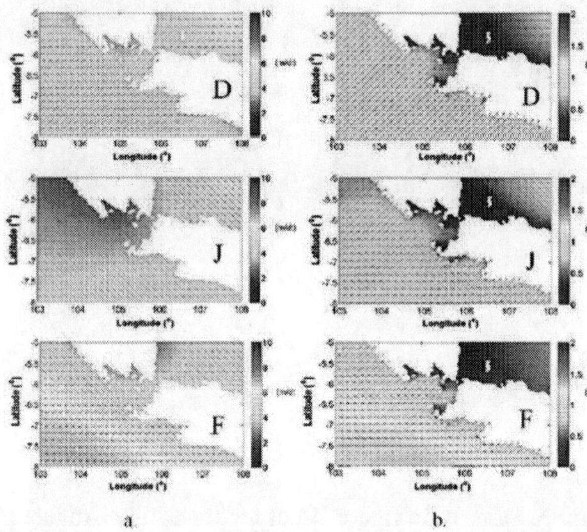

Figure 4. Figure 4. Averaged (a) wind field (m/s) and (b) significant wave height (m) for December–January–February (DJF).

During the southeast monsoon months of June–July–August (JJA), the significant wave height increases to a maximum value (1.8 m) within a year in August (figure 5). In the Indian Ocean, the significant wave height moves to the northwest in the range of 0.75–1 m. The significant wave height in the Java Sea moves to the west at about 0.5 m. Area 1 is influenced by waves that radiate northward from the Indian Ocean. Propagated wave from the Indian Ocean has a minor influence to area 2 and 3 due to inhibited area from waves coming from the Indian Ocean, in area 2 the wave effect of the Indian Ocean is very weak as the waves do not propagate eastwards into area 2. The influence of waves from the Indian Ocean in area 3 is weaker than in area 2 because the waves from the Indian Ocean do not spread eastward to area 3 and the waters in the narrow area 3 inhibit the propagation of waves from the Indian Ocean entering area 3. In addition, the formation of local wind waves in area 2 and area 3 does not occur due to the narrowness of waters in both areas.

3.2. Wind and wave

Figure 6 exposes three points in the Sunda strait. Wind speed, significant wave height, and maximum significant wave are observed monthly during a year at these three points. Time series of wind and significant wave height are shown in figure 7. It exhibits a good correspondence between both parameters except in certain months, where the significant wave height increases as the wind strengthens and vice versa, except in November at point 1 when the significant wave height reaches its maximum value. This is associated with a swell propagates from the Indian Ocean that contributes to wave energy. The significant wave height at point 1 is 1.4 m in November with a wind speed of 3 m/s, at point 2 is 0.9 m in February with a wind speed of 4 m/s, and at point 3 is 0.6 m in February with a wind speed of 4 m/s. The minimum significant wave height at the three points occurs in April with values of 0.65 m, 0.25 m, and 0.2m at point 1, 2, and 3, respectively. In addition to significant wave height, the maximum significant wave height (Hmax) displayed in Figure 8 is also calculated to determine the maximum significant wave height within a year in the Sunda Strait. Hmax in the Sunda Strait reaches about 1.25–2.6 m at point 1, 0.6–1.8 m at point 2, and 0.3 to 1.2 m at point 3. Wind direction is also suggested to play a role in contributing the wave direction (figure 9) within a year, except in January and February at point 1, in November, January, and February at point 2, in November and February at point 3. Again, this is due to swell that propagates from the Indian Ocean into the Sunda Strait.

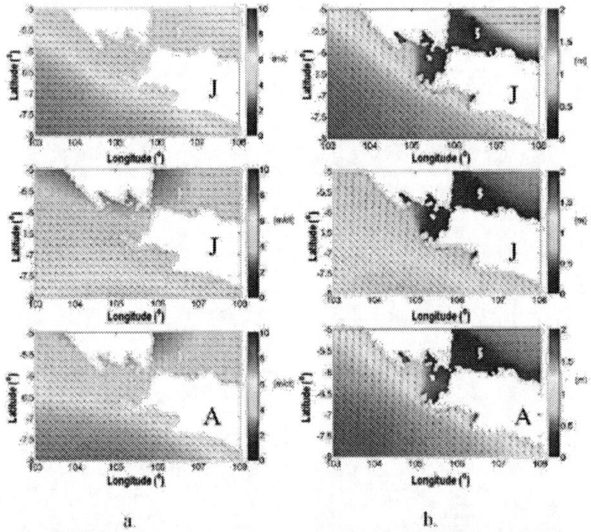

Figure 5. Same as in figure. 4 but for June–July–August (JJA).

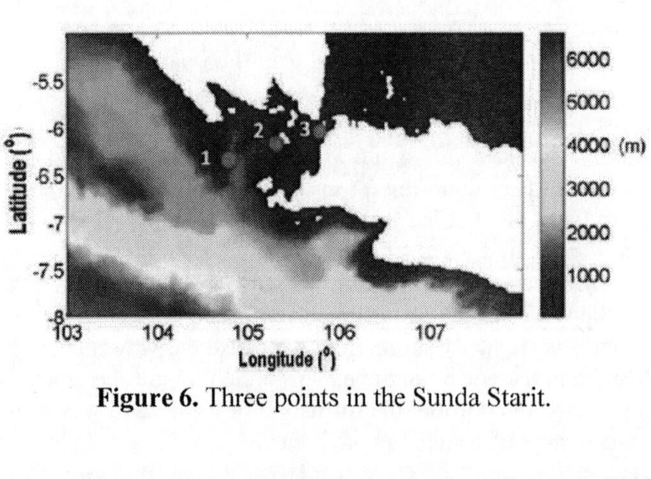

Figure 6. Three points in the Sunda Starit.

3.3. *Wave spectrum*

The wave spectrum is also studied at the three predetermined points (figure 6). Through the wave energy spectrum, the characteristics of seas and swell in these waters are analyzed by the peak of dominant energy, dominant direction, and local wind direction. The monthly spectrum at point 1 is illustrated in figure 10 where seas exist throughout the year. In March-November the local wind leads to the northwest and the peak density leads to the north, indicates the existence of seas. Likewise, in December - February local winds lead to east-southeast and the peak density of energy heads east; leading a swell to exist revealed by the peak density of energy to the north but the local wind actually leads to east-southeast. The wave spectrum in May and November shows high density levels compared to other months and denotes maximum significant wave heights while local winds are weakened in this month. To summarize, seas and swell waves occur simultaneously in December–February, May, and November. The direction of the seas to the east and the swell propagated from the Indian Ocean move north; leading to a distraction of wave direction from its local wind direction (figure 7) in January and February. Figure 11 reveals monthly average spectrum at point 2 where the local wind leads to a density peaks at the similar direction, indicating that seas occur almost throughout the year (from December–February, April–August and October). In addition, swell also exists in December-January illustrated by a peak density of energy that leads north, but local winds lead to the East-Southeast and in March-November where the local wind leads to the northwest, but the peak density of the spectral energy moves north-northeast. Similar to point 1, the results at point 2 show a high energy level of the wave spectrum in May and November and exhibit that seas and swells occur simultaneously in December-January and April-October. Monthly average spectrum at point 3 is presented in figure 12. In December-February and March-October, the wave spectrum moves in corresponding with the local wind results in an existence of seas during a year. Moreover, swell exist in January–March, May, and September-November. As at point 1 and point 2, wave spectrum in May and November at point 3 has a very high peak density of the seas energy and in January-March, May, and September–October seas to the east and swell to the north occur concurrently. Table 1 reveals the direction of the monthly seas and swell at the three points.

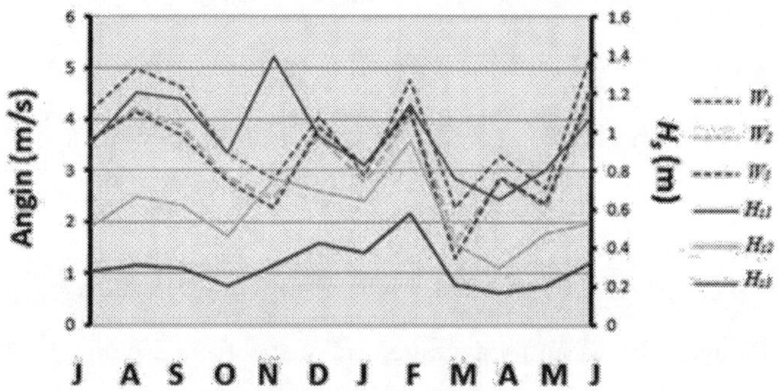

Figure 7. Time series of wind and significant wave height at the three points in the Sunda Strait.

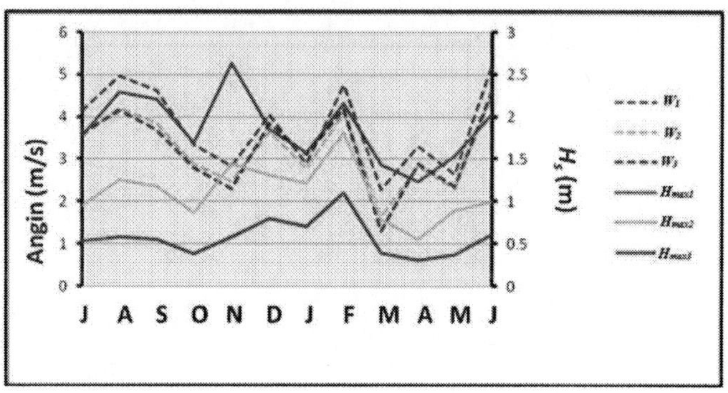

Figure 8. As in figure 7, but for wind and maximum significant wave height.

IOP Conf. Series: Earth and Environmental Science **139** (2018) 012025 doi:10.1088/1755-1315/139/1/012025

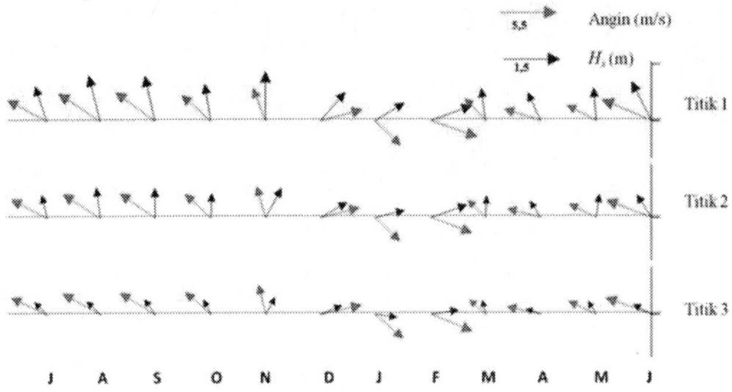

Figure 9. As in figure 7, but for wind and wave direction.

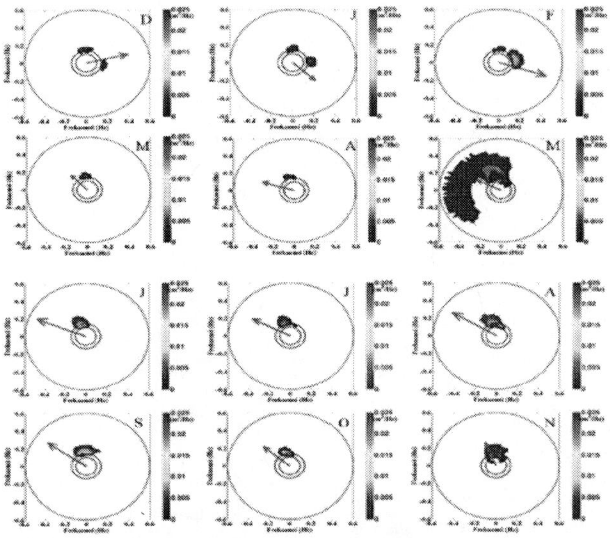

Figure 10. Monthly averaged wave spectrum at point 1

Table 1. Direction of seas and swell at the three points. N = North, NE = Northeast, E = East, SE = Southeast, W = West, NW = Northwest.

		Month											
		D	J	F	M	A	M	J	J	A	S	O	N
Point1	Seas	E	E	E	N	N	N	N	N	N	N	N	N
	Swell	N	N	N	-	-	NW	-	-	-	-	-	NW
Point2	Seas	E	E	E	-	NW	NW	NW	NW	NW	NW	NW	-
	Swell	N	N	-	N	N	N	N	N	N	N	NE	NE
Point3	Seas	E	SE	SE	NW	W	NW	NW	NW	NW	N	NW	-
	Swell	-	NE	NE	NE	-	NE	-	-	-	NE	NE	NE

Figure 11. As in Fig.10 but at point 2.

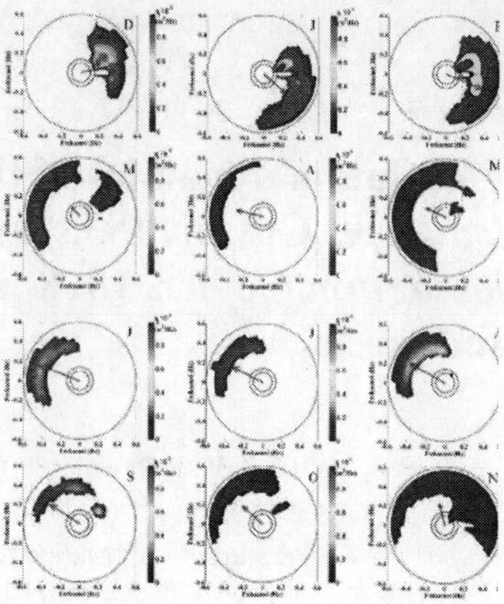

Figure 12. As in Fig.10 but at point 3.

4. Conclusion

Despite a weaker local wind, the maximum significant wave height is captured at the southern part of the Sunda Strait with its height of 2.6 m in November compared to other seasonally months. This is associated with the dominated swell from the Indian Ocean that contributes on wave energy toward the Sunda Strait. The 2D spectrum analysis that exhibits the monthly ocean wave characteristic at southern part of the Sunda Strait is dominated by seas along the year and swell propagating from the Indian Ocean to the Sunda Strait during December to February (northwest monsoon), May, and November. Seas and swell characteristics at northern part of the Sunda Strait are apprehended weaker compared to other parts of the Sunda Strait due to its location that is farther from the Indian Ocean.

References

[1] Kurniawan R and Khotimah M K 2015 *IPTEK* **26**(1) pp 19–27
[2] Kurniawan R 2012 *Karakteristik Gelombang Laut dan Daerah Rawan Gelombang Tinggi di Perairan Indonesia* (Depok: Universitas Indonesia)
[3] Rachman A 2007 *Tugas Akhir* (Bandung: Program Studi Oseanografi Institut Teknologi Bandung)
[4] Semedo A, Suselj K, Rutgersson A and Sterl A 2011 *Bull. Amer. Meteor. Soc.* p 59
[5] Ramadhan H 2014 *Tugas Akhir* (Bandung: Program Studi Oseanografi Institut Teknologi Bandung)
[6] I O C, I H O and B O D C 2003 *GEBCO Digital Atlas, Intergovernmental Oceanographic Commission dan the International Hydrographic Organization* Centenary Ed. (Liverpool: British Oceanographic Data Centre)
[7] Young I R 1999 *Int. J. Climatol* **19**(9) pp 931–950
[8] Booij N, Ris R C and L H Holthuijsen 1999 *J. Geophys. Res* **104** 7649–7666

IOP Conf. Series: Earth and Environmental Science **139** (2018) 012026 doi:10.1088/1755-1315/139/1/012026

Population dynamics of the yellowstripe scad (*Selaroides leptolepis* Cuvier, 1833) and Indian mackerel (*Rastrelliger kanagurta* Cuvier, 1816) in the Wondama Bay Water, Indonesia

R Sala[1]*, R Bawole[1], F Runtuboi[1], Mudjirahayu[1], I A Wopi[1], J Budisetiawan[2] and Irwanto[2]

[1] Department of Marine Science, Faculty of Fisheries and Marine Science, University of Papua, Manokwari, Papua Barat Province, Indonesia
[2] WWF Indonesia, Teluk Cenderawasih Site Project, Wasior Papua Barat, Indonesia
*Email: ridwansala@gmail.com

Abstract. The Wondama Bay water is located within the Cendrawasih Bay National Park and is potential for fishery resources, including pelagic fish such as yellowstripe scad (*Selaroides leptolepis* Cuvier, 1833) and Indian mackerel (*Rastrelliger kanagurta* Cuvier, 1816). Yet, information about the population dynamics of these species in the region is unknown until today. Meanwhile, the fishing activities have been quite intensive and include the dominant catches over the last ten years by traditional fishermen fishing using liftnets. Therefore, this study aims to determine some of specific characteristics of the population dynamics and fish utilization status of scad and mackerel in the waters of the Wondama Bay. Data used in this study were taken from direct observation of catch of liftnet fishery. The data then were analysed by using FISAT II to estimate the growth parameters, mortality rates, and yield per recruitment. The results showed that yellowstripe scad has the positive allometric growth, while Indian mackerel followed isometric growth. Models of fish growth were $L(t) = 22\ (1-e^{-3.0(t-0.05)})$ for yellowstripe scad and $L(t) = 27.8\ (1-e^{-4.0\ (t-0.04)})$ for Indian mackerel. The natural mortality (M) of 4.19 year^{-1}, fishing mortality (F) of 5.01 year^{-1}, and total mortality (Z) of 9.20 year^{-1} were for yellowstripe scad, and M of 4.74 year^{-1}, F of 2.52 year^{-1} and Z of 7.26 year^{-1} were for Indian mackerel. Based on the mortality rates, estimated exploitation rate for the yellowatripe scad was 54 % and the Indian mackerel was 35 %. To increase the production of catch without increasing fishing effort (fishing mortality) can be done by increasing the size of fish caught or the $Lc/L\infty$ should be greater than 0.5.

1. Introduction

The present study focuses on the yellowstripe scad (*Selaroides leptolepis* Cuvier, 1833) and Indian mackerel (*Rastrelliger kanagurta* Cuvier, 1816) that live in coastal waters of the Wondama Bay. The bay is within conservation areas of the Cendrawasih Bay National Park (CBNP). Both fishes belong to small pelagic fish that have value in supporting ecological function of coastal ecosystem. In addition, both fishes are considered economically important and consumed by most communities in Wondama. Yellowstripe scads are caught in coastal waters and live in schooling up to 50 m depth [1] and Indian mackerel schoolings are found in water up to 90 m depth [2]. Nowadays, high demand and high prices of these species encourage increased catch [3]. Data from 2008 to 2012 showed that yellowstripe scad and Indian mackerel were at the top five of fish species contributing to fish production in Wondama [4].

Utilization of fishery resources in the CBNP often leads to conflict between fishermen from outside the region and local fishermen [5]. A fishing technology approach in fisheries management in the CBNP has emphasized [6]. Based on the above facts, research on the dynamics of population becomes very important, especially from the aspects of growth, mortality, and exploitation rates of the yellowstripe scad and Indian mackerel species.

Both fishes are the dominant catch of liftnet fishery in the waters of the Wondama Bay operated by small scale fishery. Fishing with liftnets used light aids to draw fish gathered on the net. Because of the nature's attraction of the fish on light [7], it makes the fish very susceptible for being caught by the liftnet fishing gears. In addition, since the selectivity of the liftnets is low, because it used very small mesh size of net (minnow net), it makes almost every size of the fish caught by the liftnet. Consequently, young cohort fish will be caught.

Fishing on young fish or fish with size less than the first maturity size will lead to a decrease in population productivity to reproduce a new generation and reduce recruitment supply for fisheries. This may lead to disruption of the population's ability to maintain its sustainability. In addition, yellowstripe scad and Indian mackerel populations are important part of the fish communities that support the ecosystem in the Wondama Bay waters. The existence of the fish populations is important to maintain a trophic balance of food which will further maintain the balance of the ecosystem.

Concerning the declining population of small pelagic fish resources in off the Cendrawasih Bay National Park area, and efforts to obtain data on fisheries resources, study on the dynamics of the pelagic fish populations was proposed. Such study is crucial in the context of sustainable fisheries resource management [8], particularly for the management of catch size and optimum level of fishing effort.

2. Materials and Methods

2.1. Site and sample collection

This study conducted in the Wondama Bay of the (CBNP) Indonesia (figure 1). Data collection was carried out from March to April 2016. Samples of yellowstripe scad (*Selaroides leptolepis* Cuvier, 1833) and Indian mackerel (*Rastrelliger kanagurta* Cuvier, 1816) were taken from liftnet fishermen. Yellowstripe scad and Indian mackerel were bycatch of the liftnet, since the fishing targets of the liftnet in Wondama were anchovies. There were 116 individuals of yellowstripe scad and 261 individuals of Indian mackerel sampled during the survey. The fish samples were then measured their total body length (L) and weight (W). L and W of the samples were measured to the nearest centimeter (cm) and gram (g).

2.2. Length-weight relationship

The length-weight relationship was analyzed by fitting function $W = aL^b$ to length (L) and weight (W) data [9]. Regression coefficients a and b were obtained by using least square methods by firstly transforming the function into logarithmic form, $\log W = \log a + b \log L$. Statistical test was done to examine whether b value is close to 3.0 or not by using t-test, as shown in formula (1) below. If b is close to 3, it means that the species shows isometric growth. When b significantly larger or less than 3, it indicates allometric growth [9].

$$t = \left(\frac{SD_y}{SD_x}\right)\left(\frac{|b-3|}{\sqrt{1-r^2}}\right)\left(\sqrt{n-2}\right) \quad (1)$$

2.3. Growth model

Model growth of fish species was estimated by using von Bertalanffy growth function (VBGF) as follows:

$$L(t) = L\infty\,(1-e^{-k(t-to)}) \quad (2)$$

where: L(t) is the fish length at age t; L∞ is asymptotic length; K is growth coefficient; t denotes the age of fish and to is hypothetical age at which a fish would have zero length. The age *to* was estimated by using Pauly's formula [10]:

$$Log\ (-t_o) = -0.3922 - 0.2752\ Log\ L\infty\ -\ 1.038\ Log\ K \qquad (3)$$

The estimated ages at each fish length were calculated by using derivation of VBGF [11] namely:

$$t = t_0 \frac{1}{K}.Ln \frac{(1-Lt)}{L\infty} \qquad (4)$$

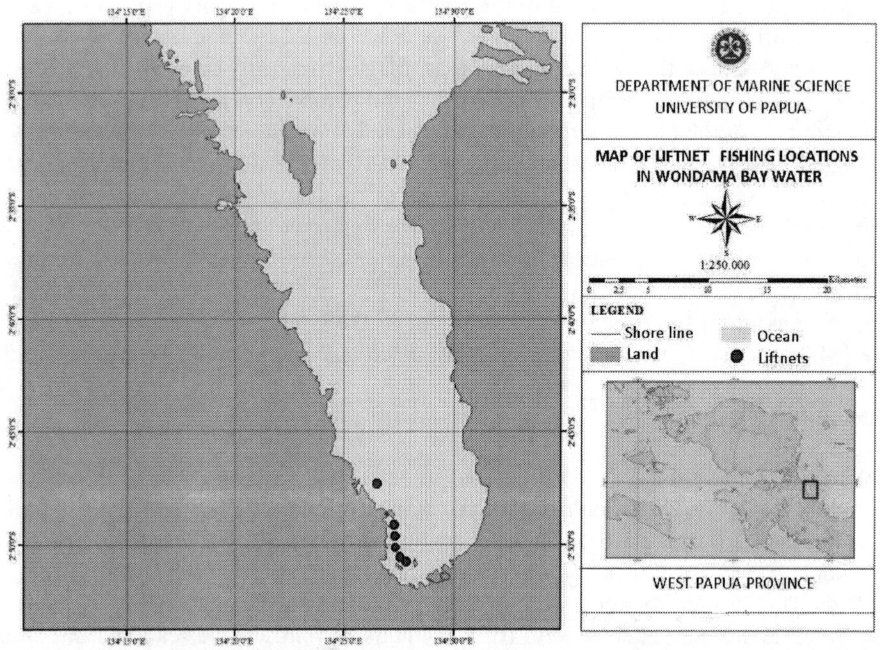

Figure 1. Map of study area in the Wondama Bay.

2.4. Mortality and exploitation rates

Total mortality rate (Z) of fish consists of natural mortality (M) and fishing mortality (F). The total mortality rate was estimated by using Beverton-Holt method which was based on the length frequency of data distribution [11], in which Z and L follow functional relationship:

$$Z = K.\frac{(L\infty - L)}{(L - Lc)} \qquad (5)$$

where Z = total mortality, L∞ = asymptotic length, K = growth coefficient, L = mean length of the fish (cm), Lc = length of fish caught (cm).

Estimation of natural mortality used empirical equation of Pauly [12] as shown in equation (6).

$$Log\ M = -0.0066 - 0.279\ log\ (L\infty) + 0.6543\ log\ (K) + 0.4634\ log\ (T) \qquad (6)$$

where M = natural mortality, L∞ = asymptotic length, K = growth coefficient, T = Mean sea surface temperature (°C). The equation shows a positive relationship between natural mortality, growth parameters, and sea surface water (SST). We used mean SST of 31 °C as suggested by [13] for water surface temperature in CBNP.

When Z and M are known, then fishing mortality rate (F) can be estimated by using formula F = Z – M. Furthermore, the exploitation rate (E) can also be estimated since E is the ratio of fishing mortality (F) and total mortality (Z) [14]. The optimal exploitation rate of fish stocks occurs when F equals M [15]. Thus, when E larger than 0.5, it indicates that the fish stock is experiencing overexploitation.

2.5. Relative yield-per-recruit (Y/R)
Relative yield-per-recruit (Y/R) was estimated by using formula [16] :

$$Y/R = E.U^{M/K}\left\{1 - \frac{3U}{1+m} + \frac{3U^2}{1+2m} + \frac{U^3}{1+3m}\right\}$$

where
$U = 1 - (L_c/L_\infty)$
$m = K/Z$
Lc = Length at fish caught
E = Exploitation rate

The analyses of data for estimating growth parameters, mortality rates and yield per recruitmet were done by using FAO-ICLARM Stock Assessment Tools II (FISAT II) [16].

2.5.1. A subsubsection. The paragraph text follows on from the subsubsection heading but should not be in italic.

3. Results and Discussion

3.1. Length frequency distribution of yellowstripe scad and Indian mackerel
Frequency distribution of yellowstripe scad and Indian mackerel caught in the Wondama Bay is shown in figure 2. Length of yellowstripe scad spreaded between the range of 10 and 24 cm, whereas the Indian mackerel ranged between 15 and 26 cm.

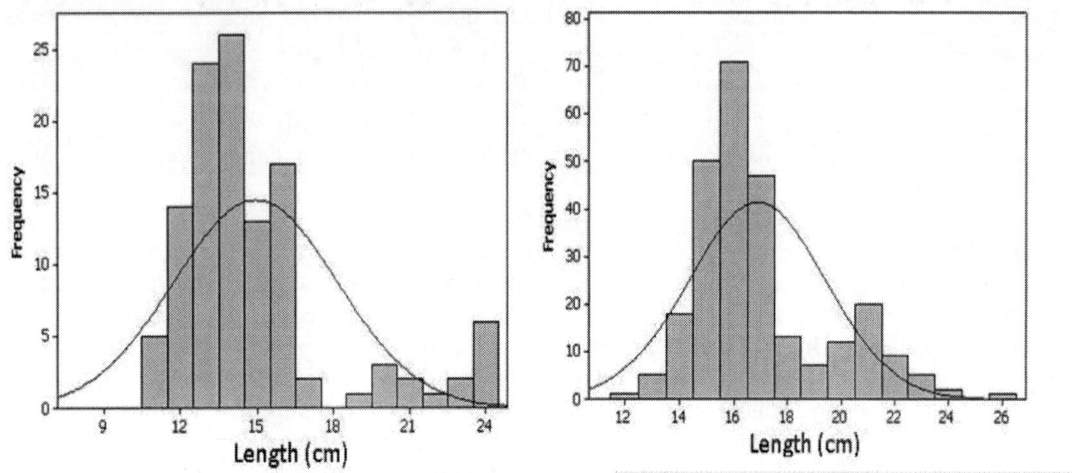

Figure 2. Size composition of yellowstripe scad (left) and Indian mackerel (right) in the Wondama Bay.

Both length curves tended to skew to the left, which showed that more portions of fish caught were at small sizes (young fish). On average, the length of yellowstripe scad is 14 cm and Indian mackerel is

16.88 cm, as shown in table 1. Fishbase (fishbase.org) records the length at first maturity (Lm) for the yellowstripe scad is 11.9 cm and for the Indian mackerel is 19.9 cm. Based on these data, the scads captured by using liftnets in the Wondama Bay were generally larger than Lm or have been in spawning ages. On the other hand, most the Indian mackerel taken using liftnets in the Wondama Bay waters were at the sizes less than Lm or immature fish.

Table 1. Mean length and weight of yellowstripe Indian scad and mackerel in theWondama Bay

No	Type	Length (cm) Average and Standard deviation (n)
1	Yellowstripe scad	14.90 ± 3.20 (116)
2	Indian mackerel	16.88 ± 2.519 (261)

3.2. Length-Weight Relationship of yellowstripe scad and Indian mackerel

The length-weight relationship (LWR) of yellowstripe scad caught in the Wondama Bay followed the equation $W = 0.003L^{3.41}$ with the coefficient of determination between length and weight (R^2) of 0.91. The LWR for Indian mackerel followed the equation $W = 0.009L^{3.02}$ with R^2 is 0.74. The relationship curves, which is in logarithmic form of L and W, is shown in figure 3.

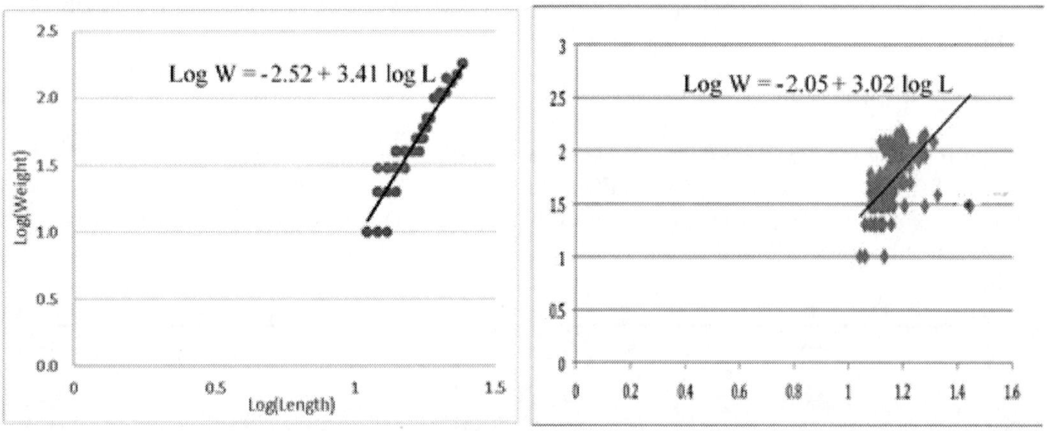

Figure 3. Length-weight relationship of yellowstripe scad (left) and Indian mackerel (right) in the Wondama Bay.

The above-mentioned pattern of growth-length relationship shows a regression coefficient (b) for yellostripe scad is 3.41 (greater than 3) which means having a positive allometric growth pattern, or growth of weight faster than growth of length. This is in contrast to the growth pattern of yellowstripe scads in some places in Indonesia that shows negative allometric and isometric growth pattern [17]. The relationship pattern of Indian mackerel exhibit that b coefficient is 3.02, indicating that the fish had an isometric growth pattern, which means growth of length is proportionally to growth of weight. This growth pattern is different from that found in the waters of the Bay of Bengal, has a negative allometric growth pattern [18], and also different from that found in the western Indonesian waters [2] that tends to be allometrically positive pattern [9] found several things that affect the difference in rates of weight growth and length growth, including differences in habitat, time, season, size, and growth phases (i.e. at the beginning of the growth phase tends to increase in the length faster than other dimensions).

3.3. Growth models for yellowstripe scad and Indian mackerel

The curves of length growth models for yellowstripe scad and Indian mackerel are shown in figure 4. The yellowstripe scad and the Indian mackerel have asymptotically long parameters of 22.0 cm and 27.8 respectively. Both fish species have large values of growth coefficient (K), i.e. 3.0 for yellowstripe scad and 4.0 for Indian mackerel. The large K values indicates that both fish have very rapid growth to achieve asymptotic length.

The growth parameter values for both fish species are slightly different from those observed in some waters in Indonesia. For yellowstripe scad, $L\infty = 18.9$ cm and K = 1,1 $year^{-1}$ are observed in Riau waters and $L\infty = 22$ cm and K = 1.2 $year^{-1}$ are found in the Java Sea [19]. For Indian mackerel, $L\infty = 23.9$ cm and K = 2.76 $year^{-1}$ are founf in Java Sea [20], $L\infty = 25.8$ cm and K = 1,630 $year^{-1}$ are observed in Java Sea (Pekalongan) [19]. The species occupying different habitat environments will exhibit different patterns of life history characteristics [21]. So, the differences in growth parameters, particularly the growth coefficient and K of between yellowstripe scad and Indian mackerel might be related to differences in the quality of the aquatic environment, including the availability of food. The Wondama Bay waters possibly has good water conditions and food availability to support the growth of fish.

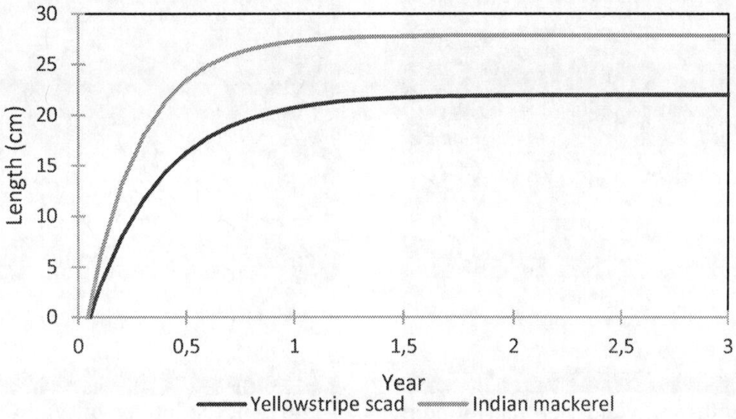

Figure 4. Growth curves of yellowstripe Indian scad and in the Wondama Bay.

3.4. Mortality rate of yeallowstripe scad and Indian mackerel

Mortality rate (Z), natural mortality (M), fishing mortality (F) and rate of exploitation for yellowstripe scad and Indian mackerel in the Wondama Bay can be seen in table 2. Natural mortality of yellowstripe scad is lower than the fishing mortality while natural mortality of Indian mackerel is higher than fishing mortality. The natural mortality rate of yellowstripe scad is higher than that in Tambelan (Riau) with a value of 0.64 [22]. However, the natural mortality rate for Indian mackerel was is similar to that observed in the Java Sea (M = 4.4) [20]. The rate of natural mortality of fish is influenced by environmental conditions like SST, and life history of the fish such as fish i.e. growth rate *(K)*. The increase in SST has affected the existence of marine ecosystems (including fish population) especially in the tropic [23]. The average SST in the CBNP isabout 31 °C [13] which possibly caused a high natural mortality of fish in that region.

The important thing to note from the table 2 is the value of the rate of exploitation (E), which describes how the portion of biomass resources that have been taken by fishermen. The value of the rate of exploitation of yellowstripe scad is 0.54 and Indian mackerel is 0.35. These values suggested that yellowstripe scad has been exploited exceeded the optimum level. Therefore, action should be given to control fishing activities on this species so that this resource can be sustainable. Different conditions for mackerel, where exploitation rate is 0.35, therea are many oppurtunities to catch this species.

Table 2. Model of growth of yellowstripe scad and Indian mackerel in the Wondama Bay.

No	Species	Z ($year^{-1}$)	M ($year^{-1}$)	F ($year^{-1}$)	E ($year^{-1}$)
1.	Yellowstripe scad	9.20	4.19	5.01	0.54
2.	Indian mackerel	7.26	4.74	2.52	0.35

3.5. Yield per recruitment

Results of analysis of yield per recruitment (YPR) on yellowstripe scad and Indian mackerel are presented in figure 5. Based on figures, the two fish species have relatively equal values of YPR. For instance, at size $Lc/L\infty = 0.5$ and $E = 0.5$, Indian mackerel had YPR approximately 0.0456 grams/recruit and yellowstripe scad has YPR of 0.0424 grams/recruit. The YPR values explain the expected yield in every one recruitment of fish [11]. For example, 0.0456 grams/recruit of YPR for Indian mackerel explain that recruitment of one Indian mackerel would give a yield of 0.0456 grams. The amount of yield per recruitment can be improved only when the minimum size of fish allowed to be caught is large (i.e. $Lc / L\infty > 0.5$) and/or E is greater than 0.5.

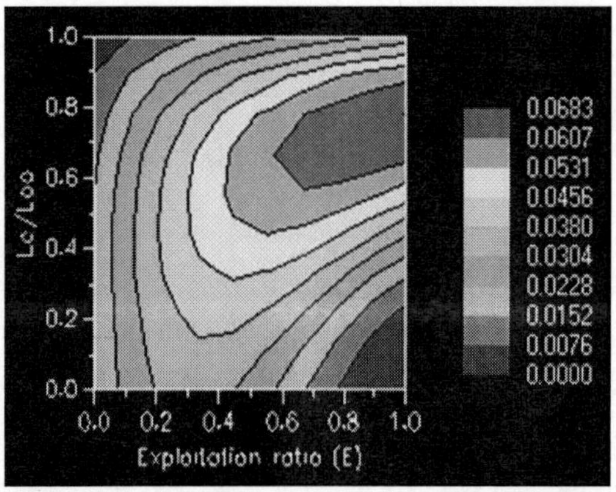

Figure 5. Isolepths of yield per recruitment (gram/recruit) for yellowstripe scad.

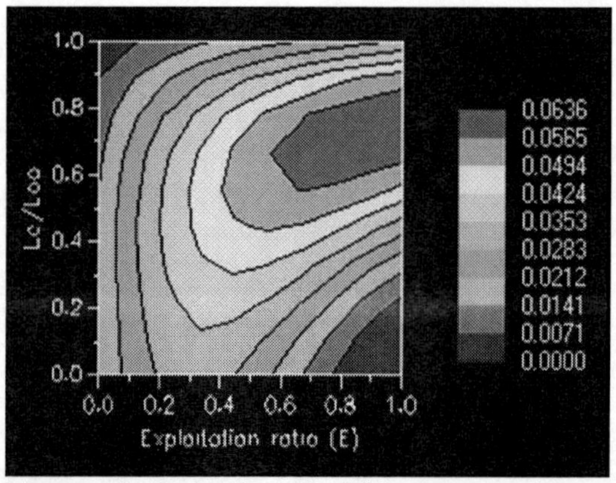

Figure 6. Isolepths of yield per recruitment Indian mackerel.

However, for the purpose of resource conservation of yellowstripe scad and Indian mackerel population, it is suggested to set L_c larger than the minimum length at first maturity of respective species to provide opportunity for the fish to reproduce before being caught. This requires an increase in the selectivity of fishing gears (liftnets) used to catch fish in the Wondama Bay. Also, the management control of fishing effort (e.g increasing the number of liftnets) should be carried out strictly and should not exceed the optimum level of effort.

4. Conclusion

- Yellowstripe scad caught in the Wondama Bay demonstrates a positively allomeric growth of body shape, whereas Indian mackerel shows an isometric growth.
- Estimation of life history for yellowstripe scad is 22.0 cm, $K = 3.0$ $year^{-1}$ and $M = 4.19$ year-1. For Indian mackerel, L_∞ is 27.8 cm, $K = 4.0$ $year^{-1}$ and $M = 4.74$ $year^{-1}$.
- The exploitation rate for yellowstripe scad is 0.54 and Indian mackerel is 0.35.
- To increase catch, it is suggested that for yellowstripe the size should be larger than Lm, which is larger than 11.9 cm, whereas for Indian mackerel should be larger than 19.9 cm.

Acknowledgements
We thank the Site Project Cendrawasih Bay National Park of WWF Indonesia, for helping with accommodation and local transportation. Also, we thank Management Body of Cenderawasih Bay National Park for issuing the permit letter, and fishermen who assisted in the field data collection.

References

[1] Allen G R and Erdmann M V 2012 *Reef fishes of the East Indies Volume I* (Australia: University of Hawai'i Press)
[2] Pauly D, Cabanban A and Torres Jr F S B 1996 *ICLARM Stud. Rev.* **23** 135–216
[3] Bawole R and Sala R 2014 *Determination of total allowable catch of pelagis and demersal fishes* Pemerintah Daerah Teluk Wondama (Unpublished Document)
[4] BPS 2013 *Wondama in figures* (Wondama: Biro Pusat Statistik)
[5] Bawole R 2012 *JMHT* **18**(2) 110–117
[6] Bawole R, Rumere R, Mudjirahayu and Pattiasina T P 2013 *JMHT* **19**(1) 63–73
[7] Ragesh N, Sajikumar K K, Remya R, Sasikumar G, Koya K P S and Mohamed K S 2014 *Mar. Fish. Infor. Serv., T & E Ser.* **219** 21–23
[8] Rahman M M and Hafzath A 2012 *J. Biol. Sci.* **12**(8) 426–432
[9] Froese R 2006 *J. Appl. Ichthyol.* **22**(4) 241–253
[10] Froese R and Binohlan C 2003 *J. Appl. Ichthyol.* **19** 376–379
[11] Sparre P and Venema S C 1998 *Introduction to tropical fish stock assessment. Part 1. Manual.* (Rome: FAO Fisheries Technical Paper) p 407
[12] Pauly D 1980 *J. Cons. Int. Explor. Mer.* **39**(2) 175–192
[13] Bawole R 2012 (Bogor: Institut Pertanian Bogor)
[14] Craig J F ed 2015 *Freshwater fisheries ecology* (New Jersey: John Wiley & Sons) p 920
[15] Gulland J A 1983 *Fish Stoc Assesment:AManual of Basic Method* (New Jersey: John Wiley & Sons) p 273
[16] Gayanilo F C, Sparre P and Pauly D 2006 *FAO-ICLARM Stock Assessment Tools II (FISAT II). Revised Version User's Guide.* (Rome: FAO Computerized Information Series (Fisheries) Rome Food and Agricultural Organization of the United Nations) p 168
[17] Dobo J and Supeni E 2015 *Pro. of The Second Annual Scientific Seminar Fisheries Development* (Tual: Politeknik Perikanan Negeri Tual)
[18] Mustafa M G and Ali M S 2003 *Bangladesh J Fish. Res.* **7**(2) 159–168
[19] Dwiponggo A, Hariati T, Banon S, Palomares M L D and Pauly D 1986 *ICLARM Tech. Rep.* **17** p 91
[20] Sudjastani T 1974 *The species of Rastrelliger in the Java Sea, their taxonomy, morphometry and population dynamics* (Colombia: University of British Columbia) p 147
[21] Adams P B 1980 *Fish. Bull.* **8**(1) 1–12
[22] Yonvitner and Fahmi 2012 *Biodiversitas* **13**(4) p 200–204
[23] Cheung W L, Watson R and Pauly D 2013 *Nature.* **497** p 365–368

IOP Conf. Series: Earth and Environmental Science **139** (2018) 012027 doi:10.1088/1755-1315/139/1/012027

The effect of monsoon variability on fish landing in the Sadeng Fishing Port of Yogyakarta, Indonesia

D Subarna[1]

[1] Centre for Atmospheric Science and Technology, LAPAN, Indonesia
*Email: dadang.subarna@lapan.go.id

Abstract. The volume of landing fish of the Sadeng Fishing Port within certain months showed an increase from year to year, especially during June, July and August (JJA). While in other months the fish production was low. The purpose of this research was to understand the influence of monsoon variability on fish landing in the Sadeng Fishing Port. Data were analyzed descriptively as spatial and temporal catch. Data were namely catch fish production collected from fishing port, while satellite and HYCOM model during 2011–2012 period were selected. The wind data, sea surface temperature (SST) and chlorophyll-a were analyzed from ASCAT and MODIS sensors during the Southeast Monsoon. The result showed the wind from the southeasterly provide wind stress at sea level and caused Ekman Transport to move away water mass from the sea shore. The lost water mass in the ocean surface was replaced by cold water from deeper layer which was rich in nutrients. The distribution of chlorophyll-a during the Southeast Monsoon was relatively higher in the southern coast of Java than during the Northwest monsoon. The SST showed approximately 25.3 °C. The abundance of nutrients indicated by the distribution of chlorophyll-a around the coast during the Southeast Monsoon, will enhance the arrival of larger fish. Thus, it can be understood that during June, July, and August the catch production is higher than the other months.

1. Introduction

The majority of seafood production, especially the catch fish is usually found in the tropics area between 10 °N and 10 °S [1]. Indonesian territory extends from 90-141 °E. This area is an important area for the productivity of tropical seafood. Natural fish stock from the Indonesian waters including the Exclusive Economic Zone (EEZ) is estimated around 6.4 million tons/year, with 63.5 % captured annually which consists of 4.14 million tons from the Indonesian Seas and 1.26 million tons from the Indonesia EEZ [2]. Pelagic fish plays an important role for the fishermen economy in Indonesia, i.e. about 75 % of the total fish stock or 4.8 million tons/year are pelagic fishes [3].

Due to the unique position of the Indonesia territory, Indonesia experienced a strong response to large climate phenomenon like the Australia-Indonesian Monsoon (AIM) and El Niño-Southern Oscillation (ENSO). The AIM and ENSO, as well as the interplaying phenomena between these two, are influenced by ocean-atmosphere interaction in the Pacific and Indian Oceans. Indonesia plays an important role in the global thermohaline cycle and global climate circulation, and also as the center of biodiversity, especially fish resources [4]. Other phenomena that may affect ocean color have been studied in Indonesian seas are ENSO, IOD [5, 6, 7], MJO [8], Waves and also the current of Kelvin and Rossby [9, 10,11]. All these ocean-atmosphere interactions affect the variability of SST, sea level anomalies, wind, and ocean color.

The Indonesian Seas are the only seas in the planet that connected two major oceans i.e the Pacific Ocean and Indian Ocean. This connection is known as Indonesian Throughflow (ITF) [12, 13]. Due to the coastal geographic influence and topographic variation of the ocean floor, the ITF variability is thought to be related to the marine color variability. Besides ITF, the ocean color in the Indonesian seas is strongly influenced by the AIM. During the Northwest Monsoon (Decemer–February: DJF), river runoff may also affect the ocean color, not only through the delivery of the nutrients but also through the increase of particulate matter and colored dissolved organic matter (CDOM), especially in western Indonesia [3, 14].

Monsoon is a reversal of seasonal wind patterns caused by temperature gradients between land and sea [15]. Monsoon winds play an important role in generating currents, waves and SST. During the Southeast Monsoon season, the southeasterly winds from Australia generated upwelling along the southern coasts of Sumatra, Java, and Nusa Tenggara Island Chain [13]. The reverse condition occurred during the Northwest Monsoon season. An understanding of the spatial and temporal evolution of upwelling is an important factor in coastal fisheries. According to [16] an area falls into the category of a monsoon region if: 1) a major wind direction shifted at least 120° between January and July, 2) persistent wind direction at least 40 % of the time in January and July, 3) Average windspeed over 3 m/s in January or July, 4) fewer than one cyclone-anticyclone change every 2 years in January or July.

Sea surface dynamics induced by monsoon winds have strong correlation with fish behavior and seasonal abundance and distribution of fishes. The study of marine processes in the Indonesian Seas, especially in the southern sea of Java can help fisheries community to estimate fish stock. A research on seasonal variability of pelagic fish production and processes that played important role in the sea around Java Sea has conducted [3]. Using over 20 years of fish catch data collected from several fish ports and ocean color data of 7 years, and the annual catch pattern was determined by monsoon activity [3]. Therefore, monsoon greatly influenced the emergence of warm and nutrient-rich surface currents in the Java Sea, surface water transport and upwelling in the Sunda Strait, the southern sea of Java which starts from the Bali Strait. Processes in different seas for each region greatly affected the fish distribution and catches.

The purpose of this research is to investigate the influence of monsoon variability on the catch production in Sadeng. In this study, the ocean color will be expressed as chlorophyll-a.

2. Materials and Methods

2.1. *Study area*

The study area in Sadeng, as shown in figure 1. Sadeng is a fishery port located in the Sadeng Bay, flanked by two villages namely Songbanyu and Pucung Villages. The distance from the provincial capital is about 85 km. The Sadeng Fishery Port is one of the work units of the regional technical implementation unit and is the headquarters of the regional fishery port which has work area along the coast of Yogyakarta. As a center for economic activities, the Sadeng Fishery Port is a gathering place for fishermen, fish traders, fishery entrepreneurs and other business entities that support economic activities in Sadeng, Songbonyu and Girisubo.

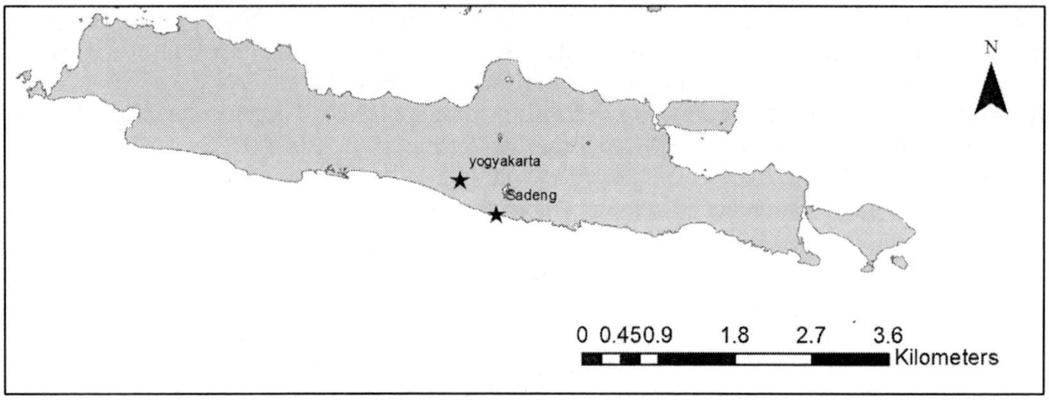

Figure 1. Research location of catch fish production in Sadeng.

2.2 Data Collection

SST and ocean color data used in this research were obtained from MODIS. The data were processed using a python script or SeaDAS software that can be obtained from http://oceancolor.gsfc.nasa.gov/seadas/. The MODIS sensor detects the emitted and reflected radian in 36 channels which sweep the visible spectrum up to the infrared spectrum. More information related to the instruments can be obtained from NASA MODIS webpage (http://modis.gsfc.nasa.gov/about/specifications.php). Wind data were obtained from the ASCAT (Advanced Scatterometer) sensor on the METOP-A satellite and can be downloaded from http://manati.star.nesdis.noaa.gov/datasets/ASCATData.php/. The marine model data were obtained from HYCOM+NCODE and can be downloaded from http://tds.hycom.org/thredds/catalog.xml. Fish production data were obtained from Sadeng.

Table 1. Development of Sadeng PPP Production 2010-2014. (Source: UPTD PPP DKP Daerah Istimewa Yogyakarta, 2015).

Year	Production (kg)	Production Value (IDR)
2009	1,639,000	13,620,186,000
2010	1,430,000	11,904,152,000
2011	961,170	10,896,980,000
2012	993,460	13,967,523,000
2013	914,490	13,073,328,500
2014	3,563,490	20,947,587,200

2.3 Data analysis

Daily SST and chlorophyll-a data for period of 2011–2012 were composed into monthly data using IDV (Integrated Data Viewer) software. The monthly composite images were then overlaid with sea surface wind.

3. Results and Discussion

3.1 Results

The result showed that during JJA, the upwelling firstly appeared in the southern Bali with low SST migrating westward towards the southern coast of West Java, as shown in figure 2. The migration path depends on the seasonal evolution of the wind. The upwelling diminished due to a wind reversal associated with the emergence of the Northwest Monsoon (DJF) and the propagation of the Indian Ocean's Kelvin wave. The interannual variability of upwelling in the southern Java is associated with ENSO. During El Niño, the upwelling of Java-Sumatra extends to near the equator and lasts longer until November.

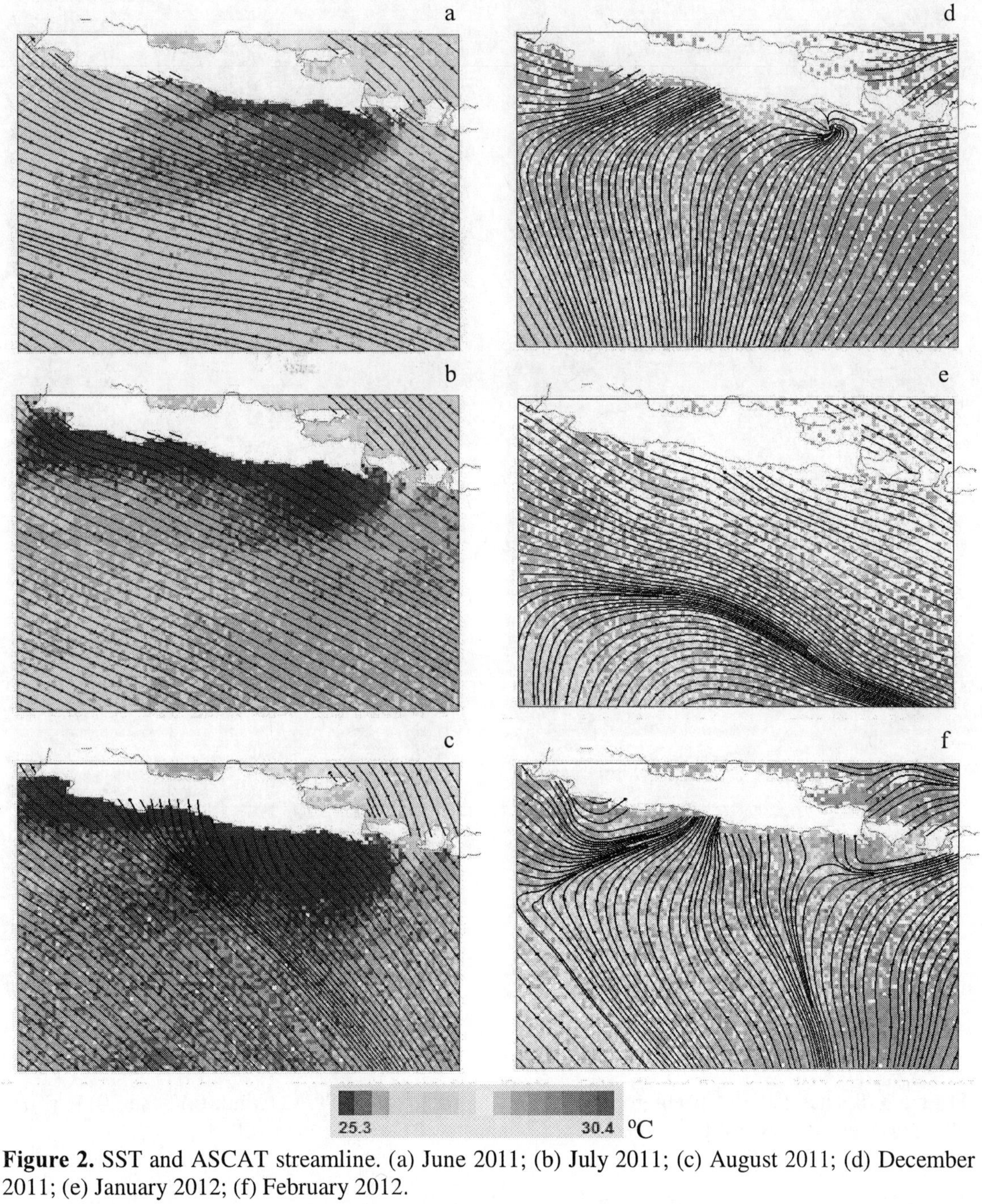

Figure 2. SST and ASCAT streamline. (a) June 2011; (b) July 2011; (c) August 2011; (d) December 2011; (e) January 2012; (f) February 2012.

IOP Conf. Series: Earth and Environmental Science **139** (2018) 012027 doi:10.1088/1755-1315/139/1/012027

The intensity of upwelling is in line with the wind speed. It has been suggested that the upwelled water mass from the subsurface is rich in nutrients. This is indicated by the high concentration of chlorophyll-a as shown in figure 3.

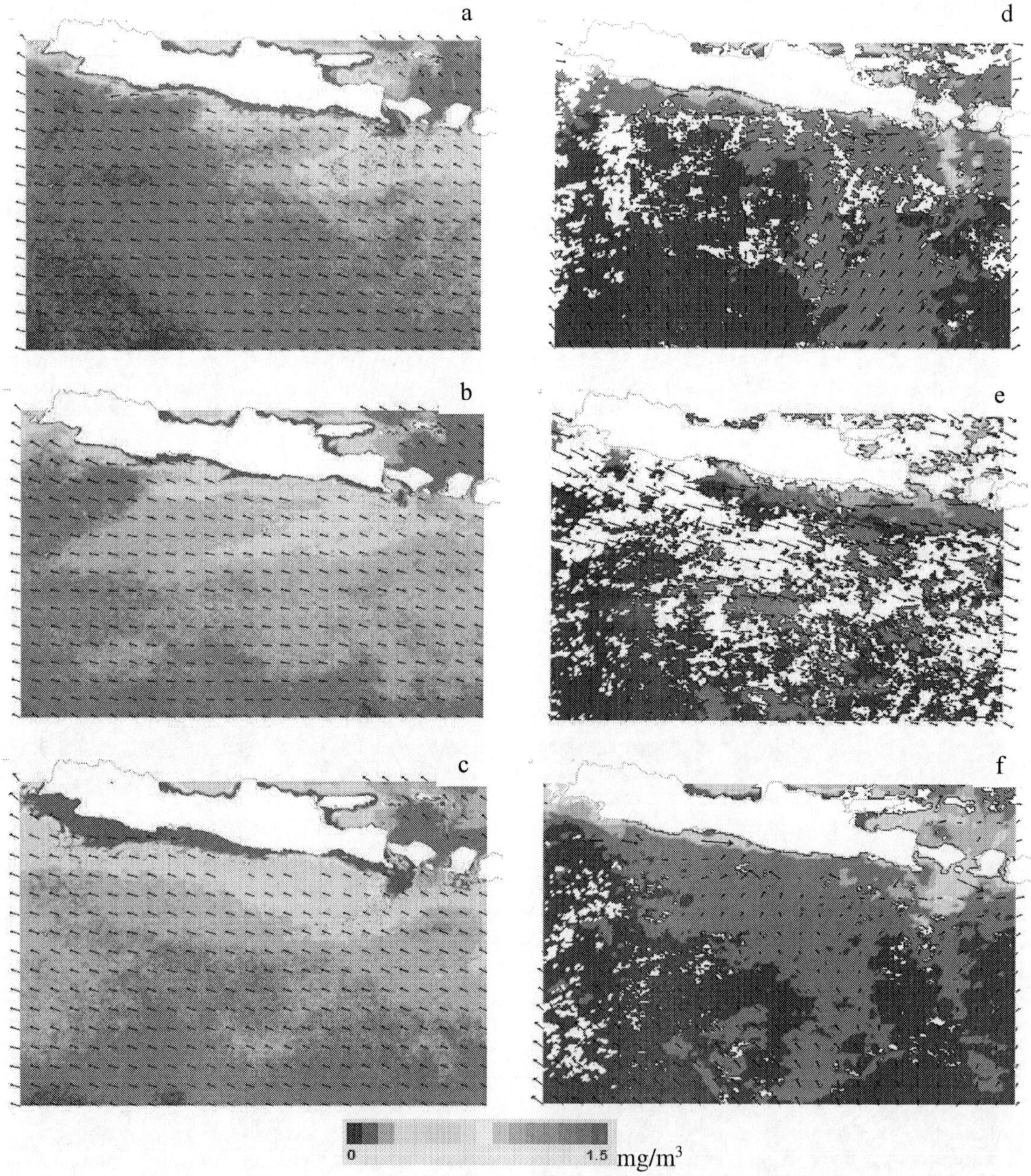

Figure 3. Spatial patterns of chlorophyll-a superimposed with ASCAT wind. (a) June 2011; (b) July 2011; (c) August 2011; (d) December 2011; (e) January 2012; (f) February 2012.

The opposite condition occurs during DJF. The concentration of chlorophyll-a in the south of the Java is low which implies low fish catch. This condition is due to the wind blows toward the beach and forms a vortex. As a result eddies formed and endangered the fishermen.

If we plot the variability of SST and chlorophyll-a, it clearly seen that the variability of both parameters follows the monsoon variability, as shown in figure 4. The low SST coincides with the higher chlorophyll-a concentration that occurs during JJA. Similarly, the higher SST corresponds to the lower chlorophyll-a concentration.

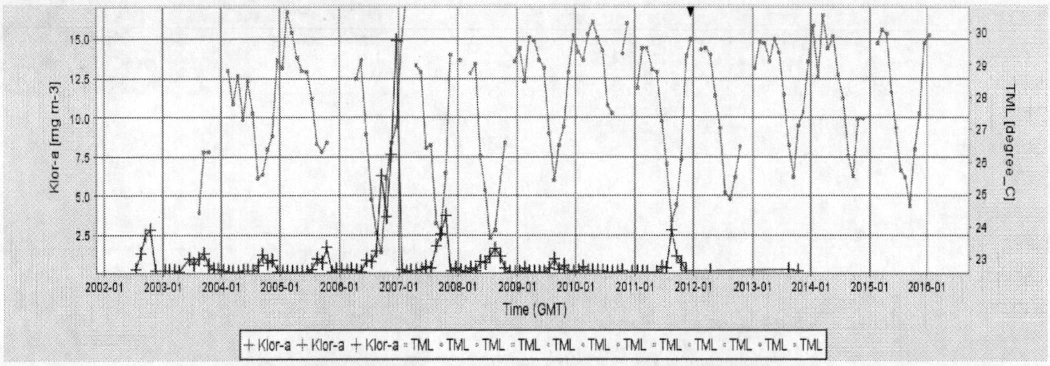

Figure 4. Temporal variability of SST and chlorophyll-a.

The results of HYCOM model are shown in figures 5 and 6 respectively for JJA and DJF included with a cross-section of the coastline around Sadeng, Yogyakarta offshore as far as 780 km (indicated by line + ---------■) in figures 5 and 6. In June the model was unable to capture upwelling. Thus no visible appearance of cold water pools on the surface and cross sections showed that the thermocline layer was at an average depth of 60 m around the coast. In July and August, the model shows the emergence of cold water pools that originated in the southern Bali that spread westward towards the Sunda Strait. From the cross section, it can be seen that the thermocline layer moves upward approaching the sea surface. Model also shows that the upwelling intensively take place in the southern Java during JJA.

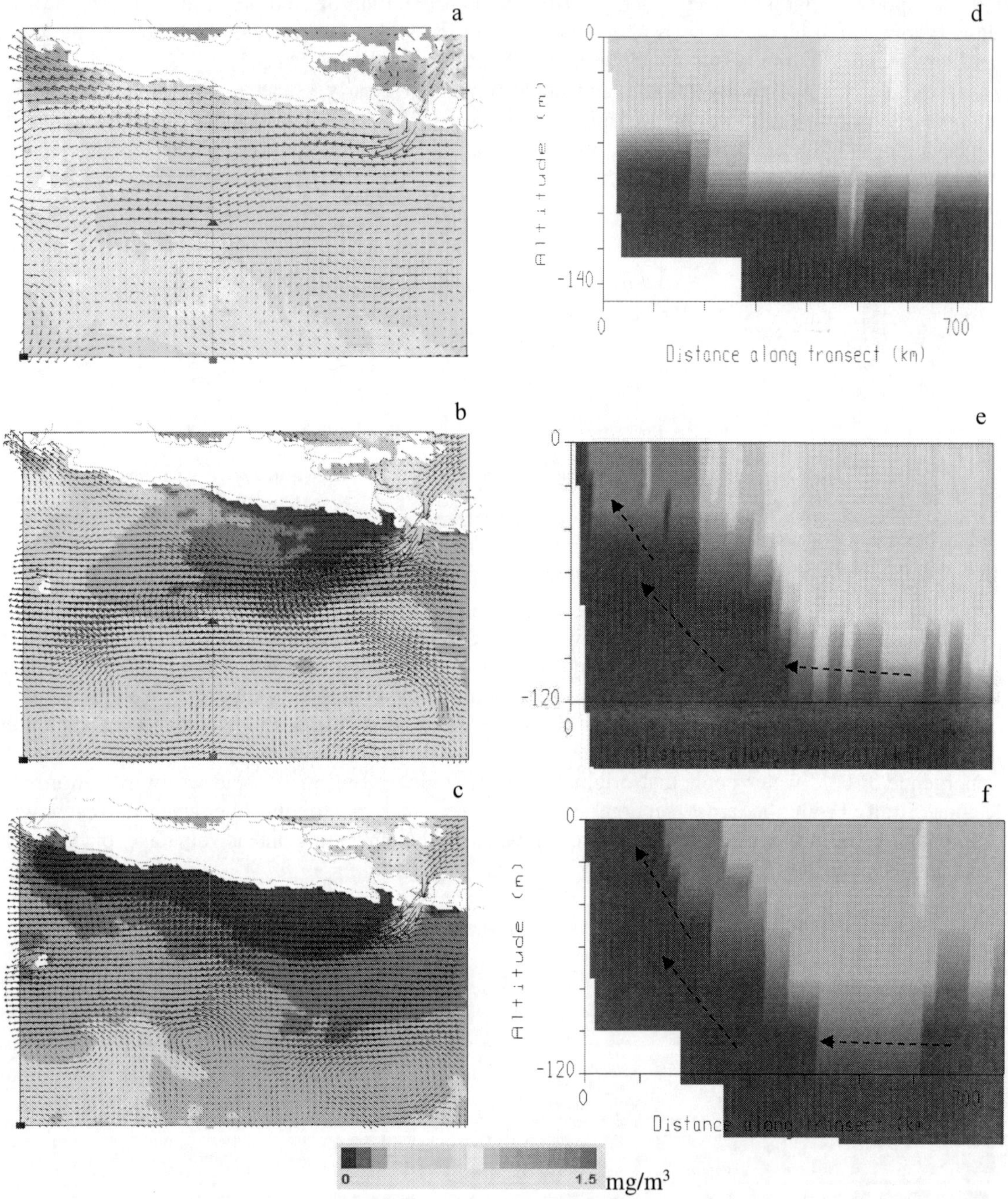

Figure 5. SST and sea surface current from the HYCOM model (a, b, c) and cross-section of ocean temperature (d, e, f) from the coastline (0) to 780 km offshore during JJA. (a, b) June 2011; (c, d) July 2011; (e, f) August 2011.

During the Northwest Monsoon, the SST is relatively high (> 26.3 °C) and the thermocline layer is relatively deep about 60 m as shown in figure 6.

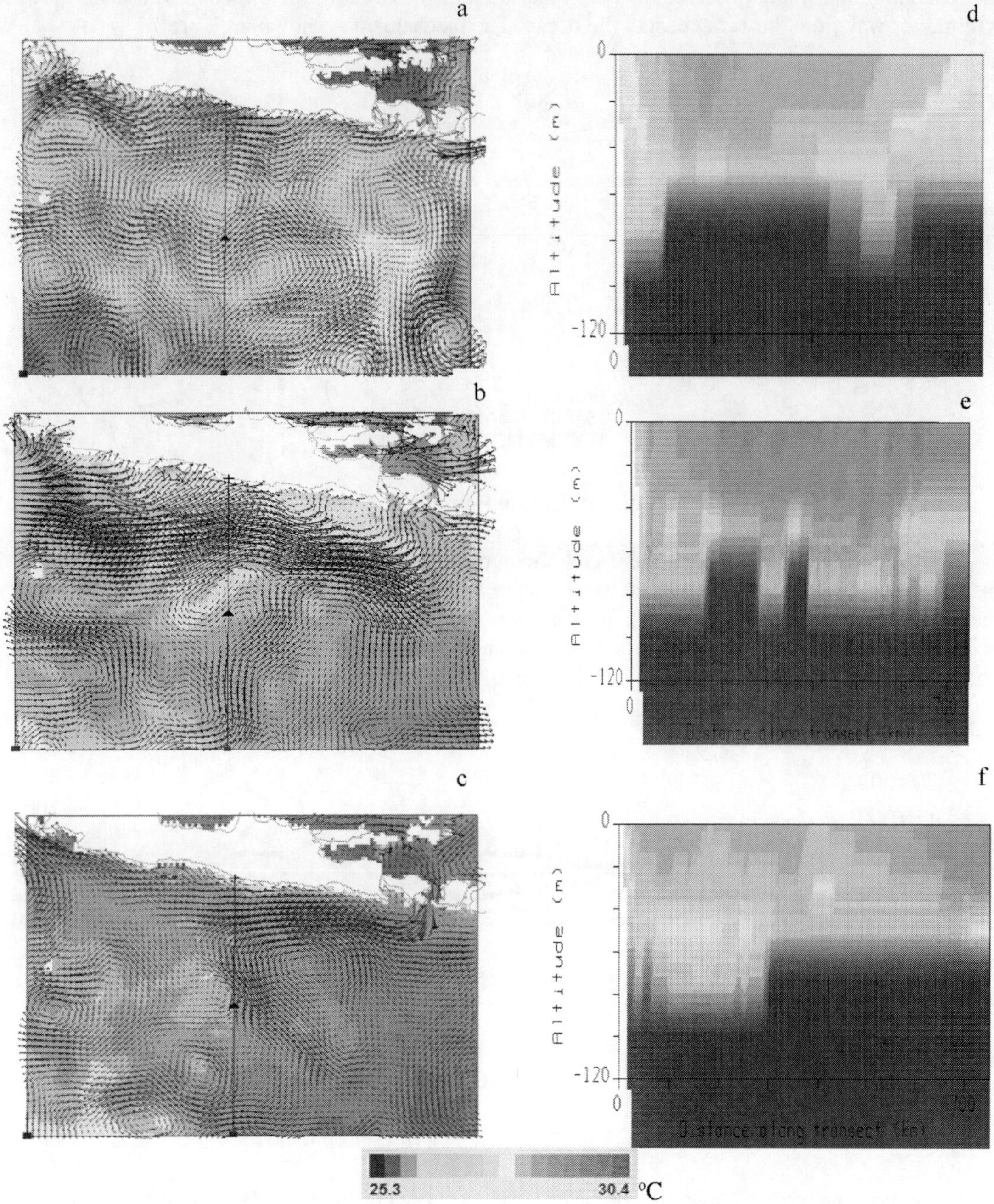

Figure 6. SST and sea surface current (a, b, c) and the monthly mean ocean temperature (d, e, f) of DJF. (a, b) December 2011; (c, d) January 2012; (e, f) February 2012.

3.2 Discussion

SST is an important indicator for coastal upwelling event. Cooler SST than the surrounding area is indicated by the coastal upwelling processes through the uplifting of the cooler water mass from the deeper layer to the surface. Monsoon wind can generate current, wave and wind stress. During JJA southeasterly winds interact with sea levels through wind stress will generate Ekman transport. Since the location of the southern sea of Java is in the south of the equator, the direction of Ekman transport will move away from the Java coast to the ocean. The mechanism is shown in figure 7.

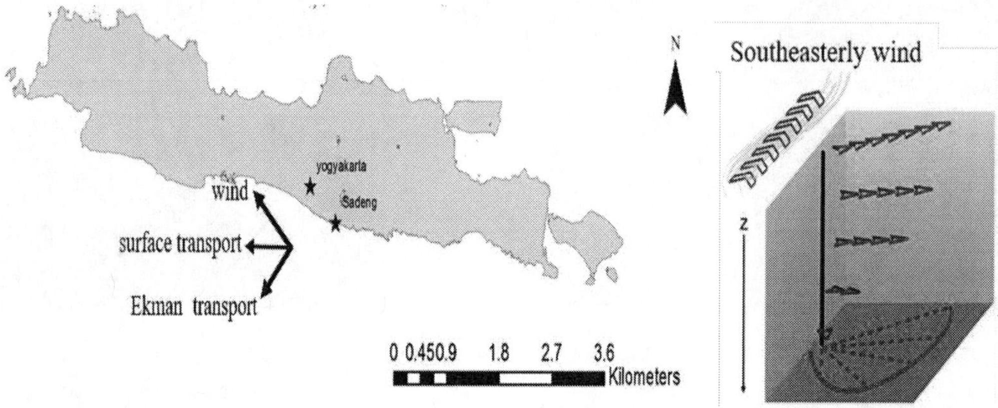

Figure 7. The mechanism of Ekman transport in the southern sea of Java.

Due to the Ekman's transport, there is a vacuum of water masses around the coast. The water is then replaced by the water mass from deeper layer with a cooler temperature of 25.3 °C than that in the surface. This water mass is rich in nutrients and is important for the marine food web. The occurrence of coastal upwelling and the higher of chlorophyll-a in the southern Java during JJA is in a good agreement with the increased of fish landing in the region (figure 8 and figure 9).

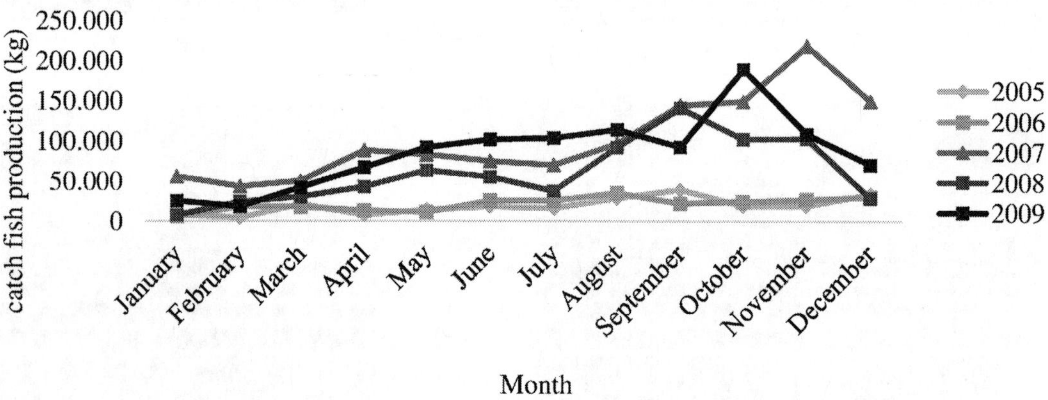

Figure 8. The montly fishcatch data from 2005–2009 [17].

There were interesting things happened in 2010, as shown in figure 9. The production of the catch is relatively higher in other months. 2009/2010 is the year of El Niño, it is anticipated that the annual upwelling variability across the coasts of Java and Sumatra is coupled to ENSO through ocean-atmosphere teleconnection [12]. The mutually reinforcing process between ENSO and AIM can generate a more intensive upwelling process and increase the abundance of chlorophyll-a around the southern coast of Java.

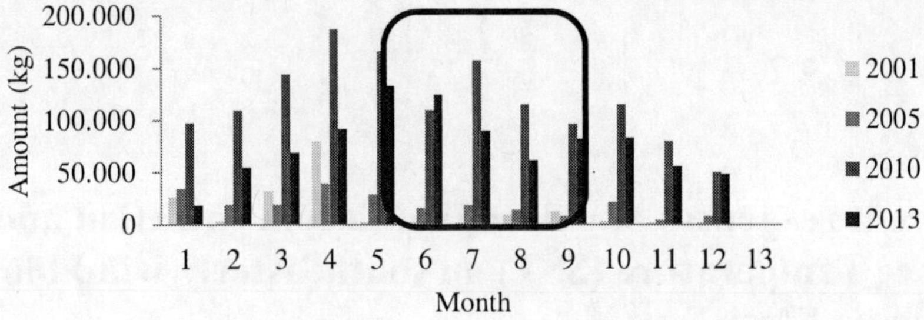

Figure 9. Monthly fish catch at the Fishery Seaport Sadeng [18].

4. Conclusions

The influence of Southeast and Northwest monsoons on the variability of SST and chlorophyll-a have been discussed in the present study. During JJA, the SST in the southern coast of Java was relatively low (25.3 ºC) compared to adjacent sea (26.32 ºC). The concentration of chlorophyll-a during this period is relatively high compared to DJF. This phenomenon is a result of the coastal upwelling caused by the Ekman transport that is triggered by the wind stress. Futhermore, our results show that higher fish catch is observed during the upwelling season.

Acknowledgments

The author would like to thank the NASA JPL, NOAA, and NOAA ESRL, PSTA LAPAN, PEMDA Yogyakarta, for providing access to the data that enabled this research.

References

[1] Longhurst A 1993 *Deep-Sea Res.*I **40** 2145–2165
[2] Agency for Marine and Fisheries Research (AMFR) 2001 *Report on Indonesia Fish Stock Assessment National Fish Stock Assessment Commission Indonesia* (Jakarta: Ministry of Marine and Fisheries Affairs)
[3] Hendiarti N, Suwarso, Aldrian E, Amri K, Andiastuti R, Sachoemar S I, and Wahyono I B 2005 *Oceanogr.* **18** 112-123
[4] Veron J E N 1995 *Corals in Space and Time: The Biogeography and Evolution of the Scleractinia* (New York : Cornell Univ. Pr.) p 321
[5] Saji N H, Goswami B N, Vinayachandran P N, and Yamagata T 1999 *Nature* **401** 360–363
[6] Webster P J, Moore A M, Loschnigg J P, and Leben R R 1999 *Nature* **401** 356–360
[7] Feng M, and Meyers G 2003 *Deep Sea Res.* II **50** 2263–2284
[8] Madden R A, and Julian P R 1994 *Mon. Weather Rev.* **122** 814–837
[9] Ffield A, and Gordon A L 1996 *J. Phys. Oceanogr.* **26** 1924–1937
[10] Sprintall J, Gordon A L, Murtugudde R, and Susanto R D 2000 *J. Geophys. Res.* **105** 17217–17230
[11] Susanto R D, Gordon A L, Sprintall J, and Herunadi B 2000 *Geophys. Res. Lett.* **27** 1499–1502
[12] Susanto R D, and Marra J 2005 *Oceanogr.* **18** 124–127
[13] Susanto R D, Moore T S II, and Marra J 2006 *Geochem. Geophys. Geosyst.* **7** 1–16
[14] Asanuma I, Matsumoto K, Okano H, Kawano T, Hendiarti N, and Sachoemar S I 2003 *J. Geophys. Res.* **108** 1–14
[15] Slingo J 2003 Overview *Encyclopedia od Atmospheric Sciences* ed Holton J R, Curry J A, Pyle J A (New York: Academic Press) pp 1365–1370
[16] Ramage C S 1971 *Monsoon Meteorology* (New York: Academic Press) p 296
[17] Rahmi T A 2010 *Thesis* (Bogor: Bogor Agricultural Institute)
UPTD [Unit Pelaksana Teknis Daerah] 2015 *Profil Pelabuhan Perikanan Pantai (PPP) Sadeng* (Yogyakarta: Maritime and Fisheries Affairs)

IOP Conf. Series: Earth and Environmental Science **139** (2018) 012028 doi:10.1088/1755-1315/139/1/012028

Different responses of chlorophyll-a concentration and Sea Surface Temperature (SST) on southeasterly wind blowing in the Sunda Strait

A Wirasatriya[1,2]*, Kunarso[1,2], L Maslukah[1,2], A Satriadi[1], R D Armanto[1]

[1] Department of Oceanography, Faculty of Fisheries and Marine Science, Universitas Diponegoro, Semarang, Indonesia
[2] Center for Coastal Disaster Mitigation and Rehabilitation Studies, Universitas Diponegoro, Semarang Indonesia

*E-mail: aninosi@yahoo.co.id

Abstract. During southeast monsoon, along the western coast of Sumatra Island and southern coast of Java Island are known as the coastal upwelling areas denoted by the occurrence of Sea Surface Temperature (SST) cooling and chlorophyll-a blooming. Located between Sumatra and Java Islands, Sunda Strait waters may give different response to the southeasterly wind blowing above. Using SST and chlorophyll-a data obtained from daily MODIS level 3 during 2006–2016, this study demonstrated the evidence on how bathymetry and topography modified the effect of southeasterly wind on the spatial variability of SST and chlorophyll-a. All datasets were composed into monthly and monthly climatology. The area in the center of Sunda Strait had the lowest chlorophyll-a concentration and the warmest SST during the peak of upwelling season. The deep bottom topography and the absence of barrier land prevented the generation of wind driven coastal upwelling. However, the chlorophyll-a concentration in this area had the highest correlation with the wind speed which means that the variation of chlorophyll-a concentration in this area was highly depended on the variability of wind. On the other hand, the areas with shallow bathymetry and in front of Panaitan and Java Islands had higher chlorophyll-a concentration and cooler SSTs.

1. Introductin

Upwelling is known as the lifting of water mass from the deep water column to the sea surface. In the sea surface, upwelling can be indicated by lower SST and higher chlorophyll-a (Chl-a) concentration since the lifting of water mass carries cold water and rich in nutrient which is required for the growth of phytoplankton. The high Chl-a concentration during upwelling period is often followed by the increase in fisheries productivity [1, 2, 3]. Thus, SST and Chl-a are important indicators for determining the fishing ground and season.

The area along southern coast of Java and Sumatra is well known as the upwelling area occurs during the southeast monsoon [4, 5, 6]. The southeasterly wind blowing along the southern coast of Java Island and Sumatra Island generates Ekman transport which carries water mass away from the coast. The water mass loss in the coastal area is then filled by the water mass from the deeper layer. Thus, the upwelling along the coast of Java and Sumatra Islands is classified as the coastal upwelling.

The Sunda strait is a narrow channel separating the Java and Sumatra Islands. As a whole entity, the Sunda Strait has been investigated by [6] which demonstrated as one of the three strongest upwelling areas along the southern coast of Java and Sumatra in 2006. The positive Chl-a anomaly of greater than 3 mg/m^3 extended offshore for more than 200 km from October to November 2006. Nevertheless, the smaller scale pattern of Chl-a concentration and SST in the south side of the Sunda Strait mouth was neglected in the previous studies. With the width of mouth passage of about a hundred kilometers, varied bottom topography and the existence of small islands, the southeasterly wind blowing over the Sunda Strait may give various impact on the pattern of Chl-a concentration and SST distribution. It has been known that, wind speed is the most important factor for the development of coastal upwelling along the southern coast of Java and Sumatra Islands [1, 2, 3]. In the present study, we demonstrated on how the same speed of southeasterly wind can give the various impact on the Chl-a and SST distribution in a relatively narrow area i.e. the Sunda Strait.

Because of the wide-ranging spatial and temporal coverages, satellite remote sensing has often been used to investigate the ocean surface conditions in the Indonesian Seas [1, 2, 4, 5, 6, 7, 8, 9]. The ocean surface dynamics like upwelling, SST front, and eddies can be monitored by using satellite remote sensing technology. Utilizing the advantage of remote sensing technology, the present study investigates Chl-a and SST variability in the Sunda Strait.

2. Materials and Methods

The main dataset used in this study were MODIS Aqua Lv3 for Chl-a and SST with spatial resolution of 0.04° × 0.04° and observation period from 2006 to 2016 [10]. For SST, we used MODIS SST 11 μm which cover the observation both for day and night. These datasets are developed and distributed by the Ocean Biology Processing Group. The algorithm for generating MODIS SST is described in [11] while for MODIS Chl-a is in [12, 13]. These datasets have been examined and validated against *in situ* data to ensure the best accuracy [14, 15, 16, 17, 18].

To investigate the formation mechanism of Chl-a and SST pattern in the Sunda Strait, we used surface wind data obtained from six hourly ERA interim data with grid interval of 0.125° [19]. We composed SST, Chl-a, and surface wind parameters into monthly and monthly climatology following [20]. We also calculated Ekman Depth following this equation [21]:

$$\boldsymbol{D_E} = \frac{\mathbf{4.3}\boldsymbol{W}}{\sqrt{\mathbf{sin}|\boldsymbol{\emptyset}|}} \quad (1)$$

where D_E is the Ekman depth (m), W is wind speed (m/s) and ϕ is latitude (°).

To obtain the bottom topography, we used Shuttle Radar Topography Mission (SRTM 30) with the spatial resolution of 30 seconds or equal to one km. SRTM 30 data is developed by National Aeronautics and Space Administration and can be accessed at http://topex.ucsd.edu/.

3. Results and Discussion

3.1. Spatial distribution of Chl-a and SST during upwelling season in the Sunda Strait

The spatial pattern of Chl-a, SST, and surface wind in the Sunda Strait during September is shown in figure 1. We show the period of September since the peak of upwelling season occurs in this month. During September, surface wind blew southeasterly. This condition was favourable for coastal upwelling generation. Outside the strait mouth, wind speed was about 6–7 m/s and decreased entering the strait.

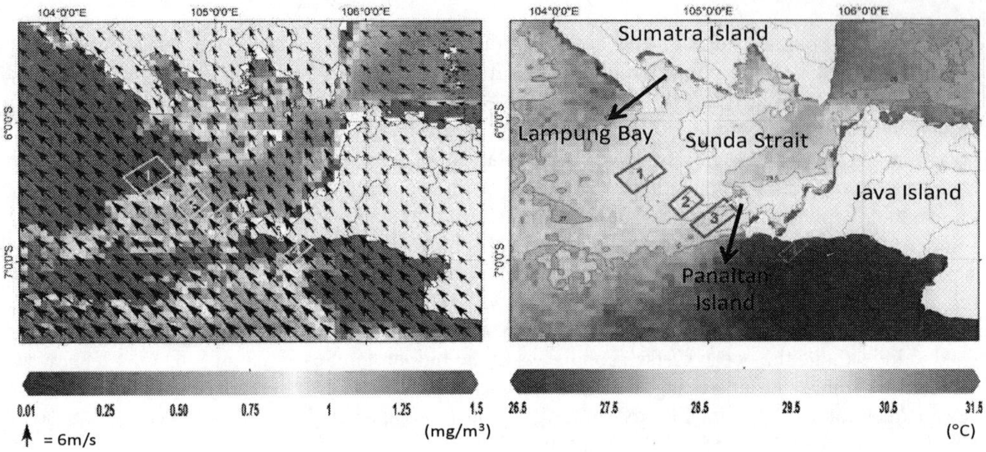

Figure 1. Distribution of Chl-a concentration and surface wind vector (left) and SST (right) in the Sunda Strait in September (2006–2016). Boxes in the figure denote the area for temporal analysis shown in figure 2. Detail coordinates of the areas are described in text.

The upwelling area was observed along the southern coast of Java Island until southwestern coast of Sumatra Island indicated by the high Chl-a (SST) bands of greater than 1 mg/m^3 (less than 27.5 °C).

Focusing at the mouth of the strait, we can easily capture the distinctive spatial pattern of Chl-a and SST. The area in the center of the Sunda Strait shows the lowest Chl-a concentration and the warmest SST during the peak of upwelling season. On the other hand, near the tip of Sumatra Island and in front of the Panaitan Island and Java Island exhibit higher Chl-a concentration and cooler SSTs. The absence of barrier land in the center of the Sunda Strait prevents the formation of coastal upwelling. However, a contradictive fact is found at the southern tip of Sumatra Island. Although there is no barrier land, the Chl-a concentration (SST) was high (low). For further analysis, we chose 4 areas at the mouth of the Sunda Strait, which represent the different characteristic of Chl-a and SST distribution in the Sunda Strait, to investigate their temporal variation as shown in figure 1. The areas are:

1. Southern tip of Sumatra Island, the corners are as follows: 104.616 °E, -6.242 °N; 104.722 °E, -6.364 °N; 104.510 °E, -6.518 °N; 104.396 °E, -6.394 °N.
2. The center of the Sunda Strait, the corners are as follows: 104.744 °E, -6.598 °N; 104.878 °E, -6.473 °N; 104.963 °E, -6.570 °N; 104.824 °E, -6.693 °N.
3. In front of the Panaitan Island, the corners are as follows: 105.104°E, -6.559 °N; 105.192 °E, -6.656 °N; 104.974 °E, -6.838 °N; 104.884 °E, -6.747 °N.
4. In front of the Java Island, the corners are as follows: N: 105.454 °E, -6.964 °N; 105.589 °E, -6.837 °N; 105.648 °E, -6.884 °N; 105.508 °N, -7.020°N.

3.2. Temporal variability of Chl-a and SST in the Sunda Strait

The temporal variation of Chl-a, SST, and wind speed in 4 areas is presented in figure 2. Figure 2a shows that the minimum (maximum) wind speed was about 1 m/s (6 m/s) which occurred in April (September). It also obviously shows that the wind speed variabilities among four areas were almost similar both in pattern and amplitude. The variation of Chl-a and SST followed the variation of wind speed. From January when the wind blew northwesterly to Jun when the wind speed was minimum, the Chl-a concentration (SST) was low (high). This is due to the absence of coastal upwelling which only generated under the condition of strong southeasterly wind. Despite the different amplitudes, the maximum (minimum) Chl-a concentration (SST) was reached during the southeast monsoon in

September. Thus, there was a different response on Chl-a and SST for each area toward the southeasterly wind blowing in the Sunda Strait.

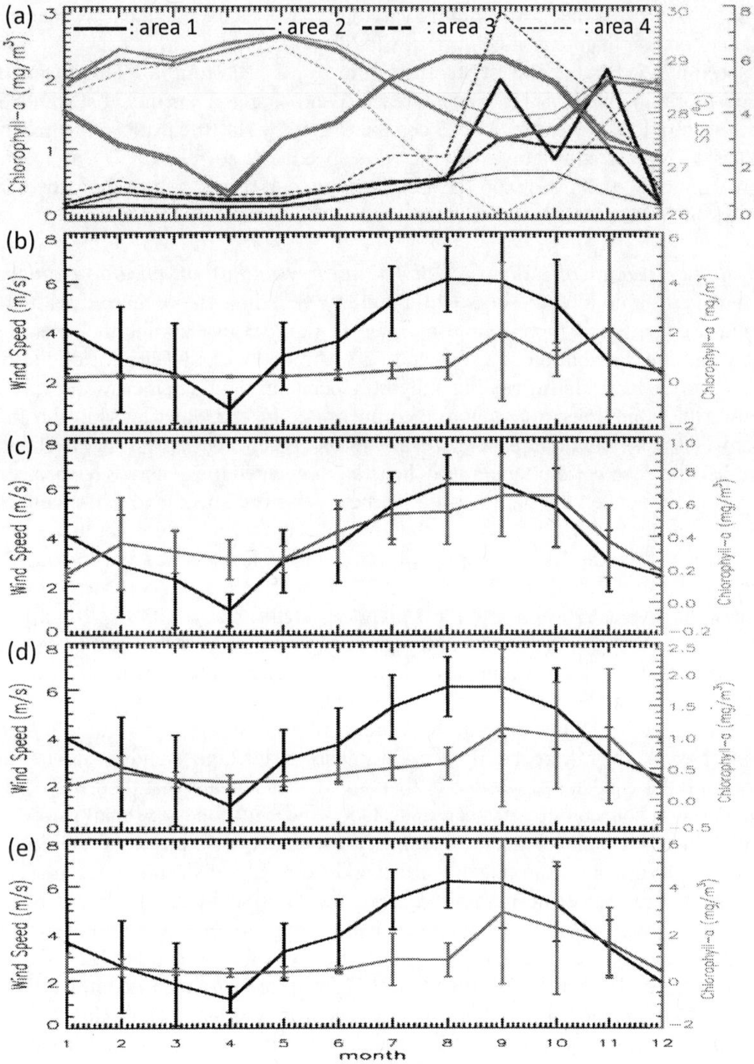

Figure 2. (a) Time series of wind speed, Chl-a, and SST in four areas as shown in figure 1. Adjusted scale of time series graphic in figure 2a and standard deviation for Chl-a and wind speed in (b) area 1, (c) area 2, (d) area 3, and (e) area 4.

The different responses of Chl-a and SST amplitudes in 4 areas toward the same wind speed blowing in the Sunda Strait are described as follows. As explained in the previous section, during the southeast monsoon the area 4 had the highest Chl-a concentration i.e., about 3 mg/m^3 and the lowest SST i.e. about 26 °C; area 2 had the lowest Chl-a concentration i.e. 0.5 mg/m^3 and highest SST i.e. 28 °C; and area 1 and 3 was in between. There was almost no fluctuation of Chl-a concentration in area 2. Furthermore, figure 2b-e shows the adjusted scale of wind speed and Chl-a axis with their standard deviation for area

IOP Conf. Series: Earth and Environmental Science 139 (2018) 012028 doi:10.1088/1755-1315/139/1/012028

1 to 4. The standard deviation of wind speed for 4 areas were only about 2 m/s which means that from 2006 to 2016 the monthly wind speed blew in the Sunda Strait were relatively stable. During the southeast monsoon, the standard deviation of Chl-a in area 1, 3, and 4 was high, i.e. maximum of about 4 mg/m^3, 1.5 mg/m^3 and 2 mg/m^3 for area 1, 3, and 4 respectively. This means that the monthly Chl-a concentration in area 1, 3, and 4 was highly variable. After scale adjustment, we can see the variation of Chl-a in area 2 clearer than presented in figure 2a. The variation of Chl-a in area 2 followed the variation of wind speed and it has the lowest standard deviation i.e. 0.2 mg/m^3. This means that although the amplitude of variation is not as large as in area 1, 3 and 4, area 2 has the most stable monthly variation of Chl-a concentration. Moreover, it has been confirmed by statistical analysis, the Chl-a concentration in area 2 has the highest correlation with the wind speed. The correlation of Chl-a and wind speed for 4 areas were 0.35, 0.80, 0.60, and 0.66 for area 1, 2, 3 and 4, respectively. Thus, the Chl-a concentration in area 2 was highly depended on the wind speed.

3.3. Relation between Chl-a and SST distribution with the bottom topography

Back to the question on how the high Chl-a and low SST in area 1 occurred without the existence of barrier land, we used bathymetry data presented in figure 3 to examine the relation between bottom topography and the distribution of Chl-a and SST. Areas 1, 3 and 4 are located at the shallow areas. Despite no barrier land, the upwelling may still occur in area 1, since water mass loss due to the southwestward Ekman transport cannot be compensated by the water mass in the shallow area. Thus, the deeper water mass should replace that water mass loss through upwelling mechanisms.

In contrast, the deep bottom topography in area 2 prevented the generation of coastal upwelling. The physical process regulated the high correlation between wind speed and Chl-a concentration in area 2 may be related to the wind generated mechanical mixing instead of coastal upwelling. With the wind speed of about 6 m/s, the Ekman depth in area 2 in September was only about 73 m. This is not comparable with the depth of area 2, i.e. about 2000 m (figure 3b). As shown by many researchers at various seas, the concentration of nutrients increases significantly with depth until about 1000 m and remain stable after that [22, 23]. Thus, although this mixing process still can entrain the nutrients from the deeper layer to the surface, it cannot reach the richest nutrient layer which is located near the bottom of the sea. Thus, the amplitude of Chl-a variation in area 2 was lower than the other three areas.

The previous section shows that although the amplitudes of Chl-a variation in area 1, 3 and 4 were higher than area 2, they were highly variable denoted by high standard deviation. Moreover, the correlation with the wind speed was lower than area 2. This means that there are other possible factors influence the Chl-a concentration instead of surface wind. Since they are shallow and located near land, freshwater discharge which bring nutrients from the land may also influence the Chl-a concentration in these 3 areas. The variable anthropogenic effect and freshwater discharge influenced by precipitation to this three areas may cause the monthly Chl-a concentration in these 3 areas become variable. As discussed by [1, 2] the anthropogenic factor such as trading harbour, agricultural rice field, and fishpond activities, all together contribute to the nutrient dynamics of the Lampung Bay which is located in the northern part of area 1. These possibilities are left for future studies. In contrast, these factors may be neglected for area 2 since it is deep and located far from the land. Thus, this indication supports the statement in the previous section i.e. the variability of Chl-a in area 2 was strongly depended on the variability of surface wind.

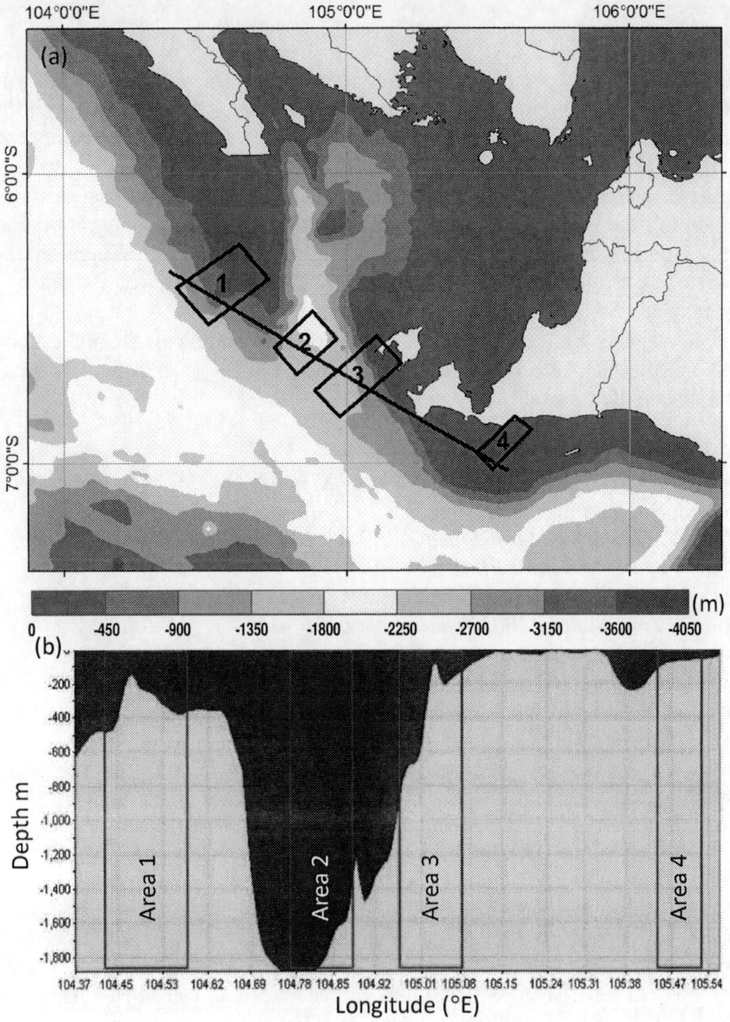

Figure 3. (a) Bathymetry of the Sunda Strait, (b) depth profile of the black line in figure 3a.

4. Conclusion

The detailed features of Chl-a concentration and SST in the Sunda Strait during the southeast monsoon has been investigated by using long-term observation of MODIS data (2006-2016). The results are concluded as follows:

1. Although the speed of surface wind is the same, the different characteristics of bottom topography and the existence of barrier land caused different distribution of Chl-a and SST in the mouth of the Sunda Strait.
2. The area with shallow bathymetry in the southern tip of Sumatra Island and the areas in front of the Panaitan and Java Islands demonstrates higher Chl-a concentration and cooler SSTs. Southeasterly wind generated the coastal upwelling in these areas may entrain more nutrients from the sea floor to the surface than the deep area without barrier land in the center of the Sunda Strait.

IOP Conf. Series: Earth and Environmental Science **139** (2018) 012028 doi:10.1088/1755-1315/139/1/012028

3. The variability of chlorophyll-a concentration in the areas with shallow bathymetry and in front of the Panaitan Island and Java Island indicates lower correlation with the wind speed. This means that there are other possible factors influence the chlorophyll-a concentration instead of surface wind.
4. The area in the center of the Sunda Strait depicts the lowest Chl-a concentration and the warmest SST during the peak of upwelling season. The deep bottom topography and the absence of barrier land prevented the generation of wind driven coastal upwelling. Surface wind may only generate mechanical mixing instead of upwelling which made the intensity of the Chl-a blooming and SST cooling was less than the other areas.
5. The Chl-a concentration in the center of the Sunda Strait reveals the highest correlation with the wind speed which means that the variation of Chl-a concentration in this area was highly depended on the variability of wind speed.

References

[1] Sachoemar S I, Yanagi T, Hendiarti N, Sadly M and Meliani F 2010 *Coast. Mar. Sci.* **34**(1) 82
[2] Sachoemar S I, Yanagi T, and Aliah R S 2012 *Coast. Mar. Sci.* **37**(2) 75
[3] Friedland K D, Stock C, Drinkwater K F, Link J S, Leaf R T, Shank B V, Rose J M, Pilskaln C H and Fogarty M J 2012 *PLoS ONE* 7(1) 1371
[4] Susanto R D, Gordon A L and Zheng Q 2001 *Geophys. Res. Let.* **28**(1) 1559
[5] Susanto R D and Marra J 2005 *Oceanography* **18**(4) 124
[6] Iskandar I, Rao S A and Tozuka T 2009 *Int. J. of Rem. Sens.* **30**(3) 663
[7] Susanto R D, Moore I I T and Marra J 2006 *Geochem. Geophys. Geosyst.* **7**(5) 1
[8] Setiawan R Y and Kawamura H 2011 *IEEE J. Sel. Topics Appl. Earth Observ. Remote Sens. (JSTARS)* **4**(1) 241
[9] Setiawan R Y and Habibi A 2011 *IEEE J. Sel. Topics Appl. Earth Observ. Remote Sens. (JSTARS)* **4**(4) 944
[10] Esaias W E *et al.* 1998 *IEEE Transactions on Geoscience and Remote Sensing* **36**(4) 1250
[11] Brown O B and Minnett P J 2009 (http://modis.gsfc.nasa. gov/data/atbd/atbd_mod25.pdf)
[12] O'Reilly J, Maritorena S, Mitchell B G, Siegel D, Carder K L, Garver S, Kahru M and McClain C 1998 *J. Geophys. Res.* **103** 24937
[13] Hu C, Lee Z and Franz B 2012 *J. Geophys. Res.* **117**(C1)
[14] Moore T S, Campbell J W and Dowell M D 2009 *Remote. Sens. Environ.* **113** 2424
[15] Zhang C Y, Hu C M, Shang S L, Muller-Karger F E, Li Y, Dai M H, Huang B Q, Ning X R and Hong H S 2006 *Remote. Sens. Environ.* **102** 250
[16] Ghanea M, Moradi M, Kabiri K and Mehdinia A 2015 *Adv. Space Res.* **57**(1) 10
[17] Lee M A, Tzeng M T, Hosoda K, Sakaida F, Kawamura H, Shieh W J, Yang Y and Chang Y 2010 *Terr. Atmos. Ocean Sci.* **21**(4) 727
[18] Qin H, Chen G, Wang W, Wang D and Zeng L 2014 *Int. J. Remote Sens.* **35**(11–12) 4315
[19] Dee D P *et al.* 2011 *Q.J.R. Meteorol. Soc.* **137** 553
[20] Wirasatriya A, Setiawan R Y and Subardjo P 2017 *IEEE J. Sel. Topics Appl. Earth Observ. Remote Sens.* **10**(12) 5513
[21] Pond S and Pickard G L 1991 *Introductory Dynamical Oceanography, 2nd edition* (New York: Pergamon)
[22] Hayase K and Shinuzoka N 1995 *Mar.Chemis.* **48**(3-4) 283
[23] Sunda W G 2012 *Frontiers in Microbiology* **3** 1

IOP Conf. Series: Earth and Environmental Science **139** (2018) 012029 doi:10.1088/1755-1315/139/1/012029

Local government units initiatives on coastal resource management in adjacent municipalities in Camarines Sur, Philippines

A Z Faustino[1*] and H L Madela[1*]

[1] Central Bicol State University of Agriculture, Calabanga Campus Sta. Cruz, Calabanga, Camarines Sur, Philippines

*Email: daines_711@yahoo.com; helen.madela@cbsua.edu.ph

Abstract. This research was conducted to determine the local government units (LGUs) initiatives on coastal resource management (CRM) in adjacent municipalities in Camarines Sur, Philippines. The respondents of this study are 100 fisherfolk leaders in the municipalities of Calabanga, Tinambac and Siruma. Descriptive, comparative and evaluative methods of research were employed and a survey questionnaire was used as the primary tool in data gathering. On the test of difference, the computed F-value of 12.038 and p-value of .001 revealed a very high difference in the implementation of CRM initiatives in the adjacent municipalities. The respondents in this study live below the poverty threshold. The intrusion of commercial fishers and the use of active fishing gears inside the 15-km municipal waters significantly affect the marine habitat while fishpond conversion kills the natural cycle in the mangrove forests. However, the FOs membership in the Municipal Fisheries and Aquatic Resources Management Council empower them to engage in governance which can be a venue for them to recommend policies related to CRM. As a result of this study, a CRM monitoring and evaluation model was crafted to guide the LGUs in the review, revision and crafting of CRM programs.

1. Introduction

The Philippines, with a general coastline of 18,000 sq km, is composed of 7,100 islands [1]. Its 2.2 million square kilometers territorial waters is considered its Exclusive Economic Zone (EEZ) [2]. Its coastal ecosystems namely coral reefs, seagrass beds and mangroves, said to be among the most biologically diverse in the world, are interconnected forming an intricate web of food chain essential to the survival of living organisms found in it. However, these resources are continuously exploited and it depleted the fishery resources of the Country. This brought the national government agencies (NGAs), non-government organizations (NGOs) and academic institutions to act on the rapid degradation of the coastal resources [3].

Coastal management in the Philippines started as early as the 1980s, even before the Earth Summit and the United Nations Conference on Environmental Development (UNCED) in 1992 which concretized the concept of Integrated Coastal Management [4]. Implementation of coastal management in the country is a result of various economic, environmental and governance concerns affecting the poor fishing communities. Most of these initiatives were started by academic institutions and non-government organizations (NGOs) such as the Siliman University and Haribon Foundation. Both of these institutions/organizations applied community-based approaches and showed a great contribution to the success of coastal management and one of which is the island-based community fish sanctuary in Apo Island in Negros Occidental. The engagement of the fishing community in governance brought them to be in the front-line of environmental protection and management of their resources [5]. Since most of the initiatives are school-based and NGO led, LGUs participation in this was limited to permitting these groups to implement CRM program, participation in the planning process and provision of manpower to assist the program implementers.

This set-up had changed when the Local Government Code (LGC) or Republic Act 7160 (RA7160) was enacted into law [6]. The LGC gave the LGUs the autonomy to manage the affairs of their respective municipality and the resources within their territorial jurisdiction. This devolution of powers, from a centralized flow of government to devolved function, gave the LGUs the authority to protect and conserve their coastal environment. This mandate of the local government was further cemented upon the enactment of the Fisheries Code of the Philippines or Republic Act 8550 (RA 8550) in 1998 [7]. The salient features of RA 8550, to name a few, are (a) extending the municipal waters from 4.83-km (3 nautical miles) to 15-km (9.32 nautical miles); (b) creation of the Fisheries and Aquatic Resources Management Council (FARMC) which will recommend necessary and appropriate policies and ordinances related to fisheries development and (c) designating competent deputy fish wardens from fisherfolk organizations who will partner with the local police in patrolling and surveillance operations. These provisions under RA 8550 gave the LGUs and the communities the opportunity not only to explore but also to preserve their coastal and marine resources. Furthermore, this change brought the local government to develop programs on coastal management and eventually formulating CRM plans to address the continuing and worsening situation in the coastal environment. Since then, the LGUs became a partner implementer of the different NGAs, NGOs and academic institutions in CRM.

Numerous studies have already been conducted on coastal resource management (CRM) and fisheries development in the Philippines focusing on its impacts on the environment and the economic well-being of the fishing communities and most of these programs though are initiated by NGAs, NGOs and academic institutions. Seldom that the LGUs initiatives are reviewed through the 'eyes' of the fisherfolk. It is in this permise that the researcher wishes to assess the LGUs initiatives on CRM in the adjacent municipalities in Camarines Sur, Philippines. The researcher also hopes to provide a CRM Monitoring and Evaluation (M & E) model which can serve as guide to the LGUs and the fisherfolk organizations to strengthen their CRM programs.

2. Methods

The respondents in this study were purposely selected in the municipalities of Calabanga, Siruma and Tinambac, in the Province of Camarines Sur, Philippines. Three (3) coastal communities with active fisherfolk organizations (FOs) in each municipality were chosen and its elected leaders, as 'gatekeepers' of their respective FO, were identified as respondents of the survey. They are the best source of information needed in the study as they are the most aware of the initiatives being done by the LGUs on CRM. A total of 100 respondents were interviewed using a survey questionnaire which was developed by the Researcher. Focus group discussions (FGDs) and key informant interviews (KIIs) with the key personnel of the Office of the Municipal Agriculturist (OMA) in each municipality were conducted to validate the responses of the respondents.

This study used descriptive-evaluative and descriptive–comparative research methods employing survey technique. Descriptive research, as explained by [8], is a collection of data describing some phenomenon that is may or may not be quantifiable such as close-scales, open-ended survey questions, observations and interviews. Descriptive-evaluative, on the other hand, evaluates or measure the results against some known or hypothesized standards. Descriptive method is most applicable in determining the profile of the respondents regarding age, sex, household size and position in the organization. It is likewise appropriate for identifying the fishery-related issues and concerns along economic, environment and governance. The descriptive-evaluative method, on the other hand, was used to determine the CRM initiatives of the LGUs and its level of implementation in the adjacent

IOP Conf. Series: Earth and Environmental Science **139** (2018) 012029 doi:10.1088/1755-1315/139/1/012029

municipalities, while the difference in the level of implementation was analyzed using the descriptive-comparative method of research. It further looks into the factors that caused the difference.

3. Results and Discussion

The demographic profile of the respondents, concerning age, most of them are nearing their senior years and only a few youngsters are involved in the organizational operations. It is also worth mentioning that females are becoming active in the formerly male-dominated fishing industry. Females are well represented in the organization and given positions not only as Secretaries but also as Presidents and Board of Directors. Based on the interview with the fisherfolks, positions in the organizations are given to older members because they are more experienced and most trusted by the LGUs. The younger generation, however, is given responsibilities in the different working committees becoming their training ground in preparation for their eventual handling of leadership positions in the future. It is true that most often the adults are given responsibilities in community building, it is also worthy to note that the youth's engagement in the development of communities is crucial as they are the second liners of the seniors. To encourage youth involvement in the community is vital to understand the influences, motivations, obstacles and feedback that they receive from the community [9].

The LGUs initiatives and its level of implementation differ from one municipality to another. In community organizing, most of the LGUs initiatives have implemented evaluation and planning session, regular meetings and in holding general assemblies. What is ironic though, they have only fairly implemented in the formation of fisherfolk organizations. The value of community commitment and participation in decisions making and in its implementation, resource management consider not only the biophysical aspects of resource management but the social, economic and legal implications. Moreover, the consideration of how these would contribute much to the organization of communities to improve livelihoods and protect coastal resources [10]. Related to this, capacity building programs are pre-determined by the LGUs. Thus the conduct of training needs assessment (TNA) to identify appropriate thematic training and seminars were reasonably implemented while monitoring of the FOs organizational development and the actual conduct of training were moderately implemented. Through thematic some training courses, stakeholders gained the capacity to promote interaction among state interests, government and their communities since they live in connection with the occupations and dynamism of the coastal area. Moreover, capacity building based on the needs of the people is the basis for natural resource protection and for improving the quality of life of coastal dwellers [11].

Initiatives on environmental protection and management are focused on patrolling activities, the establishment of protected areas and in the rehabilitation of the mangrove forests and the coral reefs. These indicators were implemented by the LGUs with the hope that these will rehabilitate the profoundly poor condition of the coastal environment. The basic sector could do the protection and management of the resources as long as they are capable and had been devised on the concepts and principles of paralegal formation process, tactical sessions, constituency building and networking [12]. This will prepare the basic sectors in facing their adversities during patrolling and surveillance activities in the field.

Rehabilitation efforts are usually based on the research conducted by LGU-commissioned research institutions such as BFAR. Results of these studies are said to be accessible in the database of the LGUs and this indicator was moderately implemented. Although formation and training of local researchers from the communities were only fairly implemented, it is still worth mentioning that there is an attempt by the LGUs to promote participatory research in the municipalities. There is only need to enhance further the capacities of the locals on participatory research methods. The common aim of participatory research is to change social reality by insights into everyday practices that are obtained through collaborative research on the part of scientists, practitioners, service users and other stakeholders [13].

Provision of livelihood projects is crucial in CRM because it will augment the income of the fishing families and at the same time divert the fisherfolks from fish capture thus providing enough time for the coastal resources to regenerate. Provision of capitalization, technical assistance and in strengthening the capacities of the FOs in operationalizing livelihood projects was moderately implemented. However, the market linkage has been fairly implemented since the LGUs have difficulty in finding better markets where they can introduce and sell the local products. Hence, the FOs maximize the opportunities to get possible buyers during aqua-trade and agri-trade fairs. One of the major dilemmas of using rural commons for industrial uses in developing countries relates to the failure of the state to evolve an agreement on distributing property rights the local and industries sector in a manner that sustains livelihoods and ecosystems [14].

Regarding market linkage, although the LGUs are fairly rated, they implement concerning partnering with different NGAs and NGOs. These organizations usually have implemented or currently implementing CRM projects in their municipalities. According to [15], coastal protection underscores the need for community-based processes that transcend understanding of conservation measures but also invests in sustainable, operative and trustful working relationships, as well as the urgency of interdisciplinary approaches in ensuring effective design and implementation of fishery management tool. It is emphasized that for CRM to succeed, all stakeholders must be committed and engaged.

The difference of LGUs CRM initiatives implementation level revealed an F-value of 12.038 and a p-value of 0.001 (table 1). This indicates that the implementation level of the given initiatives was very highly significantly different.

Table 1. One-way analysis of variance of the CRM initiatives among LGUs in the adjacent municipalities in Camarines Sur, 2017.

Sources	Sum of Squares	df	Mean Square	F	Sig.
Between groups	9.632	2	4.816	12.038	0.001
Within groups	6.001	15	0.4		
Total	15.634	17			

Since there is a very highly significant difference, Scheffe's test was made to determine where lies the difference. Among the comparisons, Calabanga and Tinambac (MD = 1.54000; Sig. = 0.003) and Tinambac and Siruma (MD = 1.56333; Sig. = 0.003) displayed significant difference. On the other hand, the comparison between Calabanga and Siruma was found to be non-significant.

Table 2. Scheffe's test of the significant differences of CRM initiatives in the adjacent municipalities in Camarines Sur, 2017.

(I) Municipalities	(J) Sum of Squares	Mean Difference (I-J)	Sig.	Stat. Sig.
Calabanga	Tinambac	-1.54000**	0.003	Highly significant
	Siruma	0.02333	0.998	Not significant
Tinambac	Calabanga	1.54000**	0.003	Highly significant
	Siruma	1.56333**	0.003	Highly significant
Siruma	Calabanga	-.02333	0.998	Not significant
	Tinambac	-1.56333**	0.003	Highly significant

The fishery sector is considered as the poorest sector in the country according to the National Anti-Poverty Commission (NAPC). This was shown in the results in the economic condition of the respondents that the highest monthly earnings of fisherfolk range from Php 5,501 to 6,500 only, which is below the expected income to meet their basic needs. According to the Philippine Statistics Authority, its 2015 census said that a family of five must earn Php 9,064 monthly to meet its food and non-food needs. In 2012, fisherfolk along with the farmers and children consistently were posted in the substandard from other nine basic sectors in the Philippines [16].

The common fishery-related issues and concerns in the coastal areas are that of commercial fishing and the use of active gears in the municipal waters. Most of those violators are from their neighboring towns and as far as Quezon Province. On the other hand, fishpond conversions threaten the biodiversity of the mangrove forests in the three municipalities. Although some respondents said that they do not tolerate this illegal conversion, there are still some residents who insisted on constructing illegal fish ponds. Aside from the above environmental issues, waste disposal in the coastlines and beaches are becoming a regular thing in the coastal communities. Although the LGUs have already a solid waste management ordinance, there are still households who throw their garbage wherever they like.

Fisherfolk organizations are well represented in the Municipal Fisheries and Aquatic Resources Management Council (MFARMC). They are mandated by law to provide recommendations to the Sangguniang Bayan (SB) to address fishery issues and concerns that affect the lives of the fisherfolks (Sec. 73, RA 8550). Most of those issues that are being addressed in the MFARMC are managing protected areas and illegal fishing inside the municipal waters. Most of their recommendations are considered and enacted into law by the SB. In a governance context, no single organization has surveillance across the variety of interests involved. However, the challenge remains as for how to reach beneficial social goals by negotiation and to balance these overlapping [17].

In consideration of the results stated herein, a CRM Monitoring and Evaluation (M & E) model was developed to guide the LGUs and to strengthen its initiatives on CRM. M & E are two different management tools that are closely related, interactive and mutually supportive through routine tracking of project progress; monitoring can provide quantitative and qualitative data useful for designing and implementing project evaluation exercises which can refine strategies and can be further developed or improved [18].

IOP Conf. Series: Earth and Environmental Science **139** (2018) 012029 doi:10.1088/1755-1315/139/1/012029

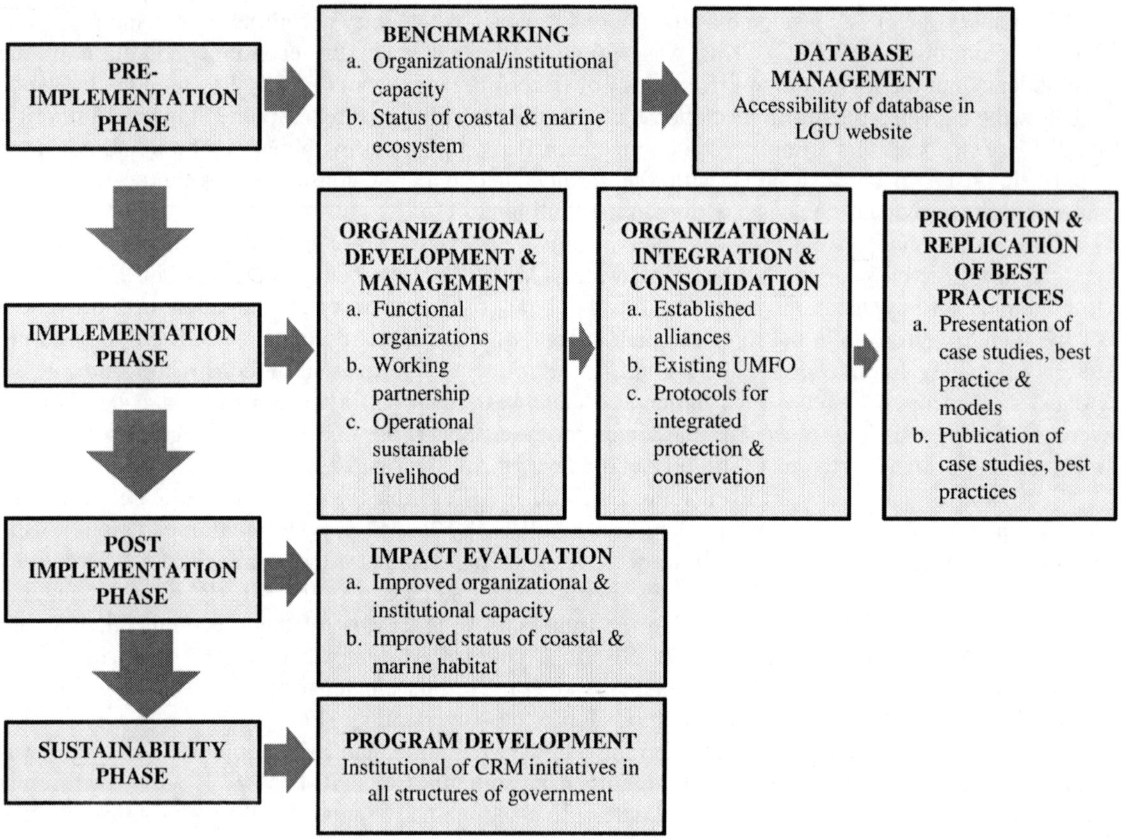

Figure 1. CRM Monitoring and evaluation model for the three (3) adjacent municipalities in Camarines Sur, Philippines.

4. Conclusion

There is a need to strengthen the initiatives of the LGUs on the various components of Coastal Resource Management. The low or moderate level of implementation as revealed in the study can be attributed to the lack of scientific and baseline data such as community needs assessment which can be used as a basis for identifying appropriate strategies to resolve the varying issues and problems encountered by the coastal communities. Community consultations are also important to make the initiatives more relevant and responsive to the needs of the fisherfolks. The LGUs' CRM program implementers must study CRM responsive activities to compliment NGOs initiatives for a sustainable CRM among municipalities.

References

[1] White A T and Trinidad A C 1998 *The Values of Philippine Coastal Resources: Why Protection and Management are Critical* (Cebu City: Coastal Resource Management Project)

[2] Aliño P M 2011 http://innri.unuftp.is/pdf/Philippine%20Fisheries.pdf Accessed in February 2017

[3] Ferrer E M and Nozawa C M 1997 *Community-Based Coastal Resources Management in the Philippines: Key Concepts, Methods and Lessons Learned* (Diliman: University of The Philippines)

[4] United Nations Sustaibanle Development www.sustainabledevelopment.un.org/content/documents/Agenda21.pdf

[5] Department of Environment and Natural Resources, Bureau of Fisheries and Aquatic Resources of the Department of Agriculture, and Department of the Interior and Local Government 2001

Philippine Coastal Management Guidebook No. 1: Coastal Management Orientation and Overview Coastal Resource Management Project of the Department of Environment and Natural Resources (Cebu City: Department of Environment and Natural Resources) p 58

[6] Republic Act 7160 1991 An Act Providing for a Local Government of 1991

[7] Republic Act 8550 1998 An Act Providing for the Development, Management and Conservation of the Fisheries and Aquatic Resources, Integrating all Laws Pertinent Thereto, and for Other Purposes

[8] Hubbard R 2016 https://www.quora.com/What-is-descriptive-evaluative-research Accessed on 3rd April 2017

[9] Brennan M A, Barnett R V and Baugh E 2007 *J. Extension* **45**(4) 4FEA3

[10] Rivera R and Newkirk G F 1997 *Ocean Coast. Manage.* **36** 73–95

[11] Echevarria L, Gomez A, Piriz C, Quintas C, Tejera R and Conde D 2013 *ICZM* **13**(4) 445–456

[12] Balaod Mindanaw 2013 *Community-based Paralegal Formation Process and Training Toolkit A Part of the Project Protecting the Watershed Forest Reserve through Legal Empowerment and funded by the Foundation for the Philippine Environment* (Cagayan de Oro City: Balay Alternative Legal Advocates for Development in Mindanaw Inc.) p 237

[13] Bergold J and Thomas S 2012 *Forum Qualitative Social Research* **13**(1) Art. 30

[14] Thomson K 2009 *Ocean Coast. Manage.***52** 586–592

[15] Velez M, Adlerstein S and Wondolleck J 2014 *Mar. Policy* **45** 171–181

[16] Bersales L G S 2014 *Fishermen, Farmers and Children Remain The Poorest Basic Sectors* (Phillipine: Philippine Statistics Authority) 20140704-SS2-01

[17] Dwyer J and Hodge I 2016 *Environ. Sci. Policy* **66** 1–10

[18] USAID 2017 www.ehproject.org/PDF/phe/8-phe_monitoring.ppt Accessed on 27th June 2017

IOP Conf. Series: Earth and Environmental Science **139** (2018) 012030 doi:10.1088/1755-1315/139/1/012030

The vulnerability of fishermen's community and livelihood opportunity through drought and seasonal changes in border area of Indonesia-Timor Leste

A D Jayanti[1], W Fitriya[1], E Setyobudi[1], S A Budhiyanti[1], Suadi[1] and S J Kune[2]

[1] Department of Fisheries, Faculty of Agriculture, Universitas Gadjah Mada, Jl. Flora Bulaksumur, Sleman, Yogyakarta, Indonesia
[2] Universitas Timor, Kota Kefamenanu, Timor Tengah Utara Regency, Nusa Tenggara Timur, Indonesia
*Email: anes.dwijayanti@ugm.ac.id

Abstract. Communities that live in coastal areas in Indonesia are affected by the ecosystem degradation because their livelihoods majority depends on ecosystem's services. Fishermen in Timor Tengah Utara Regency depends on their livelihood on fish catches and crops. TTU Regency is known as a place with drought. Agriculture sector and fisheries play the central role of communal livelihood. This research was conducted to gain information and baseline study to support the intervention scheme reducing the vulnerable level of coastal communities. This research was conducted in Insana Utara, Biboki Moenleu and Biboki Anleu District. The social-ecological and statistic descriptive analysis were undertaken and involving 53 fishermen, 4 women groups, 11 clan's elder and staffs of local government as the respondents. The data shows that the majority of the fishermen are small-scale fisheries commercial fishermen and possess a high level of vulnerability. The factors that are mostly affected the fishermen livelihood is the job diversification as farmers which is primarily supported by the crops and rely on the rainfall. The vulnerable context of fishermen in TTU can be reduced by optimizing and enhancing communal institution capacity and increasing the cooperation among the stakeholders and government also women participation.

1. Introduction

Coastal communities invariably depend on their livelihood from ecosystem services. Not only in economic aspect but ecosystem also mainly related to several, social and cultural values inside the community. Depressing ecosystem quality has commonly led them to poverty and interfere with their culture [1]. However, every community has the resilience context as well, which is the opposite of the vulnerability context. There are three aspects of vulnerability; exposure, sensitivity and adaptive capacity [2]. Exposure is the level which an ecosystem, is determined by environmental conditions. Sensitivity is the levels of dependence on natural resources and the technologies used to harvest resources. Adaptive capacity is a latent characteristic that reflects people's ability to anticipate and respond to changes and to minimize, cope with and recover from the consequences of change [3].

Timor Tengah Utara Regency is located between east and west region of Timor Leste and passed by an international road that connects it. Geographically, this current situation is a strategic location to be developed and could increase the economic opportunity for the communities around it. Regardless its strategic locations, the inhabitant that live in the village around the international road, which is coastal areas, still live in poverty. The majority among them are fishermen and farmers. Dwelling in the border area is considered as a challenging task. The fishermen in the border area are potential to have border conflict to claim their fishing ground with another country or to distribute and market their catch. Located in border area without sufficient infrastructure support is a form of vulnerability. In 2014, there was no fishing port for the fishermen, and due to the proximity to the market is quite far, the fishermen throw away the catches in the sea. The situation will get worse if the condition of the fishing port and the facilities for cold chain system in the neighbor country better, the resource could be easily smuggled and landed in the mentioned port.

There is no report mentioned before described coastal community structure, stratification interaction and vulnerability level. This information is to support the next policy or program that will be implemented in that area to overcome poverty and to increase their livelihood opportunities, particularly to strengthen the fishery sector considering that the northern coastal areas of TTU have the susceptibility to water availability for agricultural land management, but have unutilized fisheries resources.

2. Methods

2.1. Study area

This research was carried out in the Timor Tengah Utara Regency, located at the north end of NTT between Timor Leste, Indonesia (figure 1). The study sites were three districts, namely Insana Utara, Biboki Moenleu and Bibooki Anleu. Those District were selected due to the majority among the dwellers were fishermen. The area has a rural character, dominated by fishing and agriculture as major economic activities. The research sites were selected according to accessibility, high dependence on marine resources and differences regarding predominant occupations, ethnic composition.

Figure 1. Research location in Timor Tengah Utara Regency, NTT.

2.2. Questioner and sampling design

Opportunistic sampling was used in this study to select participants for semi-structured interviews [4]. To reduce the risk of bias inherent in opportunistic sampling the interviewees were chosen throughout the entire area of the village to get the better understanding of vulnerability level in the community. The social-ecological analysis was undertaken and involving 53 fishermen, 4 women groups, 11 clan's elder and staffs of local government as the respondents. Quantitative questionnaire surveys were preceded by qualitative assessments of the study villages to obtain background information and a better understanding of local conditions and to refine study questions. In qualitative data assessment, the focus lies on the quality of statements, meaning the analysis of what and how regarding a certain situation.

Besides the sampling and preparing the questioner to the fishermen, field observation and focus group discussion were held as the introductory research. This process involved traditional leaders of each clan who dwell in the coastal area, traders, local government, fish processor woman group and shop or outlet owner. Due to limitations of time available for this study, Participatory Rural Appraisal (PRA) techniques through semi-structured interviews, focus groups or (non-participant) observation, which mostly generated qualitative data, could collect only limited quantitative data was done as the introductory and to validate the collected data.

3. Results and Discussion

3.1. Fisherman profiles

Fishing activity in North Coast of Timor Tengah Utara (TTU) Regency is characterized as a traditional fisheries. Regardless of the fishing gears, the fishermen in TTU that observed in five villages indicate that they are have been undertaking professions as a fishermen during much time. They were starting their profession as a fishermen 21 years ago in average and majority among them was inherited by their parents. Most of the fishermen graduated from elementary school as high as 73.58 %. Most of the respondents' age is 42 years old and based on the further interview they are willing to inherit their skills to their children.

Table 1. Number of fisherman of each village and it's characteristic.

Village	Fisherman	Age (average)	Experiences as Fisherman (average)
Oepuah (Insana Utara)	7	39.9	18.7
Oesoko (Insana Utara)	26	42.7	23.2
Liusoesoko (Biboki Monleu)	9	43.6	31.0
Oesoko kolam dapur (Biboki Monleu)	6	37.7	21.0
Lius kolam dapur (Biboki Anleu)	4	46.0	18.8
Kolam tua (Biboki Anleu)	11	42.2	17.0
n	53	42.0	21.6

Table 2. Education level of fisherman in Timor Tengah Utara Regency.

Education Level	Fisherman	%
Illiterate	1	1.89
Elementary	39	73.58
Junior High School	10	18.87
High Senior School	4	7.55

As much as 34 % of fishermen possessed boat and based on the area study, the boat that majority under fishermen's possessions is a traditional boat that made from wood and constructed approximately 7 m length and 1 m wide and only 2 % using a machine. Another 66 % of the fishermen was boat-less and using nets to catch fish and work as a labor in other boats that they have to divide their catch equally (50 %) after it reduced by operational cost. The composition of fishing gears that are used in TTU Regency was beach seine, gillnet, mini purse seine and long line with percentage orderly 45.3 %, 49.1 %, 3.8 % and 1.9 % (table 2). Works as a fishermen in TTU is not a primer job. Their primer job mainly as a farmer and fishing activity as a side job that they are performed during extra time, yet gradually fishing becomes main activity because farming activity cannot be done when the dry season comes. During dry season vulnerability of the fishermen increasing, because it is happening at the same time with the fishing lack season.

Table 3. Fisherman ownership of boats and fishing gear.

	Yes (%)	No (%)
Boat	34.0	66.0
Beach seine	45.3	54.7
Gill net	49.1	50.9
Mini purse seine	3.8	96.2
Long line	1.9	98.1

3.2. Catch and price dynamic during peak and low catch season

The fishermen in the site research have a unique counting system for the catch. They do not use kilograms scale to weighing their catches but using basket scales instead. One basket could be estimated as 50 kg. According to the data that were gathered from the fishermen, the effort and the catch disperse variably. It shows that effort of fishermen to gain catches different one another. Based on the linear analysis, it does not show that the increase of effort or time is in line with the increasing of catches. Based on figure 2, the uncertainty of catches occurs, where the increase of effort does not show a significant increase of the catch in a week. The optimum fishing trip hour based on this study which provides the optimum return was around 20 hours/week during low season. However further analysis of the cost-revenue analysis needs to be done in peak, low and moderate season as well. Based on figure 3, there are significant differences in price, catches and income during a week for fishermen within peak, moderate and lack fishing season. The income at peak season is higher respectively fourth times than lack season, yet the price of fish lower as high as more than 100 % compare to the lack season.

There were three seasons of fishing, where lack of fish and where there are hard to catch a favourable fish. As a comparison, in Spermonde Islands South Sulawesi, there were differences between peak and low season it is also related to the wet season which usually the catch was lower than the dry season. However, the differences between the low and peak season were different compared to what had happened in TTU. They caught 28.62 kg/week with six days works which were valued 327.12 USD during low season and 38.64 kg/week which is valued 412.92 USD during the peak season [6]. Meanwhile, in TTU the fishermen caught 21 kg/week during the low season and valued 74.27USD and 198.9 kg/week which valued 337.75 USD. We revealed that there was significance during the low season that the catches were very small and during the peak season the volume of catches is high, but the value was cheap. It also regards to the commodity that is caught which mostly mackerel or sardine where it is considered as a commodity that has a regular price. Whether it is on low or in peak season the vulnerability of the fishermen were still occur.

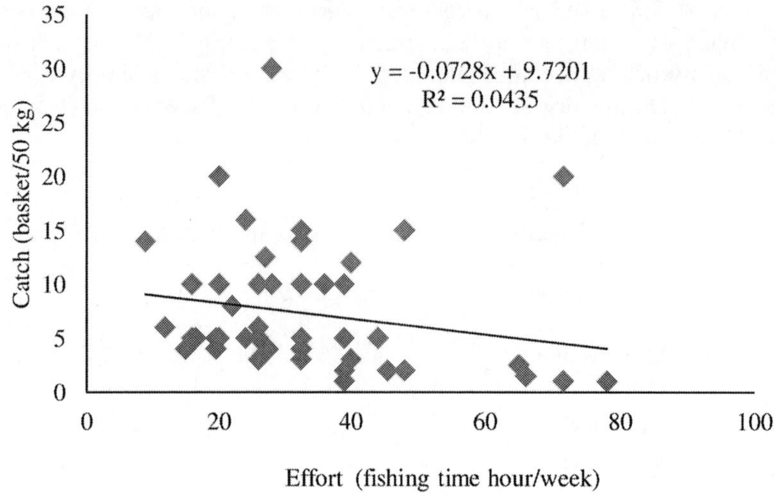

Figure 2. Average catches and effort in Timor Tengah Utara Regency.

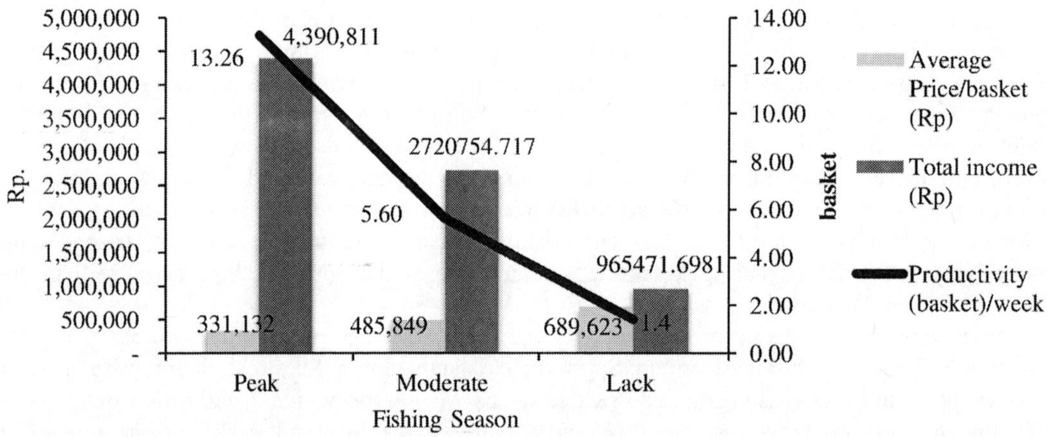

Figure 3. Fisherman's production and income during peak, moderate and lack season.

3.3. Patron-client system in TTU

Patron-client systems are typical for small-scale fisheries especially in tropical fisheries in Southeast Asia [7]. It has widespread due to the function of the patrons provide the link between the fishermen and the buyer [6], where usually fishermen stay in the remote area and apart from the market. Patron-client relationships in coastal communities which possess high vulnerability level have profound implications for fishing household livelihood security and their vulnerability by providing economic income through fishing or fish trading and social security in times of hardship and by contributing flexible and potentially exploitative credit-debt systems [8]. Fishermen in TTU Regency performs patron-client relationship 13 %. The majority of the fishermen are individual fishermen that poses their fishing gear and needs no boat. Those fishermen perform no interaction with the middleman and 77 % of the share harvesting system is 100 % for the fishermen. It is showed that commercial fishing in TTU regency is small-scale fisheries. It was not long time ago from the first catching that informing fishermen in TTU were classified as subsistence fishermen. Fishing activity was densely undertaken in

Kupang or other areas which has proximity to market. Fishermen in TTU Regency initially create small boat and function it as assets to fulfilled food to adopt drought when the dry season comes. It is learned that the elder of fishermen in TTU Regency was not increasing their fishing gear instrumental even though they use to fulfill their economic needs.

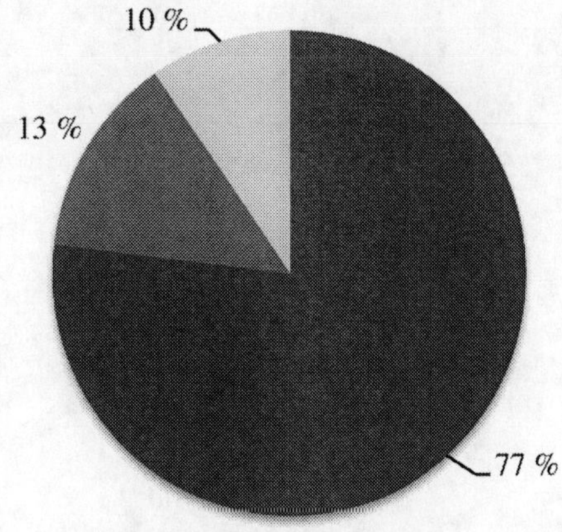

Figure 4. Share harvesting system in TTU Regency.

3.4. Vulnerability through the drought season

Based on Indonesian Agency for Meteorology, Climatology and Geophysics (BMKG figure 5), NTT was categorized as a location, which has a low to medium rainfall each year, further experienced long drought and has a problem with fresh water supply during several months in a year [5]. There are some evidences that rainfall and climate affected the catch [17]. The catch in TTU were usually occurred mostly in the peak season during August-September. However, those typical were changing since 2015. The drought season which usually starts in June which is indicated by the decrease of rainfall (figure 5) also indicate the start of fishing time. However, within three years it was changing. At present, during the drought community in the coastal area don't fish either farm. It harms their livelihood, and they have to adapt to this change.

4. Discussion

Vulnerability context is acknowledged as the adversity for a system or community to compromise and adapt to the changing that potentially reduce their resilience [3]. Also described as a stage of the exposure, sensitivity and adaptive capacity [2, 3, 11, 12, 13, 14, 15]. For the coastal community, It also regards those exposure, sensitivity and adaptive capacity are interpreted as the threat to the marine resource as a habitat change, fishing and climate impacts [2, 13, 16]. Some of the coastal communities in Indonesia do not always depend exclusively on fishing but often prefer to diversify their livelihood portfolio to comprise activities such as farming and construction in addition to fishing [9]. It also were occurred in TTU inversed way. The native fishermen possess dual existence as a fisherman and a farmer and start from the previous time; their main job is a farmer. Related to the aspect of household capital assets base and vulnerability context, fishermen in TTU, are minimum. Human capital indicates that fishermen in TTU area possess low accessibility to the fish resources due to many aspects, firstly lack boat and fishing gear. The fishermen came from Celebes who were Buton Tribe, yet the native (Dawan Tribe) was a beginner and knew a slight fishing technique. Other access is their skill to catch and operate the fishing gear and compromise with the season. Physical capitals that support fishing activity also low occurred such as availability of facilities to support cold chain system, facility for landing and handling the catch and to process it. The traditional way for the

IOP Conf. Series: Earth and Environmental Science **139** (2018) 012030 doi:10.1088/1755-1315/139/1/012030

fishermen was directly dry their catches. Developing product when catches are high can be undertaken to increase added value. Fish price at peak season falling up to 50 %.

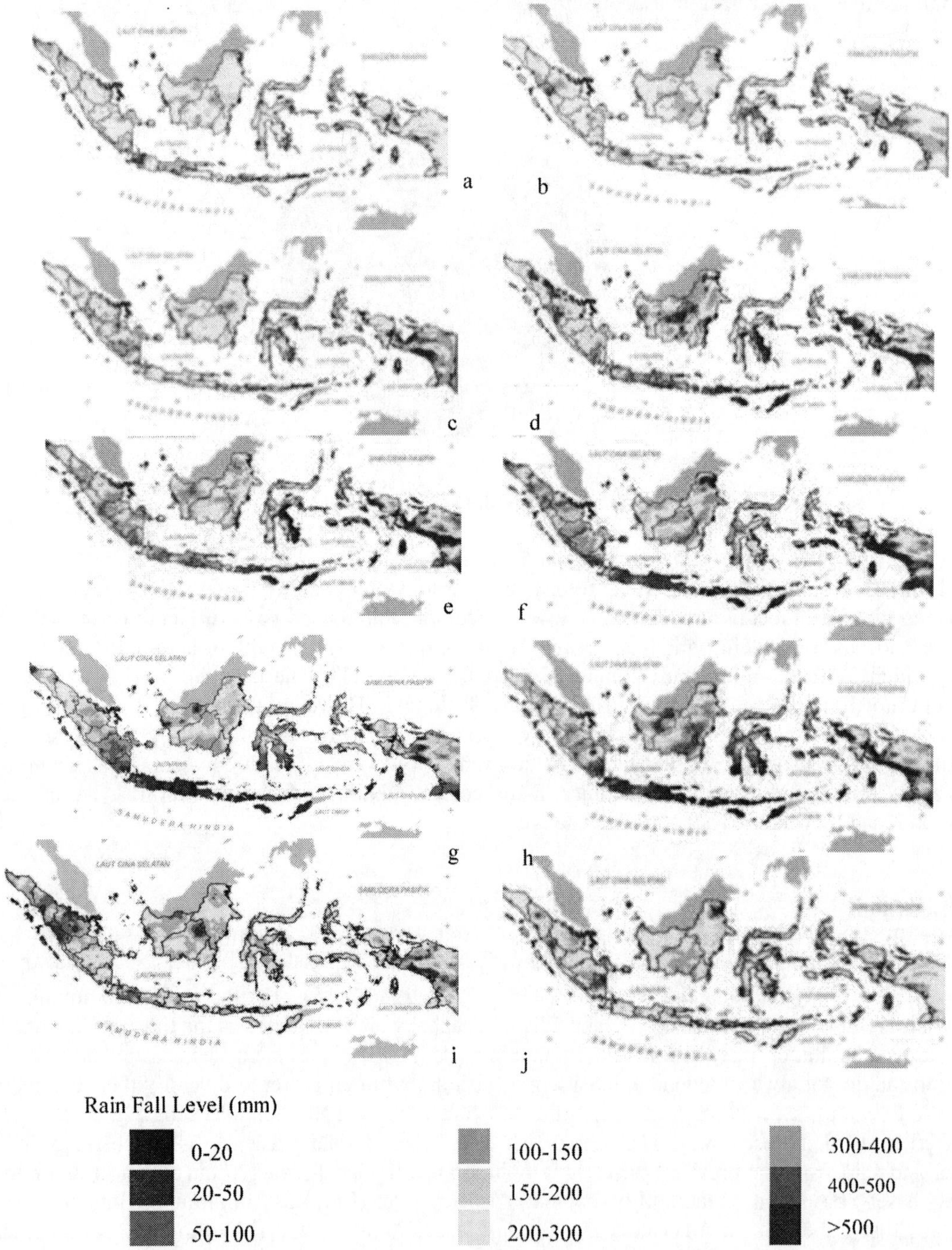

Figure 5. The rainfall rate in Indonesia in 2017, a. Febuary, b. March, c. April, d. May, e. June, f. July, g. August, h. September, i. November, j. December. Source Indonesian Agency for Meteorology, Climatology and Geophysics [5].

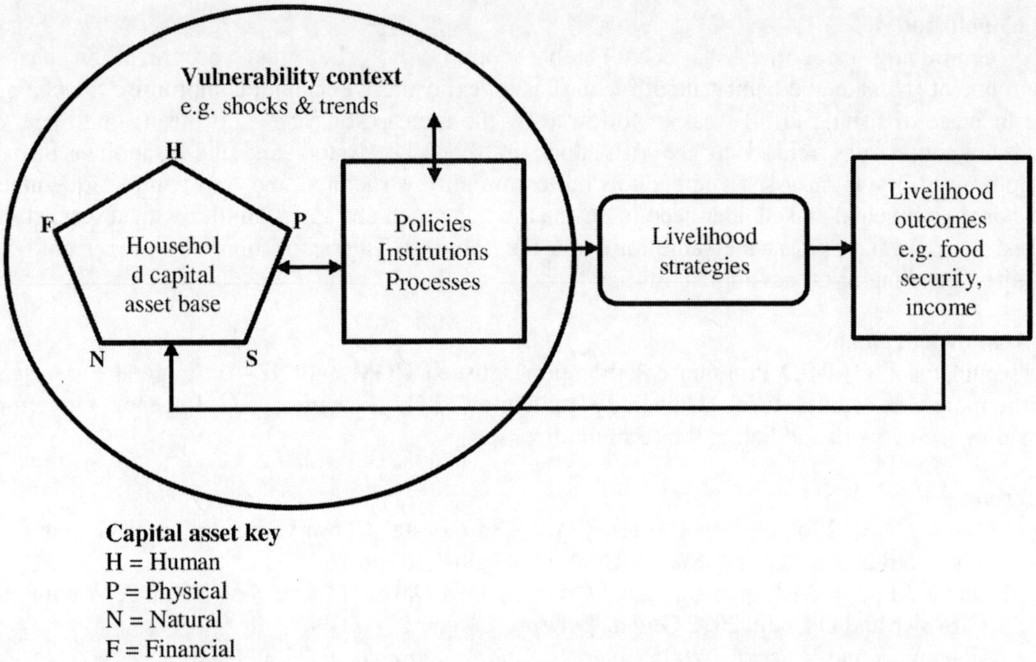

Figure 6. The Sustainable Livelihoods Framework showing the household assets and policies, institutions and processes components as existing together in the vulnerability context.

Increasing the other livelihood opportunity requires collaborations between authorities and local users. To smooth the marketing, fishermen, and stakeholder can rely on efficient information flow, involvement from government and nongovernment stakeholders which has good leader performance and a clear, structure and function, the active participation of both government and non-government stakeholders [10]. To assist and support the community, structured and transparent Governmental and nongovernmental stakeholder need to devise the fishermen and their families with several skills to extend their fishing skill and at the same time add the value of fisheries resource by proper cold chain facilities, handling or processing facilities. The facilities establishment in the border area, even more in Eastern Indonesia is a core agenda of Indonesia's national development plan. It should be followed by the capacity of the local government and also the beneficiaries communities.

The infrastructure's establishment such as road system, trading ports and fishing ports in border areas especially in the field of the fishery has been improved, since 2015, the construction of infrastructures such as roads and seaport has begun. However, there are some obstacles in the process of development of this infrastructure. The problem is the community's readiness to adapt to the changes itself. The development is expected to accelerate economic activity in the border areas and increase the sovereignty of the nations; mostly the community in the border areas has a chance for the better life. However, it also could be an exclusion potential for the local community when the economic growth lured many people from another area to join in the economic activities and left the local community behind. Therefore, the adaptive capacity of the local communities or the fishermen needs to be improved. Adaptive capacity needs to be increased as well with involved them in several productive meetings and training to protect the ecosystems on which their livelihoods depend on and making sure that there are no people in the community that are left behind towards the fast-growing access development. Lastly, there is a need to concern on the effort to repair the marine ecosystem and a water-efficient farming system to enriched their alternatives livelihood.

5. Conclusion

TTU community categorized as a vulnerable community. They just recognize the changing environment (climate and habitat modification) as an exposure, yet cannot compromise the changing. The increase of fishing effort was not followed by the increase of catches. It means that there were environmental issues related to the fish stock in the TTU waters and the fishermen have not compromised it yet. To assist and support the community, structured and transparent Governmental and nongovernmental stakeholder need to devise the fishermen and their families with several skills to extend their fishing skill at the same time add the value of fisheries resource by proper cold chain facilities, handling or processing facilities.

6. Acknowledgement

We would thank CARED Program collaboration between UGM and NZ-Aid to fund this research. Furthermore, we would thank University of Timor, TTU Fisheries, and Regency Government providing the network and link to the communities.

References

[1] Diegues A C 2005 Marine Protected Areas in Coastal Conservation in Tropical Developing Countries (Amsterdam: Marine Conf. On Multi-Use) pp 16
[2] Cinner J E, T R McClanahan, N A J Graham, T M Daw, J Maina, S M Stead, A Wamukota, K Brown and O Bodin 2012 *Global Environ. Chang.* 22 12–20
[3] Adger W N and Vincent K 2005 *Comptes Rendus Geoscience* 337 399–410
[4] Kemper E A, Stringfield S and Teddlie C 2003 Mixed methods sampling strategies in social science research *Handbook of Mixed Methods in Social and Behavioral Research* 2nd Ed. ed Tashakkori, A and Teddlie C (Thousand Oaks: Sage Publications) p 273–296
[5] Badan Meteorologi Klimatologi Geofisika (BMKG) 2018 http://www.bmkg.go.id diakses tanggal 15 Januari 2018
[6] Minarro S, Forero G N, Reuter H and vanPutten I E 2016 *Mar. Policy* **69** 73–78
[7] Ruddle K 2011 *Hum. Organ.* **70**(3) 224–232
[8] Platteau J P and Nugent J 1992 *J. Dev. Stud.* **28** 386–422
[9] Cullen L C 2007 *Dissertation* Department of Biological Sciences, University of Essex United Kingdom
[10] Wever L, M Glaser, P Gorris and D F Schulte 2012 *Ocean Coast. Manage.* **66** 63–72
[11] Daniella F S, P Gorris, W Baitoningsih, D S Adhurif and Sebastian C A F 2015 *Mar. Policy* **52** 163–171
[12] Allison E H, Perry A L, Badjeck M C, Adger N W, Brown K, Conway D, Halls A S, Pilling G M, Reynolds J D, Andrew N L and N K Dulvy 2009 *Fish Fish.* **10** 173–196
[13] Smit B and Wandel J 2006 *Global Environ. Chang.* **16** 282–292
[14] Parry M L, Canziani O F, Palutikof J P, van der Linden P J and C E Hanson 2007 *Climate Change 200: Impacts, Adaptation and Vulnerability* (Cambridge: Cambridge University Press) pp 976
[15] Turner B L, Kasperson R E, Matson P A, McCarthy J J, Corell R W, Christensen L, Eckley N, Kasperson J X, Luers A, Martello M L, Polsky C, Pulsipher A and A Schiller 2003 *A Framework for Vulnerability Analysis in Sustainability Science* (USA: Proceedings of The National Academy of Sciences of The United States of America) **100**(14) 8074-8079
[16] Gallopín G C 2006 *Global Environ. Chang.* **16** 293–303
[17] Milton D A, Satria F, Proctor C H, Utama A P A, Prasetyo M F and Andria 2014 *Cont. Shelf Res.* **91** 247–255

Public perspective towards marine litter in West Aceh City

I Kusumawati[1]*, M Setyowati[1], E Riana[2] and T Prartono[2]

[1] University of Teuku Umar Jl. Alue Peunyareng, Ujong Tanoh Darat, Meulaboh, Indonesia

[2] Bogor Agricultural University Jl. Raya Dramaga, Bogor, 16680, Indonesia

*Email: ikakusumawati@utu.ac.id

Abstract. Marine litter or marine debris is a man-made solid material discarded, abandoned or lost in coastline or into the sea. To reduce the amount of marine litter in the ocean, raising public awareness is an important way. One of the contributing factors on marine litter is the lack of understanding within the community, but to identify how people notice the problem is required adequate research literature. The purpose of this study is to examine the awareness of West Aceh community on marine litter along western coastal area. The research objectives; 1) to evaluate societal perception towards marine litter; 2) to examine the urgent indicator of public awareness in West Aceh City. This study will employ a survey approach by distributing questionnaires to 383 respondents. It was found that respondents show low awareness on marine litter according to statistical data, but there are some rooms to manage in order to raise the level of public awareness. It concludes that sense of responsibility could be enhanced by involving public in any activities for preventing and eradicating marine litter. Education aspect is also important to increase public understanding about the threats of marine debris on environment, human health and economic income.

1. Introduction

The ocean is an extraordinary resource, covering nearly three-quarters of the Earth. The ocean supports fishing industries and coastal economies, provides recreational opportunities and serves as a nurturing home for a multitude of marine plants and wildlife [1]. However, the ocean is currently under a considerable threat. The seeming vastness of the ocean has prompted people to overestimate its ability to safely absorb our wastes safely.

Marine litter is any persistent, manufactured or processed solid material discarded, disposed of or abandoned in the marine and coastal environment. Marine litter consists of items that have been made or used by people and deliberately discarded or unintentionally lost into the sea or coastline including such materials transported into the marine environment from land by rivers, drainage or sewage systems or wind. For example marine litter consists of plastics, wood, metals, glass, rubber, clothing, paper etc. This definition does not include semi-solid remains for example mineral and vegetable oils, paraffin and chemicals that sometime contaminate sea and shores [2]. In general, marine debris is any object on the surface of the sea, the sea and the coast as an impact of human activities [3]. Marine debris has become an international problem considering that debris can flow and enter any country without being impeded by national borders or political ideology and has been widely discussed in various literatures [4, 5].

Currently, the highest waste contributor is the Asian country which has long coastline and wide waters area. Indonesia became the second highest polluting country after China. Indonesia produced 3.2 million tons of plastic waste in 2010, with around 1.29 million tons of that ending up in the ocean.

IOP Conf. Series: Earth and Environmental Science 139 (2018) 012031 doi:10.1088/1755-1315/139/1/012031

The problem of marine debris is inseparable from the lack of community role in maintaining hygiene, especially the culture of our people who have not been aware to not litter and low communication between stakeholders to manage waste also become problems related to the issue of marine debris [4]. Inadequate management, infrastructure and lack of community knowledge about the impact of debris on the environment are also the problems in marine debris management [6].

Community behavior about marine debris is strongly influenced by knowledge, environmental insight, attitude to the issue and motivation to find a solution to the problem [5]. Therefore, increasing community awareness is important effort to reduce marine waste through behavioral change of people. This is the only powerful way to reduce the amount of marine debris on the coast [8]. Increasing community awareness can be an opening step in combating marine debris to then integrate with approaches to achieve a clean sea.

The issue of marine hygiene has not been a major political agenda and has not been the concern of the majority of the people. Researchers [9] said that politicians and the public have not paid attention to the level of waste plastic problems at sea. As a result, infrastructure, management and policies concerning the problem of marine waste are still very limited.

Panglima Laot of West Aceh attempted to encourage the social community in order to make various efforts to overcome the marine debris issue in Aceh. According to Panglima Laot, waste disposed into the sea will be a poison that can kill marine biota, so awakening public awareness is necessary for marine environment [10]. Therefore, researchers believe this research with title public perspective towards marine litter in West Aceh City is essential to conduct. This research aims to evaluate social perception towards marine litter and to examine the urgent indicator of public awareness in West Aceh City.

2. Materials and Methods

2.1. Study areas and research design

The study is conducted in West Aceh regency, one of regency in Aceh Province Indonesia. Figure 1 showed map of Aceh Province. Researcher uses quantitative descriptive approach in this research to explain the research problem that focused on research objectives. This design is expected to provide an overview of public perceptions on marine debris.

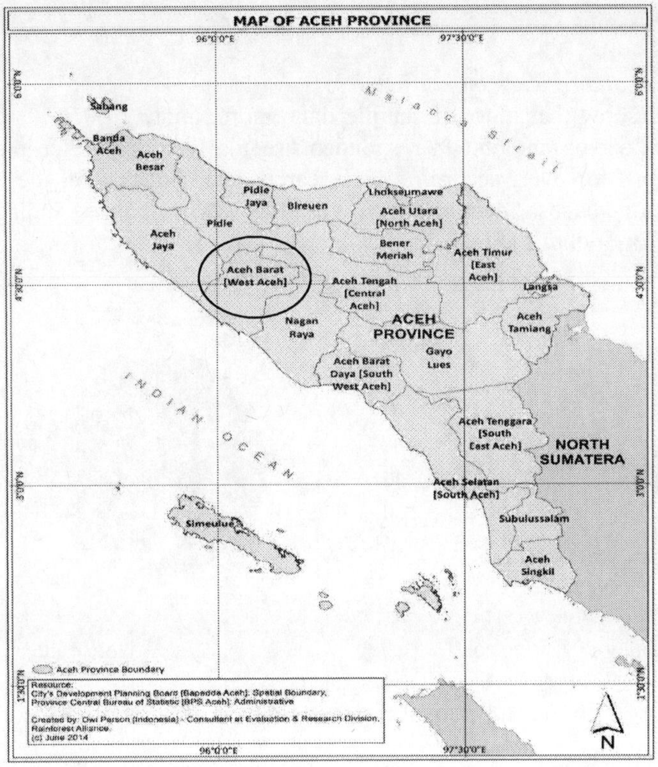

Figure 1. Map of Aceh Province, West Aceh Regency with studied areas shaded.

2.2. Data collection

The data is collected by questionnaires survey (self-administrative or face-to-face) from January 2017 to June 2017. The population is all residents of West Aceh Regency around 190,244 people [11]. The sample was taken by using random sampling approach. To determine the sample size, researchers use [12], with 95 % confidence level and confidence interval five. A total of 383 samples were selected in this study. Researcher met the respondents randomly by visiting beaches, cafes, government offices, schools and universities. This approach is used in order to get the adequate respondents without the demographic background.

2.3. Questionnaire design and data analysis

The questionnaire was designed in three stages. First, preliminary questionnaire was drafted based on previous researches. Second, the preliminary questionnaire was pre-tested with 40 respondents. Pre-test of questionnaire were conducted by face to face and got 95 % responses. After the revision by pre-test of respondents, the final questionnaire is confirmed and comprised 12 parts adapted from previous study [4]. The surveys of questionnaires were conducted in person contained close-ended questions. The questionnaire section contains three groups of questions as follow:

- Personal experience of visiting the coast.
- Personal opinion on debris amount due to people behavior.
- And others marine important issues.

The responses will be analyzed through statistical approach by providing the basic information of the descriptive data. Therefore, researcher will employ SPSS 24 to run descriptive statistical data in order to acquire adequate evidence of public awareness.

3. Results and Discussions

3.1. Public awareness and concern on marine debris

Respondents' replies show that most of sample data approximately 60.3 % have frequent visits to coastal area in a week and around 65.6 % respondents regularly noticed litter in most of their visit and every visit (figure 2). Moreover, according to research data, littering on the beach is triggered by visitor's bad behavior dispose debris directly as 92.5 % responses indicate that beaches are contaminated with a lot of debris (figure 3).

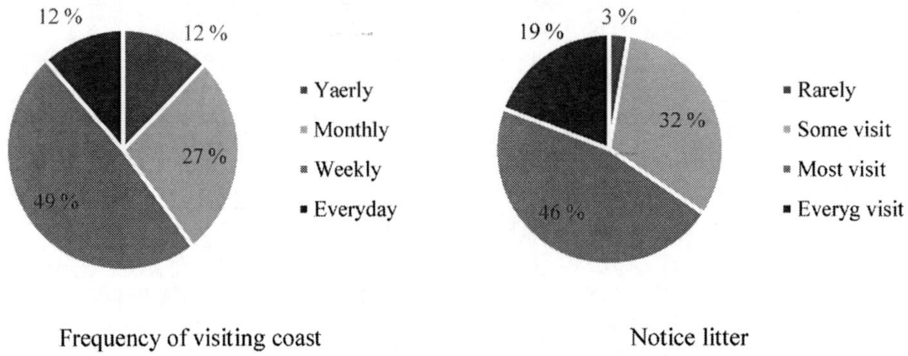

Figure 2. Personal experiences of visiting the coast.

Similarly, the ocean is polluted by debris because of human activities on the beach which abandon trash [13]. People's incompetency in managing debris becomes the core of marine debris problem.

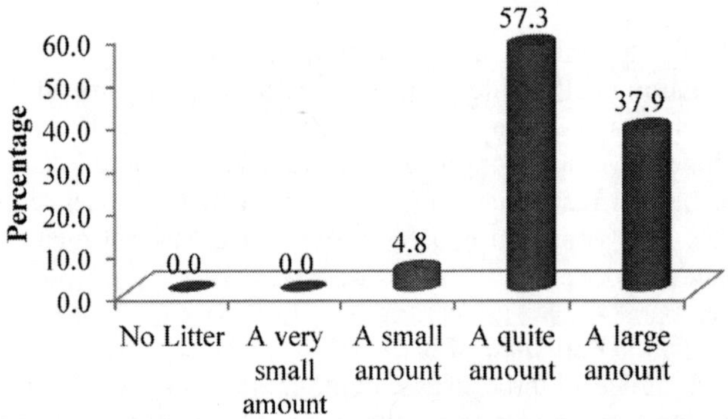

Figure 3. People behaviour in direct litter on the beach.

Regarding with people's knowledge on the amount of debris across the world referring to its type, most of respondent do not aware how many debris is according to their category. Metal has a small number of debris around 2.89 % according to previous study [14], but the majority respondents answered that metal contributed 20–40 percent of debris in the sea. Relatedly, the result shows that most respondents believe that there are more than 10 percent cloths or textile materials contaminate the ocean while the previous study shows that there are only less than live percent of cloths contaminating marine environment [14].

However, regarding the amount of plastic debris, respondents perceive that plastic is the most abundant material and ubiquitous waste materials. The sample data indicate more than 70 percent respondents feel plastic has a huge contribution in marine litter with more than 60 percent. This statistical data is supported by many studies which show that plastic give 60–80 percent in whole marine debris quantity [15] and it was the major component of debris [14, 16].

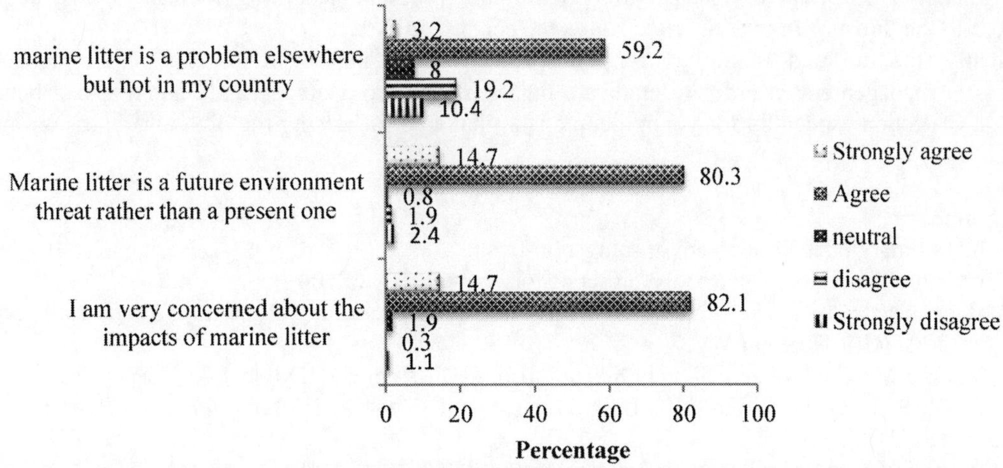

Figure 4. Marine important issues.

Based on figure 4, there are 59.2 percent of respondents who consider that marine litter is not a problem in Indonesia nowadays. Meanwhile other respondents disagree about marine litter is not a problem in Indonesia with 10.4 percent strongly disagree and 19.2 percent disagree. This finding is contradicted with some reports. Latest study reports that Indonesia is the second top countries which mismanage plastic waste approximately 3.22 MMT (Million Metric Ton) each year [17]. This situation should be a massive problem for Indonesia people, but the lack of information make society do not aware what the world facing with now. Marine debris has become a pervasive pollution problem affecting all of the world's oceans [18]. Marine litter which caused by mismanaged waste material of beachgoers has become a serious environmental problem due to its ability to move following the current sea with no border [19].

The marine debris problem is perceived by respondents as shown on figure 4 with 82.1 percent respondents feel concerned in impacts of marine litter and 14.1 percent respondents are strongly agree. The result signs that respondents aware about the problem of marine debris and worry about the situation which caused by the issue. On the other hand, respondents assume that marine litter is not the problem today, yet it will be a serious problem in the future. There is 80.3 % respondent expect the issue will be the future problem. In contrast, a report said that some evidence proof that marine debris is one of the world's most persistent pollution problems affecting the oceans [13].

3.2. The urgent Indicator of public awareness in West Aceh City

It is clearly shown that marine litter problem in West Aceh City is related to human behavior in littering. Behavior problem is the foundation of marine debris issue and to manage this problem needs strong will and effort from every stakeholder particularly government to educate people. To reduce the amount of marine debris, changing the attitude is an important factor [5].

4. Conclusion

People of Aceh aware of the marine litter situation and they believe that bad behavior is the problem to solve for eradicating the amount of debris. Furthermore, plastic is still the major debris in the ocean as the research finding indicates that abundant plastic waste was easy to find on the beach.

However, most of respondent do not perceive the amount of debris according to its category. People think the other debris such as clothes and metal contribute only a small amount. People also believe that Indonesia still in safe situation regarding marine debris. It indicates that most of respondent has no any information about how people of Indonesia dangerously contaminate the ocean. Unfortunately, the threat of marine litter is a future problem according to respondents, yet they concern to the threat of marine debris towards the ecosystem.

Finally, the full understanding of marine debris and human behavior is a requirement for a successful management in order to eradicate marine debris from our ocean. Changing bad behavior through education especially for young people is a vital step for raising the high awareness on marine litter.

References

[1] https://marinedebris.noaa.gov/turning-tide-trash
[2] http://publications.jrc.ec.europa.eu/repository/handle/JRC58104
[3] Beachwatch 2002 *The annual UK beach litter survey report: Marine: Conservation Society 2004* (UK: Ross-on-Wye) pp 92
[4] Veiga J M, Vlachogianni T, Pahl S, Thompson R C, Kopke K, Doyle T K, Hartley B L, Maes T, Orthodoxou D L, Loizidou X I and Alampei I 2016 *Marine Pollution Bulletin* **102**(2) 309–315
[5] Storrier K L and McGlashan D J 2006 *Marine Policy* **30** 189–196
[6] Author 2014 *Plastic Debris in the Ocean: The Characterization of Marine Plastics and their Environmental Impacts, Situation Analysis Report* (IUCN: Switzerland)
[7] Hartley B L, Thompson R C and Pahl S 2015 *Marine Pollution Bulletin* **90** 209–217
[8] Rees G and Pond K 1995 *Marine Pollution Bulletin* **30**(2) 103–108
[9] Depledge M H, Galgani F, Panti C, Caliani I, Casini S and Fossi M C 2013 *Marine Environ Res.* **92** 279–281
[10] http://diliputnews.com/read/10272/peringati-hari-lingkungan-hidup-pantai-aceh-barat-bebas-sampah.html
[11] http://bappedajakarta.go.id/?page_id=1131
[12] http://www.surveysystem.com/sscalc.htm
[13] http://www.un.org/Depts/los/consultative_process/consultative_process.htm
[14] OSPAR 2007 *OSPAR Pilot Project on Monitoring Marine Beach Litter: Monitoring of Marine Litter on Beaches in The OSPAR Region* (OSPAR Commission: London)
[15] Browne M A, Crump P, Niven S J, Teuten, E L, Tonkin A, Galloway T and Thompson R C 2011 *Environmental Science and Technology* (in press)
[16] Derraik J G B 2002 *Marine Pollution Bulletin* **44** 842–852
[17] Jambeck J R, Geyer R, Wilcox C, Siegler T R, Perryman M, Andrady A, Narayan R and Lavender Law K 2015 *Marine pollution* **347** 768–71
[18] http://www.greenpeace.org/international/en/publications/reports/plastic_ocean_report/
[19] Coe and Rogers 1997 *Marine Debris* (Springer: New York)

IOP Conf. Series: Earth and Environmental Science **139** (2018) 012032 doi:10.1088/1755-1315/139/1/012032

Challenges in managing freshwater fishery resource through Lebak Lebung Auction approach: a case study in Pangkalan Lampam District Ogan Komering Ilir Regency

B Muslimin[1,2*] and Suadi [1]

[1] Doctorate Program in Agricultural Sciences, Faculty of Agriculture Universitas Gadjah Mada, Bulaksumur, Caturtunggal, Kec. Depok Sleman, DIY, Indonesia
[2] Aquaculture Study Program, Faculty of Agriculture, Universitas Muhammadiyah Palembang, JL. Jenderal Ahmad Yani 13 Ulu Palembang, South Sumatera, Indonesia
*Email: boby.m@mail.ugm.ac.id

Abstract. Responsible management of fishery resources has been a tradition of Ogan Komering Ilir (OKI) of South Sumatera for generations. It was recorded that since 1630 the Dutch Government had made auction policy for managing water territory in OKI Regency as an effort to preserve nature and to avoid the conflict of inland water ownership. Currently, the community-based management model has been adopted by local governments into formal regulations, known as Regional Regulation on Auction of Flood Water Swamp and Rivers (*Lelang Lebak Lebung dan Sungai* or L3S). This paper describes the success factors and the threats for the failure of the L3S management model in OKI Regency, based on a case study in Pangkalan Lampam District. The study showed that the management mechanism through the L3S system had been well instituted and become a well-established management practice. The management model is in line with the principle of co-management and the approach has become critical success factor in L3S management. However, ecological, economic and social aspects influence the sustainability of such fishery management model. Besides, L3S management model faces limited data and information related to fish stocks, which result in difficulties in determining the total allowable catch.

1. Introduction

As a country, which waters dominate its geography, there are numbers of challenges in managing its ocean and water resources including the fishery resources. Fish is one of the renewable resources but endangered due to over-exploitation. The threat of extinction is caused by reckless management [1]. The characteristic of common-pool resources with joint ownership system also adds the complexity of its management. This complexity is caused by the two characteristics of the commons, namely (1) free-riding problems, in which the difficulty lies in issuing or managing the resources users with physical barrier or rules and (2) sub tractability problems, where sharing the resources or obtaining the same results out of the resource's utilization is difficult [2]. Two solutions to solve such resources management problems: (1) privatizing the resources ownership and (2) strong government control [3]. However, some other research showed the favorable outcome of resources management by society, which differs from the institutions [3, 4, 5]. Nevertheless, it needs to understand the factors or conditions for the success or failure of the common-pool resources management regime [6]. The joint-ownership pattern between the government and society through decentralization management with co-management system is seen as institutional innovation for joint-ownership resources management [7, 8, 9].

Water and fishery resources management system, so-called Lebak Lebung (flood water swamp) becomes one of the co-management patterns of the commons. Since a long time ago, the society and government came to realize the importance of utilizing limited Lebak Lebung fish resources, so it has to be managed sustainably. Such tradition is a way-of-life sustained by the government, which subsequently legitimized in recent Peraturan Daerah or Local Government Regulations. Since 1630, Ogan Komering Ilir Regency (OKI) managed its Lebak Lebung through an auction system. It mainly aims to cease the ownership conflict of waters and fishery resources as well as the regional revenue sources. Such auction system has been long known as *Lelang* (auction) *Lebak Lebung dan Sungai* (L3S). In 2015, 12 districts were conducting L3S which total 332 auction objects (Lebak Lebung) were offered, but only 260 objects in demand. With a long history and the dynamics of resources management policy until now, what is the current condition of such resources management model and what are the main challenges for the sustainability of such management model? These questions will be addressed in this paper, including the formulation of L3S or joint-ownership recommendation of resources management model policy for the future.

2. Methods

2.1. Time and location

This research was conducted in September to December 2016 in Sub-District of Pangkalan Lampam. The observation sites include 14 sites in four villages, namely Pangkalan Lampam, Deling, Sungai Bungin and Riding Village, which are parts of the public water of Lubuk Lampam. The location of field study and schedule were decided at the time of fishermen held an auction.

2.2. Data collecting and data analysis method

The primary data were collected through in-depth interview with key informants using purposive sampling techniques on the auctioneers during research. There were eight main respondents selected, consisting of two auction winners, two fishermen and three staff of Marine and Fishery Offices of OKI Regency. Secondary data was also collected from literature research, Pangkalan Lampam Districts data, Marine and Fishery Offices of OKI Regency, Statistics Indonesia of OKI regency as well as South Sumatera Province. The data then further analyzed with descriptive methods to describe the current condition of fishery resources, management and challenges of the L3S management model.

3. Results and Discussion

3.1. Description study site

OKI regency, geographically, has larger water area compared to the land, especially the common waters such as swamp (including floody swamp) and river. Lubuk Lampam one of the flood swamps area, which is the main source of the rural livelihood, both in fisheries and agricultural-related activities. This water called flood swamp area because it is affected by the tide of Lempuing River creek, which is connected with Komering River [10]. The Lubuk Lampam area is 1,200 ha, consisting Lebak Kumpai river 965 ha, the forest about 213 ha, the main river about 18 ha and Lebung (deep swamp) about four ha [11]. The Lubuk Lampam is also known as the fish reserve area, because of its potential to be reservation area for local fishes to reproduce.

The management of fishery resources in OKI regency mainly is based on the Fisheries Law or called Undang-Undang No. 45, 2009 [12], particularly on the principles of benefits, justice, unity, partnership, independence, equity, integration, efficiency, preservation and sustainable development. In line with these principles, the local government issued Regional Regulations (Perda) of OKI Regency No. 18, 2010 about L3S management, which subsequently renewed into Perda No. 14, 2015. This rule regulates the L3S management and the allowed or disallowed fishing equipment.

3.2. *Fish production*

Total fish production based on target species in Lubuk Lampam experienced fluctuation. In 1992, there were 63 species from 19 fish families found, but in 2008 there were only 48 species. However, in 2013 there were 63 fish species found [11]. The fish production based on the number of catch shows a declining trend. In 1997, the fish production reached 93 ton, while in 2012 it fell to 81 ton [11]. The Marine and Fishery Offices of OKI Regency recorded the number of catches using the L3S method in 2016 as 311 ton (table 1).

Table 1. Recorded the number of fish catches using L3S method in 2016.

Fish species	Amount of fish caught (kg)	Percentage (%)
Helostoma temmincki	59,164	19
Ophiocephalus micropeltes	55,042	17.7
Anabas testudineus	42,712	13.7
Ophiocephalus striatus	31,238	10.0
Others	27,432	8.8
Clarias gariepinus	24,254	7.8
Rasbora sp.	16,324	5.2
Pangasius sp.	15,250	4.9
Oreochromis nilotocus	15,250	4.9
Trichogaster trichopterus	13,592	4.4
Barbonymus schwanenfeldii	8,650	2.8
Mytus sp.	1,570	0.5
Monopterus alabus	600	0.2
Hemibagrus hoevenii	300	0.1
Total	311,378	100

3.3. *Types of fishing gears*

In general, the fishers utilize 10 types of fishing gears in Lubuk Lampam of OKI regency. In Pedamaran District, fishers use *kilung empangor* (fike net), *lulung-empang* (filtering barriers), *jaring insang tetap* (set gill nets), *jaring insang hanyut* (drift gill nets) and *pancing tajur* (hook and line), *jala* (cast nets), *bubu* (portable traps), *rawe* (long line).

Table 2. The fishing gear depends on the ebb and flow of the water tide in Pedamaran Disctrict.

Fishing Equipments	Amount (pcs)	Periods
Filtering barriers	100	Flood tide
Drift gill nets	400	Between periods
Hook and line	1.2	Flood tide
Cast nets	90	Low tide
Portable traps	400	Between periods

The utilization of the fishing gear depends on the river tide. As seen in table 2, fisher in Pedamaran District use several fishing gears because of various types of ecosystem in the district, such as reeks and swamp. Meanwhile, in Pangkalan Lampam District fishers only use few types of fishing gears, because the water area mostly consists of Lebak and Lebung (floody and deep swamp area). Based on the interview with fishery extension workers, there are five types of fishing equipment that are dominantly used based on the ebb and flow of the tide in the study site. During high tide on October to March, the fishers use *pancing tajur* (hook and line) and *jala* (cast net). However, if the water only reaches the medium height in March to May, the fishers use the *lulung empang* (filtering barriers) and *pancing tajur* (hook and line). Meanwhile, during the dry season on May to October, the main fishing gears are drift gill nets, cast nets and traps. Nevertheless, there is also challenge during this dry season that is the occurrence of environmentally unfriendly fishing practice using *setrum* (electric fishing gears). Such method directly causes the decline of fish stock in Lebak Lebung. Regional regulation No. 18, 2010 and No. 29, 2010 have stated the recommended fishing gears such to trap, gillnet and hook and line. The rule also depicted forbidden fishing gears such as *tuguk* (not selective fishing gears), cyanide and electricity. Also, those regulations also control the type and size of fishes, which are not allowed to catch (restriction of catching, carrying and selling fishes with the size of 2.5 cm to 20 cm) in table 3.

Table 3. Prohibited size of fishes to catch.

Common names	Species	Total length (cm)
Snakehead	*O. Striatus*	< 15
Giant snakehead	*O. micropeltes*	< 20
Ocellated snakehead	*O. pleurophothalamus*	< 15
Splendid snakehead	*O. maruloides*	< 10
Emperor snakehead	*O. maruloides*	< 10
Bangka snakehead	*O. bankenensis*	< 10
Kissing gourami	*Helostoma temmincki*	< 2.5
Siamese gourami	*Trichogaster pectoralis*	< 2.5

Fishing activities during March to May where the water level reaches medium height are deemed to be the most productive as the fishermen could obtain the maximum results, especially for snakehead (*Channa* spp.). During this season the catch could approximately reach 200–240 kg of snakehead. According to the interviews with the *pengemin* (the fisher who win the auction) and fishery extension worker, there were few fish species with bountiful numbers within the last two years, mainly *sapu-sapu* (*Hyposarcus pardalis*), *lundu* (*Arius maculatus*) and *catfish* (*Clarias* spp.). On the other hand, the rare species were *baung* or local catfish (*Hemibagrus nemurus*), *tapa* (*Wallago leeri*) and *selincah* (*Belontia hasselti*). Based on this information, the group of introduced carnivore species such as catfishes started to dominate fish caught and such trend could be another challenge for future management of fishery resources.

3.4. Lelang lebak lebung dan sungai (L3S) model

The local government has regulated mechanisms for managing floody swamp area and river (L3S) in OKI Regency, in which the people of OKI regency could have a chance to join the auction. The auction winner (*pengemin*) will be given rights to utilize the waters and fisheries for one year according to the existing regional (*Perda*) and regent's regulations. In the Perda of OKI regency No. 14, 2015 [13] article 18, it is stated that the auction winner or *pengemin* is obliged to pay from the highest bid and additionally five percent as permit retribution to catch fishes in the claimed area to the treasurer. In Article 28, *pengemin* is also obliged to deliver reports of their fishing activities to the Office Fisheries and Marine Affairs every month, including a) type of fish, b) amount/weight, c) selling price and d) the fishing gears. Upon the end of managing the water area, *pengemin* is obliged to do restocking at least five percent from the auction values, which is managed by the local district office and to be reported to the Offices during restocking. *Pengemin* should also prioritize in employing the people from the local village around the auction object.

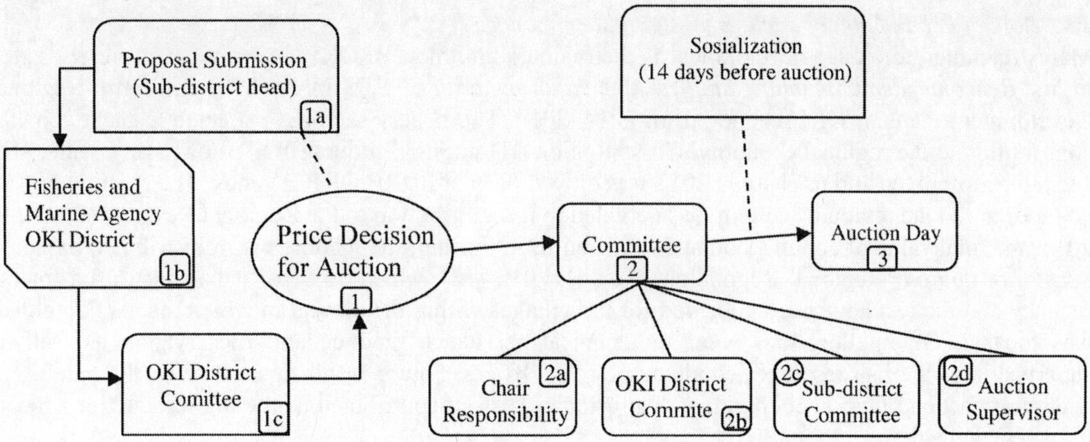

Figure 1. Standard Price Determination and Appointments committee Rural, Peraturan Bupati No. 29 tahun 2010 [13]: 1) price decision for auction determine by 1(a) proposal submission of sub-district head, 1(b) fisheries and Marine Agency OKI district and 1(c) OKI district committee; next 2) committee actor consists of government 2 (a), 2(b), 2(d) and non-government 2(c) and 3) the auction held in sub-district.

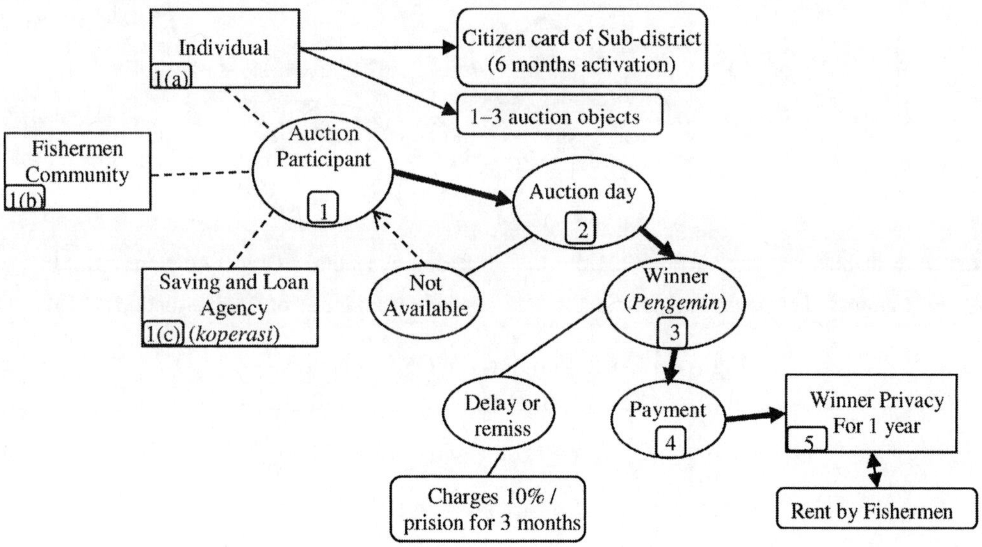

Figure 2. Auction Mechanism: 1) The auction participant could be individual 1(a), fishermen community 1(b) and *Koperasi* 1(c). The participant must be had citizen card of sub-district area with chosen minimal one auction objects and maximum three auction objects. After auction day (2), the winner called *Pengemin* must be complete the payment (4) which is can be use auction area for fisheries activity during one years and fishermen could be rent auction area from *Pengemin*, but if *Pengemin* doesn't complete the payment, *pengemin* will be charged 10 % or threats retained for three months.

3.5. Role of Lebak Lebung auction in regional economy

Mainstreaming the village development is the main commitment of OKI Government as clearly stated in its *Rencana Pembangunan Jangka Menengah Daerah* (RPJMD) or Medium-Term Regional Development Plans of OKI regency from 2014–2019. The fishery sector is expected to make a better contribution to the regional economy. Gross Regional Domestic Product (PDRB) of fishery sector still need to be improved and its share in 2012 was only 7.79 % of PDRB OKI Regency. The existing of L3S is important to the regional economy, as depicted in its contribution to the Regional Revenue (PAD). In 2015, the total value of action stood at 5.4 billion IDR. Further, the income was redistributed based on *peraturan daerah* (Regional Regulation) No. 14, 2015 [12], which 30 % of it has been used for the regency and the rest have been allocated for the villages within the district in which the auction object was located. Other allocations went to technical assistance practice and preservation as well as supervision of the fish resources conducted by the Offices. Figure 1 and figure 2 show the number of auction object and their auction values. Pampangan District contributed as the highest auction objects (62 sites) and values.

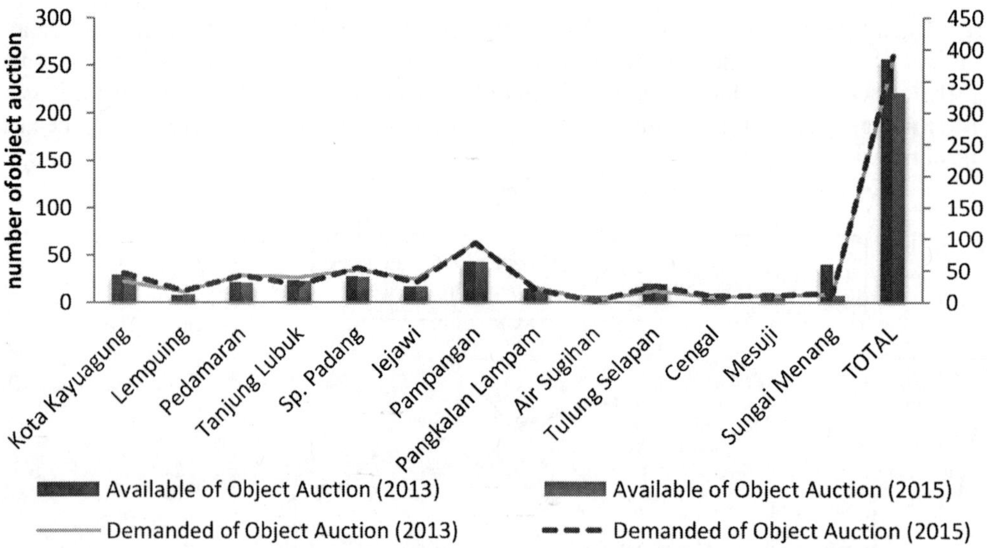

Figure 3. Auction objects in OKI Regency.

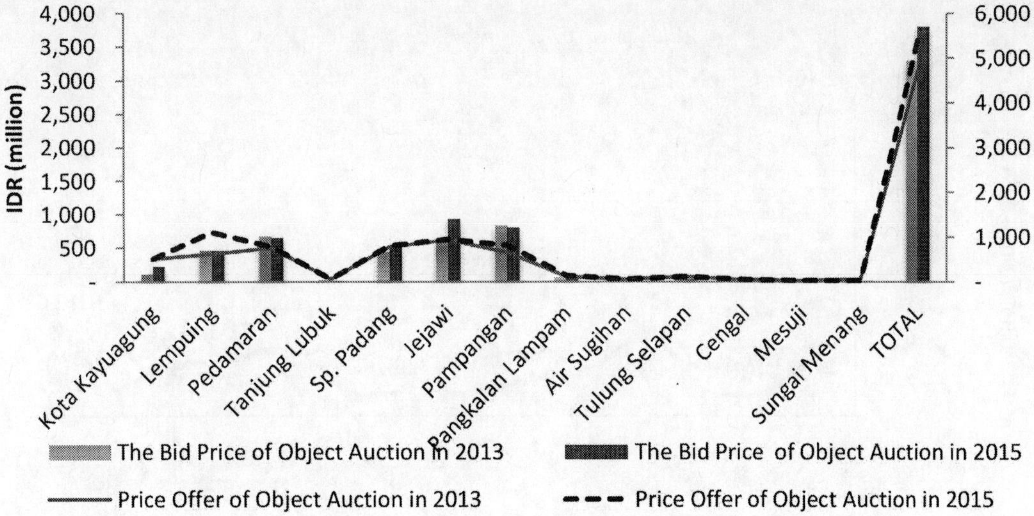

Figure 4. Auction value in OKI Regency.

As shown in figure 3 available for object auction was decreased, from 385 in 2013 became 332 in 2015. It happened in Sungai Menang District, from 59 and became 10. It assumed the lowest demand for object auction causing low cash results. Otherwise, Price offer of object-action shown in Figure 4 increased from 3.1 billion rupiahs in 2013 became 3.6 billion rupiahs in 2015. The bid price of Auctioned objects in figure 4 also experienced an increase from 4.9 billion IDR in 2013 to 5.7 billion IDR in 2015. The increase occurred in several sub-districts Sold objects used as regional income of OKI Regency. The increase occurred in several districts, namely Kota Kayu Agung, Sp. Padang, Jejawi and Pangkalan Lampam. There were 20 objects in 2015 with 13 objects in-demand, while in 2016 there were 14 objects which all became in-demand. The detailed study at micro level was conducted in Pangkalan Lampam District. However, there is an interesting fact of the auction value in Rasau Gemuk Bagian Laut in 2015, where it was bid for 42.4 million IDR; while in 2016 it decreased to 9.55 million IDR (figure 5 and 6).

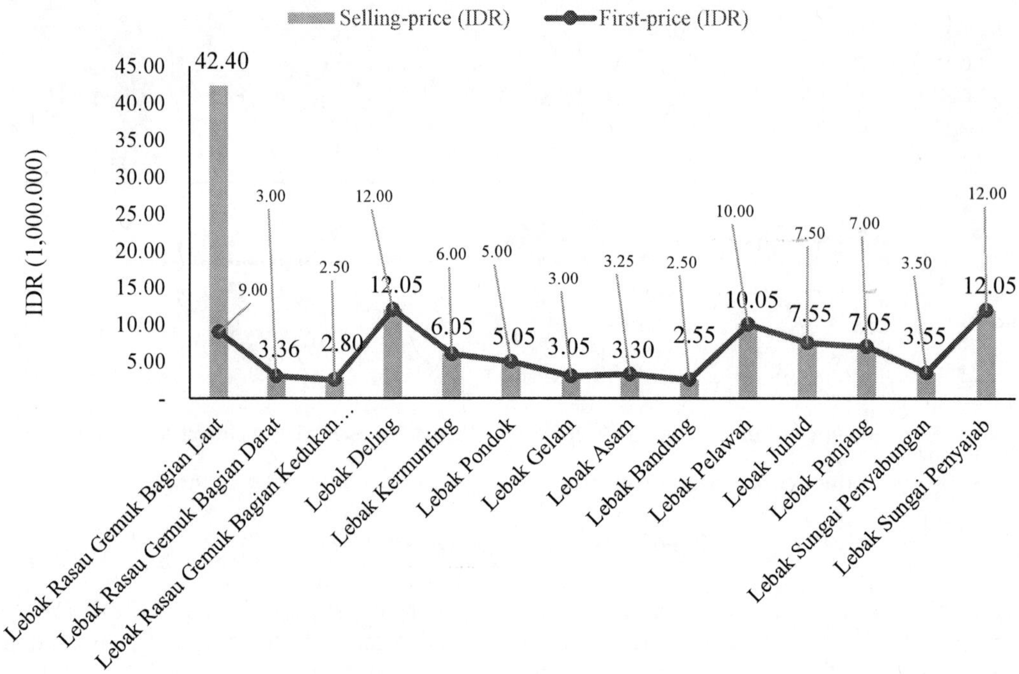

Figure 5. Auction value in Pangkalan Lampam District (2015).

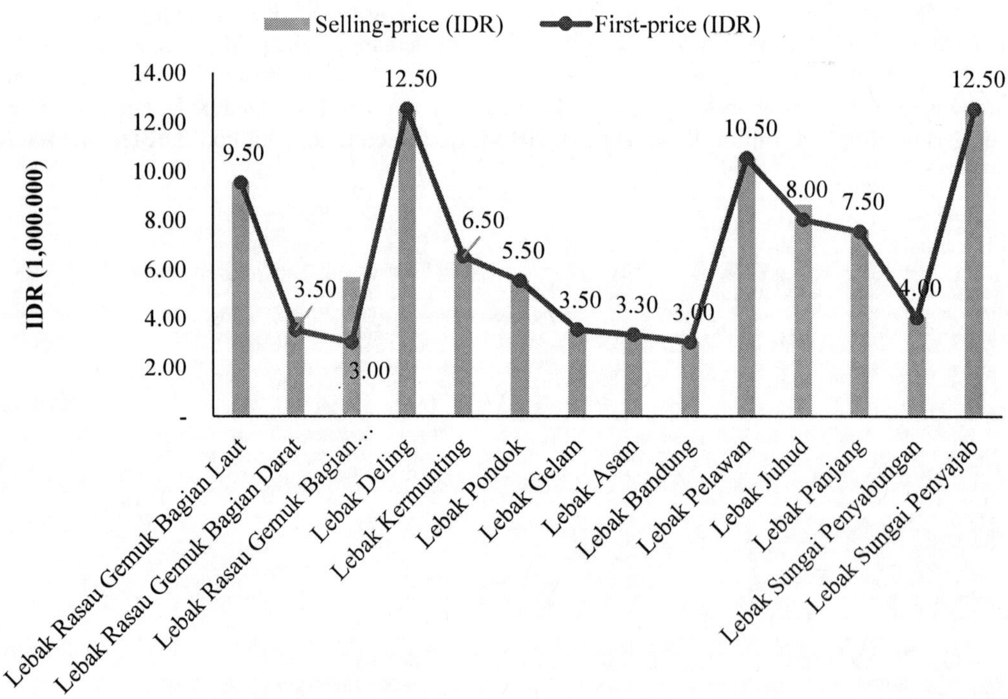

Figure 6. Auction value in Pangkalan Lampam District (2016).

3.6. The challenges of L3S model

The effort of managing the common-pool resources such Lubuk Lampan in OKI might be categorized as fisheries co-management. This approach has been practiced hereditarily in OKI Regency through the auction process. Under the scheme of water resources management in Pangkalan Lampam District, the managerial group is involved and given responsibility according to the auction mechanism. The L3S is an accommodative model for a joint-ownership scheme with co-management. However, seen from the characteristics of the local people, who are also the participants, the main issue to participate in L3S is the capital access and capacity. The capital issue makes it possible for outsiders to cooperate and make a special deal with local people. For example, the fishes caught by the local fishermen should be sold to *pengemin* with pre-determined price or the fishermen are given a chance to fish with particular payment terms. To participating in the auction, local fishermen as a *pengemin* candidate are required to possess a capital to register and pay the obtained auction object. Local government gives a chance to local people to participate either individually, by cooperation or group, with the prize of maximum amount auction object, which is three objects. The capital ownership access permits for outsiders to undertake capital loan with a particular deal. There is no specific criterion for local people in accessing capital from the local government, so they have the same chance in that process. Therefore the clear rule to involve the local in the process of an auction will be needed to enhance the opportunity of local to gain benefit from their available resources.

The current L3S management model is also facing the availability and sustainability of the fish stock. In fact, the fish recruitment process could occur naturally as well as artificially with the help of humans. Provision of a fish conservation area is one of the efforts to preserve fish stock, by deciding the spawning area and nursery area, each year, the OKI Regency Government stipulates these areas to forbid any fishing activities. The determination of nature reserve areas for research and conservation area for the fishes as well as allowed and disallowed fishing equipment have been explicitly ruled under *Perda* No. 14 year 2015 [12] which is ruler modification (*Perda* No. 18 year 2010 [15]) about management of L3S. Periodically, the OKI Regency Government also undertakes to restock for fish recruitment every year. Through the auction scheme, five percent of the auction object payment will be allocated for restocking. According to fishermen and *pengemin* or auction winners, restocked fishes are not coming from local species, but an introduced species with good adaptability and fast-growing such as catfishes and *nila* or tilapia. This further causes local endemic fishes to compete in nature hardly and it could be a threat of extinction. The restocking of non-native fish seeds and the scarcity of local/indigenous fish species, particularly seed impacted on the difficulty of obtaining endemic fish with economic values.

The L3S management model in fact also faces the challenges related to approach to e estimate the state of fish stock for auction. The estimation of maximum numbers of yield, whether through the assessment of Maximum Sustainable Yield and Maximum Economy Yield is hard to conduct because of the lack of data. Meanwhile, such estimation is critical because it could be the anchor of fixing the standardized price in the future. Also, based on the local regulation, it is stated that before initiating the auction process, the committee will determine the auction object value in the district level, usually based on the previous year's determination. The unavailable precondition in estimating the fish stock and obstruction in collecting information about the caught fishes by *pengemin* causes difficulties in determining the auction object by the indication of size fish stock. Few other challenging issues are depicted on:

- OKI Regency has made regulation for conservation zone of fish, but in practice, the regulation still allowed for fish restocking with non-native (alien) species.
- Auction winner must donate five percent of revenue for fish stocking, but hard to be monitored and conducted.
- Auction winner supposed to report the catch data (volume by species and fishing gears), but fish catch data not available.
- The auction price determined by previous year auction and didn't by fish stock condition.
- The auction mechanism did not consider Maximum Sustainable Yield and Maximum Economic Yield as Policy.

Difficulty accessing capital for local people third party funds as other options for accessing capital with deals and agreement between them.

4. Conclusion

The L3S model in OKI Regency has been employed for long, ruled explicitly under the local government regulations, involving the community participation. Such approach became a success factor in managing water area within the context of joint-ownership, which applies co-management principle emphasizing in equal opportunities for local people to participate in the auction process. The challenges, on the other hand, are laid on the approach to estimate fish stock and its state, the capacity of the local community to participate and controlling the auction winner to obey the rules. Thus, determining the total amount of caught fishes by *pengemin* or auction winner, evaluating the numbers of caught and the study of fish stocks and their state are needed to enhance the better allocation of fish stock and sustainability of fish resources. Determining the species of fish to be restocked is also necessary and makes a priority to the local/native or endogenous species to avoid the declining of original species of Lubuk Lampam. Therefore the advance study of local species and effort to domesticate will be needed.

References

[1] Widodo J and Suadi 2006 *Pengelolaan Sumberdaya Perikanan Laut* (Gadjah Mada University Press: Yogyakarta) p 252
[2] Ostrom E, Dietz T, Dolsak N, Stren P C, Stonich S and Weber E 2002 *The Drama of The Commons* (National Academic Press: Wasington DC) p 534
[3] Hardin G 1968 *J. Science New Series* **162** 1243
[4] Berkes F 1989 Common Property Resources: Ecology and Community-Based Sustainable Development (John Wiley & Sons Ltd) p 320
[5] Ostrom E 1990 *Governing the Commons: The Evolution of Institution fo Collective Action* (Cambridge Univ. Press: London) p 298
[6] Acheson J 2000 *J. Human Ecology.* **28** 145
[7] Nielsen R J, Degnbol P, Viswanathan K, Ahmed M, Hara M and Abdullah N M R 2004 *J. Marine Policy* **28** 151
[8] Carlsson L and Berkes F 2005 *J. Environmental Management* **75** 65
[9] Pomeroy R S and Rivera-Guieb R 2006 *Fishery Co-management: A practical Handbook* (CAB International Publishing: UK) p 258
[10] Jubaedah D 2015 *Pengelolaan Rawa Banjiran Berbasis Analisis Dinamika Fluktuasi Muka Air di Lubuk Lampam, Sumatera Selatan* (Bogor: Respitory of Bogor Agrciultural University)
[11] Eko P, Mohmmad M K, Ismudi M and Endi S K 2013 *Jurnal Kebijakan Perikanan Indonesia* **5** 57
[12] Peraturan Daerah Kabupaten Ogan Komering Ilir 2015 *Nomor 14 2015 Tentang Perubahan atas Peraturan Daerah Nomor 18 tahun 2010 tentang Pengelolaan Lebak Lebung dan Sungai Ogan Komering Ilir*
[13] Peraturan Bupati Ogan Komering Ilir. 2010 *Nomor 29 2010 Tentang Petunjuk Teknis Pelaksanaan Lelang Lebak Lebung dan Sungai. Ogan Komering Ilir*
[14] Undang-undang nomor 45 2009 *Tentang Perubahan atas Undang-Undang Nomor 31 Tahun 2004 Tentang Perikanan Indonesia*
[15] Peraturan Daerah Kabupaten Ogan Komering Ilir 2010 *Nomor 18 2010 Tentang Pengelolaan Lebak Lebung dan Sungai dalam Kabupaten Ogan Komering Ilir*

IOP Conf. Series: Earth and Environmental Science **139** (2018) 012033 doi:10.1088/1755-1315/139/1/012033

Livelihood analysis of floating net cages fish farmers at Sendang Village Sub-district of Gajah Mungkur Reservoir of Wonogiri Regency

Z N A Nissa[1], Suadi[1]* and Sukardi[1]

[1] Departement of Fisheries, Faculty of Agriculture Jl. Flora, Bulaksumur, Universitas Gadjah Mada, Yogyakarta, Indonesia
*Email: suadi@ugm.ac.id

Abstract. Floating net cages farming (KJA) is one of the main livelihood resources in Sendang Village, Gajah Mungkur Reservoir. The purposes of this study were to determine the livelihood asset of fish farmers, the problems of livelihood asset management and the utilization strategy to support aquaculture businesses also farmers livelihood. The study showed that the natural capital, provides easiest way to farmer in fish cultivation. The fish farmers also have good technical capabilities in fish cultivation and the product has high demand and value which is essential for farmers livelihood. The main problems faced by small-scale farmers and all large-scale farmers were transition period, and the rise of cost price which sometime cause the failure in the business. The strategies to deal the problems include technological adjustment, managing the pattern of stocking of tilapia seeds and income source diversification. There were differences in dealing the rise of cost. The small-scale farmers borrow from the bank, while medium-scale farmers use their savings. Another difference of livelihood strategy was the management of financial capital. However, various strategies were still required to increase the livelihood of fish farmers and could address the vulnerabilities in the cultivation of KJA as a common pool resources.

1. Introduction

Global tilapia production is expected to almost double from 4.3 million tons to 7.3 million tons between 2010 and 2030, according to World Bank [1]. World tilapia production in 2015 is about 5,576,800 metric tons which has price about seven USD·kg^{-1}. China is world's largest producer of tilapia farming and Indonesia's tilapia production is in the second stage after China [2]. The development of tilapia farming provides benefits for rural communities in the form of job opportunities, sources of income and collective-based socio economic activities. The fisheries sub-sector in Wonogiri Regency has a positive growth rate. One of the factors were caused by the development of fish farming in the form of floating net cage farming (KJA) in Gajah Mungkur Reservoir.

Sendang Village, Sub district of Wonogiri, Wonogiri Regency is a village that directly adjacent to Gajah Mungkur Reservoir. Sendang Village is a fishery business center that has various potential such as processing and marketing fish, catching fish and cultivation fish in floating net cage farming (KJA). According to Department Animal Husbandry, Fishery and Maritime of Wonogiri Regency said that the cultivation fish in floating cage farming (KJA) with 176 units of KJA is able to produce an average of 900 tons of fish per year [3]. The tilapia cultivation pattern of KJA fish farmers is usually spreading 69.19 kg/plot tilapia seed in one period. The rotation take is about four months [4]. Tilapia cultivation in KJA at Sendang Village has been growing for approximately 30 years. This of course have impact on ups and downs of the fish farmer's number and other stakeholders of fisheries. The ups and downs of fish farmer's number are caused by many factors, such as environmental factor. When the weather is deteriorating or the intensity of rainfall is high, that causes upwelling resulted in mass fish mortality. In addition, the ups and downs of the number of cultivators are influenced by the price of production factors such as dynamic of price like feed prices, seeds, scarcity of seed stock, highly dependence on factory feed, fluctuations in tilapia fish demand that seasonal also affect the sustainability of the business of tilapia fish farmers in Sendang Village District, Wonogiri.

One of the reasons fish farmers maintain in their business because high of demand of fish both from inside and outside the region. Demand of tilapia fish in local markets is quite a lot from the owners of restaurants that located in along the rim of the reservoir. Outside demand of tilapia fish came from Yogyakarta, Magelang and surrounding areas. The outside demand also included the export market. The strategy to optimize the potentiality of the region from each fish farmers is of course different. Differences of this strategy are influenced by differences in problems and challenges faced and the ability owned by each fish farmer household in managing their capital. Cultivation is very dependent on the availability of various capital or resources, including social capital, human capital, economic financial capital, natural resource and environment capital and physical capital (infrastructure) [5]. These capitals are often modified by the role of social relations (such as gender, economy class, age, ethnicity and religion/race), institutional influences (rules, customs, customs, markets) and vulnerabilities (such as natural disasters and conflicts, economic, fluctuating prices, population growth and population issues and changes in technology and macro policies) [5]. Department of Fishery make policy and strategy to develop the scale of micro and small business of fishery such as providing technical and managerial assistance to improve micro business to be feasible as well as *bankable* in long term and existence of financial guarantee system to operationalization of micro business activity [6]. The development of fish farming business in Sendang Village shows a good improvement in terms of production and number of fish farmers. The number of fish farmers is likely to increase. This is supporting by the easy access to the financial capital through the Bank with the KUR (People's Business Credit) with low interest rates

The approach used in analyzing the strategies applied by the fish farmers to keep the business going is a Sustainable Livelihood Approach (SLA). Allison & Ellis argue that the concept of livelihood is able to reconcile important factors that affect the vulnerability (problem) and strength of individual or family survival strategies [7, 8]. Therefore, the sustainability livelihood approach is important to be studied in order to provide an overview of vulnerability or problems to the livelihood capital faced by cultivating in KJA at Sendang Village, Gajah Mungkur Reservoir along with ideas that can be attempted to overcome the vulnerability based on livelihood capital which are owned. Based on the identification of the problem, this study aims to determine livelihood assets owned by KJA fish farmers in Sendang Village, Wonogiri Regency, to identify their problems in the livelihood capital management and to understand livelihood strategy to support cultivation in fish cages farming.

2. Materials and Methods

2.1. Research framework

This research was conducted from April to May 2016 in Sendang Village, Wonogiri Sub-district, Wonogiri Regency. The basic method used in this research is descriptive method. The research framework depicts in figure 1.

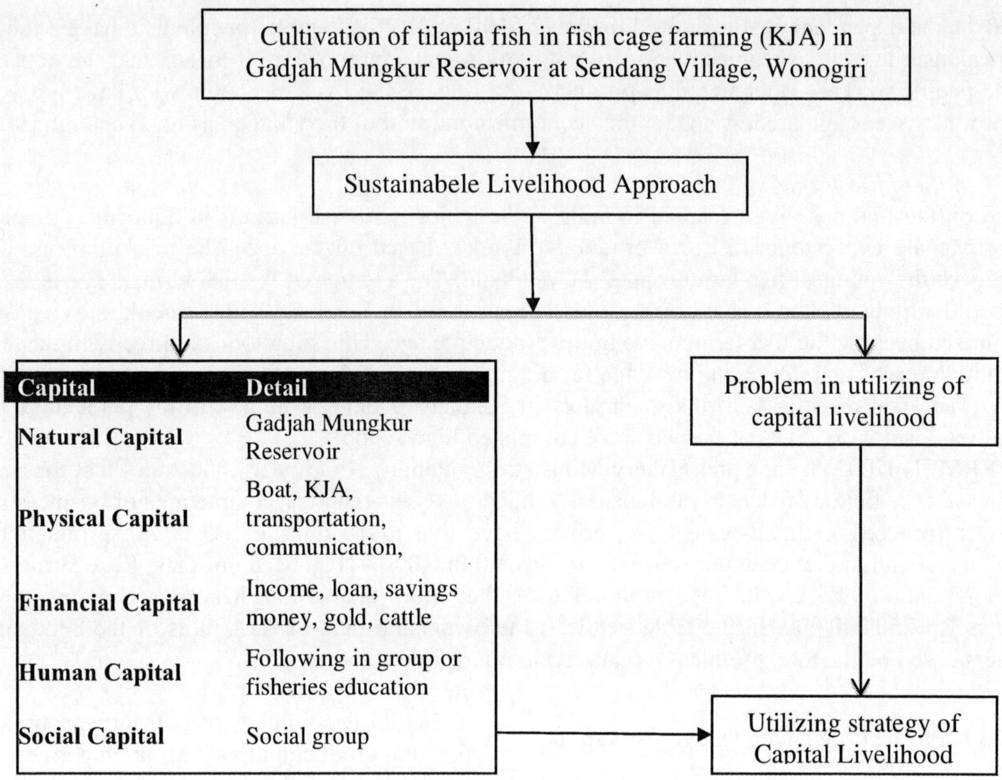

Figure 1. Consideration framework of sustainable livelihood approach.

2.2. Method of data collection
This study used random sampling method interviewing 35 fish farmers as respondents. The sampling was based on Wonogiri Fisheries Department data that had selected randomly. The selected respondents were fish farmers who agree to participate. The collected data consisted of primary and secondary data. Primary data were gathered by interviews, observation and documentation. The primary data such as of the identity of the fish farmers (i.e., the name, the age and their education level), availability and access to livelihood capital of KJA fish farmers and household activities of KJA fish farmers were collected. Also, secondary data were collected by exploring literature studies and collecting data that available in Sendang Village and Fisheries, Animal Husbandry and Maritime Department of Wonogiri Regency, such as village monographs.

2.3 Data analysis
Data were analyze based on the four important components that have been included in the questionnaire such as fish farmers profile, ownership and access to each capital (natural, physical, financial, human and social). Afterward, it was enriched by analyzing the problems or challenges and survival strategies which was applied by the fish farmers by utilizing their livelihood capital. The data result was presented as a descriptive statistic which is in the tabulation that has been averaged and grouped according to existing groups or levels following the final results obtained. Also, some photograph and proxy approach were used to support the results.

3. Results
Sendang Village is a village that has easier access to Gajah Mungkur Reservoir (fishery zonation) so that KJA's business has been growing in Sendang Village. Sendang population dependency ratio is

51.70 %, the dependency ratio showed that every 100-productive people (people who have a job) have the responsibility to 52 peoples who were not working or unproductive. It means that the dependents of the people who are categorized as productive to finance the lives of people who have not yet and not productive enough are low so that the economic condition of the villagers is good enough [9].

3.1. Profile of fish farmers

Based on the findings that presented in table 1, the majority of fish farmers in Sendang Village were majority male and categorized as a productive worker based on the age. The minimum age of fish farmers or the youngest fish farmers aged 21 years old. The average of the fish farmers age is about 36 years old farmers. The aquaculture business opportunity in Sendang Village could survive and be maximized because the fish farmers are in the productive age. This situation is indirectly indicated that the cultivation of tilapia floating net cage farming (KJA) in Sendang Village has been growing for 50 years. The level of education of fish farmers in Sendang Village is high with the percentage of the education level of 51.43 % of farmers have completed high school.

PERMEN KP (Maritime and Fishery Ministry Regulation) No. 5 year 2009 states that the scale of business in the field of fish cultivation is determined by assets (capital, volume/area of business units), turnover (proceeds). Small-scale KJA farmers have two to 10 units of KJA with production 0–2 tons/harvest and the revenue per year reaches Rp. 60,000,000.- [10]. Medium scale KJA farmers have 10 to 20 units of KJA, with total production reaches more than 2 tons/harvest and annual revenue reaches Rp 250,000,000. In the table below is the ownership data of KJA plots of the Sendang fish farmers based on the total production (harvest) in one period (table 2).

Table 1. Profile of Sendang village fish farmers.

Parameters	Frequency	Percentage (%)
Sex of respondents		
Male	33	94
Female	2	6
Education level		
Primary School	4	11.43
Junior High School	10	28.57
Senior High School	18	51.43
Diploma/Bachelor Degree	3	8.57
Age of Villagers (years)		
Less than 20	0	0
21–30	8	23
31–40	22	63
> 40	5	14
KJA Farming Experience (years)		
< 5	18	51.43
5–10	13	37.14
> 10	4	11.43

Tabel 2. The number of fish farmers based on the total production (harvest) in on period.

Total production (kg)	Fish farmers (person)	Percentage (%)	Percentage (%)
Small Scale			
less than 500	19	54.29	
500–1,000	10	28.57	91.43
1,000–1,500	3	8.57	
Medium Scale			
≥ 2,000	3	8.57	8.57
total	35	100	100

3.2. Analysis of the utilization of livelihood capital

The percentage of livelihood capital used to support the cultivation of KJA by KJA fish farmers in Sendang Village, Wonogiri Sub-district, Wonogiri Regency is presented in table 3. The utilization of livelihood capital was used to face the problems that occur in the management of KJA cultivation business. Based on table 3, there were some strategies which were applied by Sendang Village fish farmers. The most dominant livelihood capital which is utilized as a source of survival by KJA fish farmers in Sendang Village, both small and medium scale fish farmers is the natural capital by utilizing the waters of Gajah Mungkur, Wonogiri Reservoir.

Table 3. Utilization of livelihood capital by KJA fish farmers in Sendang Village.

Type of Capital	Small Scale		Medium Scale	
	Number of respondent	% respondent	Number of respondent	% respondent
Natural Capital				
Reservoir (public waters)	32	100.00	3	100.00
Physical Capital				
Handphone	32	100.00	3	100.00
Internet	10	28.57	3	100.00
Motorcycle	32	91.43	3	100.00
Pick Up	0	0.00	2	66.67
Diesel machine	25	71.43	3	100.00
Human Capital				
Skill in fisheries	32	100.00	3	100.00
Fiheries illumination	8	22.86	1	33.33
Financial Capital				
Loan	32	91.43	1	33.33
Saving money	21	60.00	3	100.00
Small scale business	6	17.14	2	66.67
Social Capital				
Processing fish group	10	28.57	0	0.00
Fish farmers group	25	71.43	2	66.67

Based on the findings that can be seen in figure 3, natural and financial problems is an important issue among the farmers of KJA tilapia in Sendang Village. Natural problems or vulnerabilities that faced by KJA fish farmers in Sendang Village were the impacts of the transition period. The strategy that has been done was a strategy that leads to minimize the risks posed such as technological manipulation, the reducing capacity of several plots of KJA, arranging the pattern of stocking and diversification of job. Financial problems such as lousy harvest, reduced revenues, fluctuation of prices and macro policies (i.e., fuel prices) affect production factor prices (i.e., increase in feed prices, seeds and other operational costs).

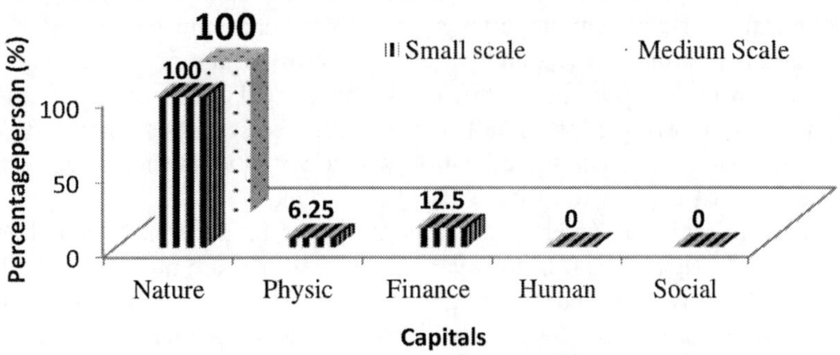

Figure 2. Capital livelihood vulnerability intensity.

Most of all fish farmers have other jobs as alternatives income source. Below is the list of profession/job of the fish farmer in Sendang Village (figure 3). Based on figure 3, people in Sendang Village have the primary job as a fish farmer (77.14 %). Meanwhile, the remaining 22.86 % has various other jobs (i.e., laborers/masons or bricklayer, farmers, headman of the village, fish sellers, middleman (fish seller), drivers of truck and boat driver for tourism). Respondents who have the primary job as a fish farmer, they provide plenty of time to manage their cultivation.

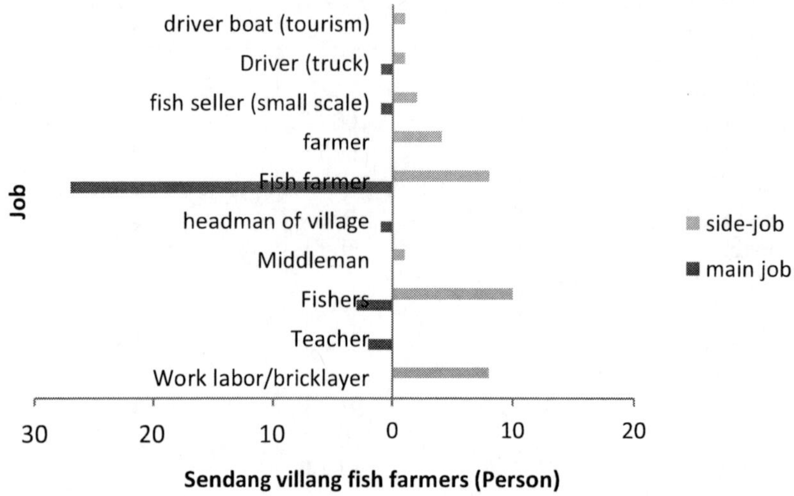

Figure 3. Number of fish farmers by their job (livelihood).

Based on (table.4), The small-scale fish farmer has vulnerability or problems in the utilization or management of physical capital, which is the possession of diesel machine and boats. However, the small-scale fish farmers face these vulnerabilities by utilizing the social capital they possess such as their presence and participation in groups. Fish farmer groups facilitate access to borrow supporting facilities for the cultivation. Fish farmers who have interest in following counseling/training were few. Fish farmers who want to learn about fish cultivation and have physical capital (internet, television or radio), they can access it from it. That is a fish farmer's strategy when they didn't follow the counseling/training/education from Wonogiri Fisheries Department.

4. Discussion

In general, the strategy undertaken by KJA tilapia fish farmers in Sendang Village was not significantly different from the strategy undertaken by KJA fishermen in Jatiluhur Reservoir in a study conducted by [11] and fish farmers in Klaten by [12]. The ability of the fish farmers manages or utilize assets owned which can be a supporting factor of livelihood [11]. Tilapia fish farmers in Sendang Village were divided into two groups: small and medium scale fish farmers. Each fish farmers certainly has a different strategy in capital utilization to deal with the problem. The research results by [12] that was done in 2013 divides the tilapia fish farmers into several groups ranging from micro, small, medium and large scale. In general, the strategies were undertaken by small-scale tilapia fish farmers in Janti, Ponggok, Nganjat and Sidowayah Klaten Villages are raising capital by applying for loans, utilizing groups to buy feed and access government aid. The strategy undertaken by middle-scale farmers is to make a concerted effort, adding business networks, marketing products out-of-town [12]. Comparing the results of [12] research, the strategy undertaken by the tilapia fish farmers in Sendang Village was reasonably similar. This tendency happens because of the rural life habit which tends to be homogenous. Other research also states that one of the characteristics of rural society is to have homogenous characteristics [13].

IOP Conf. Series: Earth and Environmental Science **139** (2018) 012033 doi:10.1088/1755-1315/139/1/012033

Table 4. Vulnerability and strategy issue that done by fish farmers.

Capital	Small scale fish farmers				Medium scale fish farmers			
	Problem	**% respondent**	**Strategy**	**% respondent**	**Problem**	**% respondent**	**Strategy**	**% respondent**
Natural	Oxygen reduction	78.12	watering by diesel machine	100	Oxygen reduction	100	Watering by diesel machine	100
	Upwelling	56.25	Give up	77.78	*Upwelling*	66.67	Give up	66.67
			Do nothing	22.22			*Do nothing*	33.33
	Shallow water	21.21	Emptying some plots of KJA	100	-	-	-	-
	Bird pest	21.21	Use bell	100	-	-	-	-
Physical	Don't have boat	3.12	Get a loan	100	-	-	-	-
	Don't have diesel machine	21.87	Get loan from fish farmer group	100	-	-	-	-
Financial	Hard to get seed	71.87	Decrease of seed	100	Hard to get seed	33.34	Buy from other region	100
	Increase seed price	59.37	Depend on group	71.43	Increase seed price and feed	33.33	Buy directly from feed and seed agency	observation
			Get a debt (transfer nota)	100			Depend on savings	100
	Low production	18.75	Get a debt	40.62	Low production	33.33	Diversification job	66.67
			Diversification job	31.25			Arrange seed pattern	100
			Arrange seed pattern	37.5	-	-	-	-
	Fluctuation in demand market	21.88	Follow the low of market	observation	Fluctuation in demand market	33.33	Sell to outside market	66.67
Human	Less interest in fisheries education/training	observation	Nothing strategy yet	-	Less interest in fisheries education/training	Observation	Learn from others sources	observation
Social	Less in presence and being active in fish farmers group	observation	Do nothing	observation	Less in presence and being active in fish farmers group	Observation	Do nothing	observation
	Price competition	25	Do nothing	100	Price competition	33.34	Sell to outside market	100

According to the data (table 2), the business opportunity of aquaculture in Sendang Village can be developed and maximized because the fish farmers are in the productive age. The level of education of fish farmers in Sendang Village is high which affects the ability of the fish farmers adapting and receiving new knowledge or technology in the management of their aquaculture business. Formal education is significantly related to the field of competence of farmers such as the marketing of business results, harvest and postharvest handling. The situation reflects that the higher formal education of the farmers, the higher level of competence they have [14]. It also said that diversification strategies could develop optimally. It was determined by some factors such as household composition, age and educational level of the farmer and his/her spouse. Moreover, the strength of social capital and social networks, the 'openness' of the farmer towards outside world, the effort to looking for an information and sources of knowledge and a professional attitude of the farmer are also key factors for developing their farming [15]. The length of the fish farmers doing the cultivation of KJA in Sendang Village is less than five years, indicated that all the growth in fish farms due to new entrants. Based on the results of interviews, the increase of fish farmers number is triggered by the high opportunities for the cultivation of KJA tilapia fish and easy access to get the bank loan. The success of fish farmers in the cultivation business is closely related to the competence of agribusiness owned by fish farmers in managing the fish farming business.

The average fish farmer does not have only one job or not just be a fish farmer, but also has multiple jobs or other jobs to avoid unemployment. Participation of fish farmers and their wives in a social group such as cultivation group and fish processing product group is a survival strategy that they have been doing. Joining into fish processing product group were fish farmer's wife strategy to increase the added value and to fulfill the demand. This activity ensured the continuity of fish farming business KJA in Sendang Village. These were a livelihood strategy that utilizes social capital or connection among fish farmer's wife. The access to credit is essential for small farmers.

Besides social capital, the access to credit is important for small farmers. The smaller fish farmers prefer to get a loan from the bank while the medium scale fish farmer prefers to use their saving money when they face challenges in finance. That is a reality or social facts that researcher found in the field because the small-scale fish farmer who had not enough saving money. The easy access to get a debt/loan from the bank was one factor that encourages small-scale farmers to get a loan from the Bank. It was a rational action according to Weberian theory was an instrumentally rational action (an action that taken consciously based on consideration and consciousness to reach the purpose of that action and the means used to achieve the objectives). Considering the objectives to be achieved (greater profits/maintaining business cultivation) and tools used to achieve goals (by borrowing from the bank) [16]. Fishery Department of Wonogiri Regency has made policy and strategy to develop the scale of micro and small business for fishery and aquaculture such as providing technical and managerial assistance to improve micro business to be feasible for a long-term, and the existence of financial guarantee system to the operationalization of micro business activity [6]. So that, they can avoid the harvest failed. If it happened so, they would use their side-job to pay interest when there is a severe harvest. It has been stated that the contribution of the new activities (off-farm) to the total household income varies from a marginal to a substantial addition to the primary income, but, whatever its size, this additional income can stabilize household finances, which would otherwise be strongly influenced by price fluctuations and seasonal cycles. Off-farm employment can stabilize the household finances, providing more than a quarter of household income systems [11].

5. Conclusion

The Gajah Mungkur Reservoir, the fish farmer ability, and the potential market were a basic capital that had a vital role in maintaining fish cultivation business in KJA in Sendang Village especially to perform survival strategy due to the disadvantages trends and shocks for Sendang Village fish farmers. The main problem faced by 78.12 % of small-scale and all large-scale fish farmers was the seasonal transition. Another problem was the increase in production factor prices (i.e., increase in feed prices, seeds and other operational costs) (35.41 % of respondents). Capital utilization strategies conducted by small and medium scale fish farmers were technological manipulation, arranging the pattern of stocking and having an alternatives job, especially when severe harvest. There was a difference in strategy when factor prices rise between small and medium-scale cultivators. Small-scale farmers chose to borrow from the Bank or have a loan, while medium-scale cultivators only use their savings. Another difference of livelihood strategy between small and medium scale fish farmers was the financial capital management which was exactly in time of hard to get tilapia seeds. Small-scale fish farmers typically reduce capacity/volume per plot while medium-scale fish farmers sufficient capital and networks can afford to buy from other regions.

References

[1] http://www.fao.org/docrep/019/i3640e/i3640e.pdf Accessed in 2017 October 30
[2] https://www.was.org/documents/MeetingPresentations/AQ2016/AQ2016_0005.pdf Accessed in 2017 October 30
[3] Department of Animal Husbandry Fisheries and Maritime Wonogiri Regency 2010 *Fisheries Condition in Gajah Mungkur Reservoir* (Wonogiri: Wonogiri Government Publishing)
[4] Ardi I 2013 *Media Aquac* **8**(1) 24–25
[5] http://www.zef.de/module/register/media/2390_SL-Chapter1.pdf Accessed in 2016 October 3
[6] Department of Fisheries and Marine 2016 *Banking Credit Schemes for the Development of the Marine and Fisheries Business Sector*. (Jakarta Government: Directorat General P2HP)
[7] Allison E H and F Ellis 2001 *J Marine Policy* **25** 377–388
[8] Ellis F 1998 *J. Development Studies* **3**(1) 1–38
[9] Sendang Headman 2015 *Sendang Village Profile* (Wonogiri: Wongiri Government Publishing)
[10] Ministry of Fisheries and Marine Regulation of Ministry of Fisheries and Maritime of Indonesia 05/MEN/2009 about Business Scale in Cultivaton Fish Indonesia
[11] Purwandari M I N 2014 *J. Solidarity* **3**(1) 56–62
[12] Febriana A V *Thesis* 2013 Universitas Gadjah Mada Yogyakarta
[13] Yuanita A 2015 *Thesis* Universitas Gadjah Mada Yogyakarta
[14] Manyamsari I and Mujiburrahmad 2014 *J. Agrisep* **15**(2) 58–60
[15] Meert H, Huylenbroeck G V, Vernimmen T, Bourgeois M and Hecke E V 2005 *J. Rural Studies* 81–97
[16] Damsar and Indrayani 2016 *Rural Sociology Introduction* (Jakarta: Kencana)

IOP Conf. Series: Earth and Environmental Science **139** (2018) 012034 doi:10.1088/1755-1315/139/1/012034

Techno-socio-economic analysis of losses in capture fishery: a case study in Pelabuhan Ratu, Sukabumi, West Java Province

A Nurhayati [1*] and A H Purnomo[2]

[1] Faculty of Fishery and Marine Science, Padjadjaran University, Jl. Raya Bandung Sumedang km 21, Hegarmanah, Jatinagor, Kabupaten Sumedang, Jawa Barat, Indonesia
[2] Research Center For Marine and Fisheries Product Processing and Biotechnology, Jl. KS Tubun, Petamburan IV, Slipi, Jakarta Pusat, Indonesia
*Email: atikah.nurhayati@unpad.ac.id

Abstract. This research was aimed at analyzing factors influencing capture fisheries losses, focusing on technical, social and economic aspects at Pelabuhan Ratu. A case study was undertaken, through a survey involving 40 respondents. These respondents represented groups of fishers, collectors, middlemen, processors and consumers. The questions delivered in the survey was adapted from the Exploratory Fish Loss Assessment Method (EFLAM). Based on this research, the fish loss was detected in Palabuhan Ratu, which amounted to 4.25 % at the fisher level and 5.12 % in the following supply chains, due to some factors. It was found that among the technical factors, the most influential ones were handling of landed fish, fish sortation, fish size, fish shelf life and season. Among economic aspect, factors with the most significant influence were fish price fluctuation and price level; meanwhile, among the social factors, those that had the most significant influence was the revenue distribution system. Based on this, the relevant policy implication of this research was the need for effective programs which covers the development of cold chain and distribution facilities and infrastructure, and an improvement in skills and knowledge of fish derivative product processors.

1. Introduction

Among Indonesian's prominent production centers, there is Palabuhan Ratu in the Sukabumi District of West Java Province. In 2016, the Palabuhan Ratu's capture fishery production was recorded at 3,839,057 kg [1]. The main species of fish loaded in Palabuhan Ratu landing place are mainly *Euthynnus affinis, Thunnus obesus* and *Katsuwonus pelamis.* However, high yield losses have reduced the value to only Rp. 106,224,971,000. Fishery loss or Post–Harvest Fish loss (PHF) is a decrease in the number of fishery resources that can be consumed, occurring in a marketing distribution chain, regarding of inputs of capture fisheries production, post–harvest handling, processing and marketing (collecting traders, wholesale and retail). The losses caused by the shrinkage of the post–harvest result in capture fishery reach 30–40 % [2]. The resulting losses of catching fisheries include the physical condition of the fish catch, quality and financial losses [3]. One disadvantage can occur at any stage in the supply chain, from catching or harvesting to the consumer [4]. The resulting fisheries post–harvest losses occur in all sequences of fisheries process and are related to the characteristics of fishery products. The series of fisheries activities include preproduction, production, management and marketing carried out in a fishery business system.

At the pre–production stage, the resulting fisheries post–harvest losses associated with the characteristics of fishery products are seasonal. The level of the capture fisheries production, in general, is highly fluctuating, depending on the seasonal pattern. In peak season, supply to fishery products increases, so they are often unabsorbed by the market priced lower by the consumer. It is a typical case of losses incidence at the pre–production stage. At the stage of production to marketing, the characteristics of a perishable fresh fishery product further increase the risk of losses if proper handling is not carried out until the stages of product distribution to the final consumer. Fishery products are also voluminous, so the handling is not easy to implemented.

The above situation underlines the need for handling losses, not to increase the value of the catch but also to improve the supply of fish for the fulfillment of consumption needs and to implement a better management system. As mentioned by New Partnership for Africa's Development the decrease in losses is closely linked to the management systems and resource wealth generation [3]. The most obvious way to increase fish supply, even without increased landings, is to reduce losses of what is currently caught, it is further that attention should be given to reducing losses [5].

Efforts to decrease or prevent losses do not escape the condition of some factors, which can be categorized as techno-socio-economic factors. With this in mind, the purpose of this research to analyzed these factors. Which affect the shrinkage of capture fisheries, with a focus on three social, which are technical social and economic aspects.

2. Methods

The market for fish is assumed to be perfectly competitive where everyone can participate and the seller is a price taker. Consider a fish value chain player that seeks to maximize profit [6]. The methodological approach used in this research is case study using data taken with survey technique involving selected respondents through random sampling. The respondents represent the group of fishermen (10 respondents), collectors (10 respondents), intermediate traders (10 respondents), processors and consumers (10 respondents), which totals 40 respondents. The questions in the survey were adapted from the Exploratory Fish Loss Assessment Method (EFLAM) approach while data processing was conducted using multiple linear regression.

These methods have been developed and adopted in many countries especially in Africa [7, 8, 9, 10, 11]. (Based on rapid and/or participatory rural appraisal (RRA and/or PRA), EFLAM is planned and performed to generate qualitative and indicative quantitative data [12, 11].

$$TFL = VPL + LRQL$$

Total Financial loss = Value of physical loss (kg) + loss in revenue due to quality loss (Rp)

Before running the regression analysis, validity and reliability tests were done, resulting in the following. Validity criteria, all factors that were found valid (coefficient value > 0.300) [13]. Meanwhile, based on the Cronbach's reliability criteria ($\alpha > 0.6$) [14]. It was found that the instrument used in the analysis is reliable ($\alpha = 0.820$). It was found that the selected independent variables were statistically able to explain 72.3 % variation in the dependent variable (fish losses), the rest (27.7 %) belonging to the non-observed independent variables. These selected and thus observed independent variables were: type of fish, fish quality, price, size of fish, shelf life, season and sorting.

3. Results and Discussion

3.1. Descriptive account of factors potentially influencing handling practices and fish losses

The problem of fish losses in Indonesia, including in Palabuhan Ratu was related to the existing natural conditions in the area and its surrounding. Sukabumi District, where Palabuhan Ratu is located, is astronomically located at 6°57'–7°07' South Latitude and 106°22'–106°33' East Longitude and is of the influence of tropical climate, with an average temperature of 23.5 °C and precipitation of 2,806 mm. The district, with an area coverage of 412,799.54 ha has a coastline length of approximately 105 km so that fisheries become a essentials part of the livelihood of people in the region [1]. Nine of 45 sub-districts in Sukabumi District is a coastal sub-district directly facing the Indian Ocean. The sub-districts are Tegalbuleud, Cibitung, Surade, Ciracap, Ciemas, Simpenan, Palabuhan Ratu, Cikakak and Cisolok. Such a high temperature and humidity of tropical climate as well as the fisher's dependency on the fishery make proper fish handling so challenging.

Fishermen in the coastal sub-districts perform a relatively centralized fish landing, in Palabuhan Ratu Landing Place (Tempat Pendaratan Ikan, TPI). The local fishermen shared roles with those from other provinces in supplying the catch to this Palabuhan Ratu. Local fishermen used boats/fishing vessels, motorboats of 5–30 GT size; on the other hand, fishermen from other provinces engaged capture with regional coverage using relatively larger vessels of 15–60 GT. Fishing equipment that was used were varied, i.e., fishing line, gillnet, chart, payang, longline, purse seine and trammel net. Size-wise, the existing vessels that land fish in Palabuhan Ratu can be categorized into: motorboats sized < 5 GT (58 %), motorboats sized 6–10 GT (8 %), motorboats sized 11–20 GT (3 %), motor boats sized 21–30 GT (4 %), motorboats sized > 30 GT (1 %) and non-engine boats (16 %). Most of the fish landed in Palabuhan Ratu was auctioned in fresh condition. This condition was related to short fishing operation time, i.e., half-day fishing. Longer-trip, which spent more than two days at the sea, catch which were landed by only a few boats, i.e., purse seiners, squid netters and longliners.

From the social factor, the condition that potentially influences the fish loss cases were human resources. In general, educational achievement of market players in Palabuhan Ratu was low. Most fishers, traders, processors and consumers only finished primary school (35 %), followed by junior high school (25 %), high school (25 %) and university (15 %).

Poor technical and social circumstances caused terrible fish handling and processing condition in Palabuhan Ratu. Fish products from processors in Palabuhan Ratu and the surrounding areas were limited to traditional ones. The average processing methods comprised smoking, salting, sun-drying and fermentation. Fisher's wives and relatives, who predominantly did the processing, sold the processed fish on a weekly basis at a local market and gave the sales revenues to their husband. These included fresh fish (40 %) smoked fish (20 %), salted fish (25 %) and another fish processing (15 %). In the case of handling of fresh fish, only a few of the market players in Palabuhan Ratu applied a cold chain system while others apply inadequate method and equipment. Such a condition results in an insufficient distributional coverage; fish products from Palabuhan Ratu were marketed only to near locations such as Bandung, Bogor, Jakarta and Sukabumi.

Table 1. Summary of fish handling practices and circumstances in Palabuhan Ratu.

No.	Market player	Activities associated to handling		
		Technical	Economic	Social
1.	Fishers	• Fishing following seasonal pattern, taking whatever available within an average length of trip of respective type of vessel (and gear) • Sorting and grading based on species (or species group), size and quality	• For selling of the catch at the auction place, while women participate more in creating profit through processing and retailing. • Optimizing the capacity of the available fish hold and fishing trip to	• Networking and communication among fishers, particularly regarding where to catch and where to sell the fish • Extending the fishing trip whenever necessary (for particular boat

		• The use of cold storage by a small number of fishers	catch as much as possible, seeking a higher level of revenue	sizes, up to 20–30 days)
2.	Trader	• First traders collect auctioned fish and store it in less hygienic and proper equipment with inadequate cooling material • Transporting fish to next market nodes on foot, by bike, motorcycle or mini trucks, depending on the distance.	• Fresh and processed products are sold on local markets in the and nearby districts • Selling prices are based mainly on the type of product and the quality of raw material	• Similar to fishers, traders build effective networking and communication so that collective massive economic losses can be managed • Most processors link themselves to certain traders or fishers in order to secure supply of good fish with the best prices
3.	Processors	• Processors apply simple technologies to meet local markets • Fish of different grades are processed for varied types of product and segmented qualities	• Profit is pursued by target market segmentation • Most local consumers opt to consume more fish when supply is high and easily convert to substitutes when there is a shortgae	• Most processors link themselves to certain traders or fishers in order to secure supply of good fish with the best prices
4.	Consumers	• Consumers are likely to prefer fresh fish than processed ones	• Most local consumers opt to consume more fish when supply is high and easily convert to substitutes when there is a shortgae	• Most processors link themselves to certain traders or fishers in order to secure supply of good fish with the best prices

IOP Conf. Series: Earth and Environmental Science **139** (2018) 012034 doi:10.1088/1755-1315/139/1/012034

3.2. The resulting condition: rate of fish loss
With the condition of factors as described above, it was found that the rate of fish loss in Palabuhan Ratu is high. The high loss rate is contributed by eight species namely frigate tuna (*Euthynnus affinis*), yellowfin tuna (*Thunnus obesus*), skipjack (*Katsuwonus pelamis*), white sardinela (*Sardinella*), bigeye hairtail (*Trichiurus lepturus*), marlin, short mackerel (*Rastrelliger faughni*) and squid (*Loligo* sp.).

Table 2. Fish loss at fish landing and the next chains in Palabuhan Ratu Sukabumi.

No	Fish species		Fish Loss Average	
	Species	Scientific name	Landing (%)	Next chains (%)
1.	Frigate tuna (*tongkol*)	*Euthynnus affinis*	4	5
2.	Yellowfin tuna (*tuna*)	*Thunnus obesus*	3	3
3.	Skipjack	*Katsuwonus pelamis*	4	5
4.	White sardinela (*tembang*)	*Sardinella*	4	6
5.	Bigeye hairtail (*layur*)	*Trichiurus lepturus*	4	5
6.	Makaira (*marlin*)	*Istiohoridae*	5	5
7.	Short mackerel (*kembung*)	*Rastrelliger faughni*	5	6
8.	Squid	*Loligo* sp.	5	6
	Fish loss average (%)		4.25	5.12

$$Total\ financial\ loss\ as\ a\ \% = VPL + LRQ$$

Table 2 showed that the fish catch landed in Palabuhan Ratu lost its amount since the first node of the chain approximately 4.25 % while in the next chain it amounts to 5.12 %. It was comprised physical and value losses. By the time the catch was landed at the port and went to the auction place, a number types of damage can be spotted which ranged from few scratches, fragmentation to total impairment. Some discarded directly as wastes while others were sorted based on (group of) species, size and quality and then priced accordingly.

In the next nodes of the chain, losses occurrence attributed to several types of mishandling including processing without proper chilling and or inefficiency in handling and processing steps. During processing, traditional techniques such as meatball and fish floss making, the fish loss were accounted for discarding non-process portion. It is worth noting here. However, those processing activities using such traditional technologies can generate value-added up to 1.25 %.

3.3. Statistical analysis
Simultaneously, these variables selected for the regression were found influencing the level of fish losses with a confidence level of 95 % (Fcalc 4.25). Table 3 shows the results of partial analysis of the variables.

Table 3. Result of regression analysis.

Variable	Parameter estimate	Significance
Species of fish	2.11	S
Quality of fish	2.31	S
Price of fish	1.75	S
Size of fish	2.14	S
Shelf life of fish	2.10	S
Season	2.01	S
Sortation	1.73	NS

Referring to this statistical result and considering the observed chronological sequence of this variables (factors), the relationships among these factors and their effect on losses can be depicted as in figure 1. In general, this figure suggests that there are some factors like season, where interventions are impossible and some others which partially or simultaneously can be intervened for the best expected output, which in this context is the level of losses.

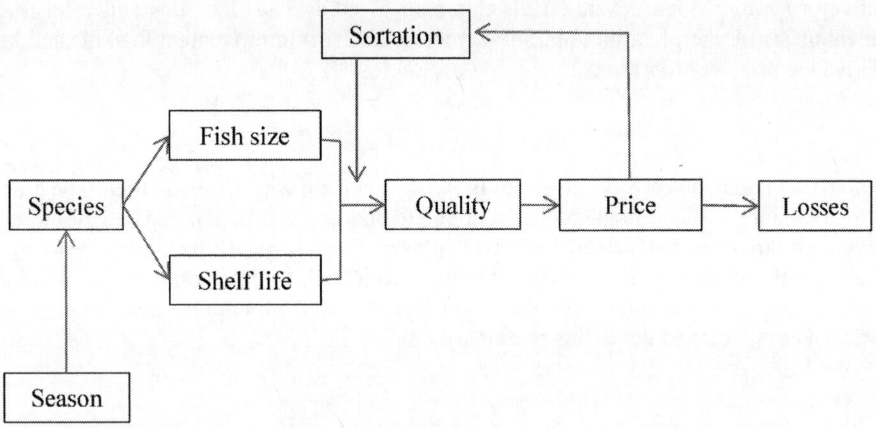

Figure 1. Level of fish losses.

3.4. Synthetis of the result

To get the best quality and to minimize losses, actually there is a standard called HACCP (Hazard Analysis Critical Control Point), that should have been applied in all stages in the handling and processing in Palabuhan Ratu. This standard equips managers with scientific, rational and systematic reference to identify, control and monitor fish processing activities such that fish quality can be maintained. However, field observation which has been presented showed that technical, economic and social environment did not allow such a standard to perfom well. As described, fish landed in the auction place of Palabuhan Ratu are caught using various types of fishing vessels that operate various types of gear and exercise different lengths of the trip and this brings consequences that implicate technical difficulties in fish handling. As it was noted during the research, fish quality has a broad range of quality disparity. Moreover, these fish belong to many owners (fishers) such that immediate collective handling is very difficult. The difficulties are capitalized as these fishers have a somewhat limited awareness of the need for proper handling of their catch. Low educational achievement undoubtedly contributes to an unfavorable social condition. As shown in table 1, only a few fishers apply cold chain system. While cold storage facility was also limited. This social factor simply complicates efforts to maintain the good quality of fish and to minimize losses. The social and technical problems could be solved if significant economic incentives were accruing from practicing proper handling and processing. However, the existing economic circumstances also could not support the necessary incentives. Processors, for example, considered the revenue margin which implements better handling and or processing would not be significant.

All of these problems represents cyclical chains, where each factor had interactions to influence one another. Given this, interventions become relevant. The options of interventions may be based on the result of the following statistical analysis. Factors listed in the results of the statistical analysis (table 1) can be grouped into external ones, where control is not possible and internal ones, to which interventions can be considered. The statistical regression (table 1) shows that among the technical factors, the most influential ones were (i) handling of landed fish, (ii) fish sortation, (iii) fish size, (iv) fish shelf life and (v) season; among economic aspect, factors with the most significant influence were (i) fish price fluctuation and (ii) price level; meanwhile, among the social factors, those that had the most significant influence was the revenue distribution system.

In line with the results of field observation presented earlier, it was identified that interventions to internal factors in brief can be described as facilities that will enable market players to exercise efforts aiming at saving the value of fish. Based on this, a relevant policy implication of this research is needed for effective programs which cover: (i) development of cold chain facilities and infrastructure, (ii) development of distribution facilities and infrastructure and (iii) improvement in skills and knowledge of fish derivative product processors.

4. Conclusion

This research found that the case of fish losses in Palabuhan Ratu was relatively high. The case involves some factors/variables which are interdependent one to another, a situation that justifies interventions. The interventions are those that can enable market players to exercise efforts aiming at saving the value of fish. In general, the interventions, which comprise technical, social and economic types should be targeted directly to most influential variables as can be deduced from both the detailed presentation and the statistical analysis carried out in this research.

References

[1] District Marine and Fisheries Service in Sukabumi 2016 *Fisheries and Maritime Data of Sukabumi District*

[2] Wibowo S, Utomo B S B, Syamdidi and K Rinta 2014 *Proc. of The 3rd International Seminar of Fisheries and Marine Science* (Pekanbaru: Universitas Riau) 59–66

[3] New Partnership for Africa's Development (NEPAD) 2013 *NEPAD Annual Review on International Partnership for African Fisheries Governance and Trade (PAF)* (Africa: New Partnership for Africa's Development) p 30

[4] Wibowo S 2013 *Pemahaman Susut Hasil Pascapanen dalam Industri Pengolahan Hasil Perikanan: Pelatihan Peningkatan Pemahaman tentang Evaluasi Susut Hasil pada Industri Pengolahan Hasil Perikanan* (Jakarta: Research Center for Marine and Fish Product Processing and Biotechnology)

[5] Akande G R 2001 *Application of Load Tracking in The Distribution Chain of Smoked West African Sardines Sardinella maderensis: A Case Study of Magbon-Alade Fishing Community, Lagos, Nigeria* (Rome: FAO Fisheries Report No. 712) 87–89

[6] Hussain W 2012 *Economics and Philosophy* **28** 311–331

[7] Akande G R and Odogbo D 2001 *Field Testing of Methodologies for Evaluating Post-Harvest Losses in The Artisanal Fisheries of Dorobaga Fishing Community and Baga Fish Market in Maiduguri* (Rome: FAO Fisheries Report No. 712) 81–86

[8] Akande G R, Jeffries D and Ward A R 2001 *Loss Assessment Using Intervention Load Tracking*

(Rome: FAO Fisheries Report No. 712) 91–96
[9] Akande G R and Diei O Y 2010 *Post-Harvest Losses in Small-Scale Fisheries: Case studies in Five Sub-Saharan African Countries* (Rome: FAO Fisheries and Aquaculture Technical Paper No. 550) pp 72
[10] Eyo A A and Mdaihli M 2001 *Assessment of Post-Harvest Losses in Nigeria Fishery: The Kainji Lake Model* (Rome: FAO Fisheries Report No. 712) 65–67
[11] Ward A R and D J Jeffries 2000 *A Manual for Assessing Post-Harvest Fisheries Losses* (United Kingdom: Natural Resources Institute)
[12] Conroy M J, Miller M W and Hines J E 2002 *Wildlife Monog.* 150 1–64
[13] Kaplan, Robert M and D P Saccuzzo 1993 *Phsycological Testing Principles, Application and Issues* (California: Cole Publishing Company) pp 752
[14] Malhotra and Naresh K 2005 *Marketing Research: Applied Approach* Translation Ir. Soleh Rusyadi Maryam, MM, 4th Edition, Vol 1 and 2 (Jakarta: PT. INDEX)

IOP Conf. Series: Earth and Environmental Science **139** (2018) 012035 doi:10.1088/1755-1315/139/1/012035

Economic valuation of mangrove ecosystem: empirical studies in Timbulsloko Village, Sayung, Demak, Indonesia

T A Perdana[1, 3*], J Suprijanto[2], R Pribadi[2], C R Collet[3] and D Bailly[4]

[1] Master of Economics and Development Studies, Diponegoro University Jl. Prof. H. Soedarto, S. H., Tembalang, Semarang, Indonesia
[2] Department of Marine Science, Diponegoro University Jl. Prof. H. Soedarto, S. H., Tembalang, Semarang, Indonesia
[3] Master E2AME Agrocampus Ouest, Rennes, University of Western Brittany 3 Rue des Archives, 29238, Brest, France
[4] Department of Economics, Management, Society (EGS), Agrocampus Ouest, 65 Rue de Saint-Brieuc, 35000, Rennes, France
[5] UMR AMURE Center for Law and Economics of the Sea European Institute for Marine Studies (*IUEM*) University of Western Brittany 3 Rue des Archives, 29238, Brest, France
*Email: titoadityap92@gmail.com

Abstract. Ecosystem resilience is the capacity of ecosystems to tolerate disorders without collapsing into different circumstances qualitatively controlled by a different set of processes. A robust ecosystem is one that can withstand shocks and rebuild itself when necessary. This study aims to identify the value of use-based economy and non-use value of current economy; calculating the total economic value of mangrove resources; and provide suggestions and recommendations based on observations in Timbulsloko, Sayung, Demak. The method used is economic valuation with total economic value technique. The sampling technique used non-probability and purposive sampling method. The results showed that the direct use value of mangroves was utilized by fisherman, fish pond farmers, *branjang* catchers, oystercatchers, trap makers, shop owner, grilled fish makers and shrimp chip makers. Indirect use value was derived from function as the breakwater, beach belt and hybrid engineering. Existing value was not less than 10 % of the direct use value. The total economic value was Rp. 6,361,430,639/year or about Rp. 202,335,580.1/ha/year. It is need to improve the community awareness to mangrove ecosystem and to the role of breakwater in order to reduce risk disaster and to develop an ecotourism in the area.

1. Introduction

Indonesia is the country with the largest mangrove level in Asia (and in the world), contributing about half the regional level of the mangrove area. Other Asian countries with significant levels of mangroves are (in the order of mangroves) of Malaysia, Myanmar, Bangladesh and India, which, together with Indonesia, account for more than 80 % of the total area of mangrove forests in Asia [1].

The mangrove ecosystem is one of the most threatened on the planet. Mangroves are being destroyed at levels 3–5 times greater than the average rate of forest loss and more than a quarter of the original mangrove cover has disappeared; driven by land conversion for cultivation and agriculture, coastal development, pollution and over-exploitation of mangrove resources. Mangroves are becoming smaller and more fragmented, important ecosystems for goods and services will be reduced or lost.

The consequences of mangrove degradation will then be especially severe for the welfare of coastal communities in developing countries, especially where people rely heavily on mangrove goods and services for daily needs and livelihoods [2].

Approximately 20 % or 3.6 million ha of mangrove areas have been lost since 1980. Approximately 185,000 ha disappeared annually in the 1980s; this number dropped to some 118,500 ha/year in the 1990s and to 102,000 ha/year in the period 2000–2005, reflecting the increased awareness of the value of mangrove ecosystems [1].

Area of mangrove ecosystem in Central Java covering ± 15,184.15 ha (not including Karimunjawa Islands) [3]. The destruction of coastal ecosystems coupled with the existence of buildings that jutted into the sealeads to abrasion. The total area affected by abrasion in Central Java is 6,566.97 ha where the districts of Brebes and Demak become areas affected by abrasion is quite severe. However, arising land or accretion also occurs on the coast of Central Java, which is 12,585.19 ha. Area of the acres land that is the potential in the rehabilitation of ecosystems especially mangrove ecosystem.

At this time, the level of environmental damage that threatens coastal areas is very high. According to a study conducted by [4], mangrove area in Central Java Coastal area of 10,786 ha in which 96.9 % of them are in damaged condition. The mangrove area continues to degrade, so that the mangrove study results in 2014 only get the area of vegetation cover 5,381.15 ha [3].

In the past, Timbulsloko Village; which located on the coast of Sayung District Demak; was an agricultural land and ponds and mangroves along the coast. However, in the 1980s there was a considerable conversion of agricultural land and mangroves due to the opening of ponds. In the 2000s, erosion began to occur on the coast of Timbulsloko Village and by 2013 the village has lost about 400–1300 m of coastal areas. Temporary estimates of the causes of high erosion rates are due to increased tidal intensity, use of wave retaining embankments and loss of mangroves [5].

Timbulsloko is a dynamic area and vulnerable to the threat of environmental degradation. The high rate of mangrove forest decline, the conversion of mangrove ecosystem to farm land and the problem of sedimentation become the main factor changing the ecological and economic function of mangrove ecosystem in Timbulsloko.

This is interesting and requires a research on what makes the reasons people do not want to move and whether there is still an economic side to take from the region. The concept of resilience may be an example of a precise concept that disaster-prone areas will not fall or collapse as they can improve on their own and be helped with community awareness.

The aim of this research are to identify the value of use-based economy and non-use value of existing economy, to calculate the total economic value of mangrove, to provide suggestions and recommendations to the community and local government.

2. Research Methods

The method used in this research is economic valuation with Total Economic Value technique, this method is used to calculate the value of direct and indirect benefit from mangrove ecosystem at research location. Sampling technique using non-probability sampling with sampling technique using purposive sampling method.

The economic valuation method concerns the monetary measurement of a change in one's well-being caused by changes in the quality of the environment [6]. This measurement value is known as Total Economic Value (TEV). Moreover, economic value as measuring the maximum amount a person wants to sacrifice goods and services to acquire other goods and services [7]. The economic value of goods and services is measured by the sum of the will to pay (Willingness to Pay-WTP) of many individuals to the goods or services in question. Economic valuation is about measuring the preferences of the community. The result of the valuation is expressed in the value of money as a way of looking for confidential preferences.

TEV is the sum of the value of use-based economy and non-value-based economic value [8]. UV consists of Direct Use Value, Indirect Use Value and Option Value. Meanwhile, NUV consists of two components of value that is the bequest value (Bequest Value) and the value of existence (Existence Value).

Several methodological approaches for valuing ecosystems/resources based on the above typology presented by [9]. Most of the approaches are based on a cost-approach on the grounds that the benefits approach is relatively difficult to predict. Some of the best-known methodologies are Effect on Production (EOP) or production analysis and benefit-cost analysis or benefit and cost analysis, where benefits are described by revenue from production. The definition of total economic value can be seen in table 1 [9].

Table 1. Definition total economic value.

No	Type of Value	Definition
1	Direct Use Value	The economic value derived from the direct utilization of a resource/ecosystem.
2	Indirect Use Value	The economic value derived from the indirect use of a resource/ecosystem.
3	Option Value	The economic value derived from the potential of direct or indirect utilization of a resource/ecosystem in the future.
4	Bequest Value	Economic value derived from the benefits of conserving resources/ecosystems for the benefit of future generations.
5	Existence Value	The economic value derived from a perception that the existence of an ecosystem/resource exists, regardless of whether the ecosystem/resource is utilized or not.

3. Results and Discussion

Demak as one of the districts in Central Java lies in the coordinates of 6°43’26”–7°09’43” South Latitude and 110°27’58”–110°48’47” East Longitude. Demak is adjacent to Jepara and the Java Sea in the North, Kudus and Grobogan District in the East, Grobogan and Semarang District in the South, Semarang City in the West. The farthest distance from West to East is 49 km long and from North to South along 41 km. Administratively the area of Demak District is 89,743 ha, consisting of 14 sub-districts, 243 villages and six urban villages.

In this research, Timbulsloko village was chosen as research location with Dusun Wonorejo and Dusun Bogorame as a specific research area. Timbulsloko Village is a village located in Sayung Sub-district, Demak District. Located in the Western part of Demak, adjacent to Semarang City and the Java Sea. The population in Timbulsloko village is 3,469 people, has an area of 4.61 km^2, consists of four sub-village.

Coastal areas in Timbulsloko were formerly agricultural land, ponds and mangroves [5]. However, in 1980, there was a large-scale conversion of agricultural land and mangroves in the pond. This triggered abrasion along the coast of Timbulsloko which began in 2000. In addition to the loss of mangroves, abrasion is caused by the increased intensity of the tidal wave. All sub-villages in Timbulsloko are affected by abrasion. However, Bogorame suffered the most from other areas. Many locals work as tambak farmers, although abrasion has damaged much of the aquaculture in the village. Most are in Karanggeneng and Wonorejo. Fishermen are found to be in areas close to the sea, especially in Wonorejo

The result of the research shows that there are some activities or work that utilize or intersect with mangrove ecosystem that is the fisherman, farmer, *branjang* catcher, shrimp catcher, processed fish maker, trap maker and also warung. Most of them have main income as fishermen and fish farmers. Work as fishermen and farmers, of course, a bit much tangent to the mangrove ecosystem. For fishermen, mangroves are where small fish take shelter before reaching adult size, while for farmers is a protector of the waves.

Much attention was devoted to four villages in Sayung Sub-district, Demak District, Morodemak, Purworejo, Surodadi and Timbulsloko village. Coastal rehabilitation programs such as mangrove planting, making wave breakers, hybrid engineering or sedimentation tools installed along the coast and making beach belts. Mangrove planting program is mostly done in Timbulsloko village, Sayung Sub-district, Demak District, for example from Marine and Fisheries Department.

Mangrove is destined to be a natural fortress for coastal areas. With expanding mangrove forests, it is expected to strengthen the area of Timbulsloko Village, Sayung Subdistrict, Demak District. Public awareness should also be improved on the importance of mangrove for coastal areas. Not only useful as a natural fortress, but mangroves also have benefits that can be felt directly by the community.

Table 2. Direct use value.

No	Description	People	Income (rupiah)		
			Daily	Month	Year
1	**Direct Use Value**				
	Fisherman	20	150,000	90,000,000	1,080,000,000
	Fish pond farmer	24		8,400,000	403,200,000
	Branjang catchers	10	75,000	22,500,000	270,000,000
	Oyster catchers	10	100,000	30,000,000	360,000,000
	Trap makers	1	75,000	1,500,000	18,000,000
	Shop owner	1	100,000	3,000,000	36,000,000
	Grilled fish makers	1	50,000	1.500,000	18,000,000
	Shrimp chips maker	1	75,000	2,250,000	27,000,000
		Total			2,212,200,000

The direct use value of the mangrove ecosystem in Timbulsloko Village, Sayung District, Demak District with an area of 31.44 ha is Rp. 2,212,200,000/year or Rp. 70,362,595.42/ha/year (table 2). Direct benefits calculated from the fisheries sector, namely fisherman, fish pond farmers, *branjang* catchers, oyster catchers, trap makers, shop owner, grilled fish makers and shrimp chip makers.

The indirect use value of the mangrove ecosystem in Timbulsloko Village, Sayung Sub-district, Demak District was obtained from the cost of making breakwater, seawall and hybrid engineering (table 3). Breakwater located in Timbulsloko Village is help from the Department of Marine and Fisheries of Central Java Province. The seawall is a relief from the Environment Agency of Central Java Province. The hybrid engineering tool is the result of cooperation between the government of Indonesia and the Netherlands.

Table 3. Indirect use value and option value.

No	Description	Area	Value (rupiah)
1	**Indirect Use Value**		
	Breakwater	180 m	900,000,000
	Sea Wall	980.8 m	1,286,276,000
	Hybrid Engineering	1.2 Km	1,200,000,000
2	**Option Value**		
	Mangrove forest	31.44 ha	6,263,038.94

To get the benefits of biodiversity than the value of biological diversity mangrove in Timbulsloko Village, Sayung Sub-district, Demak District multiplied by the average value of US $ 1,500/km^2/year or the US $ 15/ha/year [10]. The total area of mangrove forest located in Timbulsloko Village, Sayung Sub-district, Demak District reached 31.44 ha. If multiplied by the biodiversity value of mangrove forest in Indonesia US $ 15/ha/year, then the value of ecosystem choice of mangrove forest reached the US $ 471 or Rp. 6,263,038.94/year.

The value of the existence of mangrove ecosystem in Timbulsloko Village, Sayung Sub-district, Demak District was estimated using contingent valuation method technique. This method is used to ask the community how much value or price given for mangrove ecosystem remains maintained. From the calculation results can be seen that the value of existence for people who are around Timbulsloko, has an existing value of Rp. 535,471,600/year with an average value of Willingness to Pay (WTP) of Rp. 471,000/family/year (table 4).

Table 4. Non-use value.

No	Description	Value(rupiah)
1	**Bequest Value**	**535,471,600**
	average value of WTP	471,000
2	**Existence Value**	
	10 % direct use value	221,220,000

Mangrove ecosystem as an existence has a very high value. Accordingly, it is estimated that the existence value is not less than 10 % of the direct use value of ecosystem mangroves [10]. Therefore it is estimated that the existence value of mangrove ecosystem in Timbulsloko Village, Sayung Sub-district, Demak District is 10 % × Rp. 2,212,200,000 = Rp. 221,220,000/year. Total economic value (Total Economic Value) is based on the results of the identification of all types of benefits from the mangrove ecosystem in Timbulsloko. The calculation result of Total Economic Value of mangrove ecosystem is Rp. 6,361,430,639/year or about Rp. 202,335,580.1/ha / year.

Suggestions and recommendations that can be given are addressed to the community are utilize the mangroves as the ingredient for edible product, as coastal protection and supporting the recreation sector. Based on observations in the field, the use of mangroves that can be directly used cannot be done continually. Some woman which mostly fisherman's wives said that the mangrove could be processed to be mangroves crackers. Seeing the building of breakwater, sea wall and also *Hybrid Engineering* structure, it can be seen that the effort done by the government to protect Timbulsloko Village area, Sayung Sub-district, Demak District is very good. It would be better if accompanied by a wider public awareness of the importance of these programs so that in the future people in Timbulsloko are more concerned about the environment and also understand about the usefulness of various programs. Community assistance is needed regularly and also learning for children about the importance of loving the coastal areas and also care for mangrove trees. Another thing that might be taken into consideration for the future is recreation. Seeing the intensity of coastal rehabilitation programs as well as research conducted on the impact of the program, it is undeniable that the region is heeded. Coastal areas and mangrove have been like can not be separated in the village area Timbulsloko, District Sayung, Demak District. So, look at it, the need for the utilization of Timbulsloko into Ecotourism area that can be adopted from the Mangrove Tourism Area Mangunharjo Village, Tugu Sub-district, Semarang City.

It can be started first and reinforced by village regulations that are integrated with the local government so that people can better care for the environment and open themselves to the concept of ecotourism. Citizens who want to enter the seaside area and enjoy the beauty of the afternoon above the structure of the breakwater can provide extra money. Then the fisherman who owns the vessel can rent out to just around the mangrove area. It can be started first and reinforced by Village Regulations that are integrated with the local government so that people can better care for the environment and open themselves to the concept of ecotourism. Citizens who want to enter the seaside area and enjoy the beauty of the afternoon above the structure of the breakwater can provide extra money. Then the fisherman who owns the vessel can rent out to just around the mangrove area.

4. Conclusion

The mangroves ecosystem provides direct use value which was utilized by the fisherman, fish pond farmers, *branjang* catchers, oyster catchers, trap makers, shop owner, grilled fish makers and shrimp chip makers. Indirect use value was the function of mangroves as the breakwater, beach belt and hybrid engineering. The option value is obtained from mangrove forest area multiplied by the US $ 15/ha. The Bequest value is obtained by using the contingent valuation method technique. Existence value is not less than 10 % of the direct use value. Total Economic Value that exist in the mangrove ecosystem in Timbulsloko Village, Sayung Sub-district, Demak District. Rp. 6,361,430,639/year or about Rp. 202,335,580.1/ha/year. Therefore, the suggestions and recommendations for the community in Timbulsloko Village are utilize the mangroves as the ingredient for edible product, as coastal protection and supporting the recreation sector.

References

[1] FAO 2007 *The World's Mangroves 1980–2005* (FAO Forestry) pp 153

[2] UNEP 2014 *The Importance of Mangroves to People: A Call to Action* ed van Bochove J et al. (Cambridge: United Nations Environment Programme World Conservation Monitoring Centre) p 128

[3] Dinas Kelautan dan Perikanan Jawa Tengah 2016 *Kerangka Acuan Kerja (KAK) Rehabilitasi dan Konservasi Sumberdaya Kelautan dan Perikanan Tahun 2016* (Jawa Tengah: DKP)

[4] Badan Perencanaan Pembangunan Daerah Kabupaten Demak 2009 *Laporan Akhir Revisi Rencana Tata Ruang Wilayah Kabupaten Demak 2010-2030* (Demak: BPPD)

[5] Astra A S, Etwin K S, Arief M H, and Maulana M B 2014 *Keterlibatan Masyarakat Dalam Pengelolaan Pesisir dan Laut. Studi Kasus: Kawasan Perlindungan Pesisir Desa Timbulsloko, Kecamatan Sayung, Kabupaten Demak* (Bogor: Wetlands International Indonesia)

[6] Sathirathai S 2003 *Economic Valuation of Mangroves and the Roles of Local Communities in the Conservation of Natural Resources: Case Study of Surat Thani, South of Thailand EEPSEA Research Report* (Singapore: IDRC.ARCHIV 108378)

[7] Fauzi A 2006 *Ekonomi Sumber Daya Alam dan Lingkungan* (Jakarta: Gramedia Pustaka)

[8] Freeman III A M 2003 *The Measurement of Environment and Resource Values Resource for the Future* (Washington DC)

[9] Barton D N 1994 *Economic Factors and Valuation of Tropical Coastal Resources* (Norway: University of Bergen) p 128

[10] Ruitenbeek H J 1991 *Mangrove Management: An Economic Analysis of Management Options with A Focus on Bintuni Bay, Irian Jaya* (Canada: Dalhousie University Printing Center)

Analysis of eco-innovation with triple helix approach: case-study of biofloc catfish farming in Yogyakarta

D Purwadi[1]* and I Nurlaily[2]

[1] Departement of Agro-Industrial Technology, Faculty of Agricultural Technology, Universitas Gadjah Mada, Kampus UGM Bulaksumur, Indonesia
[2] Faculty of Animals Science, Universitas Gadjah Mada, Kampus UGM Bulaksumur, Indonesia
*Email: didik@ugm.ac.id

Abstract. Concerning environmental into focus of innovation process will expand the number of actor involved. Eco-innovation and triple helix are often frameworks applied to analyse how environmental concern are integrated in innovation process and how different stakeholder groups are having inter relation. Case study from biofloc catfish farming in Yogyakarta is presented to demonstrate a possible approach for researching the success of triple helix frameworks. This case is considered on basic of the result of a survey among farmers, academician and government. The paper concludes the creating of full triple helix encounters problem in practice. It also includes suggestion for further research on fisheries development.

1. Introduction

Sustainable development (eco-innovations) as innovation processes have received increasing attention during the past years [1]. Eco-innovation is increasingly connected to green growth in terms of global climate change as it is to be multi-win approach [2]. In aquaculture industry using bioflocs technology as eco-innovation aquaculture is a promising alternative to overcome the problems. Moreover, application of this technology is more profitable by decreasing inorganic nitrogen waste as well as providing additional feed for cultured fish, thus increasing fish growth and feed efficiency. Biofloc technology can be applied using addition of organic carbon into the culture media to increase C/N ratio and to stimulate the growth of heterotrophic bacteria [3]. Moreover, [4] concluded that by using bioflocs technology will had high sludge accumulation after harvest, because the deposition of dead organic matter and algae in the absence of water exchange. So, it is important to implement management to control ammonia and sludge formation. Bioflocs technology could certainly be an ideal method for farmers. The biofloc-dominated no water exchange conditions need more resources to maintain adequate water quality, growth, yield and profitability [5].

The triple helix metaphor generalized innovation dynamics to non-linear interactive process among multi-stakeholders. It provides a framework for disclosing how innovations are generated by analyzing university-industry-government linkages [6]. Integration and differentiation of triple helix are important for proper institutions and infrastructure, knowledge-intensive services tend not to contribute to local synergies in the economy because of their flexibility of moving across the administrative borders of regions [7]. It is important how Region Government should active to manage this synergy to maximize regional economic development. It is in line to [8] who concluded that market–led intermediary organizations should be used as a policy tool to promote triple helix networks. Government-funded intermediaries may help in creating triple helix networks.

The problems are how eco-innovation can be developed and giving impact to economic development. Innovation needs changes in many things. It is important to synergize all of the actors or stakeholders such as firms, politicians, unions, associations and private household. In general, they can be categorized into three actors: academician, business and government or ABG.

Eco-innovations dynamics is the process that innovative firms with clear intentions toward economic and environmental benefit developed by utilizing stakeholder interactions. Eco-innovation can be categorized as: product innovation, production (process) innovation, organizational innovation, market innovation and business model innovation. The problem is how such kind of innovation can be introduced and developed, then how to apply in business and finally how can give positive impact to the business and regional economy. This paper focuses on triple helix frameworks to disclose eco-innovation activities are developed by stakeholder interactions.

The purposes of this research are: (1) to describes the eco-innovation of biofloc technology for catfish farming and (2) to analysis of the triple helix framework in order to assure that eco-innovation can gives impact to the business and regional economic development.

2. Materials and Methods

2.1. Materials

Primary and Secondary data was used for identification of the catfish farming system. Data was gathered through observation and in–depth interview to the object of this research. The objects of this research are (1) Tunas Muda Catfish Farming Yogyakarta, (2) Faculty of Agricultural Technology Universitas Gadjah Mada (UGM) and (3) Yogyakarta Special Region Government.

The need data of this research are: (1) the role of stakeholders in triple helix frame, (2) the catfish farming system before and after introducing bioflocs technology such as the catfish total production, cost production, number of manpower, kind of technology, feed utilization, density, survival rate and production management. The secondary data is research reports, policy reports and annual reports of The Ministry of Marine and Fisheries as well as Yogyakarta special region government.

2.2. Methods

This research was done through (1) stakeholders identification of biofloc catfish farming, (2) description of biofloc catfish farming and (3) triple helix frame of biofloc catfish farming. Finding and discussion was done through descriptive-qualitative analysis.

3. Results and Discussions

3.1. Eco-innovation dynamics at Tunas Muda bioflocs farming

Tunas muda is a fishery group that has a catfish culture business with 25 members. *Tunas muda* sees environmental problem in relation to water limitations under climate change and the high cost of commercial fish feed as an opportunity. Intensive innovation by using bioflock technology have proved successful, as its business catfish aquaculture has extended another product such as catfish fingerlink and tilapia.

Based on a document of fisheries minister report, we outline the changing intensions at *tunas muda* catfish farming integrating environmental problem into innovation strategy and business development. Table 1 below presents eco-innovation dynamics at *tunas muda*. Table 1 shows that at *tunas muda*, eco-innovation relates not only to their product development, but also production process and organization.

Tunas muda is as catfish farmer group success to apply the bioflocs technology. Table 1 above describes that the introduction of bioflocs technology results product innovation, process innovation and organizational innovation. Total production and revenue are increase. Feed utilization changes from extensive production with commercial fish feed to become active nitrogen bacteria and bubble blower. It is also increase in stocked density and survival rate. *Tunas muda* business activities become more efficient due to the establishment of organization farming, grading procedure check point, water quality periodic procedure analysis.

Table 1. Eco-innovation dynamics at *tunas muda*.

Eco-innovation dynamics	Before	After
Product Innovation	Catfish cosumtion size	fingerlink catfish, tilapia
Total Production	2.5 ton	Round ponds fiber 5 ton
Cost production	Rp. 32,500,000.-	Rp 60,000,000.-
Manpower	2	1
Total Revenue	Rp. 37,500,000.-	Rp. 80,000,000.-
Production Process		
Innovation	Traditional	Eco-innovation;
Technology	Extensive production	Active nitrogen bacteria, bubble
Feed utilization	Commercial fish feed	blower 7 : 3
	Practically 100 tail.m^{-3}	Increase to 200 tail.m^{-3}
Stocked density	80–90 %	90–95 %
Survival Rate	3 × 5 × 70 m^3	250 m^2
Ponds	Private household	Organization farming, grading
Production management		procedure check point, water quality periodic procedure analysis
Organization Innovation	three persons	10 persons
	Catfish aquaculture	Breeding catfish division established

Business development at *tunas muda* is the result of the interaction of several stakeholders. Figure 1 present the stakeholder's relations and linkages. As the res ult of increasing in interaction between *tunas muda* and the government, then it can improve catfish production system. The government provides support with the strengthening of funds and quality assurance. Increased fish production provides indirect income for local government in the form of taxes. Universities in this case UGM has an interest to improve transfers benefits of research that has been done. Relationship between *tunas muda* with other fishery industry such as finger link supplier, fish feed factory and another community group is mutual support to improve prosperity together.

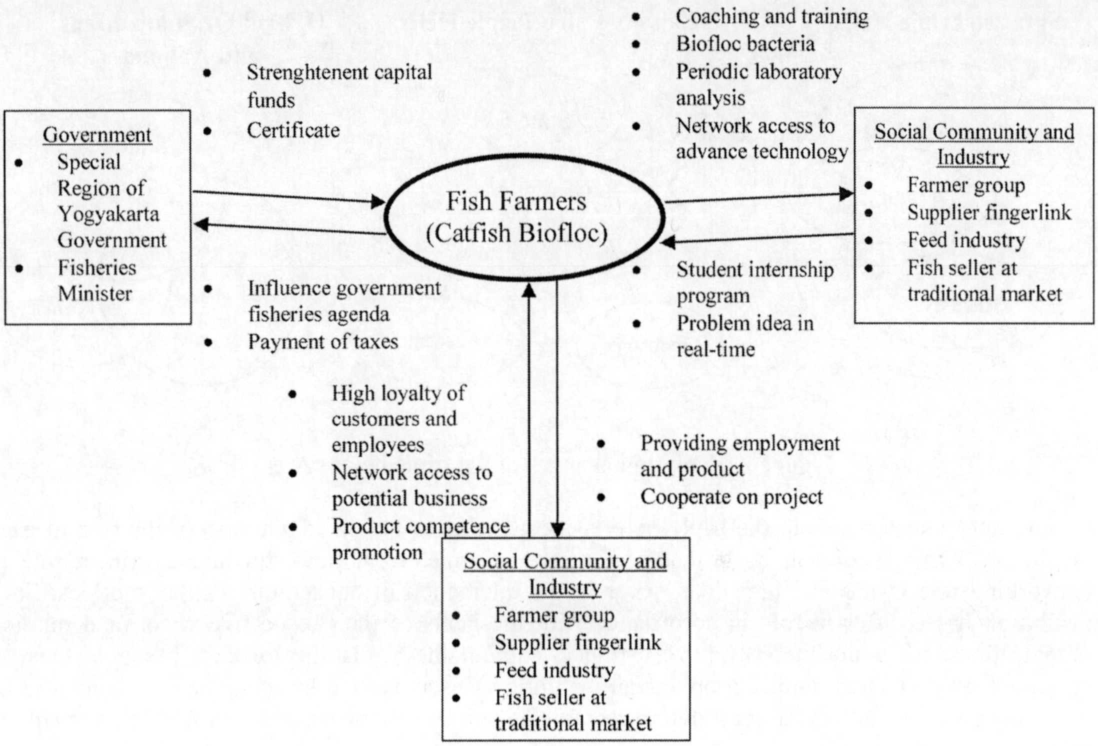

Figure 1. Triple Helix collaboration in eco-innovation dynamics at *tunas muda*.

3.2. Triple helix framework

According to agro industry system, there are three actors or stakeholders that involved in this framework system. They are academician or university entity, business entity and government entity. UGM is as academician entity, Biofloc catfish farmers is as business entity and Ministry of Fisheries and Yogyakarta Special Region Government are government entity. They are having relationship through triple helix concept or ABG (Academician, Business and Government) model.

Bioflocs technology was developed by UGM (aquaculture research) and government (Ministry of Fisheries) representative. Each stakeholder has role in this framework. UGM itself as a university organization, there are assets such as researchers or lecturers, laboratories and networks. Their assets can make the university's role as incremental cost reduction in aquaculture system, creative invention and technology transfer. The role of bioflocs farmers as the owner of catfish cultivation business is to produce catfish fingerlink and catfish consumption size. The Yogyakarta Special Region Government through Fisheries and Marine Office has a role to support by applying government regulations to implement regional economic development.

The development of aquaculture business can be improved by maximizing the role of each actor. Changes or evolutionary in the types of interactions on triple helix models occurred in recent years. The interaction of actors in the triple helix of static in which government control university and industry to laissez-faire stage in which has a balanced relationship between the three institutions; And last is the hybrid stage in which each institution has its own different roles and cuts in the same time [5]. The changes process underlying the Triple Helix system is graphically described in figure 2 below.

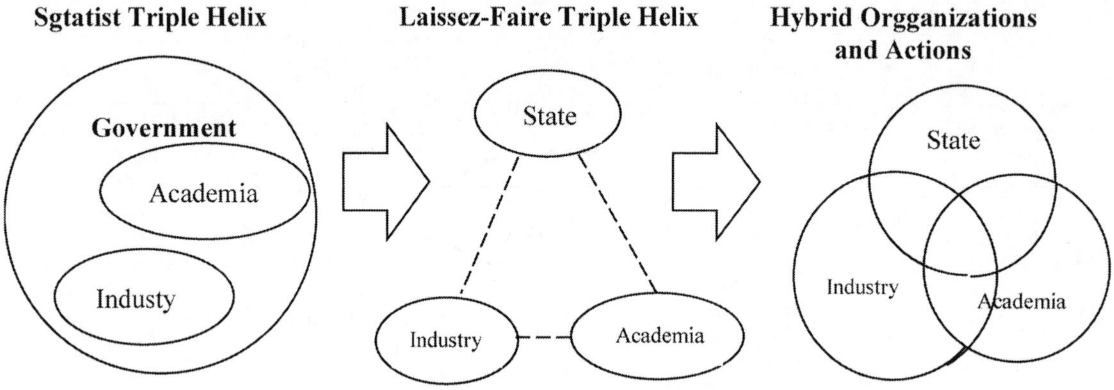

Figure 2. Evolution changes of the triple helix system.

The interaction change in the triple helix system is due to the paradigm shift in the role of each institution. The government itself tends to release its role, from powerful and dominant role to networking role Dynamic interactions occur in the interaction of aquaculture catfish bioflocs. Each institution has a different role in accordance with the goals of their respective vision and mission organizations, but simultaneously have a role to support the availability of catfish supply to meet consumer needs. Catfish farmer group as an institution was supported by government, changes to be more independent. Catfish farmer group, Academician and Government build a strategic partnership in term of new triple helix system.

Evolutionary university from university of teaching and research to entrepreneurial university, commercializing their research as start-up is the role expected by UGM. The development of eco-innovation in the field of fisheries becomes a challenge for sustainable production. The development of bioflocs as fisheries research by UGM is a synergistic role for catfish farmers to get more profit.

4. Conclusion

University, government and fish farmers have made a significant effort to promote eco-innovation as Triple Helix system in order to improve catfish production. The triple helix mechanism is a synergistic role for catfish farmers to get more profit. Business development is the result of the interaction of triple helix mechanism, by which each stakeholder simultaneously have a role to support the application of bioflocs technology, in order to maintain the availability of catfish supply to meet consumer needs.

References

[1] Etzkowitz H and Leydesdorff L 2000 *Reseach Policy* **29** 109–123
[2] Rennings K 2000 *Ecological Economics* **32** 319–332
[3] Anand P S S, Kohli M P S, Kumar S, Sundaray J K, Roy S D, Venkateshwarlu G, Sinha A and Pailan G H 2014 *Aquaculture* 418–419 and 108–115
[4] Effendy I, Deen S A and Chithambaran S 2016 *HAYATI Journal of Biosciences* **23** 106–110
[5] Bragaa, André, Magalhãesa V, Hansonb T, Morrisc T C and Samocha T M 2016 *Aquaculture Reports* **3** 172–177
[6] Etzkowitz H 2003 *Social Science Information* **42** 293–338
[7] Leydesdorff, Loet and Øivind S 2012 *Procedia - Social and Behavioral Sciences* **52** 1–4
[8] Nakwaa, Karantarat, Zawdiea G and Intarakumnerdb P 2012 *Procedia - Social and Behavioral Sciences* **52** 52–61

IOP Conf. Series: Earth and Environmental Science 139 (2018) 012037 doi:10.1088/1755-1315/139/1/012037

Economic impact studies on development project of New Yogyakarta International Airport to aquaculture in Kulonprogo Coastal

F Rachman[1*], M C Satriagasa[2] and W Riasasi[3]

[1] Department of Fisheries, Faculty of Agriculture, Universitas Gadjah Mada, Jl. Flora, Bulaksumur, Sleman, Yogyakarta, Indonesia
[2] Magister of Economic Development, Faculty of Economic and Business, Universitas Gadjah Mada, Jl. Sosio Humainora No. 1, Sleman, Yogyakarta, Indonesia
[3] Study Programme of Geography, Faculty of Science and Technology, University of AMIKOM, Jl. Ring Road Utara, Sleman, Yogyakarta, Indonesia
*Email: faizal.rachman@mail.ugm.ac.id

Abstract. New Yogyakarta International Airport (NYIA) is being constructed in Temon Sub District, Kulonprogo District. It lies on 587.2 ha area in the southern part of Java Island coastal area. Many areas of Kulonprogro coastal area are used for aquaculture of vanname shrimp. In that case, the aquaculture land needs to be cleared for the airport construction necessity and requires compensation. The value of the compensation needs to be right calculated by both sides, regarding the aquaculture land are local community assets. This study uses spatial analysis and visual interpretation. Whereas, the calculation of the acquisition value and income capital value uses Discounted Cash Flow (DCF) method. The result shows the area of aquaculture land which impacted to be cleared is 83 ha, means decreasing of shrimp ponds area in Temon Sub District, Kulonprogo District. The calculation of acquisition value for airport development indicates higher value than the income from the aquaculture activity. The results mean the aquaculture landlords do not incur losses due to the airport development project. These findings can be used for local government and related stakeholders to formulate a policy of aquaculture relocation and to estimate projection for aquaculture land suitability in Yogyakarta coastal area.

1. Introduction

Cities around the world are separated by physical distance, but individuals can travel relatively easily between cities using various forms of transportation [1]. As business markets become national and international in scale, airports are increasingly being viewed as catalysts for local economic development. Their ability to generate jobs and attract new business is being used in many locations as a justification for public investments in new airport construction and expansion [2]. The economic impact of airports and associated aviation activity is a crucial component of modern economies. However, what is also increasingly important is that this economic impact is fully understood and appreciated outside of the aviation sector [3].

Adi Sucipto International Airport can accommodate maximum 1.6 million passengers per year, but in fact, the number of passengers increasing every year. Recently, it is reported that the number of passengers has been reaching 7.2 million or almost five times than the ideal capacity. Regarding that issue, starting this year government develops a new airport in southern coast, precisely in Jangkaran Hamlet, Jangkaran Village, Temon Sub District, Kulonprogo District. New Yogyakarta International

Airport (NYIA) building was designed to accommodate up to 20 billion passengers and built above more than 587.2 ha area. One of the procedures in preparing the new airport development is the government considering the acquisition of land in the planned location, which the possessions do not only belong to palace grounds, but also belong to the local community. The land acquisition for the new airport development consists of 67 % of total land belongs to local community, 27 % belongs to Sultan and the rest belongs to government [4]. Public land procurement in Indonesia is based on Law No. 2 of 2012. Land procurement is an activity of providing land by giving fair and fair compensation to the entitled parties [5].

The procedure of land acquisition is a quite complicated process, due to many sectors involved, especially land acquisition belongs to the local community. NYIA development evicts settlements, agriculture land, rice field and fish pond in Jangkaran Hamlet. Due to the consideration of the land as a residence and workplace, the community is dejected to sell their land and migrate to another place. Fish ponds are income source for the local community, especially fish pond farmer. Fish pond is a kind of brackish aquaculture, which in Kulonprogo mostly farmers cultivate species of shrimp. Fish pond farmers in Yogyakarta mostly cultivate vanname (*Litopenaeus vannamei*) type of shrimp. One harvest time of the vanname shrimps or around four months, the farmer receives profit up to 50 million IDR. Because of the high profit of the vanname shrimp harvest, the local community tends to make the type of aquaculture as their primary income.

Some fish ponds area in Jangkaran Hamlet were evicted because they were included in planned location of NYIA development. In Indonesia, there is a particular procedure of compensation for the land owner which impacted by the public infrastructure development. According to Act Number 2 the year 2012 on land acquisition to be proposed for a public purpose, is regulated based on guidance on land right's assignment. In transferring the land right's assignment between seller and buyer, it is a must to considering protection and principle of reverence to the land affected parties. In SPI 306 the year 2013 mentions that the compensation is determined by public valuer as independent professional valuer who has a license of practice from Ministry of Finance of Republic of Indonesia and National Land Agency. Determination of the compensation has to be based on humanity, fairness, benefit, certainty, openness, agreement, participation, prosperity, sustainability and harmony. Some aspects to consider in the feasibility analysis of the project are technical, financial, economic, socio-cultural and environmental aspects [6].

The economic impact of NYIA development to aquaculture in Kulonprogo is studied by fair compensation value on land rights assessment, considering fish ponds is the local community's main income source. Profit or loss of fish pond owners was determined by comparison between fair compensation value and fish pond harvest revenue. Based on that background, aims of this research are to assess the estimation of the aquaculture acquisition area for the new airport development project and to compare among the fair compensation value and income from the aquaculture activity in Kulonprogo coastal area.

2. Methods

This research used some methods to achieve the objectives, namely remote sensing visual interpretation, calculation of fair compensation value and quantitative descriptive analysis. Remote sensing interpretation was used to estimate the fish pond area which included in NYIA planned location area and interpreted through satellite imagery PJ SPOT 6 the year 2016. The satellite imagery shows land use view on the surface of the earth clearly, moreover considering the principle of remote sensing interpretation, such as association, color and form, the form of fish pond can be recognized easily. Area of fish pond was estimated using Geographic Information System tools, ArcGIS. Fair compensation value was assessed based on Act Number 2 the year 2012, President Regulation Number 71 the year 2012 and Indonesian Assessment Standard (SPI) Number 301 and its technical guidance. Components of fair compensation value can be seen in table 1. Equation (1) shows equation for calculation of fair compensation value [7]. Assessment is defined as a combination of science and art in estimating the value of an interest in a property for a particular purpose and in a predetermined time. Also taking into account all the characteristics of the property and market conditions, including investment types on the market [7].

Table 1. Fair Compensation value assessment component.

Component of Fair Compensation Value (FCV)	
Physical compensation (NPF)	Non-Physical Compensation (NPNF)
Land (L)	Tax (Tx) [5 % from NPF]
Building (B)	Legality (Lg) [1 % from NPF]
	Time lag compensation (TL) [r = 5.92 % p.a.]

The equation for calculation of fair compensation value [8]:

$$FCV = NPF + NPNF + TL$$

$$FCV = (L + B) + (Tx + Lg) + \{(NPF + NPNF) * TL\} \quad (1)$$

Calculation of fish pond properties value uses two approaches, which are cost approach and income approach. Cost approach emphasizes fish pond properties value of replacement cost new (RCN) and income approach focuses on the cash flow of the fish pond. Values of both approaches need to be reconciled to estimate the final value of fish pond properties.

The social-economic impact was analyzed using a qualitative descriptive method by interviewing local community and field observation related to community's socio-economic of the change of land use postcondition of fish ponds to an airport. Descriptive analysis aims to describe real condition in the research location. Primary data was collected by direct interview with local people as respondents. Secondary data was collected by reviewing on related literatures. Snowball method was used to collect primary data, which the way was one key person to another one to be interviewed determined based on the previous key person. Information from interview describes the community's socio-economic condition regarding land use change of aquaculture to an airport.

3. Discussion

3.1. Physical impact on new airport development

New Yogyakarta International Airport (NYIA) development planning in Jangkaran Hamlet, Jangkaran Village, Temon Sub District, Kulonprogo District was appointed based on Decree of the Governor Number 68/KEP/2015 on Determination of Location Site for New Airport Development in Special Region of Yogyakarta on 31th March 2015. NYIA was planned to lie on the 587,2-ha area [9]. The land rights for the site consists of couple ownerships, which are local communities and Sultans of Yogyakarta Palace. NYIA development planning site can be seen in figure 1, which is attached in Decree of the Governor's appendix. In the airport, development procedure was considering a land acquisition to provide land for the infrastructure. In that planning site, kinds of land cover that impacted were settlements, rice fields, moorlands and fish ponds.

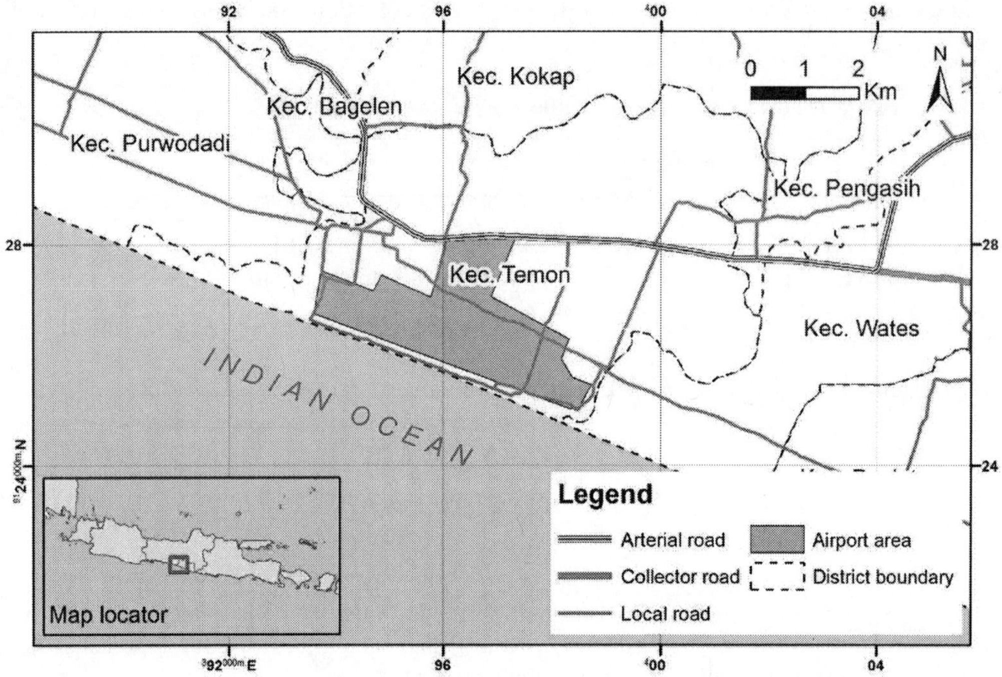

Figure 1. Map of NYIA development site.

Aquaculture sector, especially fish pond, has been very productive in the last few years, it produces very high numbers of harvest. In southern coast of Java, especially in Special Region of Yogyakarta, brackish aquaculture was a very profitable sector. Aquaculture of shrimp species, vanname (*Litopenaeus vannamei*), attracts the interest of fish pond farmers because it has more excellence than other types of shrimp, such as disease resistance, rapid growth and low feed conversion value, therefore its productivity is high on every harvest time. Furthermore, the position of Indonesia lies on equator line that having the dry and rainy season is also an essential factor for vanname shrimp to be productive along the year [9]. Based on data from Statistic Center Agency of Yogyakarta [10], the productivity of brackish aquaculture in Yogyakarta in 2011–2015 reached amounts of 3,244 tons, the highest productivity district was Kulonprogo, with productivity value was 2,581.48 tons.

Loss of fish pond area as the impact of NYIA development triggers decreasing the productivity of shrimp in Kulonprogo. Figure 2 shows NYIA development site boundary on satellite imagery PJ SPOT 6, which illustrates the site cutting fish ponds area [11]. Based on the geometric calculation using visual interpretation on satellite imagery of PJ SPOT 6, it has resulted approximately 83 ha area of the fish pond, of 313-ha total fish pond area in Kulonprogo, was impacted by NYIA development site planning. That 83-ha fish pond area consists of 379 units fish ponds with average area 2,200 m^2 per unit. Almost quarter of total fish pond area in Kulonprogo was lost as the impact of NYIA development.

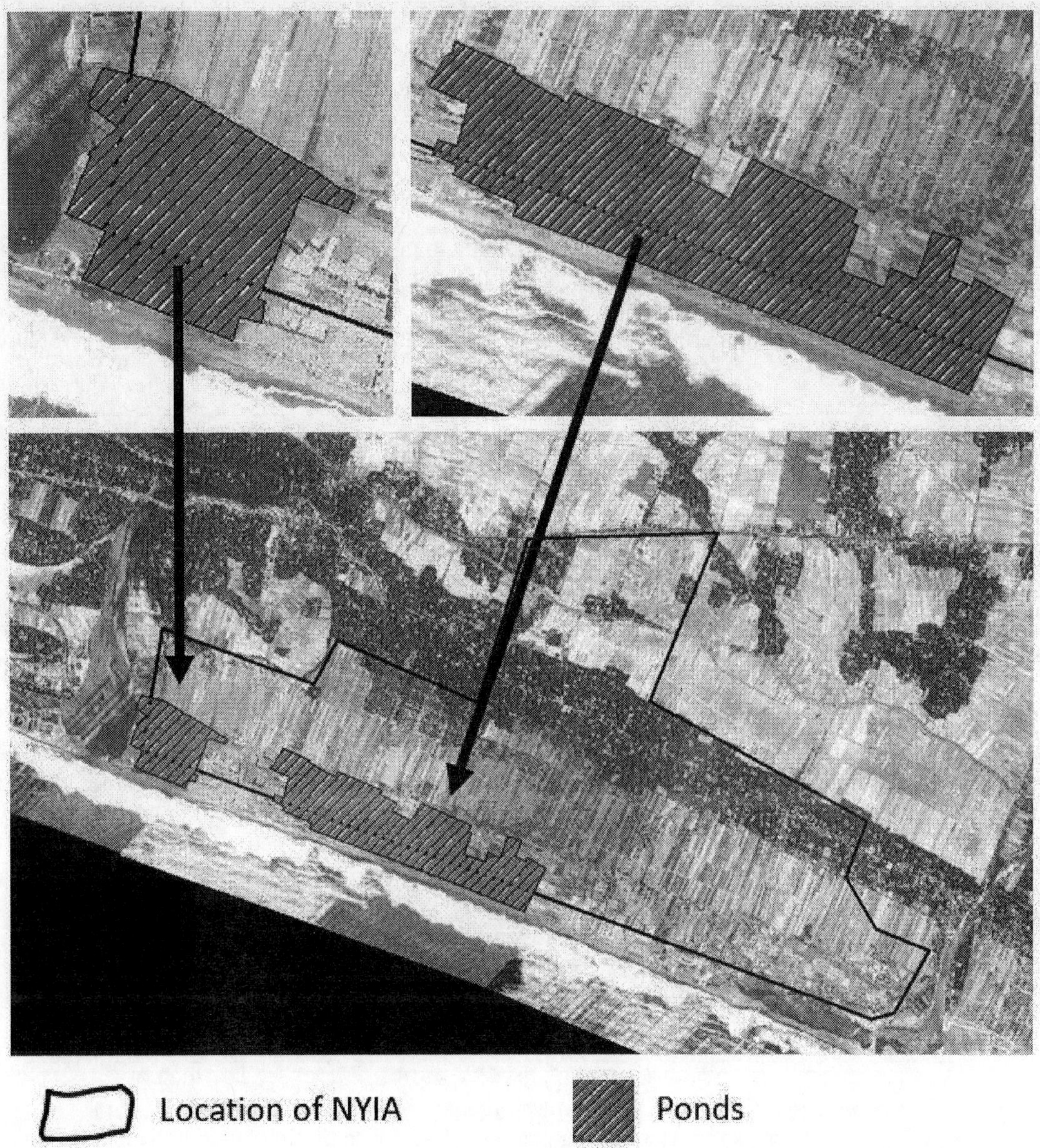

Figure 2. Impacted Area of NYIA development site (SPOT Imagery date acquisition March 2017).

3.2. Comparation fair compensation value and fish pond intrinsic value

Comparison among fair compensation value based on fish pond properties value in the area has to be assessed to determine whether the farmer was aggrieved by the land acquisition or not. Assessment of fair compensation value was calculated based on Act Number 2 the year 2012, President Regulation Number 71 the year 2012 and Indonesian Assessment Standard (SPI) Number 301 and its technical guidance. Assessment of fair compensation value consists of two main components, which are physical compensation and non-physical compensation. Result of the assessment of fair compensation value in the research area is presented in table 2.

Table 2. Fair compensation value assessment of fish pond land in NYIA airport plan area.

Component	Value (IDR)/Total Impacted Area
Physical compensation	718,421,436,225
Non-physical compensation	56,008,654,511
Fair compensation value	774,430,090,736

The result of fair compensation value shows value approximately 774 trillion IDR/total impacted area consists of 93 % physical compensation and the rest is non-physical component compensation. Nonphysical compensation is a premium privilege for landowners as unwilling to sell compensation. Calculation of fish pond properties value is presented in table 3; it shows the result the value of 92 trillion IDR.

Table 3. Reconciliation of fish pond property value assessment.

Method	Value (Rp)	Weight	Properti Value per fish pond (IDR) (2,200 m^2)	Total Properti Value (IDR) (834,301 m^2)
Income approach	245,517,663	0.85	245,655,263	92,695,808,087
Cost approach	246,435,000	0.15		

Comparison value of fair compensation and properties value can be seen in table 4. The result shows the fair compensation value is higher than the properties value, more than eight times. It because properties value consists only intrinsic value of the fish ponds, at the other side, fair compensation value considers premium value, unwilling to sell compensation, as mentioned above. It shows fish pond farmers were not aggrieved economically by the land acquisition for NYIA development.

Tabel 4. Comparison between fair compensation value and property value.

No.	Component	Value (Rp)
1	Fair compensation value	764,396,187,327
2	Property value	92,695,808,087

3.3. Socio economic impact on new airport development

NYIA development in Jangkaran Hamlet, Jangkaran Village, Temon Sub District, Kulonprogo District gave social impacts to the local community, one of them was historical value owned by the local community who already lived in the area for a long time. Fishpond activity is the source of their income, by the developing of NYIA will reduce it, as we know that fishponds in the southern part of Yogyakarta is a small-scale fisheries business [12]. Places where they were born, lived, grown and worked had to be evicted for the public facility development. Noise pollution is also one of the consequences that need to considered by the local community who live surrounding the airport. Supporting infrastructures for the new airport need to be developed by the government in the surrounding area. Nevertheless, the NYIA development also gave positive impacts, such as increasing local income, job opportunities, the use of public transportation (train and bus to airport). As the capacity for passengers in the airport increase, it will lead the tourist increasing numbers who come to Yogyakarta. It will have an impact on making the higher price of land and the most important is accelerating regional development of Yogyakarta.

Based on the assessment of fair compensation value of fish pond given by the government was very high compared to fish pond productivity, the shrimp farmers were benefiting from the land acquisition process. There are two alternatives for the shrimp farmer whose their fish pond impacted by NYIA development. The first alternative has relocated the fish ponds to other areas and the second one is they change their livelihood to supply service and trade that supporting airport services.

4. Conclusion

1. Based on geometric calculation on satellite imagery of PJ SPOT 6, the approximately 83 ha area of fish pond was impacted by NYIA development site planning.
2. Fish pond land acquisition for NYIA development shows the fair compensation value is higher than the properties value, more than eight times, means the fish pond owner was not aggrieved by the government.
3. Two alternatives for the shrimp farmer whose their fish pond impacted by NYIA development are fish pond relocation or changing livelihood to support airport service.

References

[1] Green M 2015 https://sites.duke.edu/urbaneconomics/?p=1248/ Accsessed in July 2017
[2] Weisbrod Glen E 1993 *Airport Area Economic Development Model Paper* (England: PTRC International Transport Conference)
[3] Anonim 2015a http://www.airport-business.com/2015/04/new-study-reveals-real-economic-impacts-airports/ Accsessed in July 2017
[4] Biro Komunikasi dan Informasi Publik Dephub 2017 http://www.dephub.go.id/post/read/bandara-baru-yogyakarta-siap-dibangun Accessed in July 2017
[5] Presiden Republik Indonesia 2012 *Undang-Undang No. 2 Tahun 2012b Tentang Pengadaan Tanah Bagi Pembangunan Untuk Kepentingan Umum Indonesia*
[6] Suparmoko, Sudirman D, Setyarko Y and Wibowo H S 2014 *Valuasi Ekonomi Sumberdaya Alam & Lingkungan* (Yogyakarta: BPFE)
[7] Harjanto B and Hidayati W 2014 *Konsep Dasar Penilaian Properti* Edisi 2 (Yogyakarta: BPFE)
[8] Molinos S M, Hernández S F, Sala G R and Garrido B M 2011 *Ambio* **40**(4) 408–416
[9] WWF 2014 *Seri Panduan Perikanan Skala Kecil* (Indonesia: WWF) ISBN: 978-979-1461-38-2
[10] BPS 2015 *Provinsi Daerah Istimewa Yogyakarta dalam Angka* (Yogyakarta: Badan Pusat Statistik)
[11] Anonim 2017 https://earthexplorer.usgs.gov/ Acceseed in July 2017
[12] Anonim 2015b https://jogja.tribunnews.com/ Accessed in July 2017

IOP Conf. Series: Earth and Environmental Science **139** (2018) 012038 doi:10.1088/1755-1315/139/1/012038

Technological adaptation in traditional fisheries: way to survive

A Salam

Faculty of Fishery and Marine Science, Universitas Negeri Gorontalo, Jl. Jenderal Sudirman No. 6, Kota Gorontalo, Indonesia
*Email: aziz_salam@ung.ac.id

Abstract. Traditional fisheries use traditional fishing gear and fishing boats at a scale of household and the small group of fishermen. Traditional fishing gear and boats which have been used since ancient times are still used in the era when more advanced technologies are available. What makes this simple technology able to survive today and how does it interact with new technologies? This question is tried to be answered through surveys conducted over the last few months (March-July 2017) in several fishing villages on the coast of the Celebes Sea and the Islands of Sangir in North Sulawesi. The level of mastery of local boat-making technology, the tendency of fishermen to maintain their technique known and mastered for generations and limited access to new technologies are among other factors that make traditional fisheries remain. On the other hand, the limitations of natural raw materials, the introduction of new technologies by governments or fishermen from other regions, as well as the needs of fishing communities encourage traditional fisheries to adapt technologies continually. This article also illustrates some examples of technological adaptations to the fishing boats.

1. Introduction

Boats for fishermen in everyday life are primarily economic means, yet at the same time are social symbols and cultural institutions in society. In addition to catching fish, boats serve as means of personal as well as public transportation, function as the main mean in various ceremonies of various maritime cultures throughout the archipelago. It is also a marker of particular social strata in fishing communities.

Salam and Osozawa [1] have observed the technological adaptation of all traditional boat forms in Indonesia which occurs since centuries ago, especially in the Spermonde Islands-South Sulawesi. The interaction between the original technology of the archipelago and the technology of foreign ships (Arabic, Indian, Chinese and European) since the era of trading began has led to the evolution of ship forms and their functions that are driven by the relentless adaptation of technology. The technological adaptation of boats and fishing gear can be understood as a fisherman's effort to improve the performance of their boats and fishing gear by taking advantage of the available technology while maintaining old forms and functions that are deemed usable.

Observing technological adaptations conducted independently by fishermen on boats and fishing gear gives a dynamic picture of the development of the shape and function of the boat and its gears. By understanding this dynamic, we can at least learn the patterns of acceptance and rejection, as well as modifications to a technology introduced to fishermen. Thus, it can be considered a model of technological adaptation that can easily be accepted by the community if a particular party, such as the government, is about to introduce a new technology in the field of capture fisheries. In other words, technological adaptation is a transitional path that can bridge the development of traditional fisheries toward modern technology which is a necessity if it is to expect advanced and sustainable capture fisheries in Indonesia in the future.

Technological developments in various fields directly or indirectly related to fishing activities which are involved in traditional fishing communities influence boats and fishing gear development. The influence is pervasive, either through adoption and self-adaptation by fishermen or introduced by others such as government or fishing fleets from other regions with more advanced technology. However, the acquisition of technology by fishermen through adaptation looks so slow, secondary and partial on items that are not so significant on catch performance. Since the Indonesian fishermen recognize motorization for the first time in the era of the 1970s [2, 3], there is no technological progress that stands out for the betterment of their boats. The outrigger-boat, for instance, is still with the same outboard engine and until today the predicate "traditional" is still attached to our fishermen.

How the technological adaptation in fishing boats occurs? What kind of and at what extent technologies are adopted in the traditional fishing boat technology? These two questions are attempted to be answered to understand the development of traditional fishing boats from traditional fisher's capabilities to maintain their existence.

2. Methods

This study examines the technological adaptation of boats used by the traditional fisherman along the north coasts of northern Sulawesi and in the Sangir Islands of North Sulawesi Province. This research is aimed to know the process of technological adaptation on fishing boat and adaptation of technology of traditional fishing boats.

Research on technology adaptation on traditional fishing boats and fishing gear needs to be done to obtain a knowledge base for a better understanding on how to develop boat and fishing gear for suitable for the traditional fishermen by introducing more advanced technologies in the future. The results of the research can be used as a reference to create a concept of applying advanced technology in traditional fishing boat design. If similar research is conducted in various locations with different geographical, oceanographic and fishermen's characteristics, the concept of applying technology is appropriate to local characteristics in various regions in Indonesia. Furthermore, fishing vessels and fishing gear can be built with more advanced technology and best suited to the needs of traditional fishermen in their respective localities. This concept can be useful for governments to up-grade fishermen through empowerment and development programs.

3. Results and Discussion

3.1. Boat evolution: the survival of the fittest

The evolution boat, as proved by [1], is a phenomenon that does occur. As the evolution that occurs in the living things the evolution of boats also occurs relatively slow, of course with a shorter timescale. With the passage of time, the boats used by humans are developing. The shape changes with function, capacity demands, interconnection and even subject to the human aesthetic.

Thus, the types of boats have increased and then developed into hundreds of ship types used by humans in this era. With such a rapid rate of development, countless types of extinct boats that have been found were no longer sail on the waters. In Indonesia the types of boats, especially in South Sulawesi, legendary boats that had been the backbone of the maritime axis of the archipelago such as the *phinisi* and *padewakang* now only become a symbol of the past maritime glory [4].

Boats in Indonesia have been the part of livelihood since early humans, the Australo-Melanesians who migrated from Sunda land to the Wallacea, inhabited the caves. Many prehistoric boat paintings are evident to this [5, 6]. There is a report of a boat painting in caves of the Little Kei Island in Eastern Indonesia. Reproduction of the cave imprint presented by [5] shows the boat as a dugout with additional structure and projecting bows. This painting resembles the boat painting with pictures of men, animals and the sun along with pictures of long boats called *kora-kora* found in East Timor reported by [7]. All those boats were single-bodied dugout canoes without outriggers, although some of them show the additional construction for poles and sails or other strengthening structures (figure 1) [5].

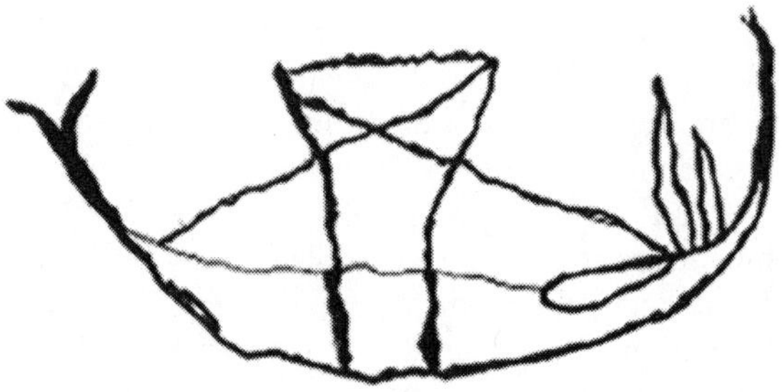

Figure 1. The boat painting on a cave wall in Ohoidertavun Village in Lesser Kei Island.

Outrigger canoes are absent from cave paintings. The outrigger boat technology is associated with the second wave of migration of the Austronesians from Taiwan to Indonesia circa 2,000 BC with more advanced technology. They might have started life in pillared houses above sea level or on their outrigger boats that roamed the waters of Nusantara.

The oldest evidence of outrigger boats is preserved on the impressive reliefs of the eighth-century Borobudur Temple [8]. However, the earliest written description is found in the fifteenth-century La Galigo cycle [9]. The boats are described as aristocratic boats used in adventures and wars to conquer neighboring states. A story in the La Galigo that narrates the creation process of the mythical boat indicates that the boat's body consisted only of a large tree that is hollowed out, a dugout boat. Further details on the manuscript describe the boats as ocean-going outrigger boats with a deckhouse structure, tripod mast, squared sails and lateral rudder. A description which very much resembles the Borobudur oceangoing outrigger ship. Outrigger boats are still in use today by small-scale fishermen across the archipelago but of course with their present appearance different from their ancestors. Hundreds of local names are used all over the country with variations in their appearance and parts.

Likewise, with the boats in the northern part of Sulawesi, some types of boats to flourish and there are also types that lead to extinction.

3.2. Traditional fishing boats on Northern Sulawesi

The north coast of Sulawesi and also Sangihe-Talaud Islands inhabited by several ethnics namely Gorontalo, Bolangitan, Minahasa and Sangir. Ethnically there is no striking difference between the boats they use in the fishery. The small boats they usually use are boats with outriggers and outboard engines mounted in the stern. There are only a few variations on how to bind the floats (*ateq*) to the transverse beam in the outrigger structures (*bahateng*) and various forms of bow and stern.

The outrigger boat used is commonly referred to as a *pelang* boat. Traditionally this type of outrigger boat is a semi dugout boat. Boat bodies are made by adding boards on the sides of the *mahera* (the base keel made from a whole piece of hollowed wood). This kind of boatbuilding technique is accepted throughout Indonesia because it is a legacy of the Austronesian boatbuilding technology.

However, there are several naming variations on boats based on size, style and place. A simple dug-out canoe with an outrigger is called *buloto*, a name that is also used for a paddle boat without an outrigger. The *pelang* boat was originally a term for a *mahera*-based boat, but with the influx of Filipino technology on Sangir fishermen, then a boat made of waterproof plywood can also be referred to as a *pelang* boat if it is made to follow the model of the outrigger boat common in this area. On the northern coast of Gorontalo, the slightly larger sized boats are called *pamo* boats that can be used to operate larger fishing gear such as gill net, and if the size is bigger, it will be called *pajeko* with larger fishing gear such as mini purse seine.

The boats used by the Sangir ethnic in the Sangihe-Talaud Islands seem to have more types. In addition to the *pelang* and *londe* boats, they use various types of boats the same kind of boats used by fishermen in the southern part of the Philippines, such as the *pambut*, *fuso* and *banka*. Large types such as the *fuso* and the *banka* are made in the Philippines territory and then purchased by Sangir fishermen. Smaller types such as the *pambut* can be made by Sangir boat builders who acquire these skills in the Philippines.

Among the types of boats is a boat *londe* is the type of boat whose numbers are already rare. This type has its own uniqueness. The current size of the *londe* boats is only about 5 m long, generally driven by oars and sometimes with a screen and operated by one fisherman. Its uniqueness mainly lies in its bow and stern. There is a kind of horn that pokes out from the bottom of the bow protruding forward and curling upwards like an elephant tusk. At the stern the horn is not so prominent but can be seen a little form backward. There is a strong supposition that this type of boat is the remnants of the ancient Sangir boat called *bininta* which is now enshrined in the symbol of the region of Sangir Islands District. Boats of the 'ancient times' *bininta* is a boat with the bow and stern both have a horn that pokes toward the front and back.

3.3. Development of traditional boat through technological adaptation

3.3.1. Simultaneous development. In this research, a development occurring on the boats used by traditional fishermen along the northern coast of Sulawesi as well as in the Sangihe Islands was observed. These developments occur simultaneously on several different levels of development. Underdeveloped types of boat are still in use according to its capacity and function along with new types of boats that have been developed with some or more changes. Thus, the previous type of boat does not necessarily become extinct. Possible changes on boats that go through technological adaptation are include changes in size, propulsion, materials and manufacturing technology, fishing gear, fishing auxiliaries and of course bringing change to the capacity and function as well as the aesthetics of the new boat.

The *buloto* used as a means to do fishing activities by using fishing rods in waters very close to the beach is a simple boat driven by oars. This type of boat is made of intact wood dredged in the interior and then formed into a boat. The outrigger float can be made of bamboo or light wood that has a buoyancy that is connected with two logs across the front and rear of the boat (figure 2).

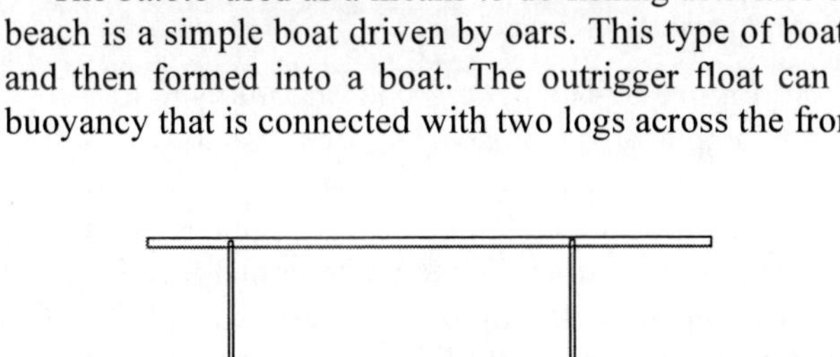

Figure 2. Buloto, the simple paddling boat with outrigger used for line fishing.

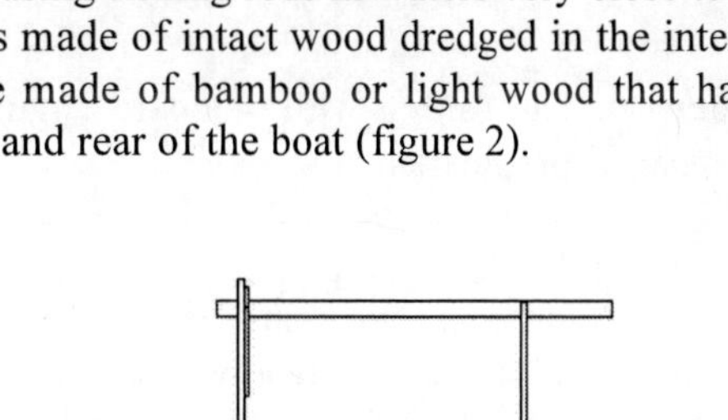

Figure 3. The small *pelang* with marine plywood for its side planks.

This outrigger paddle boat type is then developed into a small *pelang* boat of approximately the same size that is also only driven by using a paddle; its function is still the same that is for fishing with the hand line in the waters near the beach. However, this boat is made by using a *mahera* from a small tree trunk. The *Mahera* obtained from small wood does not have enough height to be used as a boat. Thus, the *mahera* is in the form of the flat keel of wood with a slight curvature. The side planks are made of marine plywood with some frame construction and side beams as reinforcement. This construction produces an outrigger paddle boat that is much lighter than the *buloto* to give comfort to the fishermen who operate it (figure 3).

The change in wooden boat making the material into plywood is a common phenomenon in northern Sulawesi. The diminishing source of wood for boatbuilding and the availability of marine plywood imported from the Philippines is a factor that allows this to happen. The reduced supply of timber is allegedly caused by the control of industrial timber plantations (HTI) by the company at locations that were traditionally the source of boat-building materials.

The *pelang* boat in figure 4 is a boat made of wood. The underside is made of a whole wooden *mahera* which is then raised with wooden planks. This boat uses an outboard engine (*katinting*) as a driving force to perform fishing operations using gill nets and line fishing in more distant waters. This type of boat is evenly distributed throughout the research sites visited. The outrigger boat that used to use sail as the main propulsion then adopted the use of the engine that could be regarded as the leap in technological adaptation by traditional boats in Indonesia in the 1970s.

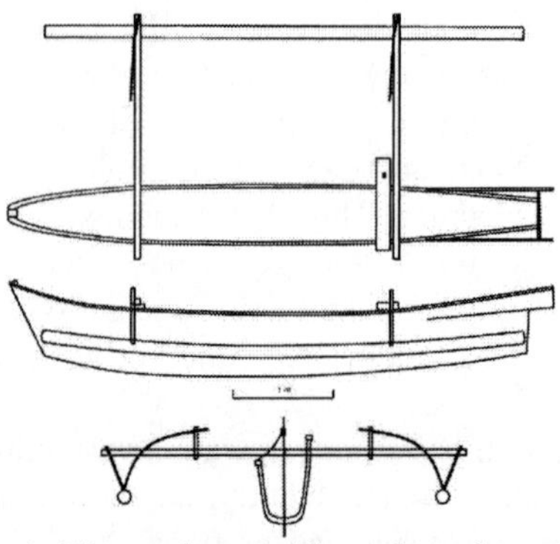

Figure 4. The *pelang* with outboard engine as its main propulsion.

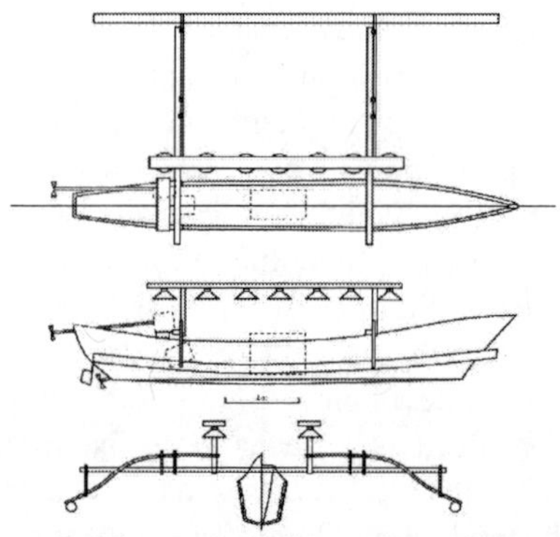

Figure 5. The *pelang* boat that turns to a lamp boat.

One of the developments of the function of the outrigger *pelang* type with an engine is to make the boat as a light boat. The boat is equipped with lights used to attract fish to gather around the boat. The boat depicted in figure 5 below is documented in Likupang at the tip of Sulawesi about half an hour from Manado City. The boat has an electric generator to turn on the lights and is equipped with an outboard engine as propulsion. The raw material is marine plywood with a length of about 7 m.

The boat operates independently collecting fish in the fishing grounds. Fishermen with purse seine who come to harvest the fish around the light boat will share the revenue of the catch to the owner of the light boat.

A very progressive development is also observed in the same village. Another light boat with larger size and capacity of lights (figure 6). This boat is made of fiberglass by local boat craftsmen who have been trained by the local government to make fiber boats. Although the design follows a modern ship design with a wide hull with good stability, it still retains the outrigger on both sides that is intended to obtain a good stability guarantee. This boat uses two internal engines mounted side by side on the stern of the ship and equipped with an electric generator.

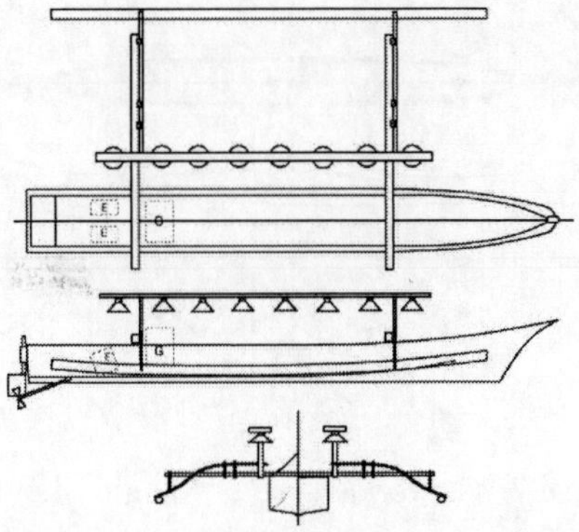

Figure 6. A bigger lamp boat made of fiberglass.

Development towards the use of fiberglass material generally occurs throughout Indonesia. Governments often provide aid in the form of fiberglass fishing boats to groups of fishermen. Even training has been done to local boat makers to make a boat from fiberglass. This has more or less helped solve the problem of scarcity of wood raw materials for boat making.

The adaptation of technology by fishermen and boat makers has shown that changes in the boat are in order to fit the boats to the needs of fishermen so that traditional fishermen can continue to maintain their existence in their efforts for livelihoods at sea.

3.3.2. Extinction. The logical consequence of technological adaptation on some types of boats which then gives better performance is that the types of boats that have been developed will be more widely used and the old types of boats will slowly be abandoned, no longer used and not made again. As a result, the old types of boats will be extinct.

In Indonesia there are many types of extinct boats either permanently or still exist in the form of a replica. *Phinisi*, *palari*, *padewakang* and *pajala* boats are some types of boats from South Sulawesi that are already extinct. *Pakur* and *sandeq* are examples of boats from ethnic Mandar in West Sulawesi that suffered a similar fate. In North Sulawesi which is the location of this study there is a type of boat that has been extinct, the *bininta* boat which already mentioned above. Meanwhile, the type of traditional boats that have been on the verge of extinction is the boat called *londe*. It has a unique shape that gives a beautiful aesthetic impression. A slim transverse profile heralds a high hydrodynamics and streamlined curvature of hull lines. At the bow there is a horn that resembles an elephant tusk arch that protrudes from the bottom of the bow towards the front and curves upward. At the stern the horn is not so prominent but can be seen a small projection backward. This boat is made of wood with side planks on its *mahera*. The outrigger beam (*bahateng*) is different on the front and back. At the front, the outrigger beam is a piece of wood that is bent at the bottom end to hold the atiq (bamboo float) without a *tadiq* (the linking piece between *bahateng* and *atiq*). While the back is a straight piece of wood linked to the *atiq* with a curved rattan (figure 7).

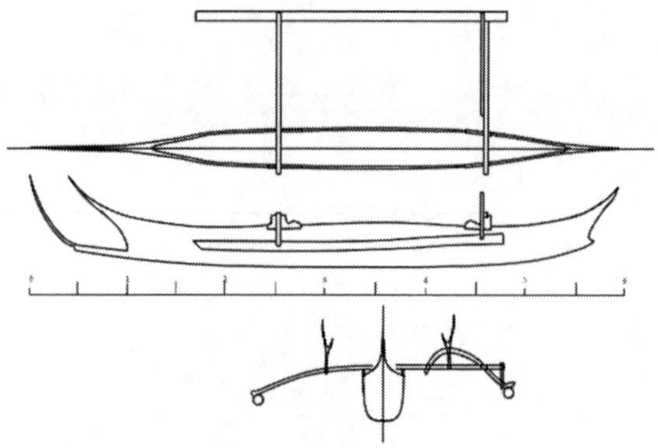

Figure 7. The *londe* type of Sangir Island.

In decades of 80s *londe* boats are still widely found along the coast through Manado City from Ranayapo, Poopo, Sario (Manado), Likupang to Batulubang near Bitung City [10]. Also found in Barangka, Tahuna, Naha and Lirung in the Sangihe Islands. The *londe* boat at that time was a sailboat used extensively in Sangir fisherman settlements (figure 8) [10]. In 2012, on Miangas Island, the outermost island on the border with the Philippines, there are still many *londe* boats [11]. However, in the same year, on the island of Marore in the Talaud Islands *londe* is no longer found which according to local residents once there are many on the island (figure 9) [11, 12].

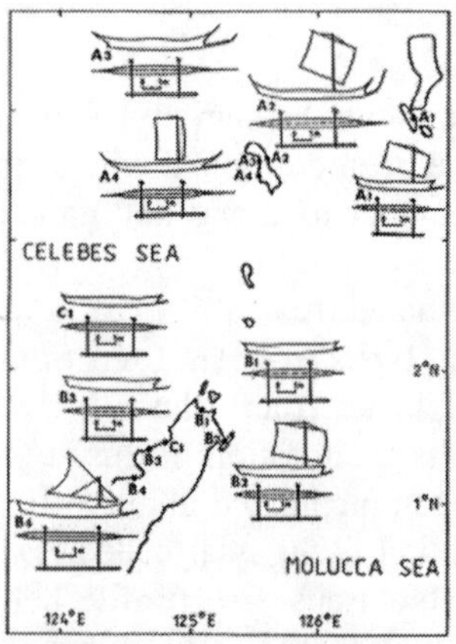

Figure 8. The existence of londe in North Sulawesi in the decade of 1980s.

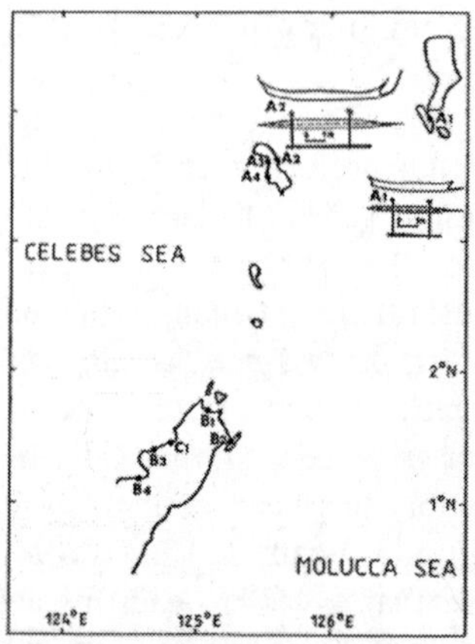

Figure 9. The existence of londe in North Sulawesi in the decade of 2010s.

The extinction of a *londe* boat with its horn characteristics is not because fishermen no longer need boats of the same size and functionality. The *londe* type began to become extinct because fishermen now no longer need the function of the horns, so they no longer order such boats. Horns were used as a place to hold to rest or to look into the sea at the time of collecting or capturing seafood by diving [10].

4. Conclusion

Changes in boat features and types in order for fishermen to maintain the existence of their fishing activities are related to the process of or the absence of technological adaptation such as: (1) boat builders adapt to new material and technology for boat building, (2) fishermen adopt types of foreign boat and technology and (3) the extinction of boat(s) as a result of lack of technological adaptation. These are the proof that technological adaptation provides a way for the traditional fishery to survive.

References

[1] Salam A and Osozawa K 2008 *J. Southeast Asian Stud.* **46**(2) 200–227

[2] Horridge A 1985 *The Prahu: Traditional Sailing Boat of Indonesia* (Singapore: Oxford University Press) p 106

[3] Hawkins C W 1982 *Praus of Indonesia* (London: Nautical Books) p 134

[4] Hadisuwarno S 1995 *Teknologi Maritim: Mengintip Kecanggihan Teknologi Pinisi No 1* (Jakarta: Himpunan Ahli Teknologi Maritim Indonesia (HATMI))

[5] Intan F S 1998 *Geologi Situs Gua Kepulauan Kei Kecil Maluku* (Jakarta: Departemen Pendidikan dan Kebudayaan)

[6] Holt C 1967 *Art in Indonesia: Continuities and Change* (New York: Cornell University Press) p 355

[7] Sukendar H 1999 *Perahu Tradisional Nusantara* (Jakarta: Departemen Pendidikan dan Kebudayaan)

[8] Lapian A B 1979 *Pelayaran pada Masa Sriwijaya* (Jakarta: Departemen P & K) pp 89–120

[9] Kern R A 1989 *Translation Series* KITLV–LIPI (Yogyakarta: Gadjah Mada University Press)

[10] Masengi W K A and Keishi Shibata 1990 *JIN* **83** 203–211

[11] Hoetagaol S M, N S A Sumampouw, J Parauba, R Tuage and M Pontororing 2012 *Studi tentang Aspek-aspek Sosial-budaya Masyarakat Daerah Perbatasan: Studi Kasus Masyarakat di Pulau Miangas* (Manado: KEPEL Press Balai Pelestarian Nilai Budaya Manado)

[12] Ulaen A J, P Nugrahini, C Setiawan, A Dukalang and Alinabur 2012 *Studi tentang Sosial Budaya Masyarakat Daerah Perbatasan: Studi Kasus Masyarakat Pulau Marore Kabupaten Kepulauan Sangihe* (Manado: KEPEL Press Balai Pelestarian Nilai Budaya Manado)

IOP Conf. Series: Earth and Environmental Science **139** (2018) 012039 doi:10.1088/1755-1315/139/1/012039

Selection of sustainable seaweed and grouper aquaculture development strategy: a case of Pulau Panjang, Serang Regency Banten Province

P A Soejarwo[1*], W P Fitriyanny[2], A Heriati[3] and A R Hakim[4]

[1] Research Center for Social Economic Marine and Fisheries, Gedung Balitbang KP I Lt. 4 Jl. Pasir Putih No. 1 Ancol Timur, Jakarta Utara
[2] Center for Integrated Research in Environmental Geoscience, Bandung Ocean Technology Research and Management, Jl. Bukit Dago Utara II No. 11, Bandung, 40135
[3] Research Center of Marine, Gedung Balitbang KP II Lt. 4, Jl. Pasir Putih No. 1 Ancol Timur, Jakarta Utara
[4] Center for Ocean Modelling Research and Operational, Bandung Ocean Technology Research and Management, Jl. Bukit Dago Utara II No. 11, Bandung, 40135
*Email: ari_permana008@yahoo.com

Abstract. Due to their high-income contribution, seaweed and grouper aquacultures are important activities in Pulau Panjang community. Determining alternative strategies in developing sustainable aquaculture for seaweed and grouper and their priority factors from theses aquaculture activities are done using TOPSIS and AHP analysis. It was found that the development strategy that must be taken is the option to maintain aquaculture activities, while, environment factor is the highest priority to maintain seaweed and grouper aquaculture in Pulau Panjang. Then three priorities are obtained from environment factor. The first is to maintain the water quality by the growth requirements of seaweed and grouper by encouraging the formation of "Environmental Community Awareness" that involved the active participation of the community to maintain quality and carrying capacity of the environment. Second is to use of natural or artificial coastal protectors (soft structure). The third priority strategy is integration and real implementation of heavy metal pollution control between government, industry sector and society.

1. Introduction

Seaweed and groupers aquaculture activities are some of the livelihoods options of coastal communities in Pulau Panjang, Serang Regency Banten Province according to [1] activities of fishery aquaculture in Pulau Panjang are in the category of sufficiently sustainable. Therefore, aquaculture activities especially seaweed and grouper are potential sectors to be developed in the region.

Furthermore, the aquaculture activities are providing more certainty for fishermen than capture fisheries activities that are very dependent on the weather and season. This situation justifies that government policy interventions to prioritize improvement production through aquaculture program are considered to be very appropriate [2]. Seaweed and grouper farming activities should be supported by factors that play a role in sustainable growth and management. The influenced factors are the environment, technology, economic and social factors [1]. The management strategy of seaweed farming activities in Pulau Panjang is located in quadrant one which is Strength-Opportunity strategy (maxi-maxi strategy) [3]. From the results of this study was obtained information that the development of aquaculture in Pulau Panjang was using the strength to take advantage of existing opportunities.

This research aims to provide a more specific alternative strategy to maintain or to optimize or to improve the development of existing seaweed and grouper aquaculture in Pulau Panjang by using TOPSIS approach. Then the AHP approach is used to determine the priority factors of the four factors (environment, technology, economic and social) by using pairwise comparison of each criterion from each factor so that the resulting strategy will be more targeted and obtain effective and efficient results [4].

2. Materials and Methods

2.1. Location and time research

This research was conducted in December 2016 until February 2017 in Pulau Panjang which is administratively located in Banten Bay, Puloampel District Serang Regency, Banten Province and presented in figure 1. The geographical location is at 6°25'18"–6°28'12"South Latitude and 106°22'9"–106°25'36" East Longitude. Primary data were obtained from the questionnaire of 66 respondents. Secondary data were obtained from the Central Bureau of Statistics (BPS), Agency of Marine and Fisheries or Ministry of Maritime Affairs and Fisheries (KKP), Pulau Panjang Sub District and other publications.

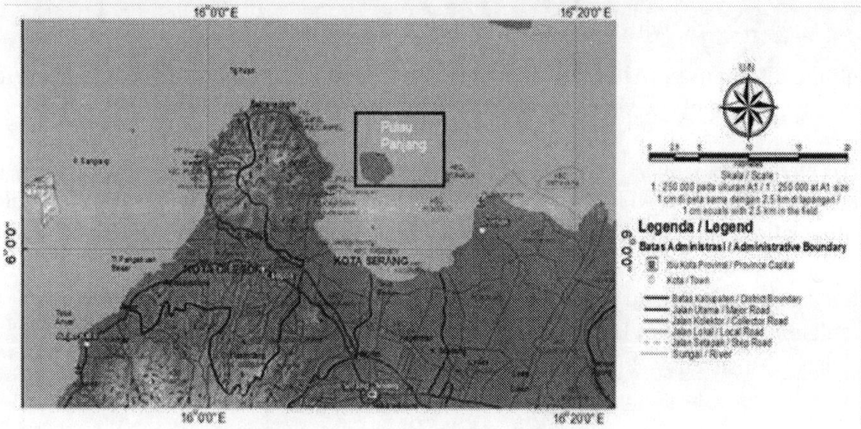

Figure 1. Research location.

2.2. The TOPSIS methods

TOPSIS (Technique for Others Reference by Similarity to Ideal Solution), initially developed by [5], is a concept where the best alternative would be the one that is nearest to the positive-ideal solution and farthest from the negative ideal solution [6].

The TOPSIS approach is used because (1) TOPSIS is rational and easy to understand; (2) easy computing process; (3) This concept allows to search the best alternative for each criterion that is described in simple mathematical form; (4) An important factor weight is incorporated into a comparison process [7].

The positive-ideal solution is composed of all best values attainable from the criteria, whereas the negative ideal solution consists of all worst values attainable from the criteria [8]. TOPSIS steps can be outlined as follows:

- Build a normalized decision matrix. The r_{ij} element resulted from the normalization of decision matrix R by the Euclidean length of a vector method is:

$$r_{ij} = \frac{x_{ij}}{\sqrt{\sum_{i=1}^{m} x_{ij}^2}} \quad (1)$$

- Build a weighted normalized decision matrix. With weight W= (w1, w2,…..,wn), then normalization of matrix V weight is:

$$V = \begin{bmatrix} w_1 r_{11} & w_2 r_{12} \cdots & w_n r_{1n} \\ w_1 r_{21} & w_2 r_{22} \cdots & w_n r_{1n} \\ \vdots & \vdots & \vdots \\ w_1 r_{m1} & w_2 r_{m2} \cdots & w_n r_{mn} \end{bmatrix} \quad (2)$$

- Determine positive and negative ideal solutions. The ideal solution is denoted A*, while the negative ideal solution is denoted A -:

$A^* = \{(\max v_{ij} \mid j \in J).(\min v_{ij} \mid j \in J'\}.\ i = 1,2,3, \ldots m\} = \{ v_{1*}, v_{2*}, \ldots v_{n*} \}$ (3)

$A- = \{(\min v_{ij} \mid j \in J).(\max v_{ij} \mid j \in J'\}.\ i = 1,2,3, \ldots m\} = \{ v_{1-}, v_{2-}, \ldots v_{n-}\}$ (4)

$J = \{ j = 1, 2, 3, \ldots$ n and j are *benefit criteria*$\}$ $J' = \{ j = 1, 2, 3, \ldots$ n dan j are *cost criteria*$\}$

- Calculate the separation. Si * is the distance (in Euclidean view) an alternative of the ideal solution defined as:

$$S_{i*} = \sqrt{\sum_{j=1}^{n} (Vij-Vj^*)^2} \text{ , with } 1=1,2,3, \ldots m \quad (5)$$

And furthest away with an ideal-negative solution.

$$S_{i^-} = \sqrt{\sum_{j=1}^{n} (Vij-Vj^-)^2} \text{ , with } 1=1,2,3, \ldots m \quad (6)$$

- Calculate proximity relative to the ideal solution

$$C_{i*} = \frac{S_{i^-}}{S_{i*} + S_{i^-}} \text{ , with } 0 < C_{i*} < 1 \text{ and } i = 1,2,3, \ldots m \quad (7)$$

- Rank the alternatives. Alternatives can be ranked based on the order Ci *. Therefore, the best alternative is one of the shortest distance to the positive ideal solution [9].

2.3. The AHP method

Analytical Hierarchy Process (AHP) is a structured technique to organize and analyze complex decisions based on mathematics and psychology [4]. It is used to determine the priority factor of the environment, technology, economic and social factors. AHP steps can be outlined as follow:

- Calculate the normalized matrix
- Calculate the maximum lamda value (λmax) and Consistency Index (CI)
- Calculate Consistency Ratio (CR) that formulated as follows

$$CR = CI / RI \quad (8)$$

where CR = *Consistency Ratio*; CI = *Consistency index*; RI = *Index Random*

If the Consistency Ratio (CR) is less than 0.10 then the result can be used to determine priority factor from seaweed and grouper aquaculture activities.

3. Results

3.1. Seaweed farming

Seaweed farming is located in the west of Pulau Panjang. The area of aquaculture is approximately 100 ha, while the seaweed plot has an area of 0.5 ha so that there are about 200 plots of the seaweed farming. In Pulau Panjang the type of cultivated seaweed is *Kappaphycus alvarezii* (cottonii) as a producer of carrageenan [10]. Seaweed farming system is using long line system by using a rope that stretched from one point to another with a length of 50–100 m. It can be in the form of loose or arranged lines in quadrilateral form with the support of buoys and anchors [11]. Seaweed life cycle is 45 days.

3.2. Grouper aquaculture

Grouper aquaculture in Pulau Panjang is conducted by floating net cage (KJA) technology, which is mounted on floating raft in the coastal area [12]. Size of floating raft is 8 × 8 m divided into four plots measuring 3.2 × 3.2 m, the raft was floated with a float and tethered with an anchor. While the size of the cage is 1 × 1 × 2 m for nursery phase and 3 × 3 × 3 m for the enlargement phase [13]. Harvest will be conducted when the weight of fish reaches 500–1,200 gram or depending on market demand. This is a measure of market demand that has high selling value [14]. Curently there are eight KJA for grouper aquaculture [15].

3.3. Result of TOPSIS approach

The expert team determines the criteria of each environmental, technological, economic and social factors for seaweed and grouper aquaculture development in Pulau Panjang. Previous experience, previous observations and research from the expert team were used to determine the criteria. From each factor, there are five selected criteria so that the total criteria used are 20 criteria.

Meanwhile, the TOPSIS calculation based on the distance of each strategy option to the shortest distance from the positive ideal solution and the furthest distance from the negative ideal solution as well as the value of relative proximity to the ideal solution (Ci) is shown in table 1.

Table 1. Distance of strategic options to ideal positive solutions (S+) and ideal negative solutions (S-) as well as the value of relative proximity to the ideal solution (Ci).

Strategy Option	Environment	Technology	Economic	Social	S+	S-	Ci	Rank
Maintain	0.2095	0.2882	0.1895	0.0677	0.1185	0.2393	0.6687	1
Increase	0.1048	0.0913	0.1327	0.1524	0.1847	0.1048	0.3621	2
Optimizing	0.0873	0.0456	0.0948	0.1863	0.2393	0.1185	0.3312	3

3.4. Result of AHP method

Of the four supported factors, each factor is compared pairwise between factors to obtain priority factor in maintaining existing seaweed and grouper aquaculture. The result of pairwise comparison between environmental, technological, economic and social factors is shown in table 2.

Table 2. The result of pairwise comparison between environmental, technological, economic and social factors.

	Environment	Technology	Economic	Social	Weight	Rank
Environment	1	1.9792	1.1465	1.376	0.3214	1
Technology	0.5053	1	1.5886	1.1993	0.2446	3
Economic	0.8722	0.6295	1	1.9834	0.2538	2
Social	0.7268	0.8338	0.5041	1	0.1802	4
						CR = 0.0884

Analysis of hierarchical structure of priority strategies to maintain seaweed and grouper aquaculture in Pulau Panjang is conducted by using AHP approach presented in figure 2.

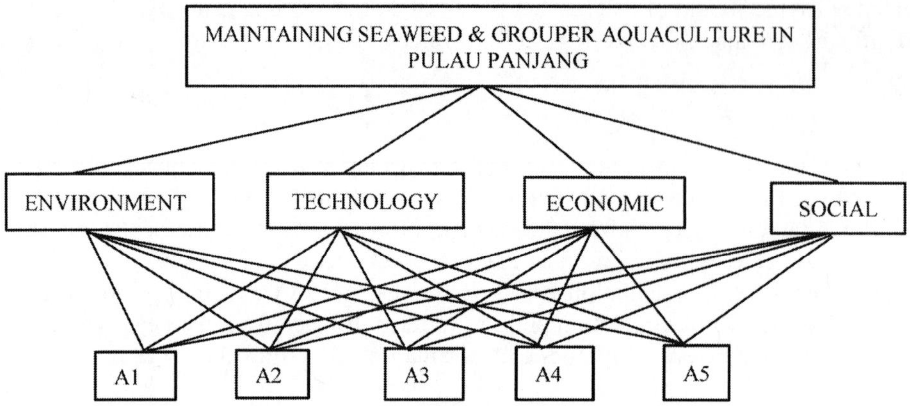

Figure 2. Hierarchy of priority strategies to maintain seaweed and grouper aquaculture in Pulau Panjang.

The alternative strategies to maintain seaweed and grouper aquaculture in Pulau Panjang by considering the priority factor of environmental is presented in table 3.

Table 3. Alternative strategy with environmental factors as a priority factor.

	Alternative	Descriptions
A1	Maintaining the water quality by the growth requirements of seaweed and grouper by encouraging the formation of "Environmental Community Awareness" where there is an active participation of the community to maintain the quality and carrying capacity of the environment	Seaweed and grouper have certain water quality threshold to be able to live and grow continuously.
A2	Integration and real implementation of heavy metal pollution control between government. industry sector and society	Pulau Panjang waters allegedly affected by heavy metal pollution from factories in the Bojonegara area which is characterized by high values of Pb and Cd.
A3	Preparation and implementation of planting season calendar of seaweed and seed grouper seedlings calendar	Seaweed disease attacks mostly occur during the rainy season and pests occur in certain predictable months so that these predictions can be used as a reference for planting seaweed and spreading grouper seeds to obtain maximum yield
A4	Using natural or artificial coastal protector (soft structure)	Water protection is essential to maintain the stability of oceanographic factors (currents, waves, winds) so that seaweed and grouper can be well developed. Planting mangroves as a natural coastal protector can also be useful as an eco-tourism location in Pulau Panjang.

A5	Provision of facilities and infrastructure of TPS (Temporary Disposal Site) and *cubluk* or septic tank for Pulau Panjang community	The availability of TPS and *cubluk* or septic tank is crucial to process the domestic waste of Pulau Panjang community before being discharged into the waters.

Meanwhile, the weight and rank of priority strategies to maintain seaweed and grouper aquaculture can be seen in table 4.

Table 4. The rank of priority strategies to maintain seaweed and grouper aquaculture.

Matriks	A1	A2	A3	A4	A5	Weight	Rank
A1	1	2.1077	3.4463	1.6158	0.8444	0.3009	1
A2	0.4745	1	2.3072	1.1734	1.6454	0.2076	3
A3	0.2902	0.4334	1	0.6077	0.6132	0.0962	5
A4	0.6188	0.8521	1.6454	1	3.1770	0.2313	2
A5	1.1841	0.6077	1.6305	0.3147	1	0.1640	4
							CR = 0.0813

4. Discussion

4.1. Strategy options for seaweed and grouper aquaculture development

From table 1 can be seen that the strategy option to maintain the existing seaweed and grouper aquaculture in Pulau Panjang has the biggest proximity (Ci) value of 0.6687. From the table it can be seen that the maintain option also has the shortest distance from the positive ideal solution (S+) with a value of 0.1185 and the furthest distance from the negative ideal solution (S-) with a value of 0.2393. So, the best strategic option to be implemented in Pulau Panjang is to maintain the existing seaweed and grouper aquaculture development. This result appropriate with interview result with respondent that the best option is to maintain the existing development of aquaculture based on their economic and social condition as well as environment and technology condition.

4.2. Identification of priority factors to maintain seaweed and grouper aquaculture in Pulau Panjang

Table 2 shows that environmental factor is a priority factor to maintain seaweed and grouper aquaculture activities with a weight value of 0.3214. Seaweed and grouper aquaculture activities need appropriate environmental and natural conditions to be able to live and grow well and continuously. The appropriate environmental condition is the initial factor of how seaweed and grouper aquaculture activities can run as expected. These conditions according to [1] and [16] can be regarded as a key factor in aquaculture activities. Therefore the selection of priority strategies to maintain seaweed and grouper aquaculture in Pulau Panjang is prepared by notice all criteria of environmental factors as a priority factor.

4.3. Recommendation of priority strategies

Table 4 shows that A1 priority strategy is the priority strategy with a weight of 0.3009. The priority strategy is to maintain water quality under the growth requirements of seaweed and grouper by encouraging the formation of "Environmental Community Awareness" where there is active participation of the community to maintain the quality and carrying capacity of the environment. Water quality must be maintained to the appropriate threshold for seaweed and grouper growth. Environment carrying capacity can be interpreted as the level of sustainable use of natural resources or ecosystems without causing damage to resources and environment [2].

Therefore, a social approach is needed to encourage the active involvement of the community in maintaining the water quality and environment, especially those directly related to the of seaweed and groupers aquaculture. The changes in water and environmental quality can have an impact on the production decrease of seaweed and grouper. This will certainly also affect the average income of seaweed and grouper farmers in Pulau Panjang in the future. Active involvement of the community can be formed by approaching influential figures in Pulau Panjang to be able to mobilize and increase public awareness of the importance of maintaining the environment carrying capacity in Pulau Panjang.

The second priority strategy is the A4 strategy, which is the application of natural or artificial coastal protector with a weight of 0.2313. Water protection to maintain the stability of oceanographic factors such as currents, waves and winds is essential for the continuity of seaweed and grouper aquaculture in Pulau Panjang. The ideal current velocity for the living conditions of seaweed and grouper is between 20–40 $cm{\cdot}sec^{-1}$ [17]. Mangrove can serve as a natural coastal shield where mangroves can serve as greenbelt to withstand waves naturally, play an important role as fauna habitat, protect the shoreline, spawning and nursery ground for some species of fish and crustaceans, natural wastewater treatment facilities and prevent pollution in coastal areas [18]. Mangrove also can support tourism potential where mangrove can become one of ecotourism area in Pulau Panjang.

The third priority strategy is A2 strategy which is integration and real implementation of heavy metal pollution control between government, industry sector and society with weight of 0.2076. Based on data on the number and type of industry located around Pulau Panjang according to [19] it can be predicted that these industries have hazardous wastes for the environment if not well managed such as heavy metal waste in steel industry, shipbuilding, sulphonation, organic waste from refined sugar manufacturing industries, heat waste resulting from the construction and operation of steam power plants and the disposal of natural gas pipelines as well as turbidity of waters from sand mining activities. From the results of the activities of these industries, [1] found that Pulau Panjang waters allegedly affected by heavy metal pollution from the factory in Bojonegara area is marked by the high value of Pb (0.207–0.214 $mg{\cdot}L^{-1}$) and Cd (0.01–0.2 $mg{\cdot}L^{-1}$). So that, the implementation of heavy metal pollution control especially for the affected locations in the waters of Pulau Panjang is needed immediately.

5. Conclusion

From the result of alternative determination strategy of seaweed and grouper aquaculture development in Pulau Panjang by using TOPSIS approach, the strategy is obtained to maintain the development of existing seaweed and grouper aquaculture.

Meanwhile, from the results of the pairwise comparison of environmental, technological, economic and social factors, it is seen that environmental factors are a priority factor in maintaining seaweed and grouper aquaculture activities in Pulau Panjang with a weighted value of 0.3214. Appropriate environmental conditions is the initial factor of how seaweed and grouper aquaculture activities can run as expected.

From the weight and rank of priority strategies in maintaining the activities of seaweed and grouper cultivation on Pulau Panjang, there are three main priority strategies. The priority strategy is to maintain water quality under the growth requirements of seaweed and grouper by encouraging the formation of "Environmental Community Awareness" where there is active participation of the community to maintain the quality and carrying capacity of the environment. The second priority strategy is the application of natural or artificial coastal protector. The third priority strategy is integration and real implementation of heavy metal pollution control between government, industry sector and society.

Acknowledgement

This research is a continuation of P A Soejarwo thesis that was supported by LPDP Scholarship Batch seven, Ministry of Finance. This research is partially funded by Botram Science Foundation, award number 2017001–A1. A Heriati is supported by Research Center of Marine, Agency of Research and Human Resources for Marine and Fisheries.

References

[1] Soejarwo P A 2016 *Thesis* Program Magister Teknik Lingkungan ITB Bandung
[2] Marzuki M 2014 *Dissertation* Program Pascasarjana IPB Bogor

[3] Soejarwo P A and Fitriyanny W P 2016 *Jurnal Kebijakan Sosial Ekonomi Kelautan dan Perikanan* **6**(2) 123–34
[4] Saaty T L 2008 *Journal Services Science* **1** 83–98
[5] Hwang C L and Yoon N 1981 *Multiple Attributes Decision Making Methods and Application* (Berlin: Springer-Verlag)
[6] Hsu P F and Hsu M G 2008 *J. Quality & Quantity* **42**(2) 181–201
[7] Wang T C and Chang T H 2007 *J. Expert Syst. Appl.* **33** 870–80
[8] Wang Y J 2007 *J. Expert Syst. Appl.* **34** 1837–45
[9] Morteza Z, Reza M F, Seddiq M M, Sharareh P and Jamal G 2016 *Journal Ocean and Coastal Management* **130** 179–87
[10] Astuty S and Diana S 2004 *Journal Agriculture* **14** 166–70
[11] BSN 2010 SNI 7579.2–2010 *Standar Nasional Indonesia* Badan Standardisasi Nasional Jakarta
[12] Sunyoto P 1994 *Pembesaran Kerapu dengan Karamba Jaring Apung* (Jakarta: Penebar)
[13] BSN 2006 SNI 01–7222–2006 *Standar Nasional Indonesia* Badan Standardisasi Nasional Jakarta
[14] WWF 2015 World *Wide Fund Indonesia* (Jakarta: WWF)
[15] Dinas Kelautan dan Perikanan Provinsi Banten 2015 *DKP Dalam Angka 2014* (Banten: DKP Provinsi Banten)
[16] Susilo S B 2003 *Dissertation* Program Pascasarjana IPB Bogor
[17] KKP 2014 *Kajian Dampak Penambangan Pasir Laut Pantai Utara Banten untuk Reklamasi Teluk Jakarta Terhadap Sumberdaya Laut dan Pesisir* (Jakarta: Badan Penelitian dan Pengembangan Kelautan dan Perikanan)
[18] Biswas S R, Mallik A, Choudhury J and Nishat A 2008 *Journal Wetlands Ecology and Management* **17** 365–83
[19] http://bplhd.co.id 2013 Accessed in 16 June 2016

Bioeconomic of profit maximization of red tilapia (*Oreochromis* sp.) culture using polynomial growth model

D Wijayanto[1*], F Kurohman[1] and RA Nugroho[1]

[1] Faculty of Fisheries and Marine Science, Diponegoro University Jl. Prof. H. Soedarto, S. H. Tembalang, Tembalang, Kota Semarang, Jawa Tengah, Indonesia
*Email: dianwijayanto@gmail.com

Abstract. The research purpose was to develop a model bioeconomic of profit maximization that can be applied to red tilapia culture. The development of fish growth model used polynomial growth function. Profit maximization process used the first derivative of profit equation to time of culture equal to zero. This research has also developed the equations to estimate the culture time to reach the target size of the fish harvest. The research proved that this research model could be applied in the red tilapia culture. In the case of this study, red tilapia culture can achieve the maximum profit at 584 days and the profit of Rp. 28,605,731 per culture cycle. If used size target of 250 g, the culture of red tilapia need 82 days of culture time.

1. Introduction

Red tilapia (*Oreochromis* sp.) is one of the major aquaculture commodities in Indonesia. Red tilapia culture is the main production of red tilapia with several types of culture methods. Red tilapia could be cultured use cages, floating cages, ponds and brackish water ponds. Red tilapia is relatively easy to adapt to the aquatic environment. However, culture of red tilapia should be careful because red tilapia is classified an invasive species in Indonesia. The growth average of tilapia production (including red tilapia) in Indonesia was 21.41 percent per year in the period of 2010–2014.

Table 1. The progress of tilapia production in Indonesia [1].

Years	Tilapia Production (Tonnes)
2010	461,191
2011	567,078
2012	695,063
2013	924,772
2014	999,615

Red tilapia culture has a strategic role in fisheries development of Indonesia, so the study of optimization to red tilapia culture is important, including profit maximization. In this research, the optimization of red tilapia culture was focused on monoculture and intensive culture. Research to learn the profit maximization of red tilapia culture could use the bioeconomic approach.

Fisheries bioeconomic is a combination of fisheries biology and economics science. The function of fish growth and the fish stock could be estimated using fisheries biology. While the maximization of profit could be estimated using economics, especially microeconomics (profit maximization). Research on aquaculture bioeconomics has been done by several researchers [2, 3, 4, 5, 6, 7]. Study of harvest optimization by [2], using the Beverton-Holt growth model and then further developed by [3, 5, 6]. A study of harvest time optimization using the von Bertalanffy growth model with the assumption of length exponent equal to three (isometric) [4, 7]. While this research using the polynomial growth model. The polynomial models have proven to be more accurate than the von Bertalanffy models in fish growth modeling [7, 12, 13].

2. Materials and Methods

This research object was focused on red tilapia culture to produce the fish consumption size. In this research, red tilapia growth model used polynomial function degree two for simplifying the model but still not ignoring the accuracy of the model. Stages of research have been including data collection and continued with data analysis using some of the equations described in the description below.

2.1. Data collection

This research used primary and secondary data. The primary data includes prices and costs by a survey to the red tilapia farmers in Magelang, Klaten and Wonosobo Regency. Secondary data includes statistical data and relevant references, including the biology of red tilapia.

2.2. Fish growth

The growth of red tilapia follows this polynomial equation:

$$W_t = at^2 - bt \tag{1}$$

Wt is the size of fish (g) at age t day, 'a' is intercept and 'b' is slope.

2.3. Costs, revenue and profit

In general, profit was the revenue minus cost. Revenue was influenced by the selling price and biomass of the harvested fish. Fish biomass was affected by the weight of individual fish and the number of fish still alive in the process of aquaculture. While the cost component includes the cost of seed procurement, labor costs, feed costs and procurement cost of facilities and production equipment. The cost component was converted to IDR per day. The quantity of artificial feed was influenced by food conversion ratio (FCR) and growth of fish biomass.

$$\pi = TR - TC \tag{2}$$
$$TR = Btb.Pi \tag{3}$$
$$Btb = Wtb.Ntb \tag{4}$$
$$Ntb = No - M.tb \tag{5}$$
$$TR = Wtb.Pi.\ (No - M.tb) \tag{6}$$
$$TC = Cp + Cb + Ctk + Cd \tag{7}$$
$$Cp = Pp.Qp \tag{8}$$
$$Qp = (Btb - Bo).FCR \tag{9}$$
$$Bo = No.Wtbo \tag{10}$$
$$Ctk = Ptk.tb \tag{11}$$
$$Cd = Pd.tb \tag{12}$$
$$tb = t - tbo \tag{13}$$

Notes:

π : profit at tb (IDR per culture cycle)
tb : culture time (days)
tbo : age of fish at initial culture (days)
TR : total revenue at tb (IDR per culture cycle)
Btb : fish biomass at tb (g)

Wtb : fish weight at tb (g per ind)
Pi : fish price (IDR per g)
Ntb : number of fish at tb (ind)
No : initial number of fish (ind)
M : average rate of mortalitity (ind per day)
TC : total cost at tb (IDR per culture cycle)
Cp : feed cost at tb (IDR)
Cb : seed cost (IDR per culture cycle)
Ctk : labor cost at tb (IDR)
Cd : depreciation cost of assets at tb (IDR)
Pp : feed price (IDR per g)
Qp : quantity of feed accumulation at tb (g)
Bo : initial fish biomass (g)
Wtbo: initial individual fish weight (gram/ekor)
FCR: feed convertion ratio
Ptk : labor cost per day (IDR per day)
Pd : asset depreciation per day (IDR per day)

2.4. *Profit maximization*

The first derivative of equation (2) to tb equal zero is be used to the profit maximization process as the first order condition or FOC ($d\pi/dtb = 0$) and the second order condition or SOC ($d^2\pi/dtb^2$) was negative. If we use equation (1) to (13), so could be solved the equation (14) and (15).

$$\pi = Btb.Pi - Cp - Cb - Ctk - Cd \tag{14}$$

$$\pi = g.tb^3 + h.tb^2 + i.tb + j \tag{15}$$

Notes:

$g = a.\ (Pp.FCR.M - Pi.M)$

$h = Pi.a.No + Pi.b.M + 2.Pp.FCR.a.tbo.M - 2.Pi.a.tbo.M - Pp.FCR.a.No - Pp.FCR.b.M$

$i = 2.Pi.a.tbo.No + Pi.b.tbo.M + Pp.FCR.a.tbo^2.M + Pp.FCR.b.No - Pi.a.tbo^2.M$
$\quad - Pi.b.No - 2.Pp.FCR.a.tbo.No - Pp.FCR.b.tbo.M - Pd - Ptk$

$j = Pi.a.tbo^2No - Pi.b.tbo.No - Pp.FCR.a.tbo^2.No + Pp.FCR.b.tbo.No + Pp.FCR.Bo - Cb$

The first derivative to equation (15) could be solved the equation (16):

$$d\pi(dtb)^{-1} = 0 = 3.g.tb^2 + 2.h.tb + i \tag{16}$$

Then, equation (16) could be used to find the optimal of tb by use the quadratic solution [8]:

$$tb_{1,2} = \frac{(-(2h) \pm \sqrt{(2h)^2 - 4.(3g).i)}}{2.(3.g)} \tag{17}$$

$$tb_1 = \frac{-(2h) + \sqrt{(2h)^2 - 4.(3g).i}}{2.(3.g)} \tag{18}$$

$$tb_2 = \frac{-(2h) - \sqrt{(2h)^2 - 4.(3g).i}}{2.(3.g)} \tag{19}$$

2.5. *The Target of fish size*

In research, we also study about harvest size base on market demand. If the market demand size is set as Wt, so by using equations (1) and (3), we can modify the equation as follows:

$$Wtt = a\ (tbo + tbt)^2 - b.\ (tbo + tbt) \tag{20}$$

Note:

Wt_t : target of fish size (g per ind)
tb_t : time culture to generate Wt_t

If Wtt could be found, so tbt also could be found. Equation (20) could be modified as follows:

$$Wtt = a.tbo^2 + 2.a.tbo.tbt + a.tbt^2 - b.tbo - b.tb \tag{21}$$

$$0 = a.tbt^2 + (2.a.tbo - b).tbt + a.tbo^2 - b.tbo - Wtt \tag{22}$$

$$tb_{t(1,2)} = \frac{-(2.a.tbo-b) \pm \sqrt{(2a.tbo-b)^2 - 4.a.(atbo^2 - b.tbo - Wt_t)}}{2.a} \tag{23}$$

3. Results and Discussion

3.1. Estimation of growth function

In this research, fish growth function could be estimated using the growth data of red tilapia in table 2. The result of growth estimation can be seen in figure 1.

Tabel 2. The growth of red tilapia [7, 9].

Age (days)	Individual weight (g)
90	24.83
100	48.97
110	73.10
120	98.00
130	124.67
140	151.30

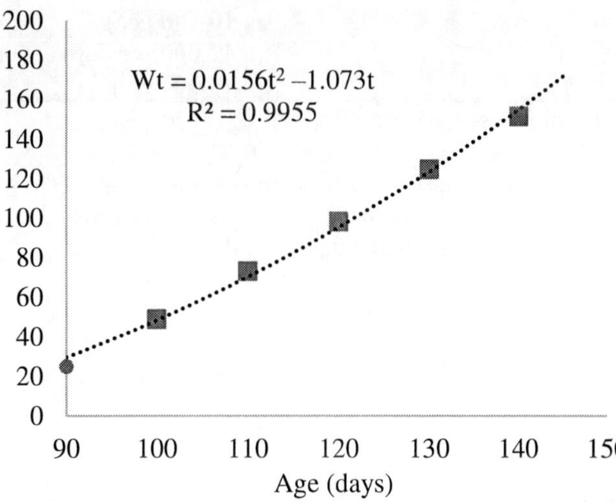

Figure 1. The growth estimation of the red tilapia.

The estimation of red tilapia growth is follow this equation:

$Wt = 0.0156t^2 - 1.073t$

$R^2 = 99\ \%$

Subject to:

$Wt, t > 0$

$Wt \leq W_{inf}$

Winf (maximum weight) is 3,744 g, Linf (max length) was 60 cm, and the relationship of length and weight follows the equation $0.02042\ L^{2.96}$ [10, 11].

In this research, red tilapia proved to be growing rapidly. On the red tilapia culture with a harvest target size of 250 g, the red tilapia fish was in a fast growth phase, so the use of polynomial growth model in this research was relatively suitable. The growth of red tilapia was slowing as it toward the weight of infinity.

The growth of red tilapia was influenced by water quality, artificial feed, seed quality, fish density, the age of fish and gender. Male red tilapia grow faster than female red tilapia. Red tilapia fish culture using floating cages can result in faster fish growth compared to red tilapia culture using ponds. It was caused by an adequate supply of oxygen to the floating cages. Dissolved oxygen was required for fish metabolism, including to support the growth of fish [7].

3.2. Optimal profit

In the application, there are diversity of red tilapia culture depend on the behavior of fish farmers. Several assumptions were be used in this research as follows the table 3.

Table 3. Research assumptions.

Assumptions	Values
a	0.0156
b	1.073
Initial weight (*Wtbo*)	20 g
Seed age (*tbo*)	84 days
Initial number of seeds (*No*)	6,000 ind
Mortality rate (*M*)	6 ind per hari (SR of 120 days is 88 %)
Fish price (*Pi*)	Rp. 17,500 per g
FCR	1
Feed price (*Pp*)	Rp. 10,000 per g
Depreciation (*Pd*)	Rp. 42,000 per days
Labor wage (*Ptk*)	Rp. 7,000 per days

Note: USD 1 equal to Rp. 13,343 at 30 April 2017

Based on the simulation, it was estimated that maximum profit at 584 days of culture with profit per cycle of Rp. 28,605,731 for 6,000 seeds. However, Indonesia fish farmers have the limited capital, so it was difficult for the fish farmer to applied 584 days of culture.

Table 4. Optimal culture of red tilapia.

Notes	Values
Profit equation	$\pi = -\,0{,}47.tb^3 + 422.tb^2 - 3.164.tb$
The first derivative of profit equation ($d\pi/dtb = 0$) as FOC	$d\pi/d\,tb = -\,1{,}404.tb^2 + 843.tb - 3.164 = 0$
Optimal time of culture (*tb*) use equation (19)	597 days (t = 681 days)
The second derivative of profit equation ($d^2\pi/dtb^2$ = negative) as SOC	$-\,2{,}808.tb{+}843 = (-\,2{,}8080{\times}597) + 843 = -\,832{,}53$ (negative, proven)
The fish weight at optimal profit	6,504 g (SR 40 %)
The fish biomass at optimal profit	15,726,572 g or 15.7 tonnes
The maximal of fish biomass	15,995,762 g (SR 35 %; tb = 651 days, t = 735 days)

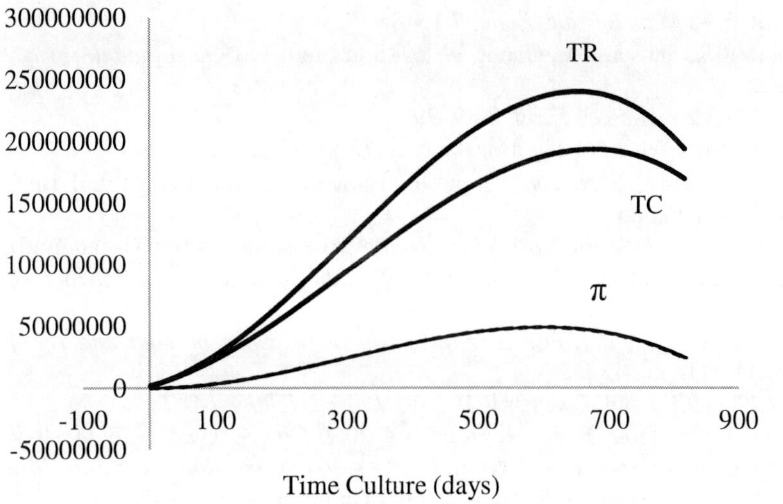

Figure 2. TR, TC and profit.

After reaching the maximum profit level, the red tilapia culture was still experiencing revenue growth. However, this revenue growth was smaller than the cost growth, so it caused a decrease in profits. We also applied this bioeconomic model to giant-gouramy and catfish culture [12, 13]. The giant-gouramy culture generated the maximal profit at 324 days and catfish (*Clarias* sp.) culture generated the maximal profit at 345 days.

3.3. Size of market

By applying equation (23), it could be estimated the culture time to produce a target of fish size. If the target size was 250 g, seed size of 19.9 g and seed age of 84 days, then the estimated culture time (tb_t) was 82 days (2.7 months). By using the equation (15), we could estimate the profit of Rp. 2,317,101 per culture cycle.

Fish farmers tend to prioritize faster turnover of capital with fewer profits than the maximum profits but slow in capital turnover. It caused by the limited capital of fish farmers and opportunity costs. If using 597 days of culture time to achieve the maximum profit, fish farmers will lose the opportunity to take seven cycles of cultures with an assumption of culture time per cycle was 82 days. Fish farmers will also had difficulty selling fish in the local market with a size too large compared to the target size.

4. Conclusion

Based on this research, we proved that this model could be used to estimate the optimal time culture of red tilapia. The profit of red tilapia culture follow the equation: $\pi = g.tb^3 + h.tb^2 + i.tb + j$. While the culture time at optimal profit to follow the equation: $tb = (-(2h) - (2h)^2 - ((4.(3g).i)^{1/2}))(2.(3.g))^{-1}$. In case of this study, red tilapia culture can generate the maximum profit at 597 days, with a profit of Rp. 28,605,731 per culture cycle. Then if the target of fish size is 250 g, it takes 82 days culture time.

Acknowlegements

Our team would like to thank for LPPM-Diponegoro University for funding this research through PNBP (RPP scheme) at the fiscal year 2016.

References

[1] KKP 2015 *Kelautan dan Perikanan Dalam Angka Tahun 2015* (Jakarta: Kementerian Kelautan dan Perikanan)
[2] Bjorndal T 1988 *Mar. Resour. Econ.* **5** 139–159
[3] Arnason R 1992 *Mar. Resour. Econ.* **7** 15–35
[4] Springborn R R, Jensen A L, Chang W Y B and Engle C 1992 *Aquacult. Fish. Manage.* **23** 639–647
[5] Heap T 1993 *Mar. Resour. Econ.* **8** 89–99
[6] Strand I and Mistiaen J A 1999 *Mar. Resour. Econ.* **13** 231–246
[7] Wijayanto D 2014 *Disertation* (Unpublish) Faculty of Economics and Business Diponegoro University Semarang
[8] Rosser M 2003 *Basic Mathematics for Economists* Second Edition (London: Routledge) pp 535
[9] Carman O and Sucipto A 2015 *Pembesaran Nila 2.5 Bulan* (Jakarta: Penebar Swadaya Publisher) pp 100
[10] Eccles D H 1992 *Field Guide to The Freshwater Fishes of Tanzania (Project URT/87/016)* (Rome: FAO) pp 145
[11] Froese R, Thorson J and Reyes Jr R B 2013 *J. Appl. Ichthyol* 1–7
[12] Wijayanto D, Kurohman F and Nugroho R A 2016 *Omni–Akuatika* **13**(1) 54–59
[13] Wijayanto D, Kurohman F and Nugroho R A 2016 *Prosiding Seminar Nasional Hasil–Hasil Penelitian Pascasarjana* (Semarang: FPK Undip Press) p 249–254

IOP Conf. Series: Earth and Environmental Science **139** (2018) 012041 doi:10.1088/1755-1315/139/1/012041

The use of preservatives consist of green tea, piper betel and potassium sorbate on boiled salted fish processing

F Ariyani[1]*, I Hermana[1] and I Hidayah[1]

[1] Research and Development Centre for Marine and Fisheries Product Processing and Biotechnology, Jl. K.S Tubun, Petamburan VI, Jakarta, Indonesia
*Email: idapoernomo@yahoo.co.id

Abstract. The main problem in boiled salted fish *ikan pindang* is mucus and mold on the surface of the fish which is produced relatively fast as well as the high level of histamine content especially when scombroid fish species are used as raw material. This study was performed to evaluate the effectiveness of various preservatives to overcome such problems. Three combinations of preservatives P1 (green tea and sorbate), P3 (green tea, piper betel, sorbate), P4 (green tea and piper betel) and P0 (no preservative/control) resulted from the previous study were used in this study. Before being used, the preservatives were tested against deteriorating microorganisms commonly found in boiled salted products, of which the result showed that all microorganisms were inhibited. The preservatives were then applied at three different stages of the process of boiled salted fish, i.e. before boiling, during boiling and after boiling. Sensory attributes and microbial characteristics of the products were then evaluated. The results showed that the performance of all tested preservatives against deteriorating microorganisms was relatively similar. It was also shown that the application before and during boiling performed better.

1. Introduction

Boiled salted fish *ikan pindang* is one of the traditional fisheries products which is popular in Indonesia as well as in ASEAN Countries. In the Philippines, this product is named *sianeng*, while in Thailand is named *pla-tu-nung* [1]. There are three kinds of boiled salted fish products available in the market, i.e. *pindang naya*, *pindang badeng* and *pindang presto*. The first two are the original *pindang* products of which the process involving heating the fish (usually scombroid fish) in solar salt (salt made from evaporation of seawater or other brine in the sun) and brine solution respectively. While *pindang presto* is a modified traditional process involving high-pressure cooking. Milkfish is commonly used as raw material and the product is vacuum packed.

In *pindang naya* processing, the fish is arranged in a bamboo basket and then boiled in 15 % brine solution for approximately 30 min. After being cooled, the product is transported and sold in the market. As the salt content is low, while the packing is not perfect, this product has a short shelf life (1–2 days). Bacterial and mould contamination during transportation and marketing contributed to the short shelf-life of the product resulting in slimy and mouldy products, stale taste and odor and even made the consumer feel itchy when scombroid fish is used.

Different methods of preservation have been used to extend the shelf-life of boiled salted fish such as low-temperature storage and proper packaging. Several natural products have also been studied to preserve boiled salted fish, such as guava leaf [2], chitosan [3], lemon grass [4] and green tea leave [5], all of which was applied as a single preservative.

Betel (*Piper betle* L) leave extract was reported as a potential antioxidant, antibacteria as well as antifungal agent [6, 7, 8, 9, 10, 11]. Betel leave extract was able to reduce histamine producing bacteria of fresh mackerel (*Rastrelliger kanagurta*) kept at ambient temperature [12]. Meanwhile, potassium sorbate was reported to be able inhibiting mold and yeast [13, 14] while sorbic acid could inhibit the growth of mold on smoked milkfish during refrigerated storage [15], and on fishloss [16].

All of the above studies used the preservatives as a single preservative. The existing study reports the used of multiple preservatives, i.e. green tea leave, betel leave and potassium sorbate to produce a better quality of *pindang naya*.

2. Materials and Methods

2.1. Materials

Commercial green tea was purchased from tea processing plant, Gambung, West Java. Betel leave was collected from Biofarmaka IPB and potassium sorbate was purchased from PT. Setia Makmur Jakarta.

The bacteria used in this study were *M. morganii* ATCC® 25830™, *K. pneumonia* ATCC®13883™, *P. fluorescen* ATCC® 13525™, *P. aeruginosa* ATCC® 10145™, *A. flavus* ATCC® 26949™ and *P. citrinum* ATCC® 9849™ collected from Food and Nutrition Culture Collection of IUC Food and Nutrition, Universitas Gadjah Mada, Yogyakarta.

Fish used as raw material of dried salted fish was skipjack (*Euthynnus affinis*) with the size of 22–25 cm long and 126–163 g weight, collected from a fish collector in Muara Baru, Jakarta.

2.2. Preparation of preservatives

Water extract of leaves was prepared according to Arambewela et al. [17]. Fresh leaves were washed to remove dust, debris and other foreign material, drained and dried at room temperature (27–32 °C). Dried leaves were then boiled in water with the proportion of dried leaves and water was 1:4 (w/v) for 4 h, cooled at ambient temperature and filtered using nylon fabrics of 30 mesh. The filtrate was packed in a dark plastic container and kept frozen until being used.

Preservatives were prepared by the formulation of green tea extract, betel leave extract and potassium sorbate to set three formulae, i.e. P1 (consists of green tea and sorbate), P3 (consists of green tea, piper betel, sorbate) and P4 (consists of green tea and piper betel).

2.3. Preparation of bacterial suspension

Each stock culture of bacteria (*M. morganii, K. pneumonia, P. fluorescen* and *P. aeruginosa*) were cultured on nutrient broth at 35 °C for overnight. After incubation, the culture was streaked on nutrient agar (slant agar), incubated at 25 °C for 3–5 days, and then stored in the refrigerator. The bacteria were then placed into a test tube filled with 10–15 mL nutrient broth, incubated for 24 h and then set the concentration of the inoculum to 0.5 McFarland's standards (ca. 10^8 cfu·ml^{-1}).

2.4. Preparation of fungi suspension

Each stock of fungi (*A. flavus* and *P. citrinum*) were cultured on Potato Dextrose Broth (PDB) media, incubated at 25 °C for 3–5 days. After incubation, the culture was streaked on Potato Dextrose Agar (slant agar), incubated at 25 °C for 3–5 days, and then stored in the refrigerator. The fungi were placed into a test tube filled with PDB, incubated for 48 h and then set the concentration of the inoculum to 0.5 McFarland's standards (ca. 10^8 cfu·ml^{-1}).

2.5. Antibacterial and antifungal assay

Antibacterial activity of the preservatives was tested against spoilage bacteria *P. aeruginosa* ATCC® 10145™ and *P. fluorescens* ATCC® 13525™ and against histamine forming bacteria *M. morganii* ATCC® 25830™ and *K. petricola* ATCC® 13883™, while antifungal activity was tested against *A. flavus* ATCC® 26949™ and *P. citrinum* ATCC® 9849™. Both antibacterial and antifungal activities were measured using disc-diffusion method [18]. Sterile paper discs (6-mm diameter) were soaked in the preservatives for 15 min and allowed to dry for 10 min. The inoculums suspension of 0.15 mL was swabbed uniformly on the surface of Mueller Hinton agar plates, allowed to stand 15 min and the disk was placed in different areas on the plate. They were incubated at 37 °C for 24 h for the test bacteria and at room temperature for 2–4 days for fungal strains. The blank disc was used as negative control, while antibiotic of chloramphenicol (30 μg) was used as positive control. The plates were duplicated in all the experiments.

2.6. Application of preservatives on boiled salted fish processing

Skipjack as a raw material of boiled salted fish was washed to remove debris and blood. The processing of boiled salted fish consisted of boiling of fish in a brine solution (with a salt concentration of 15 %) for 30 min, draining and storing at ambient temperature.

The application of the preservatives (P1, P3 and P4) on boiled salted fish processing in this study was conducted at three different stages of the process, as a) soaking solution for fish before boiling process, b) boiling solution during boiling process and c) soaking solution after boiling process. Duration of soaking was 30 min. The process of boiled salted fish without adding preservatives was used as a control.

Samples were evaluated for sensory attributes, including color and taste [19] on the day of processing, while microbial characteristics including histamine forming bacteria [20], TPC [21] and total fungi [22] were observed on the day of processing and after being stored for 3 days. The treatment experiment was conducted in three replicates.

3. Results and Discussion

3.1. Antimicrobial activity

In this present study, evaluation of antibacterial activity of various preservatives was conducted against spoilage bacteria, histamine-forming bacteria and fungi.

Antimicrobial activity of preservatives against spoilage bacteria and histamine forming bacteria was presented in figure 1 and figure 2 respectively, while that against mould was shown in figure 3.

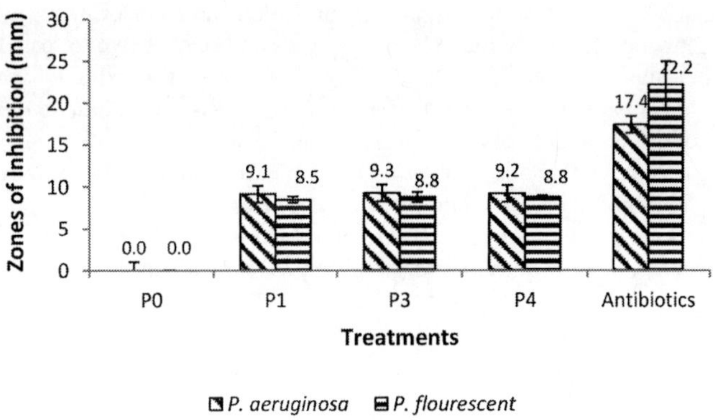

Figure 1. Inhibition zones of various preservatives on isolates of spoilage bacteria. (P0: no preservative/negative control; P1: green tea, sorbate; P3: green tea, piper betel, sorbate; P4: piper betel, sorbate).

Figure 1 shows that all preservatives were effective against spoilage bacteria *P. aeruginosa* and *P. fluorescens*, and there was no significant difference among preservatives in term of their effectiveness against these bacteria, although the effect was still lower than that of antibiotics chloramphenicol. It seems that *P. aeruginosa* was slightly more sensitive than *P. fluorescens*. Combination of green tea, betel leaves and potassium sorbate performed similar effectiveness to a combination of green tea and betel leave as well as green tea and potassium sorbate. Each component of preservatives has antibacterial activity against specific bacteria. Aqueous extract of betel leave has performed inhibition activity against *P. aeruginosa* with inhibition zones of 6 mm [23], while inhibition zone of 8 mm against the same bacteria was shown by 10 % betel leave extract [24]. In related to green tea extract, 10 % of aqueous green tea extract could inhibit *P. aeruginosa* with diameter inhibition zone of 12 mm and 9.4 mm, respectively [18, 24].

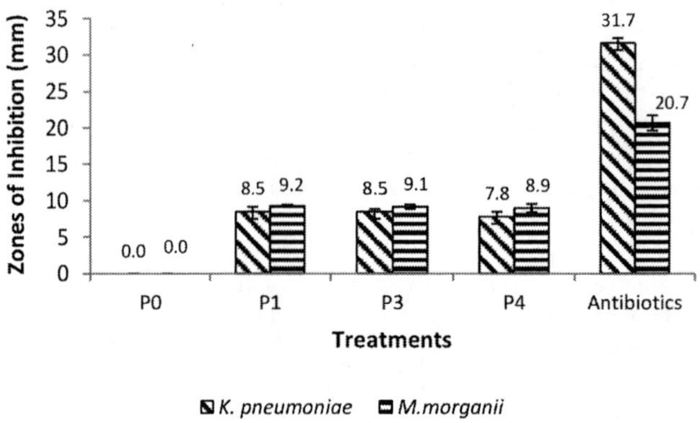

Figure 2. Inhibition zones of various preservatives on isolates of histamine forming bacteria. (P0: no preservative/negative control; P1: green tea, sorbate; P3: green tea, piper betel, sorbate; P4: piper betel, sorbate).

Related to histamine forming bacteria, the result of an experiment on disc-diffusion method indicated that all preservatives demonstrated zone of inhibition of 8.9–9.2 mm towards *M. morganii* and 7.8–8.5 mm against *K. pneumonia* (figure 2), however, their performance was still lower than antibiotics. It was shown in this study that *M. morganii* was more sensitive to tested preservatives than *K. pneumonia*. This supports the previous finding that green tea extract, one of the bioactive components in the preservatives was effective in inhibiting *Klebsiella pneumonia* [25]. Regarding betel leave, as another bioactive component present in this preservative, it was reported that nonaqueous extracts, namely ethanolic extract, petroleum extract and chloroform extract of betel leave were very effective in inhibiting *K. pneumonia* with a zone of inhibition of 23–33 mm [26]. Aqueous extract of green tea more effective toward *M. morganii* than *Klebsiella* spp., while betel leave showed comparable effectiveness towards both *M. morganii* and *Klebsiella* spp. [24].

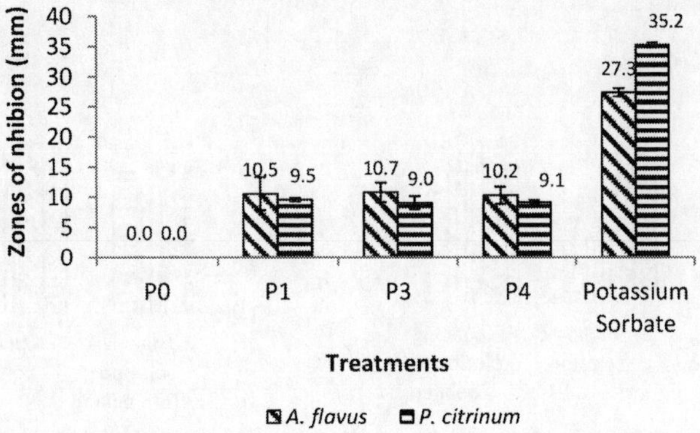

Figure 3. Inhibition zones of various preservatives on isolates of fungal. (P0: no preservative/negative control; P1: green tea, sorbate; P3: green tea, piper betel, sorbate; P4: piper betel, sorbate).

The performance of preservatives in inhibited fungi activity was shown in figure 3. *Aspergillus flavus* seems more susceptible than *P. citrinum* when treated with preservatives, on the other hand, potassium sorbate as control positive was more effective towards *P. citrinum* than that of *A. flavus*. The previous study showed that betel leave extract at 1 % completely inhibited the growth of *A. flavus* [27]. Green tea extract showed higher inhibitory effect towards *A. flavus* compared to *P. citrinum* [24], on the other hand, betel leave extract was more effective in inhibiting *P. citrinum* compared to *A. flavus*. Considering that *A. flavus* was more susceptible compared to *P. citrinum* in this study, it seems that the dominant effect of inhibiting microorganisms in this study was performed by green tea extracts.

3.2. Application of preservatives on boiled salted fish processing

3.2.1. Sensory characteristics. The results of the sensory assessment of boiled salted fish treated with various preservatives were shown in figure 4. The use of preservatives as a soaking solution before boiling as well as a soaking solution after boiling resulted in better color compared to the use of preservatives as a boiling solution. Preservatives used as boiling solution produced the darker color of boiled salted fish due to phenolic compounds present in green tea including catechin and its derivates [28] and also in betel leave such as chavicol, allylprotocatechol, chavibetol and hydroxychavicol [29]. During boiling process, these phenolic compounds are oxidized and further condensated to form components with a higher molecular weight having brown color [30].

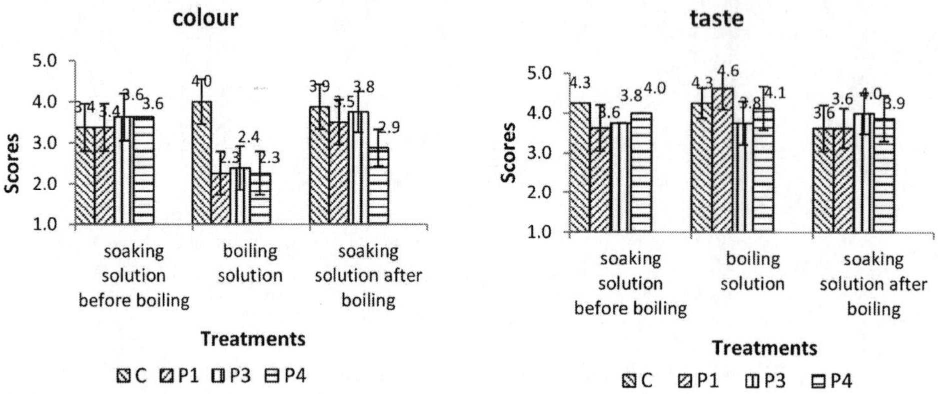

Figure 4. Sensory characteristics of boiled salted fish treated with preservatives. (P0: no preservative/negative control; P1: green tea, sorbate; P3: green tea, piper betel, sorbate; P4: piper betel, sorbate).

The taste score of boiled salted fish treated with preservatives was not much different from that of negative control/without preservatives (figure 4). Panelists rate the taste attribute of boiled salted fish in the range of 3.6–4.6 with a description of not bitter/not pungent-slightly bitter/slightly pungent. This indicates that addition of preservatives consisted of betel leave and green tea did not much effect on the taste of boiled salted fish. This could be due to the low concentration of preservatives used. It is known that sharp burning, pungent and astringent is the characteristics of betel leave taste [31], whereas green tea contained tannin have a bitter taste.

3.2.2. Microbiological characteristics

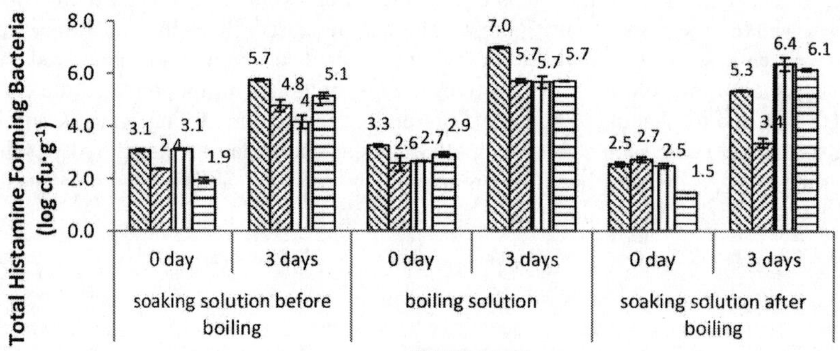

Figure 5. Total histamine forming bacteria count of boiled salted fish treated with preservatives. (P0: no preservative/negative control; P1: green tea, sorbate; P3: green tea, piper betel, sorbate; P4: piper betel, sorbate).

Based on microbiological analysis of boiled salted fish treated with preservatives, it was shown that soon after processing, total histamine forming bacteria slightly lower in boiled salted fish treated with preservatives compared to that of control (figure 5). After being stored, the histamine-forming bacteria of treated boiled salted fish was significantly lower compared to control, except for the use of preservatives as a soaking solution after boiling process. The role of green tea contained catechin might be responsible for lowering histamine-forming bacteria, as informed by a previous study [5].

Treatment of boiled salted fish with preservatives slightly decreased total bacterial count especially the use of preservatives as a soaking solution before boiling process (figure 6). However, after 3 days of storage, it was less effective since total bacterial count increased significantly reaching up more than 5 log cycle. Boiled salted fish produced by boiling in a brine solution for 30 min (*pindang naya*) could be kept only for 2 days [32].

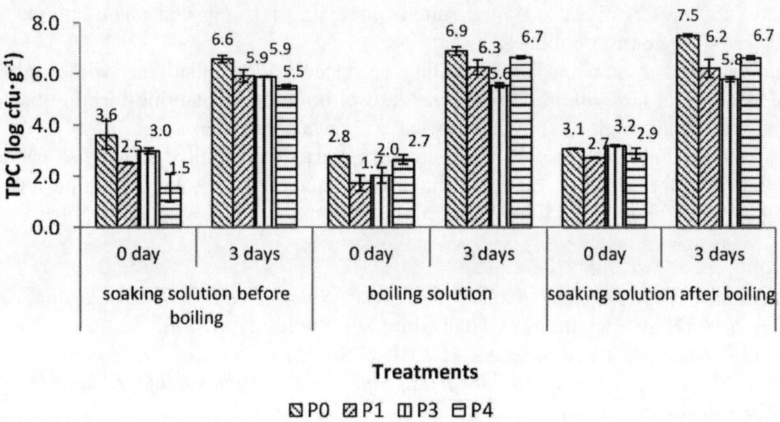

Figure 6. Total Plate Count of boiled salted fish treated with preservatives on a different step of processing. (P0: no preservative/negative control; P1: green tea, sorbate; P3: green tea, piper betel, sorbate; P4: piper betel, sorbate).

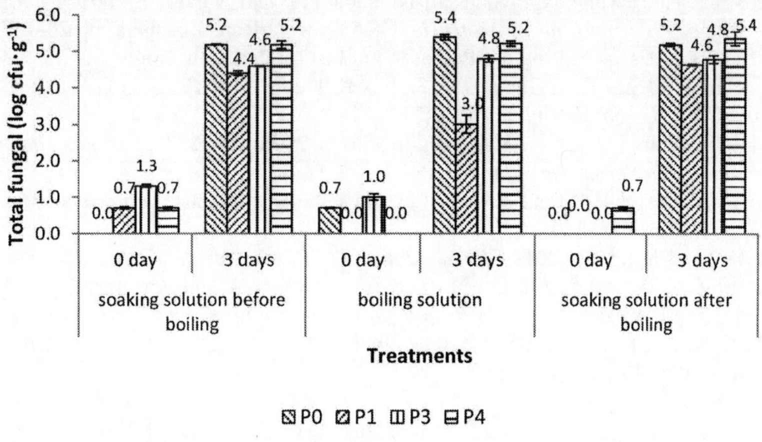

Figure 7. Total fungi of boiled salted fish treated with preservatives on a different step of processing. (P0: no preservative/negative control; P1: green tea, sorbate; P3: green tea, piper betel, sorbate; P4: piper betel, sorbate).

Figure 7 presented total fungi of boiled salted fish treated with preservatives. It is shown that boiled salted fish treated with preservatives (P1 and P3) contained lower total fungi although the products were stored for 3 days. It appears that the main active component for antifungal effects is potassium sorbate presence in preservatives P1 and P3. Potassium sorbate could inhibit fungi and yeast [13, 14]. This is in agreement with previous researchers [15], who prevented the growth of fungi on smoked milkfish by the use of sorbic acid.

4. Conclusion

In term of antibacterial and antifungal effects, the performance of all tested preservatives is relatively similar.

The use of the preservatives in boiled salted processing is suggested to be applied as a soaking solution before boiling or as a boiling solution.

Preservatives used as soaking before boiling produced boiled salted fish with slightly yellowish grey in color, no bitter taste and the lowest increase of both total histamine-forming bacteria (2.2 log number) and total fungi count (3.8 log number) after 3 days of storage.

Preservatives used as boiling solution was also suggested with boiled salted fish characteristics of brownish grey in color, no bitter taste and having an increment of total histamine-forming bacteria count of 2.9 log number and total fungi count of 4.0 log number after 3 days of storage.

References

[1] Gopakumar 1997 *Tropical Fishery Products: Some Traditional Dried And Smoke Cured Products* (New England: New Hampshire Sci. Publ. Inc.) p14–67
[2] Ariyani F, Murtin, J T and Siregar T H 2010 *JPBKP* **5**(1) 29–42
[3] Ariyani F and Yennie Y 2008 *Jurnal Pascapanen dan Bioteknologi Kelautan dan Perikanan* **3**(2) 139–146
[4] Murtini, J T, Ariyani F and Andayani F *2011 Seminar Nasional Tahunan VIII Hasil Penelitian Perikanan dan Kelautan* (Yogyakarta: Jurusan Perikanan Fakultas Pertanian UGM) pp 34 (1–8)
[5] Heruwati S H, Ariyani F, Triwibowo R, Rachmawati N and Hermana I 2009 *JPBKP* **4**(2) 161–168
[6] Jenie B S L, Andarwulan N, Puspitasari-Nienaber N L and Nuraida L 2001 *IFT Annual Meeting Technical Program Book of Abstracts* (USA: New Orleans Lousiana) pp 23–27
[7] Sugiastuti S 2002 *Thesis* Program Pascasarjana Institut Pertanian Bogor
[8] Ali I, Khan F G, Suri K A, Gupta B D, Satti N K, Dutt P, Afrin F, Qazi G N and Khan I A 2010 *Ann. Clin. Microbiol. Antimicrob.* **9**(7) 1–12
[9] Ariyani F, Amin I, Fardiaz D and Budiyanto S 2008 *Jurnal Pascapanen dan Bioteknologi Kelautan dan Perikanan* **3**(2) 157–169
[10] Himratul-Aznita W H, Mohd-Al-Faisal N and Fathilah A R 2011 *Journal of Medicinal Plants Research* **5**(6) 878–884
[11] Patel R M and Jasrai Y T 2013 *Cibtech. Journal of Bio-Protocols* **2**(2) 21–28
[12] Heruwati E S, Aryani F, Indriati N, Yenni Y, Riyanto R and Kusmarwati A 2006 *Laporan Teknis Pusat Riset Pengolahan Produk dan Bioteknologi Kelautan dan Perikanan* (Jakarta)
[13] Alrabadi N I, AL-Massad M and Gharaibeh A A 2013 *American-Eurasian J. Agric. and Environ. Sci.* **13**(11) 1497–1502
[14] Dharmadhikari M 2014 Sorbic Acid https://www.extension.iastate.edu/NR/rdonlyres/173729E4-C734-486A-AD16-778678B3E1CF/73968/SorbicAcid1.pdf Accessed on 25th Mei 2014
[15] Heruwati S H 1974 *Lembaga Teknologi Pengolahan* 37–45
[16] Indriati N and Putri A K 2011 *Seminar Nasional Tahunan VIII Hasil Penelitian Perikanan dan Kelautan* (Yogyakarta: Fakultas Pertanian Jurusan Perikanan Universitas Gadjah Mada) pp **27** 1–7

IOP Conf. Series: Earth and Environmental Science **139** (2018) 012041 doi:10.1088/1755-1315/139/1/012041

[17] Arambewela L, Arawwawala M and Rajapaksa D 2006 *International Journal of Food Science* **41** 10–14
[18] Archana S and Abraham J 2011 *Journal of Applied Pharmaceutical Science* **01**(08) 149–152
[19] Meilgaard M G V, Civille and Carr B T 1999 *Sensory Evaluation Technique* 3rd ed (New York: CRC Press Boca Raton London)
[20] Indriati N, Nata G A and Siregar R R 2008 *Prosiding Seminar Nasional Tahunan V Hasil Perikanan dan Kelautan* Jilid 3 Pascapanen Hasil Perikanan dan Bioteknologi Kelautan dan Perikanan
[21] Badan Standardisasi Nasional 2006 SNI 01-2332.2006 *Standar Nasional Indonesia* (Jakarta: Badan Standardisasi Nasional)
[22] Badan Standardisasi Nasional 2009 SNI 2332.7.2.2009 *Standar Nasional Indonesia* (Jakarta: Badan Standardisasi Nasional)
[23] Shukla R, Satish V, Singh V K, Kumar S, Gupta S and Gavani U 2009 *The Pharma Research* **01** 110–113
[24] Ariyani F, Hermana I and Hidayah I 2016 *Prosiding Seminar Nasional Hasil Litbang Pengolahan Produk dan Bioteknologi Kelautan dan Perikanan* (Jakarta: Pusat Penelitian dan Pengembangan Daya Saing Produk dan Bioteknologi Kelautan dan Perikanan Balitbang KP Kementerian Kelautan dan Perikanan) pp 209–216
[25] EL-Farmawi D, Olama Z and Holail H 2014 *Int. J. Curr. Microbiol. App. Sci.* **3**(3) 576–588
[26] Bangash F A, Hashmi A N, Mahboob A, Zahid M, Hamid B, Muhammad S A, Shah Z U and Afzaal H 2012 *J. App. Pharm.* **03**(04) 639–646
[27] Salim A B 2011 *Int. J. Acad. Res.* **3** 116–120
[28] Das S, Tanwar J, Hameed S and Fatima Z 2014 *Journal of Biochemical and Pharmacological Research* **2**(3) 167–174
[29] Bhalerao S A, Verma D R, Gavankar R V, Teli N C, Rane Y Y, Didwana V S. and Trikannad A 2013 *Journal of Pharmacognosy and Phytochemistry-An Overview* **1**(2) 10–19
[30] Haard N and Chism G W 1994 *Characteristics of Edible Plant Tissuesin Fennema O R Food Chemistry* 3rd ed (New York: Marcel Dekker Inc.) pp 943–1001
[31] Irani F 2005 *Aroma Therapy Journal* www.NAHA.ORG Accessed 12th Juli 2005 p 32
[32] Pandit I G S 2004 *Perbaikan Cara Pengolahan Ikan Pindang* (Denpasar: Jurusan Perikanan Fakultas Pertanian Universitas Warmadewa)

IOP Conf. Series: Earth and Environmental Science 139 (2018) 012042 doi:10.1088/1755-1315/139/1/012042

Chitinase activity of *Pseudomonas stutzeri* PT5 in different fermentation condition

N Chalidah[1], I N Khotimah[1], A R Hakim[1], B A Meata[1], I D Puspita[1]*, P S Nugraheni[1], Ustadi[1] and S Pudjiraharti[2]

[1] Department of Fisheries, Faculty of Agriculture, Universitas Gadjah Mada, Jl. Flora, Bulaksumur, Yogyakarta, Indonesia
[2] Research Unit for Clean Technology, Indonesian Institute of Sciences, Jl. Sangkuriang LIPI Complex Building 50, Bandung, Indonesia
*Email: indun_dp@ugm.ac.id

Abstract. This study aimed to determine the incubation condition of *Pseudomonas stutzeri* PT5 in producing chitin degrading enzyme in various pH and temperatures; to compare the production of chitin degrading enzyme in chitin medium supplemented with additional nitrogen, carbon and a mixture of nitrogen and carbon sources and to observe the production of chitin degrading enzyme in 250 mL-shake flasks and 2 L-fermentor. The parameters tested during production were chitinase activity ($U \cdot mL^{-1}$) of culture supernatant and N-acetylglucosamine concentration ($\mu g \cdot mL^{-1}$) in the medium. The results showed that *Pseudomonas stutzeri* PT5 was able to produce the highest chitinase activity at pH 6 and temperature of 37 °C (0.024 $U \cdot mL^{-1}$). The addition of 0.1 % of ammonium phosphate and 0.1 % of maltose, increased the chitinase activity of *Pseudomonas stutzeri* PT5 by 3.24 and 8.08 folds, respectively, compared to the control. The addition of 0.1 % ammonium phosphate and 0.1 % maltose mixture to chitin medium resulted in the shorter time of chitinase production compared to the addition of sole nutrition. The production of chitinase using 2 L-fermentor shows that the highest chitinase activity produced by *Pseudomonas stutzeri* PT5 was reached at 1-day incubation (0.0283 $U \cdot mL^{-1}$), which was shorter than in 250 mL-shake flasks.

1. Introduction

Chitin is the insoluble linear β-1,4 linked polymer of N-acetylglucosamine (NAG). After cellulose, chitin is the most abundant polysaccharide in nature and is primarily present in the exoskeletons of crustaceans such as crabs, shrimp, lobsters and also in various insects, worms, fungi and mushrooms [1]. Chitin is biocompatible, biodegradable and has antibacterial as well as wound-healing abilities. Furthermore, the hydrolyzed products of chitin such as oligomer, dimer and monomer have a very broad range of applications in different fields such as material science, microbiology, agriculture, waste water management, cosmetics, drug delivery system, food technology, biomedical, textile industry and much more. Chitin can be hydrolyzed to produce its oligomer, dimer and monomer fragments by a means of the chemical and biological process [2].

Generally, the chemical hydrolyzes of chitin are carried out by a strong acid such as hydrochloric acid (HCl). This procedure, however, produced low yield (below 65 %), costly and produced undesirable wastes which are one of environmental concern [3, 4]. Alternatively, the chitin hydrolysis can be performed bybiological processes, such as enzymatic reaction. Enzymatic hydrolysis of chitin can be performed by chitinase whichis specifically hydrolyzed β-1,4 linkage of NAG present in chitin polymer. Chitinase is produced by a various organism from plantsto bacteria [5].

Chitinase production by bacteria gives more advantage compared to other organisms [6]. Therefore the exploration of chitinase-producing bacteria from various environments has given considerable attention from researchers [7, 8]. Chitinolytic bacteria can be isolated by cultivating the bacteria on the selective chitin agar media. The chitinolytic bacteria will produce the clear zone surrounding the bacterial colonies [9]. *Pseudomonas* is the most prominent bacterial genus in chitinase production. It is a Gram-negative bacteria, aerobic and can be easily found in soil, water, plant and animal [10]. Chitinase activity of *Pseudomonas* has been reported else where [11, 12, 13]. Previously, we isolated *Pseudomonas stutzeri* PT5 from shrimp pond sediment located at PT. Indokor Bangun Desa, Yogyakarta, showing achitinolytic index of 1.63 and chitinase activity of 0.0002 U·mL^{-1} [14, 15].

Chitinolytic bacteria requires an optimal environment condition to produce a maximum chitinase, hence, the optimization of environmental factor such as pH, temperature and nutrient concentration in the medium (carbon and nitrogen) are necessary. Nitrogen is one of the amino acid constituent, the small unit of protein, that plays an important role in the microbial functional system, metabolism for growth and enzyme production [16]. Meanwhile, carbon helps to increase the energy source for bacteria to grow and produces enzyme [17]. Chitinase production can also be increased by a fermentation process using a fermentor because the fermentation condition such as agitation, aeration, dissolved oxygen, turbidity and foam formation can be controlled [18]. This study was aimed to determine the incubation condition of *Pseudomonas stutzeri* PT5 in producing chitin degrading enzyme, including pH, temperatures, additional nutrients and type of fermentors.

2. Materials and Methods

2.1. Chitin and colloidal chitin preparation

Chitin preparation was comprised of deproteination, demineralization and depigmentation [19]. After depigmentation process, chitin was washed with water until the pH becomes neutral, dried using theoven at 60 °C for 24 h, mashed using a blender then strained in 20 mesh sized sieve to obtain chitin powder. Colloidal chitin preparation was carried on by the addition of HCl and 4 °C cold water [19].

2.2. Inoculum preparation

Refreshment of *Pseudomonas stutzeri* PT5 from the glycerol-stock culture was done in a chitin agar medium and used as a working culture. Refreshment was performed using a streak method and incubated at 30 °C for 3–5 days. The inoculum was prepared by taking a single colony from working culture and grown in 7 mL of chitin broth incubated for 3 days at 37 °C. Chitin agar medium was prepared by mixing a 2 % colloidal chitin; 2 % bacto-agar; 0.07 % K_2HPO_4; 0.03 % KH_2PO_4; 0.05 % $MgSO_4$; 0.0001 % $ZnSO_4$ and 0.0001 % $MnCl_2$ with aquadest and sterilized [20]. Chitin broth was prepared in the same way as chitin agar without the addition of agar.

2.3. Chitinase production in various condition of incubation and additional nutrient in medium

Chitinase production was performed by inoculating a 225 µL of inoculum to 45 mL chitin broth in 100 mL. Erlenmeyer then incubated in water bath shaker for 5 days at 70 rpm. To observe the chitinase production at various temperatures, the pH medium was adjusted to six by the addition of 1 N HCl and temperature of incubation was varied into 30, 37 and 40 °C. Meanwhile, to observe the chitinase production in various pH (5, 6, 7 and 8), the temperature of incubation was maintained at 37 °C. The adjustment of pH was done by the addition of 1 N HCl or 1 N NaOH. Every 24 h, a culture filtrate was taken, centrifuged and analyzed to determine the chitinase activity. Incubation temperature and pH value of medium showing the highest chitinase activity was chosen as an optimum condition that was used afterward.

The chitinase production procedure described earlier was also performed to observe the production of chitinase in various source of 0.1 % (w/v) nitrogen (yeast extract, peptone, ammonium phosphate and ammonium oxalate) and 0.5 % (w/v) carbon (dextrose, maltose, glucose and starch) which was performed separately. Chitin medium without addition of other nitrogen and carbon source was used as the control medium. After obtaining the optimum source of additional nitrogen and carbon source in the production medium, the experiment proceeded to observe the effect of concentration. The chitinase production procedure described earlier was performed for producing a chitinase in various concentration of the best nitrogen (0.05 %; 0.1 %; 0.15 % and 0.2 %) and carbon (0.5 %; 1 %; 1.5 % and 2 %) source.

2.4. Chitinase production in 250 mL-shake flasks and 2 L-fermentor

The combination of the best source of nitrogen and carbon with the optimum concentration obtained previously was used as a production medium. The production in 250 mL-shake flasks was performed by adding 0.5 mL of inoculum in 100 mL of production medium and incubated at 100 rpm. The condition of optimum temperature and pH obtained from the previous experiment were maintained. To scale-up the chitinase production, the process was performed in fermentor jar by inoculating a 60 mL of inoculum to 1 L production medium in a 2-L fermentor then incubated for 5 days at 200 rpm agitation speed, 1 vvm aeration. Every 24 h, a culture filtrate was taken, centrifuged and analyzed to determine the chitinase activity.

2.5. Chitinase activity assay

Chitinase activity assay was performed according to [21]. The culture filtrate was centrifugated (6,000 rpm for 10 min at 4 °C). A total of 0.5 mL supernatant was added with 1 mL of 1.3 % of colloidal chitin (pH 7.4), incubated in water bath shaker at 37 °C for 30 min, boiled for 3 min and centrifugated at 10,000 rpm for 5 min. A total of N-acetylglucosamine (NAG) concentration was determined by taking 0.25 mL supernatant then added with 0.05 mL potassium tetraborate, boiled for 3 min and cooled in water [22]. The volume of 1.25 mL DMAB reagent was added then incubated in water bath shaker at 37 °C for 30 min then measured by a spectrophotometer colorimetric method, using a wavelength of 584 nm. The absorbance was compared with the standard curve of NAG. One unit (U) of the chitinase activity was defined as the amount of enzyme required to release 1 mmol of NAG from chitin per min.

2.6. NAG concentration assay

NAG concentration in the medium was determined by taking 1 mL of culture filtrate from the fermentation medium, then centrifugated (6,000 rpm at 4 °C for 10 min). A total of NAG concentration in the medium was determined according to the procedure described earlier [22].

3. Results and Discussion

3.1. Chitinase activity produced by Pseudomonas stutzeri PT5 in various pH medium and incubation temperature

The value of pH in the medium affects the environment condition for microorganisms to grow and produce the enzymes. Each species has a certain range of pH and optimum pH for growth. Microorganism needs to maintain a neutral cytoplasmic pH and for this, the plasma membrane may be relatively impermeable to protons. Drastic changes in cytoplasmic pH can harm microorganisms by disrupting the plasma membrane or inhibiting the activity of enzymes and membrane transport proteins. Prokaryotes will die if the internal pH drops below 5.0 to 5.5. External pH alterations also might alter the ionization of nutrient molecules and thus reduce their availability to the organism [23, 24]. Furthermore, the pH value of medium shows the total of H^+ ion existed in the medium. Total ion affects enzyme structure that consisted of amino acids, particularly on the hydrogen bond. Enzymes activity closely related to its structure. Changes in enzyme structure will interfere enzyme-substrate interaction, causing the alteration of enzyme activity [16].

Chitinase activity of *Pseudomonas stutzeri* PT5 culture supernatant produced in various pH on day-4 fermentation is presented in figure 1. The highest chitinase activity of 0.0028 $U{\cdot}mL^{-1}$ was reached at pH 5 followed by 0.0025 $U{\cdot}mL^{-1}$ of pH 6. The condition of pH 5 and pH 6 giving a favorable environment for *Pseudomonas stutzeri* PT5 to grow and produce the enzyme. We choose pH 6 for further experiment step because the activity was not significantly different from pH 5, moreover, the effort and resources required to prepare the pH medium of six are considered more efficient compared to pH 5. The result of enzyme activity was in accordance with the concentration of N-acetylglucosamine (NAG) in the medium, indicating that chitinase is secreted by the bacteria for hydrolyzing chitin in order to obtain carbon and nitrogen which is essential for cell growth [25]. The optimum pH obtained from this result was different from previous researchers [12], whose reported that the highest chitinase activity produced in chitin broth fermentation using *Pseudomonas stutzeri* YPL1 was obtained on the pH value of 7 with the maximum activity of 33.5 U·mL^{-1}, followed by pH 6, 8 and 5, respectively.

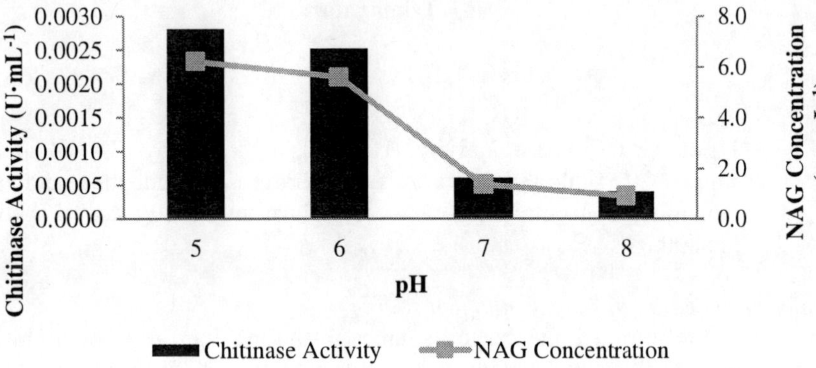

Figure 1. Chitinase activity (U·mL^{-1}) produced by *Pseudomonas stutzeri* PT5 cultured in various pH of liquid chitin medium and the remaining concentration of N-acetylglucosamine (µg·mL^{-1}) in the medium.

Chitinase activity of *Pseudomonas stutzeri* PT5 culture supernatant produced in various temperature on day-4 fermentation is presented in figure 2. The highest chitinase activity of 0.0024 U·mL^{-1} was reached at 37 °C incubation. The similar result was reported by previous researchers [26], that obtained the highest chitinase activity from *Pseudomonas pseudomallei* after 18 h fermentation at 37 °C reaching the value of 0.385 U·mL^{-1}. The maximum chitinase activity of *Pseudomonas aeruginosa* K-187 was produced at 37 °C [27]. The optimum range of temperature for *Pseudomonas* sp. to grow is 37–42 °C [28]. Temperature affects microorganism's growth by influencing the enzyme-catalyzed reactions. Beyond a certain point of higher temperature, slow growth takes place and damages the microorganisms by denaturing enzymes, transport carriers and other proteins. The plasma membrane is also disrupted as lipid bilayer simply melts and the damage is such an extent that it cannot be repaired. At very low temperature, membranes solidify and enzymes do not work rapidly. In summary, when organisms are above their optimum temperature, both the function and cell structure is affected [23, 24]. Moreover, temperature affects the reaction rate of the enzyme-substrate complex [16]. Figure 2 indicated that at a temperature of 37 °C, the reaction rate occurred faster giving the results of the highest concentration of NAG product in the medium. Chitinase production in various environment condition showed that optimum chitinase activity obtained at pH 6 and temperature 37 °C. The optimum condition of pH and temperature obtained in this stage were used to produce chitinase in the next step.

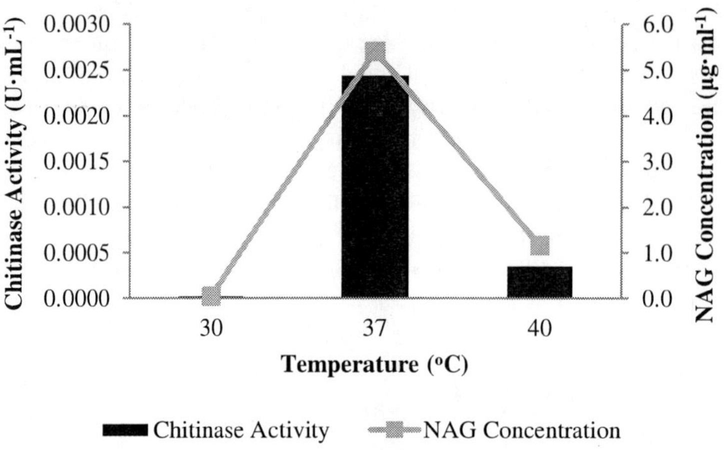

Figure 2. Chitinase activity (U·mL^{-1}) produced by *Pseudomonas stutzeri* PT5 cultured in various temperature of incubation and the remaining concentration of N-acetylglucosamine (µg·mL^{-1}) in the medium.

3.2. Chitinase production in various nutrient source

Additional of suitable nitrogen and carbon sources in the medium other than chitin, which is more complex in structure, supplies more ready-to-use nutrient which helps to increase the available energy for the microorganism to grow and produce the enzyme for degrading chitin. Figure 3 shows that ammonium phosphate produced the highest chitinase activity compare to other treatment, resulted to an activity of 0.031 U·mL^{-1} on day-3 of fermentation, meanwhile, control treatment produced highest chitinase activity of 0.0073 U·mL^{-1} on day-4 of fermentation. However, this value was lower than that reported previously [12]. They reported the superiority of ammonium phosphate compares to other inorganic nitrogen sources added to chitin medium for *Pseudomonas stutzeri* YPL-1 in producing chitinase after incubation of 84 h at 30 °C. Chitinase activity produced by *Pseudomonas stutzeri* YPL-1 in chitin medium with the addition of ammonium phosphate, ammonium nitrate, ammonium sulfate, ammonium chloride, sodium nitrate and potassium nitrate were 21.4; 13.4; 11.4; 7.2; 1.4 and 0.8 U·mL^{-1}, respectively. However, this value is lower than that of produced with the addition of organic nitrogen, namely peptone, tryptone and asparagine, with the activity of 32.8; 24 and 22.8 U·mL^{-1}, respectively. The increase in chitinase activity of *Pseudomonas stutzeri* PT5, when added with ammonium phosphate, was followed by the increase of NAG concentration in the medium. It indicates that ammonium phosphate promotes the production of chitinase by *Pseudomonas stutzeri* PT5 to degrade chitin into NAG that would be used further as nutrition for growth.

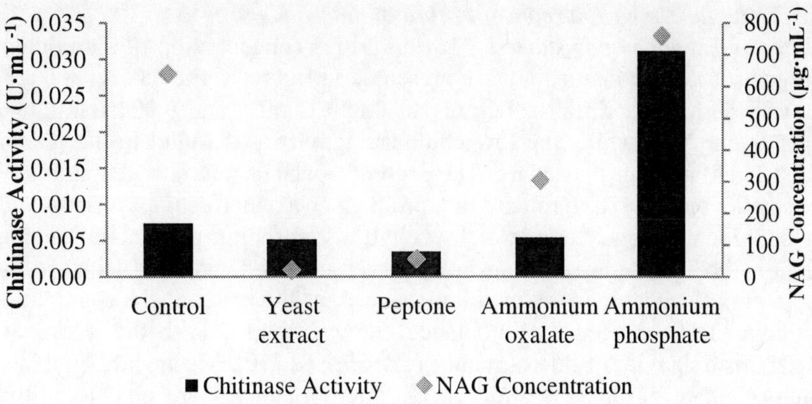

Figure 3. The highest chitinase activity (U.mL^{-1}) produced by *Pseudomonas stutzeri* PT5 during cultivation in liquid chitin medium added with 0.1 % of various nitrogen source phosphate and the remaining concentration of N-acetylglucosamine (µg·mL^{-1}) in the medium.

The highest chitinase activity produced by the addition of various carbon sources in the medium for all treatment was obtained at day-4 fermentations. Figure 4 shows that maltose addition to chitin medium produced the highest chitinase activity of 0.0173 U·mL^{-1} among the other carbon source. This result indicates that *Pseudomonas stutzeri* PT5 preferred to use maltose compared to other carbon sources that might relate to the effectivity of energy production by this carbon source. Other carbon sources may be acted as an anti-induction orinhibiting enzyme production. The preference of bacteria against certain sugar compound is controlled by the Carbon Catabolite Repression (CCR) mechanism [29]. The preeminence of maltose in inducing chitinase production in *Pseudomonas stutzeri* YPL-1 was also reported previously [12]. After 84 h incubation at 30 °C, chitinase activity produced by *Pseudomonas stutzeri* YPL-1 in basal medium containing chitin, maltose, glucose, N-acetylglucosamineand starch were 31.2; 25; 24.2; 17.8 and 16.2, respectively.

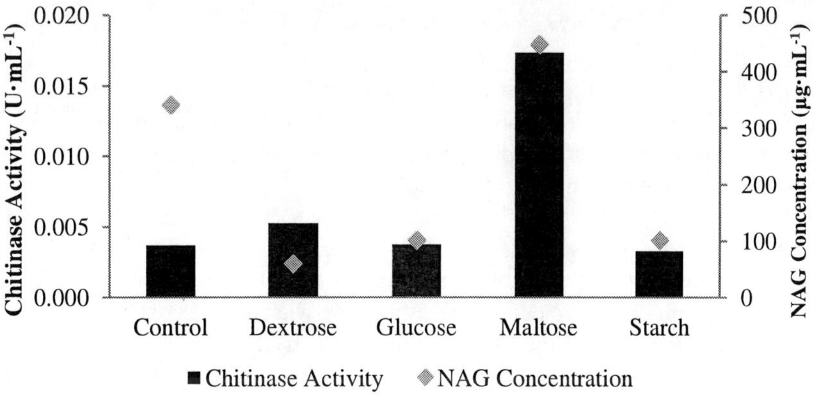

Figure 4. The highest chitinase activity produced by *Pseudomonas stutzeri* PT5 during cultivation in liquid chitin medium added with 0.5 % of various carbon source and the remaining concentration of N-acetylglucosamine (µg·mL^{-1}) in the medium.

3.3. Chitinase production in various concentration of nutrient source

Chitinase activity of *Pseudomonas stutzeri* PT5 in various concentration of ammonium phosphate as additional nitrogen sources in the medium are presented in figure 5. The 0.05 % and 0.1 % ammonium phosphate showed the highest chitinase activity of 0.030 U·mL^{-1} and 0.024 U·mL^{-1}, respectively, in day-4 of fermentation. Meanwhile, the low chitinase activity was found in the medium added with 0.15 % and 0.20 % ammonium phosphate. The presence of abundant nitrogen amount in the medium caused the synthetic process of fermentation products was ineffective. A large number of total nitrogen from additional nitrogen source or from hydrolyzed chitin products in medium (NAG) will accelerate the growth of bacteria resulting in an increase in cell biomass. However, as the biomass increases, total nutrient available in the medium reduced, as the result of the competition to obtain the nutrients. The competition caused the imbalance energy released with the intake of nutrients and eventually, metabolism activity of microorganisms decreased [30]. Meanwhile, in a low concentration of ammonium phosphate of 0.05 % and 0.1 %, microorganisms are able to utilize the nitrogen effectively resulting in an optimum growth. The optimal growth of microorganisms showed a balance between biomass cell and the availability of a nutrient in the fermentation medium.

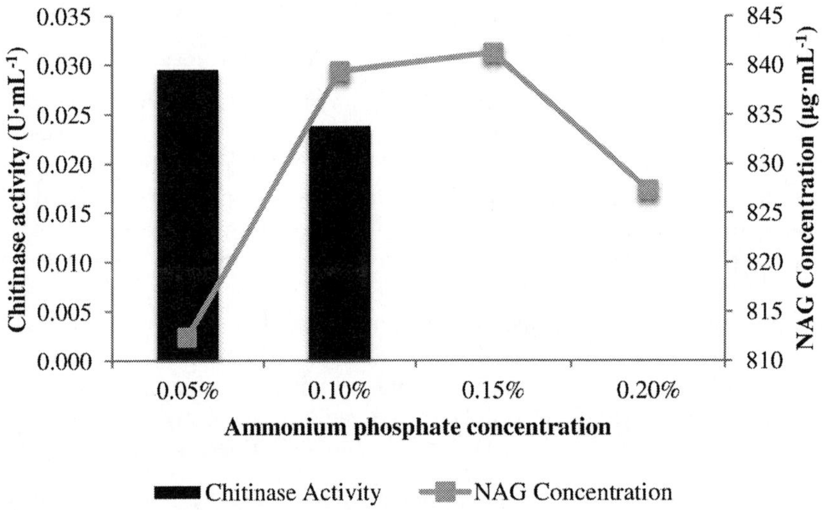

Figure 5. The highest chitinase activity (U·mL^{-1}) produced by *Pseudomonas stutzeri* PT5 during cultivation in liquid chitin medium added with a various concentration of ammonium phosphate and the remaining concentration of N-acetylglucosamine (μg·mL^{-1}) in the medium.

Chitinase activity of *Pseudomonas stutzeri* PT5 in various concentration of maltose as additional carbon sources in the medium are presented in figure 6. The highest chitinase activity produced by maltose addition in the medium for entire concentration was obtained at day-4 fermentations. The addition of 0.1 % maltose produced the highest chitinase activity 0.030 U·mL^{-1}. Maltose is a suitable carbon source for *Pseudomonas stutzeri* growth [31]. However, the reduction of chitinase activity was observed with the higher concentration of maltose. The role of maltose as a carbon source might relate to the ability of this sugar molecule to be transferred inside the cell for metabolism and growth. High concentration of maltose in the medium might cause the inhibition of chitinase gene expression. The relationship between sugar concentration and the mechanism of gene activation/shutting is controlled by the Carbon Catabolite Repression (CCR) mechanism [30].

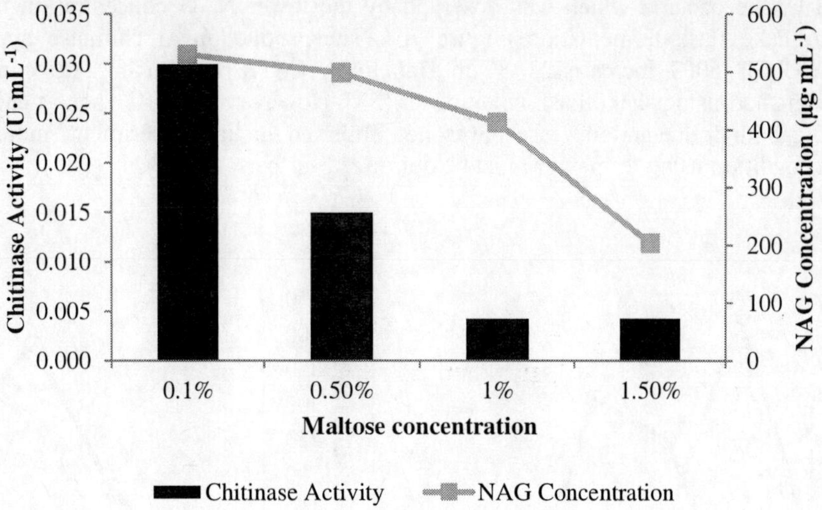

Figure 6. The highest chitinase activity (U·mL^{-1}) produced by *Pseudomonas stutzeri* PT5 during cultivation in liquid chitin medium added with a various concentration of maltose and the remaining concentration of N-acetylglucosamine (µg·mL^{-1}) in the medium.

3.4. Chitinase production in a combined nutrient source using shake flask and laboratory bench top jar fermentor

The combination of nutrient sources of nitrogen and carbon was added to the fermentation medium to produce a maximum chitinase by *Pseudomonas stutzeri* PT5. The combination of 0.1 % ammonium phosphate and 0.1 % maltose were chosen, as in this concentration (when solely added to chitin medium), the chitinase activity and NAG concentration in the medium were high. Chitinase activity in a combined ammonium phosphate and maltose as additional nutrient sources in the medium are presented in figure 7. The result shows that chitinase activity starts to increase in a day-2 of fermentation with the activity of 0.0167 U·mL^{-1} and reached the highest chitinase activity of 0.021 U·mL^{-1} on day-3 fermentations. Combined nitrogen and carbon source as an additional nutrient in chitin medium showed the lower chitinase activity compared to single additional of nutrient sources (0.030 U·mL^{-1} both for the single addition of 0.1 % ammonium phosphate or 0.1 % maltose), but shorter fermentation time of fermentation was observed. In a single addition of nutrient source, the highest activity was reached at day-4 fermentation, meanwhile, in a combined addition of nutrient source, the highest activity was reached at day-3 fermentations. Allegedly, a high concentration of substrate will become an inhibitor of the enzyme to hydrolyze the substrate. The appropriate ratio of combined nitrogen and carbon source required in chitin medium must be observed further in order to provide a nutrient balance for bacteria to grow (increase cell biomass) and to synthesize enzyme.

Microbial enzyme production in a large amount is usually prepared using a fermentor in which the factors that support microbial growth is controllable [32]. The time for *Pseudomonas stutzeri* PT5 produced the highest chitinase activity in a laboratory fermentor was shorter compared to shake flask Erlenmeyer (figure 7a). The highest chitinase activity observed in the culture supernatant of *Pseudomonas stutzeri* PT5 cultured in the laboratory fermentor (0.028 U·mL^{-1}) was reached at the first day of incubation, meanwhile, the highest chitinase activity in the shake flask erlenmeyer (0.021 U·mL^{-1}) was observed after 3 days of incubation. The shorter time of fermentation was reached using a fermentor because other factors that supported the growth of *Pseudomonas stutzeri* PT5 was controlled, such as agitation and aeration. An optimum condition for microbial growth caused the chitin hydrolyzing process occurred faster. This condition implied to the faster rate of NAG consumption by the bacteria which was observed by the lower NAG concentration in the medium compared to shake flask fermentation (figure 7b). The production of chitinase by *Lecanicillium muscarium* CCFEE 5003 increase 23 % in Bench-top CSTR Bioreactor 2 L compared to the conventional method using shake flask Erlenmeyer [33]. However, the difference in chitinase activity obtained using both fermentation vessels was not observed in this experiment, indicating that the fermentation condition using fermentor must be optimized further.

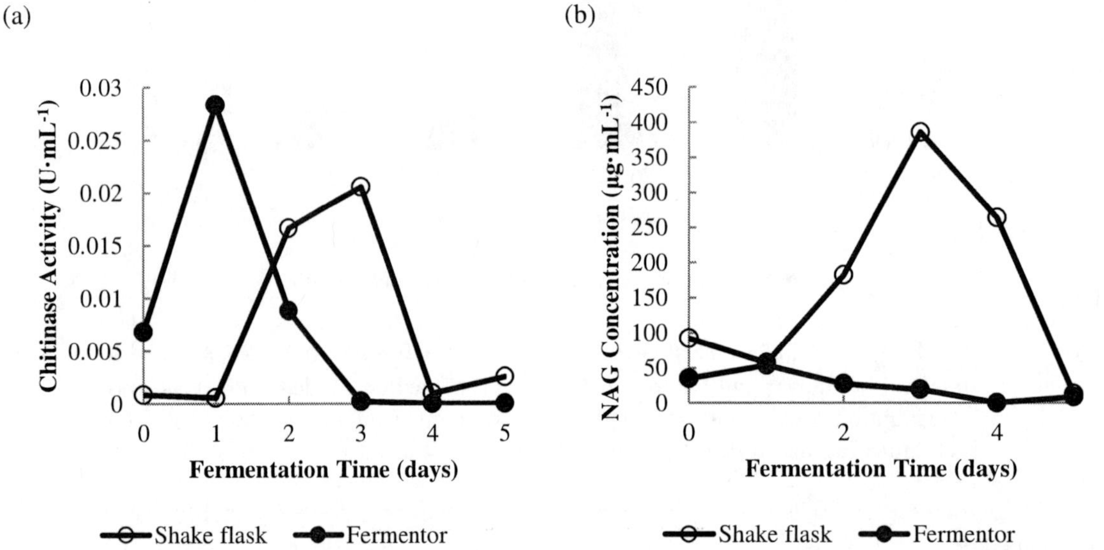

Figure 7. (a) Chitinase activity (U·mL^{-1}) produced by *Pseudomonas stutzeri* PT5 cultured in chitin medium added with 0.1 % ammonium phosphate and 0.1 % maltose using 250 mL-shake flask erlenmeyer (○) and 2 L-laboratory fermentor (●); (b) the remaining NAG (µg·mL^{-1}) concentration in the medium of shake flask (○) and fermentor (●).

4. Conclusion

This paper shows that the production of chitinase by *Pseudomonas stutzeri* PT5 was able to be optimized by the addition of suitable nutrient of nitrogen and carbon other than chitin in the medium. The combination of additional nitrogen and carbon other than chitin into medium resulted in the shorter time of chitinase production. Moreover, the production of chitinase using 2 L-fermentor was able to shorten the incubation time in order to reach the highest chitinase activity compared to 250 mL-shake flasks. It can be concluded that the production of *Pseudomonas stutzeri* PT 5 was affected by the condition of fermentation.

References

[1] Muzzarelly R A A 1977 *Chitin* (Oxford: Pergamon Press)
[2] Rinaudo M 2006 *Polymer Science* **31**(7) 603–632
[3] Ju-Hee K, Woo Jin J, Gyung-Hyun J, Joon-Seob A, Kil-Yong K and Ro-Dong P 2005 *Biotechnology Letters* **27** 7–11
[4] Sashiwa H, Fujishima S, Yamano N, Kawasaki N, Nakayama A, Muraki E, Hiraga K, Oda K and Aiba S 2002 *Carbohydrate Research* **337** 761–763
[5] Haliza and Suhartono M T 2012 *Bulletin Teknologi Pascapanen Pertanian* **8**(1) 1–14
[6] Bhattacharya A, Nagpure and Gupta R K 2007 *Critical Reviews in Biotechnology* **27**(1) 21–28
[7] Brzezinska M S, Jankiewicz U, Burkowska A and Walczak M 2014 *Curr. Microbiol.* **68** 71–81
[8] Kielak A M, Cretoiu M S, Semenov A V, Sorensen S J and van Elsas J D 2013 *Applied and*

Environmental Microbiology **79**(1) 263–272
[9] Muzarrelli R A A 1999 *Native, Industrial and Fossil Chitins* (Basel: Springer Basel AG)
[10] Siegrist J 2010 *Pseudomonas: A Communicative Bacteria* (St Louis: Sigma-Aldrich)
[11] Velusamy P, Ko H and Kim K 2011 *Agriculture, Food and Analytical Bacteriology* **1** 15–23
[12] Ho-Seong L and Sang-Dal K 1994 *J. of Microbiology and Biotechnology* **4**(2) 134–140
[13] Wang S L and Chang W T 1997 *Applied and Environmental Microbiology* **63**(2) 380–386
[14] Kholifah A 2015 *Isolasi dan Identifikasi Bakteri-bakteri Kitinolitik dari Sedimen Tambak Udang* (Yogyakarta: Fakultas Pertanian Universitas Gadjah Mada)
[15] Triwijayani A U 2016 *Identifikasi Bakteri Kitinolitik dari Sedimen Tambak Udang dan Karakterisasi Gen Kitinasenya* (Yogyakarta: Fakultas Pertanian Universitas Gadjah Mada)
[16] Lehninger A 1982 *Dasar-Dasar Kimia* Jilid I (Jakarta: Penerbit Erlangga)
[17] Pandey A 1999 *Types of Fermentations and Factors Affecting It Biotechnology: Food Fermentation* (New Delhi: Educational Publisher) pp 383–426
[18] Satiawihardja B 1983 *Buletin Pusbantepa* (Bogor: FTDC Institut Pertanian Bogor) pp 62–70
[19] Hargono H, Abdullah A and Sumantri I 2008 *J. Reaktor* **12**(1)53–57
[20] Hsu and Lockwood J 1975 *Applied Microbiology* **29**(3) 422–426
[21] Wang S L, Lin B S, Liang T W, Wang C L, Wu P C and Liu J R 2010 *J. of Microbiology and Biotechnology* **20**(6) 1001–1005
[22] Reissig J L, Strominger J L and Leloir L F 1955 *J. of Biological Chemistry* **217** 959–966
[23] Pommerville J C 2014 *Fundamentals of Microbiology* 10th Edition (Glendale Community College)
[24] Tortora G J, Funke B R and Case C L 2010 *Microbiology an Introduction* 10th Edition (San Francisco: CA Pearson Benjamin Cummings)
[25] Reinhold B and Fritz T 2002 *FEMS Microbiology Letters* **209**(2) 141–148
[26] Purkan, Azizah B, Baktir A and Sumarsih S 2014 *Molekul* **9**(2) 128–135
[27] Wang S L and Chang W T 1997 *Applied and Environmental Microbiology* **63**(2) 380–386
[28] Brooks M 2012 *Changesin Cell Structure, Morphology and Activity of Streptomyces Venezuelae During The Growth, Shocking and Jadomycin Production Stages* (Canada: Microbiology Biochemical Technology Dalhousie University Canada)
[29] Parisutham V, Kim M K, Lee S K and Ghim C M 2012 *Rewiring Carbon Catabolite Repression for Microbial Cell Factory* (South Korea: Korean Society Biochemistry and Molecular Biology)
[30] Suresh P V 2012 *World Journal of Microbiology and Biotechnology* **28** 2945–2962
[31] Lalucat J, Antoni B, Rafael B, Elena G V and Noberto J P 2006 *Microbiology and Molecular Biology Reviews* **70**(2) 510–547
[32] Mejía-Saulés J E, Waliszewski K N, Garcia M A and Cruz-Camarillo R 2006 *Food Technology and Biotechnology* **44**(1) 95–100
[33] Fenice M, Barghini P, Selbmann L and Federici F 2012 *Microbial Cell Factories* 111–112

IOP Conf. Series: Earth and Environmental Science **139** (2018) 012043 doi:10.1088/1755-1315/139/1/012043

Prebiotic index evaluation of crude laminaran of *Sargassum* sp. using feces of wistar rats

A Chamidah

Department of Fishery Products Technology, Faculty of Fisheries and Marine Science, University of Brawijaya, Jl. Veteran, Lowokwaru, Malang, Indonesia
*Email: achamidah@yahoo.co.id

Abstract. Today, prebiotics are often added to food. This compound is a food substance which could not be digested, yet benefiting the host by selectively stimulating the growth or activity of one or more bacteria in the colon to improve the health of the host. One of the foodstuffs derived from algae, which could not be digested is laminaran (β-Glucan). The relationship between microflora with the added prebiotics was tested by measuring their prebiotic index, which is supported by total sugars and laminaran levels. The results showed that total sugar content of Laminaran Acid Extract (LAE) (9.075 %) was higher than that of Laminaran Modified Extract (LME) (7.355 %), while the laminaran level of LME (42.23 %) was higher than that of LAE (30.92 %). HPLC test result confirmed the presence of laminaran. The obtained prebiotic index values of LAE and LME were 1.29 and 2.10, respectively, with a negative index score for LAE prebiotic, yet positive one for LME in terms of probiotic from *Lactobacillus* group. Laminaran extract, especially LME, deserves to be regarded as a prebiotic candidate.

1. Introduction

Prebiotic has developed rapidly as result of consumer awareness growth about a link between health and diet. Diet is a major factor in treating digestive tract of healthy human [1], since many physiological processes related to health and disease of the host starts from here, resulting in the development of the concept of prebiotics [2]. Prebiotics are nondigestible compounds having some physiological effects on humans by selective stimulation of growth or bioactivity of beneficial microorganisms that are present in the intestines or lead therapeutical [3]. Prebiotics are food that cannot be digested, which benefits the host by selectively stimulating the growth or activity of one or more bacteria in the colon to improve the health of the host [4].

The higher demand for prebiotic results in the emergence of new prebiotics, one of the potential ones is algae. Polysaccharide polymers in algae are largely (approximately 40 %) comprised of non-starch polysaccharide (NSP) [5]. Typical carbohydrates in varieties of brown algae consist of laminaran [β-(1.3)-glucan], fucoidan, cellulose, alginate and mannitol [6]. Laminaran, also called β-glucan, is one of macroalgae food reserves located within the cell (cytoplasm) [7]. It is a polysaccharide reserving in some brown algae with varied structures, similar to laminaran in terrestrial plants and mushrooms cell wall [8].

One way to evaluate a prebiotic candidate is by looking at the prebiotic index value. Prebiotic index (PI) describes the relationship between changes in beneficial microflora with those in unwanted microbes associated with the initial concentration. The PI gives a quantitative score that describes the prebiotic effect [9]. Prebiotic index of crude laminaran could be tested using methods of Sanz et al.

[10]. The purpose of this study was to evaluate the prebiotic index of Laminaran Acid Extract (LAE) and Laminaran Modified Extract (LME), which is produced from brown algae *Sargassum crassifolium*.

2. Materials and Methods

2.1. Raw materials and medium
The raw material used in this study was brown algae *Sargassum crassifolium*, clumps having talli with length from 40–100 cm. It was obtained from Garut, West Java, in February 2009 which was harvested from the wild with unknown age. Its natural habitat was at a depth of 50–250 cm of tidal area. Chemicals used were Laminaran and Laminarinase (Sigma Chemical Co), sodium acetate buffer and DNS. Various selective media (Columbia agar + cysteine HCl, and propionic acid for Bifidobacteria; Rogosa agar and glacial acetic acid supplements for Lactobacilli; reinforced clostridial agar + novobiocin and colistin for Clostridium; TSA + kanamycin, haemin, vancomycin and lake horse blood for Bacteroides) were from Oxoid Ltd.

2.2. Preparation of LAE and LME
First algae powder was defatted (using ethanol, chloroform and final solution with acetone), then extracted using 0.9 M (1:16) H_2SO_4 solution, 70 °C for 2.5 h using a water bath shaker. Then performed filtration, resulting in supernatant and residue. The supernatant was precipitated with 96 % ethanol (1:2) then centrifuged resulting in filtrate and residue (pellet). The pellets were dried and ground as LAE. The H_2SO_4 extract residue was extracted using aquadest (1:10) with the same temperature and time, and the process was the same until the pellet was produced. The pellets were dried and ground as LME.

2.3. Preparation of fermentation medium
The fermentation medium was prepared by mixing 0.3 g trypticase, 30 mL solution of trace minerals, 30 mL of buffer solution, 30 mL of macromineral and 4.02 mL of reducing solution, and then suspended in distilled water to reach a total volume of 120 mL [11]. The obtained fermentation medium was then divided into two: a-100 mL used to dissolve the substrate (LAE, LME, Inulin and control) and a-20 mL used to dissolve the inoculum.

2.4. Preparation of inoculum
The inoculum was derived from the cecum contents of Wistar rats (200–250 g). Rats were anesthetized in a closed container, which has been saturated with vapor of sodium pentobarbital (60 $mg \cdot kg^{-1}$). Surgery was then performed to the treated rats to take their cecum contents, which were then weighed by 2.0 g and added to 20 mL of fermentation medium to reach sterile inoculum concentration of 100 $g \cdot L^{-1}$. The mixture was finally homogenized for 10 min in a stomacher.

2.5. Substrate preparation
Each of substrates (LAE, LME and Inulin) was weighed as much as 0.312 g and then diluted with 25 mL of the fermentation medium. The final concentration of substrate solution was 0.0125 %. Fermentation medium without additional carbon source was used as a control. Next, 4 mL of each substrate solution and control was transferred into 15-mL-test screw tubes. The number of tubes was according to with the number of measured hours, i.e. four tubes for each treatment. Then, tubes were covered and sterilized at 121 °C for 15 min.

2.6. Measurement of prebiotic index
Sterile substrate solution and control prepared previously (4 mL) was added to 1 mL of homogenous inoculum. The mixture was then incubated at 37 °C for 48 h. Fermentation medium without an

additional carbon source (control) was used as a zero value, while medium with additional inulin as carbon source was used as a standard (completely fermented substrate).

Plating of bacterial culture was carried out at 0 h and 12 h of fermentation to observe the growth of bacteria on the medium before and after fermentation by using the pour-plate method. From each tube,1 mL of culture was taken and plated on to selective agar media for colony counting. MRS-Raffinose was used for *Bifidobacterium*, Rogosa agar was used for *Lactobacillus*, Tryptone Bile X-glucuronide medium (TBX) was used for *E. coli*, and Perfringens agar was used for *Clostridium*, while Plate Count Agar (PCA) was used for Total Count. Incubation was then conducted for 48 h at 37 °C. Both *Bifidobacterium* and *Clostridium* groups of bacteria were incubated under anaerobic conditions by adding an anaerobic gas kits (Oxoid) in an anaerobic jar. After 48 h, the number of bacterial colonies was counted, and then the prebiotic index of each substrate was calculated using prebiotic index formula [9, 10].

PI= *[(Bif a / Bif b) / (Total a / Total b)] + [(Lab a / Lab b) / (Total a / Total b)] - [(Bac a / Bac b) / (Total a / Total b)] - [(Clos a / Clos b) / (Total a / Total b)]*

Note:

- Bif: the number of *Bifidobacteria* (at sample time) / number at the time of inoculation.
- *E.coli*: the number of *E. coli* (at sample time) / number at the time of inoculation.
- Lac: the number of *Lactobacilli* (at sample time) / number at the time of inoculation.
- Clos: the number of *Clostridia* (at sample time) / number at the time of inoculation.

2.7. Test parameters

The test parameters used in this research included total sugar [12], laminaran levels [13], the content of simple sugars by HPLC and prebiotic index and score prebiotic index.

3. Results and Discussion

3.1. Total sugar

Total sugar is the total content of whole sugars in algae. The total sugar of LAE (9.075 ± 0.01 %) was significantly different ($p < 0.05$) from that of LME (7.355 ± 0.107 %) (table 1). This is likely due to the acid solution used in the extraction process of LAE, which could hydrolyze all parts of the cell including cell walls so that all carbohydrate components possibly including fucoidan are counted. In addition, the produced polysaccharides likely were still crude Laminaran, so it is possible that there were sugars other than glucose which caused the higher value of total sugar. In general, the content of total sugars in algae consists of glucose, sucrose and fructose, as well as some other content in small amounts, such as an alcohol sugar, mannitol, dulcitol and sorbitol. Meanwhile, when the residue was extracted only using water (LME), then only glucose and a small amount of mannitol obtained so that its components were relatively homogeneous.

Table 1. Total sugar and levels of laminaran.

Parameters	Crude Laminaran	
	LAE	LME
Total Sugar (%)	9.075 ± 0.01 a	7.355 ± 0.107 b
Levels of Laminaran (%)	30.92 ± 0.01 a	42.23 ± 0.39 b

The total sugar content of *S. crassifolium* was much lower from *A. nodosum*, *F. veslculosus* and *S. longicruris* by 89.6; 84.1 and 99.1 %, respectively [14]. Likewise, *L. saccharina* using Black, Yvins and News methods by 15.59; 10.60 and 55.56 %, respectively [13]. The content of a polysaccharide is usually influenced by season, population age, species and geographic location [14, 15], while based on

[16], the harvested area was responsible for the discrepancies of polysaccharide along with its structures.

3.2. Levels of laminaran

Brown algae contain polysaccharides such as laminaran, fucoidan and alginate, which are different from those of algae classes. Laminaran is estimated by measuring the concentration of its glucose produced by enzymatic hydrolysis of laminaran [13]. Table 1 shows that the level of LME laminaran is significantly higher ($p < 0.05$) than those of LAE. Higher levels of laminaran at LME were supported by yield and reducing sugar (unreported result). Compared to the levels of commercial laminaran, both LME and LAE laminaran levels (42.23 ± 0.39 and 30.92 ± 0.01 %, respectively) were lower. It was supported by the analysis result by HPLC which also showed a lower value. The low value was caused by the compared compound was patent products, with different raw materials, harvest season and habitat of growing.

The laminaran content of brown algae varies according to season ranging between 0–35 % of the dry weight [17]. *L. japonica* harvested at 1 and 2 years was also contain a very low level of laminaran (< 1 %) [15]. Laminaran structure and composition vary according to the species of algae [18]. In addition, the content and structure of laminaran of *L. cichorioides* vary greatly depending on the season and the age, thus mature *L. cichorioides* is a rich source of laminaran. On the other hand, laminaran content of *F. evanescens* depends on where it grows.

3.3. Simple sugar content of crude laminaran

Analysis of neutral sugars on crude laminaran sample generated by acid hydrolysis process, either total or partial hydrolysis, is required to determine whether the sample contains glucose. Sugar content in the samples was tested qualitatively using HPLC. It was conducted by comparing the retention time of each peak shown by the sample chromatogram with a retention time of the peak of standard compound.

A sample of products obtained by hydrolysis on commercial laminaran showed five peaks on chromatograms (figure 1) with a retention time (tR) 8.61; 10.555; 10.845; 13.414 and 15.512, respectively. By comparing the retention time of these peaks with that of standard LAE and LME compound standard, it could be expected that peak-3 (10.845 min) belong to maltose while peak-4 (13.414 min) belong to glucose. It could be seen that peak of glucose was substantially greater than that of maltose, but other peaks could not be determined because of the absence of other neutral sugar standards. It is possible that the rest peaks are other saccharides constituting polysaccharides of brown algae, such as fucoidan, alginate, funoran and manuronan considering that the test sample was crude laminaran. But the existence of these other components did not interfere with the identification of the main components.

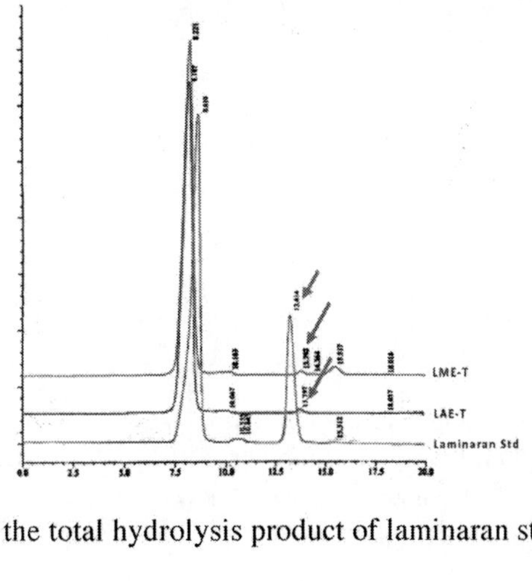

Figure 1. Chromatogram of the total hydrolysis product of laminaran standard, LAE and LME.

Analysis of neutral sugar content of LAE and total LME by total hydrolysis resulted in simple sugars monosaccharides, which were more heterogeneous than from that when only partially hydrolyzed. Investigation result on the composition of LME and LAE extracts by using HPLC showed that samples from all treatments contained glucose. This means that both extracts are laminaran compound. The content of glucose in the hydrolysate is associated with the presence of a laminaran substance in biomass [19]. This was also supported by previous study [20] showing that based on the composition of carbohydrate; glucose was the dominant component in laminaran with smaller amounts of mannit. Thus, it could be inferred that *S. crassifolium* contains laminaran.

3.4. Prebiotic index

Prebiotic index describes the relationship between changes in beneficial microflora with those in unwanted microbes associated with their initial levels. Based on Palframan equation, the increase in the population of *Bifidobacteria* or *Lactobacilli* is assumed as a positive effect, while the increase in Bacteroides and Clostridia (sub grouphistolyticum) is considered as negative effects. The number of bacterial colonies grown in the 0 h and 12 h of each treatment is shown in table 2.

As described in table 2, the number of *Bifidobacterium* colonies grown during 0 h to 12 h on a medium with a carbon source and without the addition of a carbon source showed an increase in average by 4 log cycle, whereas the *Lactobacillus* bacteria with the same conditions could only produce an increase in average by 3 log cycle. This indicates that both *Bifidobacterium* and *Lactobacillus* are equally capable of using crude laminaran (LAE and LME) and inulin as a carbon source for growth, but the ability to utilize crude laminaran of *Bifidobacterium* is better than that of Lactobacillus. Lactobacillus to take advantage of β-glucan (laminaran) must be able to break down the glycosidic bond between glucose, so glucose loses and can be used to stimulate growth. On the other hand, though *Bifidobacterium* could not utilize laminaran directly, cross-feeding possibly occurred. *Bifidobacterium* is not able to directly degrade laminaran, but the result of degradation laminaran by *Clostridium* and *Bacteroides* can be used as a carbon source for growth [8, 21]. This is the reason why the total growth of *Bifidobacterium* higher than *Lactobacillus* for all substrates.

Table 2. The growth of bacteria on various substrates at 0 and 12 hours incubation.

Bacterial	Substrate	Number of Bacteria (log 10 cells·g^{-1} feces)	
		0 h	12 h
Bifidobacterium	LAE	5.98	9.23
	LME	5.86	9.32
	Inulin	5.96	9.26
	Control	5.91	9.00
Lactobacillus	LAE	6.04	9.15
	LME	5.34	9.15
	Inulin	5.83	9.18
	Control	5.80	9.23
Bacteriodes	LAE	5.84	8.70
	LME	6.24	8.00
	Inulin	5.87	8.60
	Control	5.91	8.93
Clostridium	LAE	6.26	8.30
	LME	6.02	9.13
	Inulin	6.18	7.28
	Control	6.18	8.32

Table 2 shows that the substrate of crude laminaran (LAE and LME) was capable of supporting the growth of pathogenic bacteria, although its number is still lower than that of beneficial bacteria. This may be related to the factor of cross-feeding, where nature has set the balance between the microflora present in the colonic ecosystem. *Bifidobacteria* could not directly degrade laminaran. Only after it is degraded by Clostridia or Bacteroides, then the resulted compounds could be utilized by *Bifidobacteria*. Thus, the presence of both pathogens was still needed if their number could not exceed the balance.

Bifidobacteria and *lactobacilli* are known as organisms directly inhibit the growth of pathogenic bacteria, such as certain species of Clostridia [22]. Moreover, laminaran encourage the growth of cecal Bifidobacteria in rats [23]. Laminaran could suppress the growth of spoilage bacteria producing components in vitro [24]. Prebiotic effect of laminaran could be seen from an index value of its prebiotic. The quantitative calculation was used to facilitate fermentative analysis of prebiotics [9]. Compare of prebiotic index values of crude laminaran with that of other carbon sources and without the carbon source is showed in table 3.

Table 3. Average value of the prebiotic index of crude laminaran.

Substrate	Prebiotic index
LAE	1.29 ± 0.02 a
LME	2.10 ± 0.01 ab
INULIN	3.26 ± 0.02 b
CONTROL	1.06 ± 0.03 a

As shown by table 3, the PI value of LAE (1.29) and LME (2.10) indicated that both laminaran have the ability as a prebiotic, even if lower than inulin does (3.26). This is because inulin (β-(2,1) fructans linear) has selective colonic fermentation nature leading to real change in the microflora composition of the intestinal tract by increasing the number of bacteria which have potential to promote health and reduce the number and type of potentially harmful bacteria [25].

PI value of LME was higher than that of LAE. This is likely because the components contained in the LME are large compounds that are easily broken down by the intestinal microflora. The opposite occurred in LAE which contains mostly components that are not degraded by the intestinal microflora. This is supported by data of laminaran levels that were also higher in LME. In accordance with the mentioned formula, the provided treatment on both LAE and LME show their ability as a prebiotic although lower than the index of commercial inulin prebiotic.

The PI value of Wu-AX (water-insoluble wheat arabinoxylan unextractable) in 8 h fermentation by 2.03 ± 0.60 [26], when compared to the obtained laminaran PI value is not too different. Laminaribiose (G2) at 12-h of incubation could produce PI of 5.60 [10]. This much higher result is likely due to differences in raw materials, where the raw material used was not derived from algae but terrestrial plants, and due to different substrates used. POS (pectic polysaccharides), FOS (fructooligosaccharides) and OA (orange-albedo) at 10-h fermentation gave the higher value of PI (5.37; 6.32 and 4.9, respectively) [22]. Hence, we could conclude that the prebiotic index value of *S. crassifolium* algae is lower than terrestrial plants. However, there had been no publications found about PI values of other algae making it difficult to directly compare with the results obtained in this study.

The PI equation assumes that an increase in the populations of *Bifidobacteria* and/or *Lactobacilli* is a positive effect while an increase in bacteroides and clostridia (histolyticum subgroup) are negative [9]. If a bacterial group showed a relative increase more than the total population of bacteria, then such increase is considered > 1, but if the increase is relatively lower than total bacteria is considered < 1. This value was then added to the PI calculation to result in the overall scores. To determine the effect of laminaran prebiotics, then the PI score formula was used as displayed in table 4.

Table 4. Prebiotic index scores laminaran crude.

Medium	Bacterial	Calculation	PI Score
Control	*Bifidobacterium*	0.90 ± 0.05	< 1
	Lactobacillus	1.00 ± 0.02	> 1
LAE	*Bifidobacterium*	0.98 ± 0.01	< 1
	Lactobacillus	0.94 ± 0.04	< 1
LME	*Bifidobacterium*	0.97 ± 0.01	< 1
	Lactobacillus	1.15 ± 0.025	> 1
Inulin	*Bifidobacterium*	1.07 ± 0.035	> 1
	Lactobacillus	1.08 ± 0.031	> 1

From table 4, it could be inferred that the PI score of *Bifidobacterium* for crude laminaran (both LAE and LME) < 1, whereas the PI score of *Lactobacillus* > 1 only on the LME. In addition, the PI scores of Inulin > 1, which was simply because inulin is known as a prebiotic which has been tested. When examining PI scores from stirred pH-controlled batch cultures fermentation at pH 6.8 for 8 h of fermentation yielded HMP (high methylated pectin) and POS2 (pectic-oligosaccharide mixture) of 0.11 and 0.00 (PI score < 1), but when using LMP (low methylated pectin) and POS1 (pectic-oligosaccharide) yields PI values of 1.29 and 1.79 (PI > 1) [27]. When researching inulin from static batch culture fermentation of 1% for 5 h yields a score of 0.36 (< 1) and 24 h yields of 1.82 (> 1) score [28]. Thus, laminaran especially LME extract deserves to be regarded as a prebiotic candidate.

4. Conclusion

The resulted prebiotic index values obtained using Laminaran Acid extract (LAE) and Laminaran Modified Extract (LME) were 1.29 and 2.10, with a negative index score for LAE prebiotic and a positive score for LME prebiotic of *Lactobacillus* group. Laminaran especially LME extract deserves to be regarded as a prebiotic candidate.

References

[1] Vanhoutte T, de-Preter V, de-Brandt E, Verbeke K, Swings J and Huys G 2006 *Appl. Environ. Microbiol.* 5990–5997
[2] Vulevic J, Rastall R A, Gibson G R 2004 *FEMS Microbiol. Lett. 1.* **236**(1) 153–159
[3] Warrand J 2006 *Food Technol. Biotechnol.* **44**(3) 355–370
[4] Gibson G R and Roberfroid M B 1995 *J. Nutr.* **125** 1401–1412
[5] Lahaye M 1991 *J. Sci. Food Agric.* **54** 587–594.
[6] Dawczynski C, Schubert R and Jahreis G 2007 *J. Food Chem.* **103** 891–899
[7] Craeyveld V V, Swennen K, Dornez E, De Wiele T V, Marzorati M, Verstraete W, Delaedt Y, Onagbesan O, Decuypere E, Buyse J, de Ketelaere B, vedBroekaert W F, Delcour J A and Courtin C M 2008 *J.Nutr.* **138** 2348–2355
[8] Salyers A A, Palmer J K and Wilkins T D 1977 *Appl. Environ. Microbiol.* 1118–1124
[9] Palframan R, Gibson G R and Rastall R A 2003 *Lett. in Appl. Microbiol.* **37** 281–284
[10] Sanz M L, Gibson G R and Rastall R A 2005 *J. Agric. Food Chem.* **53** 5192–5199
[11] Goni I and Garcia-Alonso A 2000 *Eur. Food Res. Technol.* **211** 316–321
[12] Apriantono A, Fardiaz D, Puspitasari N L, Sedarnawati S and Budiyanto 1989 *Petunjuk Laboraturium Analisis Pangan* (Bogor: Dep. P.dan K. Dirjen Pendidikan Tinggi PAU Pangan dan Gizi IPB)
[13] Deville C, Damas J, Forget P, Dandrifosse G and Peulen O 2004 *J. Sci. Food Agric.* **84** 1030–1038
[14] Rioux L E, Turgeon S L and Beaulieu M 2007 *Carbohydr. Polym.* **69** 530–537
[15] Zyyagintseva T N, Shevchenko N M, Chizhov A O, Krupnova T N, Sundukova E V and Isakov V V 2003 *J. Exp. Mar. Biol. Ecol.* **294**(1) 1–13

[16] Percival E and McDowell R H 1967 *Chemistry and Enzymology of Marine Algal Polysaccharides* (New York: Academic Press) p 219
[17] Deville C, Gharbi M, Dandrifosse G and Peulen O 2007 *J. Sci. Food Agric.* **87**(9) 1717–1725
[18] Chizhov A O, Dell A, Morris H R, Reason A J, Haslam S M and McDowell R A 1998 *Carbohydr. Res.* **310** 203–210
[19] Usov A I, Smirnova G P, Kamenarska Z, Dimitrova-Konaklieva St, Stefanov K L and Popov S 2004 *Russ. J. Bioorg. Chem.* **30**(2) 161–167
[20] Yunizal 2004 *Teknologi Pengolahan Alginat* (Jakarta: Pusat Riset Pengolahan Produk dan Sosial Ekonomi Perikanan)
[21] Kuda T, Fujii T, Saheki K, Hasegawa A and Okuzumi M 1992 *Nippon Suisuti Gakkaishi* **58** 307–314
[22] Manderson M, Pinart K, Tuohy M, Grace W E, Hotchkin A T, Widmer W, Yadhav M P, Gibson G R and Rastall R A 2005 *Appl. Environ. Microbiol.* 8383–8389
[23] Kuda T, Yano T, Matsuda N and Nishizawa M 2005 *Food Chem.* **91** 745–749
[24] Kuda T, Goto H, Yokoyama M and Fujii T 1998 *Fish Sci.* **64** 583–589
[25] Roberfroid M B 2007 *J. Nutr.* **137** 2493S–2502S
[26] Vardakou M, Palop C N, Christakopoulos P, Faulds C B, Gasson M A and Nabad A 2008 *Int. Food Microbiol.* **123** 166–70
[27] Olano-Martin E, Gibson G R and Rastall R A 2002 *J. of Appl. Microb.* **93** 505–511
[28] Rycroft C E, Jones M R, Gibson G R and Rastall R A 2001 *J. of Appl. Microb.* **91** 878–887

ACE inhibition and antioxidant activity of different part of *Channa striata* prepared by various cooking method

E Chasanah[1]*, S Budiari[2], M Thenawijaya[2] and N S Palupi[2]

[1] Research Center for Marine and Fisheries Product Processing and Biotechnology, Jl. KS Tubun, Petamburan 6, Jakarta, Indonesia
[2] Food Science Program, IPB, Darmaga, Bogor, Indonesia
*Email: ekowatichasanah@gmail.com

Abstract. *Channa striata* (snakehead) extract has been known possessing positive activity, one of which is the ability to inhibit Angiotensin Converting Enzyme (ACE) activity *in vitro*. Aims of this study were to determine the effect of cooking and parts of *C. striata*, i.e. meat/fillet, gonad, skin, gill against the ACE inhibition activity and antioxidant activity *in vitro*. Heat processing methods used were direct boiling and indirect boiling and steamed at 100 °C for 10 min. ACE inhibition activity was analyzed using hippuryl-L-histidyl-L-leucine (HHL) as substrate and antioxidant activity was analyzed using DPPH method. The result shows that the higher the concentration of the extract (5 %, 20 %, 35 % and 50 %), the higher the antioxidant activity. The highest antioxidant activity was shown by gonad followed by meat extract, skin, and gill. Cooking treatment affected antioxidant activity, being the detrimental treatment were steam and direct boiling. The egg/gonad of *C. striata* showed the highest capability to inhibit ACE activity followed by meat/fillet, gill and skin. In concentration of 10 mg, extract of *C. striata* gonad was comparable to captopril, a commercial hypertension drug. While uncooked fillet showed the highest ACE inhibition activity followed by indirect boiling, direct boiling and steaming.

1. Introduction

Fish and seafood, in general, is well known as nutritious food containing high quality of protein and lipid important for health. Regular consumption of fish is important for children to combat malnutrition and for efficient prevention of several degenerative diseases for adult people. Snakehead or *Channa striata* is an indigenous freshwater fish species in Indonesia, and well known to have high nutritional value, i.e. excellence amino acid and lipid composition [1, 2, 3]. Rich in amino acid glycine and lipid arachidonic acid, *C. striata*, well known as *haruan* in Malaysia and *ikan gabus* in Indonesia, has been promoted to be used for postoperative wound healing, anti-inflammatory, antinociceptive, platelet aggregation, lowering blood glucose as well as mild antimicrobial and antifungal properties [4, 5, 6, 7]. *C. striata* also had been reported to have peptide released when the meat was hydrolyzed biologically, which was active against ACE or hypertension [8, 9].

Hypertension is becoming a global health concern including Indonesia. Hypertension or high blood pressure is responsible for one-quarter of all deaths in the world [10]. The World Health Organization reported that more than 60 % of strokes and 50 % of heart attacks are caused by high blood pressure [11], causing economic loss due to high-cost lifetime treatment. The Centre for Disease Control and Prevention (CDC) the USA has estimated that hypertension-related costs reached $ 76.6 billion in 2010 [11]. Hypertension or high blood pressure is related to Angiotensin I-converting enzyme (ACE; peptidylpeptide hydrolase, EC 3.4.15.1) activity that plays an important role in the regulation of blood pressure as well as in cardiovascular function [12]. ACE provides the powerful vasoconstrictor angiotensin II and removes the C-terminal dipeptide from the lysate of the precursor decapeptide angiotensin I. Increasing vasoconstriction and the development of high blood pressure were therefore as consequence the presence of high-level ACE in the body of patients. Extract of *C. striata* meat without hydrolysis was positively active against ACE *in vitro*, suggesting that *C. striata* extract was potentially used to combat hypertension [3].

In recent years, peptides have received more attention due to their activity against hypertension by preventing the ACE activity. Bioactive peptides derived from food proteins are safe compared to the synthetic hypertension drug but milder activity [13]. Peptides are defined as short-chain protein molecules (usually < 20 amino acid residues), therefore, in food rich in protein such as fish, peptides can be processed by hydrolyzing the protein. Bioactive peptides from biologically hydrolyzed snakehead fish myofibrillar protein and sarcoplasmic protein have been isolated [9, 14]. Using commercial protease, alcalase, two novel ACE inhibitor peptides LYPPP and YSMYPP with IC_{50} values of 1.3 and 2.8 mM were identified, respectively. Peptides from fish and shellfish have acknowledged lately as sources of natural antioxidant as well [15].

Antioxidants play an important role in human health due to their capability in scavenging the reactive oxygen species (ROS) such as superoxide anion ($O2^-$), hydroxyl radical (OH) and hydrogen peroxide (H_2O_2) that are normally produced in living organisms during metabolism. The antioxidant has been reported to have an important role in preventing cell injury [16]. ROS has been implicated taking a role in human disorders such as in diabetes, hypertension, atherosclerotic cardiovascular disease and cancer [13]. Most of the natural antioxidants reported were mostly non-protein compounds in plants.

When consuming *C. striata* for maximal health benefit, people do cooking usually by direct heating such as steaming. Cooking was also intended to enhance flavor and taste of food and inactivate pathogenic microorganisms. The objective of this work was to obtain the best tissue part of *C. striata* and the best cooking strategy to obtain *C. striata* extract with maximal ACE inhibition and antioxidant activity.

2. Material and Methods

2.1. Sample preparation and extraction

Wild *Channa striata* of 15 kg (average weight and the total length of 100–350 g and 28–34 cm, respectively) were obtained from lakes surrounding Parung, Bogor, West Java, Indonesia, and was life transported to the laboratory in a plastic bag. The fish was killed by icing them for 15 min as soon as the fish in the laboratory. Filleting was conducted to obtain fish flesh, meanwhile, gonad/egg, gill, and skin were separated. All preparation was conducted through cold chain system, and the meat, gonad, gill, and skin were frozen at -20 °C before following day's extraction.

Extraction was following Falkenberg et al. [17] with modification. Thawed meat, gill, egg/gonad and skin of *Channa striata* were minced in a food processor and 20 g of the meat was added with 60 mL of distilled water. The suspension was homogenized with Ultraturrax T25 (IKA Labortechnich) using the speed of 13,500 rpm for 5 min in ice. The suspension was centrifuged for 15 min at 10,000 g, 4 °C. The supernatant was used for treatment and analysis. The experiment was conducted in triplicates.

Fillet of *Channa striata* was heat processed by 1) direct boiling 10 min (in water), 2) indirect boiling 10 min (using a water bath) and 3) steaming at 100 °C for 10 min. Extraction was conducted as previously mention after heat treatment.

2.2. Protein content analysis

Protein content was analyzed using [18] against a standard curve of Bovine Serum Albumin (BSA) standard.

2.3. Measurement of antioxidant activity

Antioxidant activity was assessed using DPPH radical scavenging activity [19] with modification. Sample extracts at a concentration of 30 μg·mL^{-1} were tested for antioxidant activity using the 1,1-diphenyl-2-picrylhydrazyl DPPH method. DPPH (Merck) solution was prepared by solubilizing 3 mg DPPH in 5 mL methanol. Extract of 160 μL was added with 40 μL DPPH in a 96 microplate well and then incubated for 30 min in a dark room. The absorbance was read using microplate reader at 517 nm. Vitamin C was used (10 μg·mL^{-1}) as a control.

2.4. Measurement of ACE inhibitor activity

Inhibition activity of Angiotensin-I Converting Enzyme (ACE) was determined based on the formation of hippuric acid from hippuryl-L-histidyl-L-leucine (HHL) measured by UV spectrophotometer [20]. Sample (50 μL, 15 mg·mL^{-1}) added with 125 μL substrate buffer (containing 7.6 mM N-hippuric-his-leu hydrate and 608 mM NaCl in 10 mL borate buffer pH 8.3), and added with 15 μL BSA (10 mg·mL^{-1}). The mixture then incubated at 37 °C for 15 min using shaker water bath. The reaction started by addition of 50 μL ACE 50 mU·mL^{-1} and incubated for 30 min. The reaction was stopped by addition of 200 μL HCl 1 N, and after mixed using vortex, 1,140 μL ethyl acetate was added to extract hippuric acid. The mixture was centrifugated at 10,000 g for 10 min. The supernatant (1,000 μL) was dried (oven 95 °C for 75 min). After solubilized with 1,000 μL aquabidest, the absorbance was measured at 228 nm using spectrophotometer UV-Vis. The positive control used was captopril, a hypertension drug. ACE inhibition (%) was calculated as:

$$[(K-BK)-(S-BS)]/(K-BK)\times 100$$

Where K = absorbance of control; BK = absorbance of control blank; S = absorbance of sample; BS = absorbance of sample blank

All analyses were conducted in duplicates and the result was presented in an average of the duplicate measurements.

3. Results and Discussion

Our previous study showed that *C. striata* meat extract was active for ACE inhibition. Using 5 mg extract concentration, the extract was comparable with commercial hypertension drug, captopril (1–5 mg) in inhibiting ACE activity *in vitro* [3]. In this study, we used a higher concentration of the extract, i.e. 10 mg·mL^{-1}. Among tissue part of *C. striata*, egg extract showed the highest activity both for ACE inhibition and antioxidant (figure 1). In concentration of 10 mg, egg/gonad of *C. striata* showed comparable activity in inhibiting ACE as the captopril, followed by meat and gill extract and skin extract. Figure 2 showed the scavenging DPPH activity by different parts of *C. striata* extracts. Egg/gonad of *C. striata* also showed the best capability in scavenging DPPH, followed by meat and other parts.

The egg is dedicated to a new generation, therefore, the egg is composed of high-quality material and best environment condition for the egg to be grown. In the early incubation of the chicken egg, the chick embryo is exposed to considerably high oxygen tensions. Therefore, natural defense system in the form of antioxidant material effectively will develop and increase with increasing incubation time [16]. From this, it can be understood that *C. striata* egg contains the highest antioxidant naturally dedicated for optimal protection of the fish egg. Following egg, meat extract possesses the second highest antioxidant, and this might be due to the high nutritional quality of the *C. striata* as mention before. A study conducted by previous researchers [15], showed that water-soluble extracts of farmed sea bream meat (*Sparus aurata*) exhibited better DPPH scavenging activity of radical compared to the wild meat. From this result, it can be inferred that the water extract of fish such as sea bream and *C.*

striata especially in the egg and meat contain substance/material that was able to donate hydrogen and acted as electron donors and could react with free radicals to convert them to more stable products.

Cooking is intended primarily for obtaining healthy nutritious food. Fish meat becomes edible and more digestible when they are subjected to cooking. However, loss of the nutritional value of fish due to inappropriate heating will cause undesirable changes in some components of the protein and lipid. For snakehead, grilling was the best cooking for healthy eating of fish since the proximate and mineral were not affected [21]. Different fish such anchovy (*Engraulis encrasicholus*) was recommended also to be cooked by grilling and baking to obtain healthy consumption of anchovy [22].

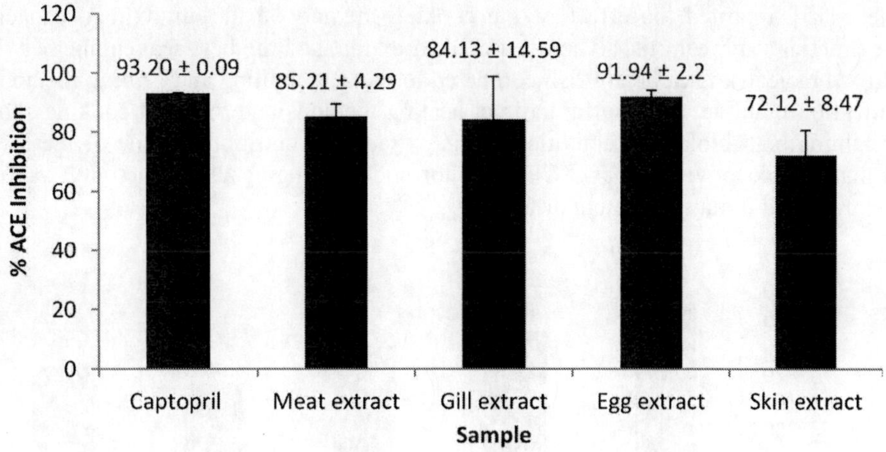

Figure 1. ACE inhibition by different *C. striata* tissue extract.

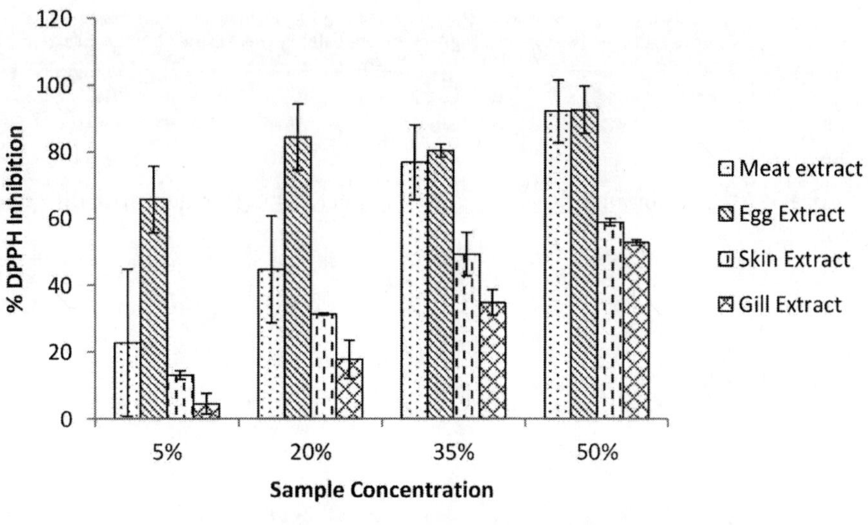

Figure 2. DPPH inhibition of different *C. striata* tissue extract.

In this study, cooking was proved to decrease biological activity of the extract against ACE inhibition. Heat processing methods used were direct boiling (in water), indirect boiling (using a water bath) and steaming, all at 100 °C for 10 min. Direct heat treatment such as steaming is practically applied when people cook snakehead for patients. Both boiling and steaming is a moist-heat cooking method; boiling uses convection to transfer heat from a hot liquid to the fish meat submerged in it; while steaming is transferred from steam to the fish meat being cooked by direct contact. When uncooked fillet accounted or 100 %, steam and direct boiling reduced the ACE inhibitor activity by about 23 % while indirect heat-treated meat lost 6 %. Accordingly, cooking treatment also affects antioxidant activity, being the detrimental treatment were steaming and direct boiling. Steaming and direct boiling reduces scavenging capability by 37 % (figure 3 and figure 4). Indirect boiling that used a water bath, i.e. the meat was in the beaker which was set having a water temperature of 100 °C, was the least affected. This might be related to the protein coagulation caused by heat treatment, where indirect heat-treated meat has undergone the least coagulation. Consequently, the water-soluble protein content of indirect heat-treated meat was the highest (table 1). A study conducted by previous researchers [15], reported the effect of the cooking method on the antioxidant capacity of water-soluble extract of sea bream fish. The oven-cooked extract had the best scavenging activity on DPPH and hydroxyl radicals of sea bream fish compared to those by boiling and grilling method. From these studies mentioned above, cooking by indirect heating methods and dry-heat cooking showed a better result, retaining both biological activities of the extract and nutritional value of the meat. It can be inferred that the bioactive such as ACE inhibitor and scavenging DPPH activity was in the water-soluble extract, and it may be protein or peptide.

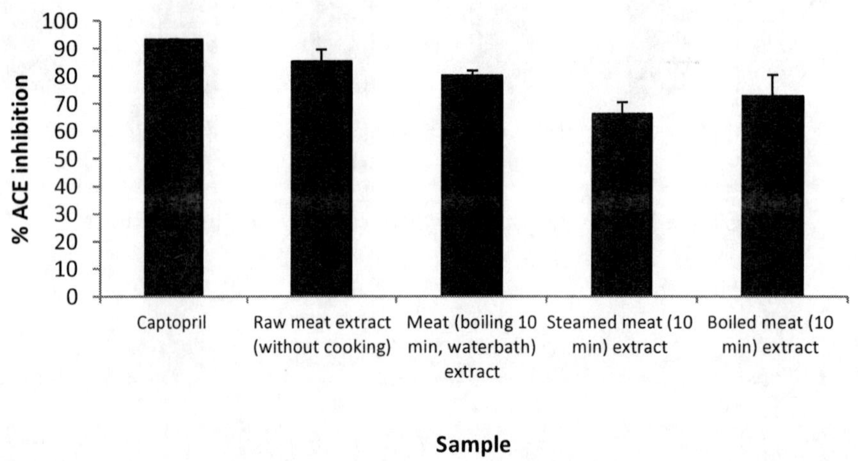

Figure 3. ACE inhibition of *C. striata* meat extract (10 mg) treated with different cooking methods.

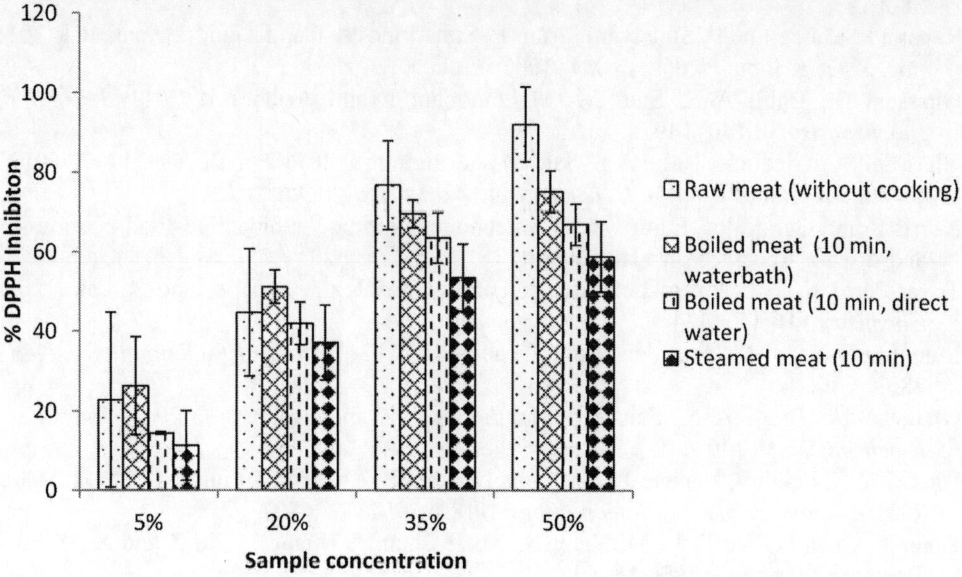

Figure 4. Effect of cooking methods on the ability of DPPH radical scavenging activity.

Table 1. The protein content of *C. striata* meat soluble fraction after heat treatment.

No	Heat treated fillet	Protein (%) ± SD
1	Boiling	4.809 ± 0.2439
2	Steaming	3.2349 ± 0.6950
3	Indirect heating (water bath)	7.6296 ± 0.6278

4. Conclusion

C. striata egg demonstrated the best activity of ACE inhibition and antioxidant activity followed by meat extract, while gill and skin part had lower activity. *C. striata* egg possessed comparable ACE inhibition activity with captopril in the same application concentration (10 $mg \cdot mL^{-1}$). Cooking was proved to decrease biological activity of the extract against ACE inhibition and scavenging DPPH activity. Steaming and direct boiling was not a good cooking method for *C. striata* as indirect boiling. Water extract of *C. striata* egg and meat might contain a source of antioxidant and ACE inhibitor substances that is potential for preventing hypertension and another related disease.

References

[1] Lay-Harn G, Chiuan-Yee L and Saringat B 2005 *Malaysian Journal of Pharmaceutical Sciences* **3**(2) 19–30

[2] Mustafa A, Sujuti H, Permatasari N, Widodo M A 2013 *International Journal of Science and Technology (IJSTE)* **2**(4)1–11

[3] Chasanah E, Nurilmala M, Purnamasari A R, Fithriani D 2015 *JPB Kelautan dan Perikanan* **10**(2) 123–132

[4] Zakaria Z A, Mat-Jais A M, Goh Y M, Sulaiman M R and Somchit M N 2007 *Clinical and Experimental Pharmacology and Physiology* **34** 198– 204

[5] Aisyatussoffi N and Abdulgani N 2013 *Jurnal Sains dan Seni Pomits* **2**(1) 2337–3520

[6] Haniffa M A K, Sheela P A J, Kavitha K and Mat-Jais A M 2014 *Asian Pac. J. Trop. Biomed.* **4**

S8–S15

[7] Rahayu P, Marcelline F, Sulistyaningrum E, Suhartono M T and Tjandrawinata R R 2016 *Asian Pac. J. Trop. Biomed.* **6**(12) 1001–1007

[8] Ghassem M, Babji A S, Said M, Mahmoodani F and Arihara K 2011 *Journal of Food Biochemistry* **38** 140–149

[9] Ghassem M, Arihara K, Babji A S, Said M and Ibrahim S 2011 *Food Chem.* **129** 1770–1777

[10] WHO 2011 *A Global Brief on Hypertension: World Health Day 2013*

[11] Norris R and Gerald R J F http://www intechopen com/books/bioactive-food-peptides-inhealth-and-disease Accessed on 11th July 2017

[12] Tovar-Pérez E G, Guerrero-Legarreta I, Farrés-González A and Soriano-Santos J 2009 *Food Chemistry* **116** 437–444

[13] Memarpoor-Yazdi M, Asoodeh A and Chamani J 2012 *Journal of Functional Foods* **4**(1) 278–286

[14] Ghassem M, Babji A S, Said M, Mahmoodani F and Arihara K 2013 *Journal of Food Biochemistry* **38** 140–149

[15] Bhouri A M, Fekih S, Hanene J H, Madiha D, Imen B, ElCafsi M, Hammami M and Chaouch A 2011 *African Journal of Biotechnology* **10**(84) 19623–19630

[16] Duan X, Ocen D, Wu F, Li M, Yang N, Xu J, Chen H, Huang L, Jin Z and Xu X 2014 *Food Research International* **56** 18–24

[17] Falkenberg S S, Mikalsen S O, Joensen H, Stagsted J and Nielsen H H 2014 *International Journal of Applied Research in Natural Products* **7**(4) 11–25

[18] Lowry O H, Rosebrough N J, Farr A L and Randal R J 1951 *Protein Measurement with The Folin Phenol Reagent* (Misssouri: Department of Pharmacology Washington University School of Medicine St. Louis)

[19] Li Y, Li X, Lee U K, Kang J S, Choi D H and Son B W 2006 *Chemistry Pharmaceutical Buletin* **54**(6) 882–883

[20] Arihara K, Nakashima Y, Mukai T, Ishikawa S and Itoh M 2001 *Meat Science* **57** 319–324

[21] Marimuthu K,Thilaga M, Kathiresan S, Xavier R and Mas RHMH 2012 *J. Food Sci. Technol.* **49**(3) 373–377

[22] Uran H and Gokoglu N 2014 *J. Food Sci. Technol.* **51**(4) 722–728

IOP Conf. Series: Earth and Environmental Science **139** (2018) 012045 doi:10.1088/1755-1315/139/1/012045

Physicochemical properties of chitooligosaccharide prepared by using chitosanase from *Stenotrophomonas maltophilia* KPU 2123

Y N Fawzya[1]*, A Rahmawati[2] and G Patantis[1]

[1] Research Centre for Marine and Fisheries Product Processing and Biotechnology, Jl. K.S. Tubun, Petamburan VI, Jakarta, Indonesia

[2] Syarif Hidayatullah State Islamic University Jakarta, Jl. Ir. H. Djuanda No. 95, South Tangerang, Banten, Indonesia

*Email: nurifawzya@gmail.com

Abstract. Study on the physicochemical properties of chitooligosaccharide (COS) prepared by hydrolysis of chitosan using chitosanase from *Stenotrophomonas maltophilia* KPU 2123 has been carried out. Hydrolysis process was conducted by reacting the soluble chitosan with 8 $U \cdot g^{-1}$ chitosan of chitosanase for 0; 8; 16 and 24 h incubation and stopped by addition of 0.25 M NaOH until reached pH 7. The COS was obtained as supernatant after being centrifugation. The liquid COS were then freeze-dried and analyzed their physicochemical properties, which comprised yield, viscosity, moisture and ash content, the degree of deacetylation (DD), as well as lead (Pb), arsenic (As) content and analyses of COS by Thin Layer Chromatography (TLC). The optimum hydrolysis time was found to be 16 h with the COS viscosity was 8.50 ± 0.87 cPs. The high COS yield was related to high ash content, i.e. 251.70 ± 77.97 % and 50.45 ± 3.19 % (db), respectively. There was lead (Pb) and arsenic (As) metals detected, i.e. 4.4 ppm and 0.1 ppm, respectively. However, they still met the requirement of Pb and As content in a commercial COS referred. Based on the COS properties, desalination process should be applied in the preparation of COS by enzymatic method.

1. Introduction

Chitooligosaccharides (COS) or Chito-oligomers are the hydrolysates of chitosan, mainly composed of β-(1→4)-linked D-glucosamine and partially of N-acetyl-D-glucosamine. In the last few decades, COS has been the subject of increased attention particularly in pharmaceutical and medicinal applications, due to their wide bio activities, such as antibacterial [1], antifungal [2], antitumor activity [3] and radical scavenging [4]. In addition, COS is non-toxic substance, water soluble and low in viscosity which makes them absorbable in the in vivo systems.

COS is generated by depolymerization of chitosan using acid, enzyme or physical hydrolysis. Enzymatic degradation is commonly preferable because the processes are relatively mild, eco-friendly and also produce more homogeneous products. Enzymes involved in COS preparation are particularly chitinolytic enzymes. In Indonesia, research on exploration of chitinolytic enzymes have been conducted in several research institutions, including isolation and screening of microbial chitinolytic enzymes, application of the enzymes for COS preparation and determination of COS bioactivities [2, 5, 6, 7, 8, 9, 10]. Most of them have been done on a laboratory scale with only a few grams of chitosan used as raw material.

Our research center has carried out studies on preparation and bioactivities of COS using chitosanases from our laboratory's bacterial collection since the last decade. The studies were done on a minimal scale (grams of chitosan as raw material) due to the exploratory research purposes and the availability of equipment and instrumentation. One of chitosanases, which is produced by *Stenotrophomonas maltophilia* KPU 2123, has been characterized and used to prepare COS. The resulting COS has been tested for antibacterial, antifungal and antitumor [2, 7]. The result showed that the COS seems to be potential for antifungal. In order to obtain other properties of COS which related to it's specification, preparation of COS at a larger scale should be carried out. This research aimed at production COS using chitosanase *Stenotrophomonas maltophilia* KPU 2123 at a larger scale and characterization of its physicochemical properties.

2. Materials and Methods

2.1. Materials

Materials used in this study were chitin and chitosan from crab shell (Sigma, with chitosan degree of deacetylation or DD more than 85 %), chitosan obtained from Bogor Agricultural University (IPB), glucosamine (Sigma) and standard COS (1–6 units) from Seikagaku Corp., Japan. *Stenotrophomonas maltophilia* KPU 2123 isolated from shrimp waste was used to produce chitosanase. Other chemicals and microbiological media components used were in analytical grade.

2.2. Methods

The research was done in four steps, including the production of chitosanase enzyme from *S. maltophilia* KPU 2123, preparation of soluble chitosan, production of COS and analysis of COS's product.

2.3. Production of chitosanase enzyme

Chitosanase was produced in a 5 L fermentor with working volume of 3 L minimal synthetic medium (MSM) contained of 0.1 % K_2HPO_4; 0.01 % $MgSO_4.7H_2O$; 0.1 % NaCl; 0.7 % $(NH_4)_2SO_4$; 0.05 % yeast extract and 0.5 % colloidal chitin as inducer [6]. The enzyme was harvested after 72 h incubation at 30 °C, 120 rpm by centrifugation at 10,000 × g, 4 °C for 15 min. The enzyme was then subjected to ultrafiltration using a 10 kDa MWCO membrane (UFP-10-E-4MA) for further concentration by 10 × of volume and assayed its activity by Schales method with glucosamine as standard [6].

2.4. Preparation of soluble chitosan

Soluble chitosan was prepared by dissolving of 50 g chitosan in 1 M CH_3COOH, stirred manually, followed by addition of 1 L of aquadest and stirred at 100 rpm. The mixture was then added with 1 M CH_3COONa until pH 6 and the volume was made 2.5 L by addition of 0.05 M CH_3COONa. The preparation of soluble chitosan was done in a 5 L beaker glass on a plate equipped with a speed adjustable agitator.

2.5. Production of COS

Production of COS was done by mixing the soluble chitosan with chitosanase at a concentration of 8 $U \cdot g^{-1}$ chitosan [2]. The mixture was then incubated for 0; 8; 16 and 24 h at room temperature. Soon after incubation, the solution was measured its viscosity using a viscometer (Brookfield Synchro-electric) with spindle no. 2 [11]. The enzymatic reaction was stopped by addition of 0.25 N NaOH until pH 7. The mixture was then centrifuged at 2,000 rpm, room temperature for 20 min. The hydrolysate containing COS was then freeze-dried.

2.6. Analyses of COS

The dried product was analyzed for its physicochemical properties, including yield, moisture content [12], ash content [13] and degree of deacetylation using Fourier Transform Infrared

Spectroscopy or FTIR [14]. The selected COS (based on the best properties) was determined for its Pb and As content [15] and analyzed the mixture by TLC using silica gel 60 F 254 chamber. Chitooligosaccharides (1–6 units) from Seikagaku Corp., Japan; D-glucosamine. HCl, as well as N-acetylglucosamine from Sigma, were used as standards.

3. Results and Discussion

The activity of crude chitosanase produced from *S. maltophilia* KPU 2123 was 0.191 ± 0.018 U·mL^{-1}, increased to 0.459 ± 0.032 U·mL^{-1} after being concentrated 10 × volume by ultrafiltration using a 10 kDa MWCO membrane (UFP-10-E-4MA). This method will separate the enzyme from other substances which have a molecular size lower than 10 kDa MWCO. They will pass through the membrane, whereas the chitosanase enzyme retained inside the membrane, and concentrated. Hence, the activity of the enzyme will increase because most of the residue which can lower its activity has been removed [7]. Crude chitosanase produced by *S. maltophilia* KPU 2123 has an activity of 0.1314 U·mL^{-1} and 1.0049 U·mL^{-1} after being concentrated using ultrafiltration [2].

Preparation of COS was carried out by addition of chitosanase into soluble chitosan conducted at pH 6 related to the optimum pH of enzyme work [7]. Hydrolysis process rapidly occurred at the beginning, indicated by a very sharp drop viscosity of soluble chitosan (table 1). The initial viscosity was high, i.e. 1,537.50 cPs which corresponded with the large molecular mass of chitosan [16]. It drastically decreased to 30.25 cPs soon after mixing with chitosanase, and then gradually declined in slow rate until 24 h incubation. This might be due to the decrease of β-(1, 4) glycosidic bond existed within chitosan as well as enzyme activity to support the degradation reaction. The similar result was investigated by other researchers [17], who studied on preparation of low molecular weight chitosan by complex enzymes hydrolysis. They further stated that the slow rate reaction was caused by inhibition of enzyme activity by the increase of oligomeric sugar concentration at a certain level. The decrease in viscosity showed that chitosan was already converted to COS's.

Chitosan used in this study is categorized as high viscosity chitosan, i.e. in the range of 800–2,000 cPs [18]. The viscosity reduction reached 98.03–99.54 %, resulted in COS viscosity 7.00 cPs after 24 h incubation. Statistical analysis indicated that incubation time 0 h gave a significant difference in viscosity with 8; 16 and 24 h incubation. Meanwhile, the 3 h incubation time gave similar product viscosity or did not show any difference in their viscosity. A commercial COS product specification required viscosity less than 10 cPs [19]. If we refer to this specification, chitosan hydrolysis in such conditions takes 16 h. GlcNAc which was produced using a crude enzyme from *Paenibacillus illinoisensis* KJA-424 reached the maximum production after 24 h of incubation [20]. At the same time, N-acetyl chitooligosaccharides were also produced in a lower concentration.

Table 1. The viscosity of soluble chitosan and chitosan hydrolysate containing chitooligosaccharides.

Incubation time (h)	Viscosity (cPs)	Viscosity reduction (%)
0	$30.25^{a} \pm 7.25$	98.03
8	$11.50^{b} \pm 2.60$	99.25
16	$8.50^{b} \pm 0.87$	99.45
24	$7.00^{b} \pm 0.00$	99.54
Soluble chitosan	1,537.50	-

The yield of chitosan hydrolysate containing COS is presented in figure 1. It is shown that the yield of COS increased with the time of incubation/enzyme reaction. The yield calculation was based on the weight of chitosan in soluble chitosan reacted with chitosanase. The high yield (> 100 %) was probably due to impurities derived from salts of enzymatic reaction results as well as inactivation using NaOH solution which was related to the high ash content produced in hydrolyzate containing

COS (table 2). Whereas, the moisture content of COS were found to be less than 10 % which meet with commercial chitosan oligomer requirement [19].

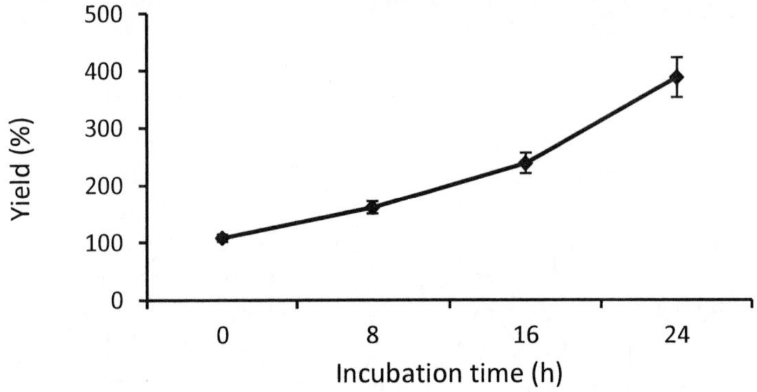

Figure 1. The yield of hydrolysate containing COS.

Ash content of COS was much higher than that of commercial COS specification (table 2). This was probably caused by salts produced during the process, such as in the preparation of soluble chitosan, enzyme addition as well as in the activation process which used NaOH solution. The salt content of COS was contributed by a hydrolysis reaction, and need a desalting process to obtain a pure oligosaccharide [21]. To reduce the salt content, desalination of COS was conducted using dialysis and gel filtration [10]. They found that those dialysis and gel filtration chromatographies were effectively removed the salt from COS with the efficiency of desalting up to 100 %.

Table 2. Moisture and ash content of chitosan hydrolysate containing COS.

Samples	Moisture content (%)	Ash content (% db)
COS – 0*	5.05 ± 2.53	58.85 ± 0.29
COS – 8*	2.50 ± 0.60	53.52 ± 5.93
COS – 16*	2.30 ± 0.54	50.45 ± 3.19
COS – 24*	3.75 ± 1.15	56.42 ± 1.38
Chitosan	11.46 ± 0.51	2.19 ± 0.12
Soluble chitosan (2 % w/v)	92.30 ± 0.02	48.46 ± 0.49
Chitosanase enzyme	98.96 ± 0.02	38.10 ± 0.36

*The number shows time (h) of hydrolysis.

The degree of deacetylation as quality parameter indicates acetyl group which can be removed from chitosan. The higher DD value, the less acetyl group contained within chitosan. The result of DD measurement showed that the DD value of chitosan hydrolysate containing COS ranged from 43.24 ± 0.88 % to 53.02 ± 3.08 %. The chitosan, as the raw material, had DD value of 42.35 ± 1.65 %. These values were much lower than that of commercial COS specification, due to the different method applied in DD measurement. To make sure that different methods result in different DD values, DD measurements have been done on Sigma chitosan. The DD value of chitosan from Sigma as listed on the label was > 85 %, however, the value based on FTIR method showed 45.05 ± 3.26 % (figure 2). Thus, the DD value of chitosan hydrolysate containing COS from 16 and 24 h hydrolysis was probably > 85 %, if they were measured as Sigma chitosan method.

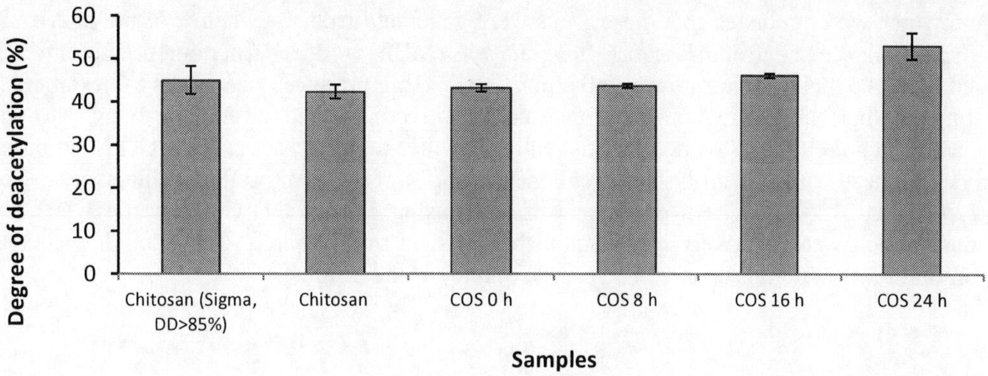

Figure 2. The degree of deacetylation of chitosan and chitosan hydrolysate containing COS.

Heavy metals may a cause of environmental pollution and cause a detrimental human or environmental effect. These metals include lead (Pb), cadmium (Cd), mercury (Hg), arsenic (As), chromium (Cr), copper (Cu), selenium (Se), nickel (Ni), silver (Ag) and zinc (Zn) [22]. They are toxic at higher concentrations. Excessive levels can be damaging to the organism. Since commercial chitosan are mostly produced from crustacean shell waste which its habitat may be subject to heavy metal contamination, the chitosan industry specifies the requirements for heavy metal content, e.g. Pb and As.

The analyzed heavy metals were lead (Pb) and arsenic (As) of soluble chitosan and chitosan hydrolysate from 0 h and 16 h (table 3). It was shown that Pb was not detected on soluble chitosan as raw material, whereas As content in this sample was 0.017 $\mu g \cdot g^{-1}$. In addition, chitosan hydrolysate was found to be positive in the two heavy metals content. The initial time of chitosan hydrolysis was found to be higher, i.e. 2,700 $\mu g \cdot g^{-1}$ and 4,400 $\mu g \cdot g^{-1}$ for Pb and As, respectively, however at 16 h hydrolysis the Pb and As content was 0.072 $\mu g \cdot g^{-1}$ and 0.102 $\mu g \cdot g^{-1}$, respectively. It was predicted that the heavy metals were generated from colloidal chitin as well as aquadest. However, the samples still meet the specification of commercial COS reported previously [19] related to Pb and As i.e. < 5 ppm and < 3 ppm, respectively. Pb may be found naturally in surface water [23]. Fresh or river water usually contains 1–10 $\mu g \cdot L^{-1}$ Pb, but this metal content in seawater is commonly lower. Industries that potentially cause Pb contamination include battery industry, foundry, fuel, cable, etc. Some arsenic compound as methylarsenic may be found in water as biological activity product. This compound will oxidize to methylarsenic acid. Organic arsenic compounds can be found in marine life and are very resistant to chemical degradation [24].

Table 3. Lead (Pb) and arsenic (As) content of chitosan and chitosan hydrolysate containing COS.

Samples	Heavy metal content ($\mu g \cdot g^{-1}$)	
	Lead (Pb)	Arsenic (As)
Soluble chitosan	nd	0.017
COS – 0*	2,700	0.072
COS – 16*	4,400	0.102

*The number shows time (h) of hydrolysis;nd not detected.
Limit detection for Pb: 0.005 ppm; for As: 0.0005 ppm.

The result of COS analysis is presented in figure 3. It is shown that there was no COS product detected in TLC result. It may be due to the low concentration of COS's under the detection limit of

IOP Conf. Series: Earth and Environmental Science **139** (2018) 012045 doi:10.1088/1755-1315/139/1/012045

TLC and the high salt or ash content which interferes with the reaction occurred in analyses of COS using TLC. Oligosaccharides move according to its molecular weight. Thin irregular spots were observed, which were predicted that these spots were generated from 1–2 degree of polymerization of COS. Crude chitosanase from *Bacillus licheniformis* MB2 produced monomer and dimer when incubated with 100 % DD chitosan, while the purified enzyme produced pentamers to hexamers [5]. It was estimated that the crude enzyme consisted of mixed enzyme both endo- and exo-type of chitosanase which are able to work synergistically. In other research, at least six COSs (monomer to hexamer) produced from hydrolysis of chitosan using 8 U·g^{-1} chitosan of chitosanase from *S. maltophilia* KPU 2123 [2]. The experiment was done using 51.4 % DD chitosan and 1.004 U·mL^{-1} chitosanase which were relatively higher than that used in this research. This fact showed that the specification of raw materials used will affect the quality of the product.

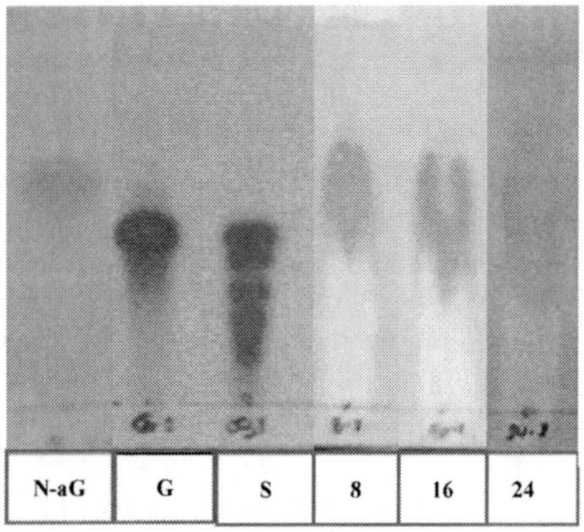

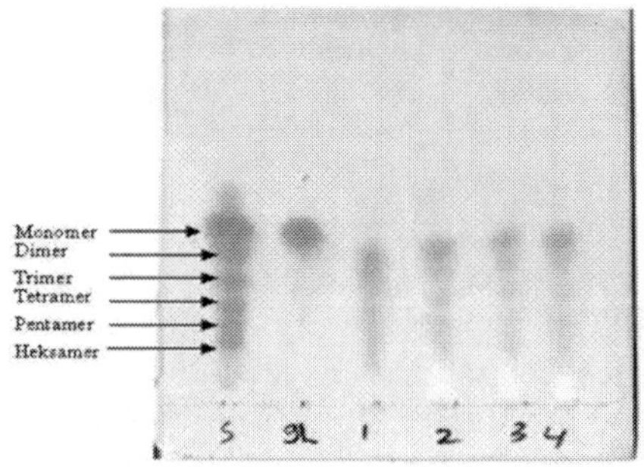

Note :
N-aG: N-acetyl glucosamine
G: Glucosamine
S: COS standard (DP 1–6)
8: COS from eight hydrolysis
16: COS from 16 h hydrolysis
24: COS from 24 h hydrolysis

G : Glucosamine
S : COS standard (DP 1–6)
1 : COS from 1 h hydrolysis
2 : COS from 2 h hydrolysis
3 : COS from 3 h hydrolysis
4 : COS from 4 h hydrolysis

Figure 3. TLC profiles of COS produced in the hydrolysis of soluble chitosan by crude chitosanase from *S. Maltophilia* KPU 2123 (left) compared to [2] result on COS hydrolyzed by the same chitosanase with different activity's stock, i.e. 0.459 U·mL^{-1} (left) and 1.0049 U·mL^{-1} (right). Chitosan's DD used was 42.4 % (left) and 51.4 % (right). Hydrolysis process was conducted on 2 % soluble chitosan using 8 U·g^{-1} chitosan of chitosanase *S. maltophilia* KPU 2123.

A good result of TLC with clear spots was found in some studies on COS production using purified enzyme or further treatment of COS to purify it, as reported on COS production using pure cellulase from *Trichoderma reesei* [7], and purified COS from chitosan using crude enzyme of *Bacillus cereus* D-11 [25].

Based on the results, related to the high ash content; ultrafiltration membrane with 5 kDa MWCO was used to reduce ash content of COS from 16 h-hydrolysis. The sample was mixed with a 25 mM acetic buffer with the same volume and passed through the membrane until equilibrium was reached (the same volume was obtained). The process was repeated in three cycles. Sampling was done for ash content analyses at each cycle (table 4). It was shown that ultrafiltration reduced ash content drastically from 50.45 % (db) to less than 1 % (db). Thus ultrafiltration can be regarded as one of the purification steps [26]. Other techniques for separation and purification of COS had also been reported, such as gel filtration [10, 27], ion exchange [28] and metal affinity chromatography [29].

Table 4. Ash content of chitosan hydrolysate containing COS before and after ultrafiltration.

No	Sample	Ash content (% db)
1	Before ultrafiltration	50.45
2	After ultrafiltration (1st cycle)	0.79
3	After ultrafiltration (2nd cycle)	0.60
4	After ultrafiltration (3rd cycle)	0.35

Commercial products of chitooligosaccharides are commonly expressed as being the salt of acetic, lactic or citric acid of the oligosaccharides [21]. In addition, glucosamine is available in the form of a hydrochloric or sulfuric salt. This is because of the expensive and complicated process of purification to reduce or remove the salt. For that reason, scaling up production of COS should be done by using: high-quality chitosan (e.g. high-DD of chitosan) and crude enzyme with high activity as well as high-quality chemicals (such as acid for preparing soluble chitosan) to minimize salt content as well as cost production of the COS.

4. Conclusion

Preparation of COS by using crude chitosanase from *S. maltophilia* KPU 2123 with the ratio enzyme/substrate of 8 $U \cdot g^{-1}$ chitosan produced COS which high in ash content and poor visualization of TLC. The enzymatic hydrolysis for 16 h produced COS with viscosity < 10 cPs which met the requirement of a commercial COS. The COS was also eligible for the specification of commercial COS related to Pb and As content. Desalination process should be done to reduce ash content which passes for a commercial specification.

References

[1] Wang Y, Zhou P, Yu J, Pan X, Wang P, Lan W and Tao S 2007 *Asia Pac. J. Clin. Nutr.* **16** 174–177
[2] Chasanah E, Riviyanti T A and Noriko N 2011 *Jurnal Penelitian Pasca panen dan Bioteknologi* **6**(2) 139–146
[3] Suzuki S 1996 *Fragrance J.* **15** 61–68
[4] Mendis E, Kim M M, Rajapakse N and Kim S K 2007 *Life Sciences* **80**(23) 2118–2127
[5] Chasanah E 2004 *Dissertation* Institut Pertanian Bogor
[6] Pratitis A 2006 *Thesis* Department Fishery and Marine Sciences Institut Pertanian Bogor
[7] Fawzya Y N, Pratitis A and Chasanah E 2009 *Jurnal Pascapanen dan Bioteknologi Kelautan dan Perikanan* **4**(1) 69–78
[8] Chasanah E, Zilda D S and Uria A R 2009 *Journal of Coastal Development* **12**(2) 64–72
[9] Baehaki A and Herpandi 2013 *Asian Journal of Chemistry* **25**(6) 3154–3155
[10] Martosuyono P, Pratitis A, Prasetya A and Prabawati E K 2014 *Squalen Bulletin of Marine and Fisheries Postharvest and Biotechnology* **9**(3) 127–136
[11] Sephanodora P and Benjakula S 1993 *Proceeding at the 8th Asia Good Conference* (Jakarta)
[12] BSN 2006a SNI 01-2354.2-2006 *Standar Nasional Indonesia* (Jakarta: Badan Standardisasi Nasional)
[13] BSN 2006b SNI 01-2354.1-2006 *Standar Nasional Indonesia* (Jakarta: Badan Standardisasi Nasional)
[14] Srijanto B, Paryanto I, Masduki and Purwantiningsih 2006 *Akta Kimindo* **1**(2) 67–72
[15] BSN 2011 SNI 2354.5 2011 *Standar Nasional Indonesia* (Jakarta: Badan Standardisasi Nasional)
[16] Yacob N, Talip N, Mahmud M, Sani N A I M, Samsuddin N A and Fabillah N A 2013 *Journal of Nuclear and Related Technologies* **10**(1) 39–44
[17] Xie H, Jia Z, Huang J and Zhang C 2011 *International Journal of Chemistry* **3**(2) 180–186

[18] Suptijah P, Salamah E, Sumaryanto H, Purwaningsih S and Santoso J 1992 *Laporan Penelitian* (Bogor: Fakultas Perikanan dan Ilmu Kelautan Institut Pertanian Bogor)
[19] Dalian Glycobio Co, LTD 2007 http://www.glycobio.cn Accessed on 12th September 2012
[20] Jung W, Alfred S, Ro-Dong P and Donald S 2007 *Carbohydrate polymers* **67**(2) 256–259
[21] Jung W J and Park R D 2014 *Mar. Drugs* **12**(11) 5328–5356
[22] Singh R, Gautam N, Mishra A and Gupta R 2011 *Indian J. Pharmacol.* **43**(3) 246–253
[23] Sudarmaji J, Mukono and Corie I P 2006 *Jurnal Kesehatan Lingkungan* **2**(2) 129–142
[24] Sukar 2003 *Jurnal Ekologi Kesehatan* **2**(2) 232–238
[25] Xing-Ai G, Zhang Y, Park R, Huang X, Zhao X Y, Xie J and Jin R 2012 *J. Appl. Biol. Chem.* **55**(1) 13−17
[26] Lopatin S A, Derbeneva M S, Kulikov S N, Varlamov V P and Shpigun O A 2009 *J. Anal. Chem.* **64** 648–651
[27] Sørbotten A, Horn S J, Eijsink V G and Vårum K M 2005 *FEB. S. J.* **272** 538–549
[28] Haebel S, Bahrke S and Peter M G 2007 *Anal. Chem.* **79** 5557–5566
[29] Le Dévédec F, Bazinet L, Furtos A, Venne K, Brunet S and Mateescu M A 2008 *J. Chromatogr. A.* **1194** 165–171

IOP Conf. Series: Earth and Environmental Science **139** (2018) 012046 doi:10.1088/1755-1315/139/1/012046

Production and characterization of the healthy brown rice milk with sodium alginate addition from brown algae *Sargassum binderi* as emulsifier

R N Latifah[1*] and F M Warganegara[1]

[1] Chemistry Department, Institute Technology of Bandung, Jl. Ganesha No. 10, Bandung, Indonesia
*Email: raisnurlatifah038@gmail.com

Abstract. Brown rice milk, a plant milk, is potential to cure beriberi disease. Alginate was used as an emulsifier and increased the nutrition of brown rice milk. Alginate from *Sargassum binderi* was extracted by nonacidic treatment. The yield of alginate was 6.25 %. The moisture and water content of alginate were 18.27 % and 12.2 %, respectively. The density and viscosity of 0.1 % alginates in aqueous solution were 1.007 g·mL^{-1} and 7.651 × 10^{-3} kg·m^{-1}s^{-1}, respectively. The characteristics peaks of alginate appeared at 3,477; 1,633 and 1,419 cm^{-1}, corresponding to hydroxyl (OH), carbonyl (C=O) and carboxyl (COOH), respectively. The best composition of alginate addition in brown rice milk was 0.2 %. The viscosity of the 0.2 % alginate in brown rice milk was 1.206 × 10^{-2} kg·m^{-1}s^{-1} and showed a texture of small particle with closer spaces between the particles. The addition of alginate in brown rice milk also inhibited the process of sedimentation of milk which showed in the first order. The nutrition composition of the best brown rice milk was 40.21 mg·mL^{-1} carbohydrate; 1.3 mg·mL^{-1} protein; 0.158 mg·mL^{-1} fat and 13.1 mg·mL^{-1} total dietary fiber.

1. Introduction

Milk is called the fifth source that makes perfect nutrition [1]. The growth of raw milk production every year reflecting the increased in consumer demand for milk. The main components of milk are proteins, fats, sugars (lactose), minerals (solid state) and water [2].

Originally, milk is referred to the product released from the mammal gland of the animal; however, the liquid extract from a plant that has a similar consistency to animal milk is also called as milk, such as soy milk and rice milk. In this paper, we called this type of milk as plant milk. The composition of plant milk is similar to cow's milk, in terms of protein, and is usually consumed as a substitution to cow's milk by those who are allergic to cow's milk. Plant milk is an alternative for people lacked lactase in their digestive tract that is not able to digest lactose contained in cow's milk. Therefore, the development of plant milk product by food industries aims to target this type of consumers.

Brown rice is an example of an ingredient that is used in the development of plant milk product. Utilization of brown rice as a beverage offers a good alternative to protein sources for a wide range of consumers, from children to the elderly [3]. The content of vitamin and mineral in brown rice was 2–3 times higher than white rice. Brown rice has a low glycaemic index (low starch and high complex carbohydrates) that can reduce the risk of type-2-diabetes [4].

In this study, we used alginate as an emulsifier in the production of brown rice milk. Moreover, alginate was used to increase the nutrition of brown rice milk and improve the texture of milk. Alginate was extracted from brown seaweeds (*Sargassum binderi*) obtained from Sayang Heulang

Beach, Pameungpeuk, Garut, West Java. Alginate is the main content of brown algal cell walls. Alginate is a polysaccharide composed of β-D-mannuronic acid and α-L-guluronate linked by 1,4-glycosidic bond. Alginate levels in brown alga are about 20–27 % of the total dry weight [5]. This research aimed to use the sodium alginate extracted from brown algae *Sargassum binderi* as an alternative emulsifier for the production of brown rice milk. Milk products are expected to have a good stability, high nutritional value and have a long shelf life.

2. Methods

2.1. Algae collection
Brown algae were collected from Sayang Heulang Beach, Pameungpeuk, Garut, West Java of Indonesia. The species were in family Sargassaceae: *Sargassum binderi*. Brown algae were washed in distilled water, dried overnight at 60 °C in an oven. The dry weights were obtained after drying overnight at 105 °C.

2.2. Extraction of alginate
The samples were suspended in 2 % (w/v) $CaCl_2$ for 2 h, washed with deionized water and treated with 40 % formaldehyde for 2 h to cross-link phenolic compounds. Alginate was extracted according to a method of Haug [6]. The samples were then washed three times with deionized water and the alginate was extracted by addition of an aqueous solution of 3 % Na_2CO_3 for 48 h. In other treatment, samples were also suspended in a mixture of 1 M Na_2CO_3 and 0.5 % EDTA, and the pH of the suspension was adjusted to pH 11 [6]. This suspension was then filtered through muslin cloth and precipitated in ethanol as the sodium salt. The precipitate was separated by centrifugation and dried overnight at 60 °C. The yield of alginate was expressed as percentage/dry weight.

2.3. Purification of sodium alginate extracts
Sodium alginate was purified according to the method of Gomez et al. [7]. An aqueous solution of sodium alginate obtained in the extraction step was directly precipitated, under stirring, by addition of ethanol until reaching a proportion of 1:1 in volume, respectively. Thus, the insoluble polymer was separated and then exhaustively washed with acetone by soxhlet for 100 h. Finally, the biopolymer was dried at room temperature under vacuum condition until constant mass was reached.

2.4. FT-IR analysis of alginate
Fourier-transform infrared spectroscopy (FT-IR) analysis of the samples was carried out using an FT-IR spectrophotometer in ITB to identify the functional groups. The dry sample was mixed with potassium bromide and pressed into a plate for measurement. The FT-IR spectrum was then recorded.

2.5. Analysis of moisture content of alginates
One gram of sodium alginate was accurately weighed and dried in a vacuum oven at 70 °C for 24 h. The sample was allowed to cool to room temperature in vacuum desiccators and re-weighed. The moisture content of the alginate sample was determined from the weight difference and expressed as a percentage of original weight.

2.6. Production of brown rice milk
Brown rice milk was prepared by mixing brown rice with water in the ratio of 1:15 (w/v), homogenized and processed to reach the consistency of porridge. Then the mixture was filtered using a calico cloth to obtain brown rice milk. Then, alginate was added to milk with the concentration of 0.05 %, 0.1 % and 0.2 % (w/v). The milk was pasteurized at 85–95 °C. Each treatment was replicated three times. Then, milk texture was observed under a microscope with 400 times magnification.

2.7. Analysis of the rate of deposition of brown rice milk
One mL of sample was put into a flask, diluted to 10 mL by aquadest. The transmitter's measurement of the suspension was performed at a wavelength of 656 nm. The transmittance results were then converted into a turbidity value with the equation below.

$$S = -log\left(\frac{T}{100}\right) \qquad (1)$$

Where:

S = turbidity
T = transmitters

2.8. Analysis of carbohydrate, protein, fat and fiber content
Nutrition analysis of milk was performed according to BSN [8], on carbohydrate, protein, fat and fiber content. Carbohydrates, protein, fat and fiber content were analyzed using Luff Schroll method, Kjeldahl method, gravimetric method and enzymatic-gravimetric method, respectively.

3. Results and Discussion

3.1. Sargassum binderi
Seaweed was the main ingredient used in this study. It was obtained from the growing area in Sayang Heulang Beach, one of the coral beaches in the southern part of Java Island. The coordinate sampling points were 7°40'12"S and 107°41' 37"E.

Figure 1. *Sargassum binderi.*

Sargassum binderi had a characteristic of the thallus (± 1.5 cm), smooth/slippery, the height reached about 60 cm (figure 1). Regular branching, opposite (left-right), main branch adjacent to each other, arises on the main short stem (± 1–2 cm) above holdfast. Leaves were oval, serrated edge, length of 5 cm, the width of 1 cm and a pointed tip. The bladder was round, rounded or pointy tip, 1 cm long-winged and 0.4 cm in diameter. The reproductive organs formed a special stalk, scratch, flatten and jagged. Leaves and bladders were shrunk.

3.2. Alginate extraction
Brown seaweeds (*Sargassum binderi*) from Indonesia were collected for extraction and purification of sodium alginate. The yield of alginate is 1.2 g from dry weight *Sargassum binderi*. The moisture and water content of alginates extracted from seaweed sample was 18.27 % and 12.2 %, respectively. The density and viscosity of 0.1 % (v/v) alginates in aqueous solution were 1.007 $g{\cdot}mL^{-1}$ and 7.651×10^{-3} $kg{\cdot}m^{-1}s^{-1}$, respectively, while the viscosity of commercial brown rice milk (without the addition of alginate) is 2.45×10^{-3} $kg{\cdot}m^{-1}s^{-1}$. The high viscosity of alginate indicates that it can be applied to thicken the consistency of brown rice milk.

Figure 2. Alginate from *Sargassum binderi*.

3.3. FT-IR of alginate

The FT-IR spectrum of sodium alginate is given in figure 3. As the alginate is a carbohydrate, the characteristic peaks of alginate appeared at 3,477; 1,633 and 1,419 cm^{-1}, corresponding to hydroxyl (OH), carbonyl (C=O) and carboxyl (COOH), respectively. This result indicates that these three peaks were appeared in FT-IR spectrum obtained from sodium alginate extracted from brown alga *Sargassum binderi* localized in Indonesia. The results of the study are in accordance with the results of previous study [9] which reported that the characteristic peaks of alginate appeared at 3,429; 1,630 and 1,428 cm^{-1}, corresponding to hydroxyl (OH), carbonyl (C=O) and carboxyl (COOH), respectively.

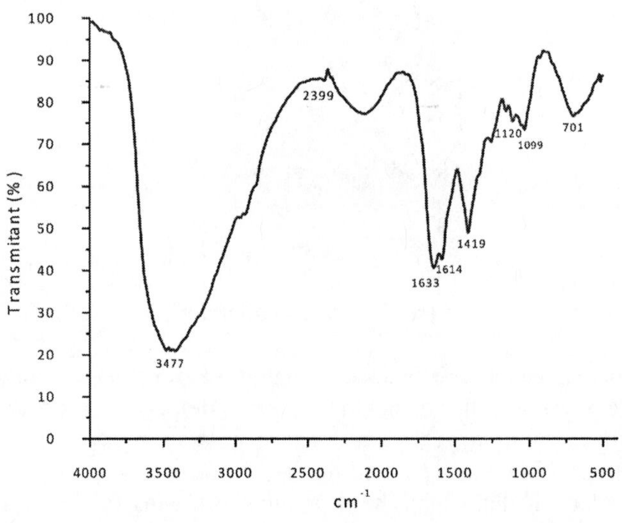

Figure 3. The FT-IR spectrum of alginate.

3.4. The rate of deposition of brown rice milk

The effect of alginate addition on the stability of brown rice milk was evaluated by measuring the turbidity value of the final product. The turbidity value was obtained by the measurement using the UV-vis spectrophotometer at λ = 241 nm. The rate of precipitation of brown rice milk with various alginate concentrations was tested kinetically by using the integral method on the order of 0, 1 and 2.

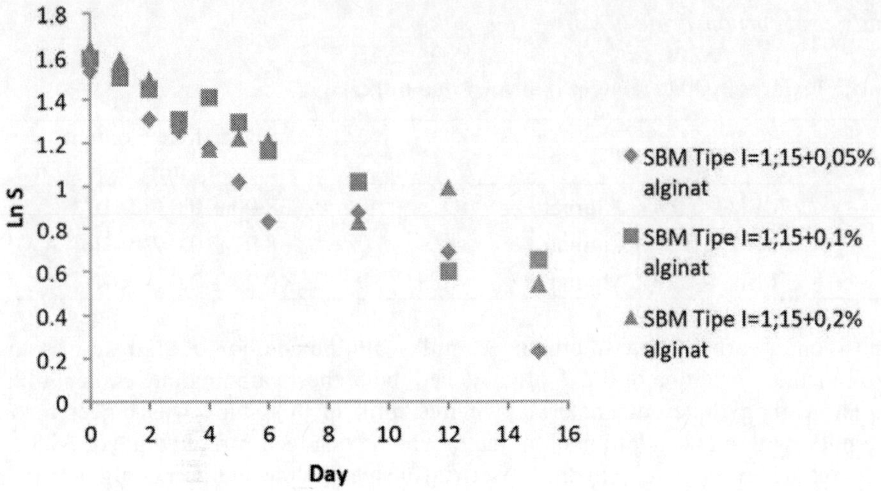

Figure 4. Variation of alginate addition to the stability of brown rice milk.

Based on the kinetics analysis shows that the rate of precipitation of brown rice with the addition of alginate was of order one. This result indicates that the addition of alginate two-fold from the initial quantity resulted in the four times increase of stability. This result also indicates that alginate could increase the stability of brown rice milk so that the resulting brown rice milk does not quickly settle or separate into two phases. Alginate interacts with brown rice milk dispersed in the liquid phase so that the milk is kept awake.

3.5. Texture of alginate

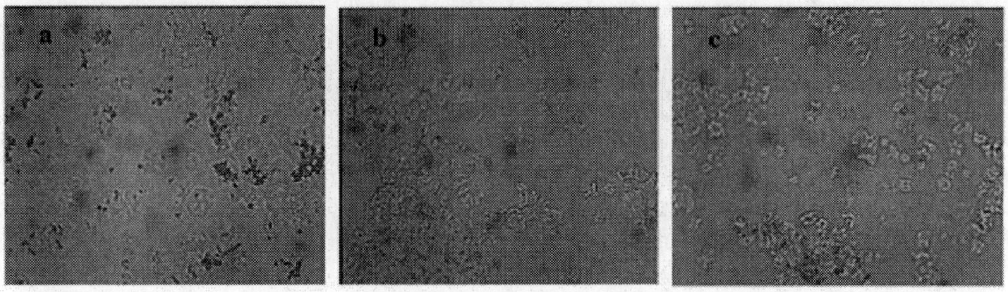

Figure 5. Texture of brown rice milk with alginate addition 0.05 % alginate (a); 0.1 % alginate (b) and 0.2 % alginate (c).

Analysis of milk texture was performed under a microscope using 400 times magnification. Brown rice milk with 0.2 % alginate addition had a denser particle constituent and clearer particle transparency properties compare to other treatments. Particles of brown rice milk with the addition of alginate formed an emulsion. This is because the effect of the addition of alginate is a process of incorporating hydrocolloid polymers or polymer chain interactions resulting in a larger and more stable aggregate polymer. When mixed with brown rice milk, the alginate polymer will form a surface-active compound that converts the suspension into a smaller emulsion resulting in a softer milk texture.

3.6. Analysis nutrition of brown rice milk

3.6.1. Analysis of carbohydrate in brown rice milk

Table 1. Carbohydrate content in brown rice milk.

Treatments	Carbohydrate Content ($mg \cdot mL^{-1}$)
Milk + 0.05 % Alginate	$38.00 \pm 0.034 \times 10^{-4}$
Milk + 0.1 % Alginate	$38.97 \pm 0.042 \times 10^{-4}$
Milk + 0.2 % Alginate	$40.21 \pm 0.048 \times 10^{-4}$

The composition of carbohydrate in brown rice milk with the addition of alginate was analyzed using Luff Schroll method. Addition of 0.2 % alginate had the highest carbohydrate content (40.21 $mg \cdot mL^{-1}$) (table 1). The carbohydrate content of brown rice milk in this study was higher than commercial brown rice milk (without the addition of alginate) which contained carbohydrate of $38.3 \pm 0.047 \times 10^{-4}$ $mg \cdot mL^{-1}$. Increased levels of carbohydrates were influenced by the addition of alginate in milk because alginate was a carbohydrate polysaccharide.

3.6.2. Analysis of protein composition in brown rice milk

Table 2. Protein content in brown rice milk.

Treatments	Protein content ($mg \cdot mL^{-1}$)
Milk + 0.05 % Alginate	$1.01 \pm 0.045 \times 10^{-5}$
Milk + 0.1 % Alginate	$1.20 \pm 0.033 \times 10^{-5}$
Milk + 0.2 % Alginate	$1.30 \pm 0.027 \times 10^{-5}$

Determination of protein content was done by the Kjeldahl method. This method principally determined the total nitrogen contained in the sample. The addition of alginate in the processing of brown rice milk affected the amount of protein in milk. This is shown by the increasing amount of protein in milk with the addition of higher concentration of alginate (table 2). The increase of protein content in brown rice milk was influenced by the interaction that occurs between alginate monomers through hydrogen bonds. The presence of these hydrogen bonds trapped and bound the protein contained in brown rice milk. The protein content of brown rice milk in this study was higher than commercial brown rice milk (without the addition of alginate) which contained protein of $0.97 \pm 0.033 \times 10^{-5}$ $mg \cdot mL^{-1}$.

3.6.3. Analysis of fat composition in brown rice milk. Fat content in brown rice milk was analyzed by gravimetric method. Table 3 shows that the fat content in brown rice milk increased as the addition of alginate into milk. The highest fat content was in the addition of 0.2 % alginate with a fat content of 0.158 $mg \cdot mL^{-1}$.

Table 3. Fat content in brown rice milk.

Treatments	Fat content ($mg \cdot mL^{-1}$)
Milk + 0.05 % Alginate	$0.124 \pm 0.022 \times 10^{-4}$
Milk + 0.1 % Alginate	$0.128 \pm 0.035 \times 10^{-4}$
Milk + 0.2 % Alginate	$0.158 \pm 0.038 \times 10^{-4}$

Alginate can increase the dissolved nutrient content of brown rice milk. Alginate forms a hydrogen bond that is a carboxyl group with a hydroxyl group of glycerol in an edible film. The existence of this interaction causes an esterification reaction in the process of fat hydrolysis causing the molecule is nonpolar which can be extracted by nonpolar solvent also in hexane. So the fat content produced from brown rice milk with the addition of alginate is increasing. The fat content of brown rice milk in this study was higher than commercial brown rice milk (without the addition of alginate) which contained fat of $0.97 \pm 0.033 \times 10^{-5} mg \cdot mL^{-1}$.

3.6.4. Analysis of total food fibers in brown rice milk. Dietary fiber is a component of plant tissue that is resistant to the hydrolysis process by enzymes in the stomach and small intestine [10]. Fiber is the main component of cell walls in various vegetables and fruits. Chemically the cell wall consists of several types of carbohydrates such as cellulose, hemicelluloses and pectin. Fiber is a polysaccharide that linked to the β-glucoside bond that will not be degraded by α-amylase and also stomach acid. The content of fiber in brown rice milk with the addition of alginate is presented in table 4.

Table 4. Total food fiber content in brown milk.

Treatments	Mass of total fiber/mL milk	Mass of Crude fiber	Mass of Food fiber
Milk + 0.05 % Alginate	13.5 mg	7.8 mg	5.7 mg
Milk + 0.1 % Alginate	13.6 mg	6.7 mg	6.9 mg
Milk + 0.2 % Alginate	18.9 mg	5.8 mg	13.1 mg

Table 4 shows that the highest fiber content was in brown rice milk with the addition of 0.2 % alginate. The addition of alginate increased the total fiber content in milk. This indicates that the fibers digested by human digestive enzymes are more numerous and can be used as energy sources.

4. Conclusion

Sargassum binderi was collected from Sayang Heulang Beach, Pameungpeuk, Garut, West Java with a yield of alginate of 6.25 %. The characteristics peaks of alginate appeared at 3,477; 1,633 and 1,419 cm^{-1}, corresponding to hydroxyl (OH), carbonyl (C=O) and carboxyl (COOH), respectively. The best composition of alginate addition in brown rice milk was 0.2 % (w/v) in 100 mL of the brown rice milk.

References

[1] Moccance and Widdownson 1991 *The Composition of Foods* 5th ed (Cambridge Royal Society of Chemistry: Royal Society of Chemistry and Ministry of Agriculture Fisheries and Food)
[2] Miller G D, Jarvis J K and Mcbean L D 2000 *The Importance of Milk and Milk Product in the Diet,in: Handbook of Dairy Foods and Nutrition* Second Edition (USA: CRC Press)
[3] Indriyani F, Nurhidajah and Suyanto A 2013 *Jurna lPangan dan Gizi* **4** 27–34
[4] Garrow J S 2000 *Human Nutrition and Dietetics* (USA: Churchil Livingstone) pp 324–356
[5] Bertagnolli C and Stevan S K 2014 *Journal of Carbohydrate Polymers* **111** 619–623
[6] Haug A 1964 *Thesis* Noewegian Institute and Technology Trondheim

[7] Gomez C G, Lambrecht M V P, Lozano J E, Rinaudo M and Villar M A 2009 *Int. J. Bio. Macromal.* **44** 365–371
[8] BSN 1992 SNI 01-2782-1998/Rev.1992 *Metoda Pengujian Susu Segar* (Jakarta: National Standardization Agency)
[9] Latifi AM, Ehsan S N and Hamid B 2015 *Journal of Applied Biotechnology Reports* **2** 251–255
[10] Srujana G 2011 *International Journal of Pharma and Bio Science* **2** 139–143

Extracting of yellowfin tuna (*Thunnus albacares*) fish skin gelatin as influenced by alkaline concentration and extraction times

E M Mafazah[1*], Y Pranoto[2] and A Rohman[3]

[1] Department of Food Science and Technology, Faculty of Agricultural Technology, Universitas Gadjah Mada, Jl. Flora, Bulaksumur, Yogyakarta, Indonesia
[2] Department of Food Science and Technology, Faculty of Agricultural Technology, Universitas Gadjah Mada, Jl. Flora, Bulaksumur, Yogyakarta, Indonesia
[3] Department of Pharmaceutical Chemistry, Faculty of Pharmacy, Universitas Gadjah Mada, Sekip Utara, Yogyakarta, Indonesia
*Email: eli_faza@yahoo.com

Abstract. Yellowfin tuna (*Thunnus albacares*) skin is a by-product of the fish processing industry. Yellowfin tuna skin was used as materials for gelatin because it contains collagen. This study was conducted to determine the characteristics of yellowfin tuna skin gelatin (YSG). In this study, YSG was obtained by the extraction process in 0.01 M NaOH solution for 12 h. The results show that the yield of YSG was 12.49 % dry basis, with the gel strength of 291.73 g Bloom, viscosity of 8.6 cPs and melting point of 35 °C. The gel strength and melting point of YSG were lower than commercial bovine gelatin (CBG), which was 309.7401 g Bloom and 36.33 °C, respectively. However, the viscosity of YSG was higher than the CBG which was 5.6 cPs.

1. Introduction

Gelatin is defined as a denatured protein derived from collagen by thermal hydrolysis and the thermoreversible transformation between sol and gel determines the property of gelatin. Gelatin has been widely used in the food, pharmaceutical and photographic industries [1]. The worldwide production of gelatin in 2005 of which 44.9 % were from pigskin, 27.9 % from bovine hides, 27.2 % from bones and a few percent from other parts [2].

In fish industries, 30 % of the waste is in the form of bones and skins [3]. The study of gelatin from fish by-products, such as skin and bone, has increased for the replacement of mammalian resources [4]. Fish gelatin can be used to replace porcine and bovine gelatin which are a major constraint for consumers with the religion and socio-cultural background. For example, porcine gelatin which is prohibited for Muslims or Jews and bovine gelatin which is not consumed by Hindus and also usually not acceptable to Jews and Muslims. Moreover, frequent occurrences of bovine spongiform encephalopathy (BSE) affects the safety issue of bovine gelatin.

The most common use of gelatin is its thermally reversible gelling properties with water, as for example, in the production of table jellies. An aqueous solution of a few percent gelatin forms thermally reversible gels with water and the gel melting temperature (< 35 °C) is below body temperature, which gives gelatin products unique organoleptic properties and flavor release [5]. The thermal reversibility of this process gives the gelatin gel it's unique "melt in mouth" quality. Other gelling agents such as starch, alginate, pectin, agar and carrageenan are all polysaccharides from plant sources, but their gels lack the melt-in-the-mouth quality, elastic properties of gelatin gels. Gelatin is notable for its gelling properties and clean flavor profile. The gelatin gel has been described as having a sparkling and clear appearance with clean melt in the mouth texture that has yet to be duplicated by any polysaccharide [4].

The degree of collagen conversion into gelatin is related to the severity of both the pretreatment and the warm water extraction process, as a function of pH, temperature, and extraction time [6]. This research aimed to know the effect of NaOH concentrations in extraction process and extraction times on the quality of gelatin made of yellowfin tuna (*Thunnus albacares*) skin (YSG) to meet the standard of gelatin in food industry, to determine the characteristics of YSG and to compare YSG with commercial bovine gelatin (CBG).

2. Materials and Methods

2.1. Materials

Yellowfin tuna (*Thunnus albacares*) skin was provided by Indo Marlin, Pelabuhan Perikanan Muara Baru Jakarta. Mammalian gelatin extracted from the skin bovine was purchased from Sigma Chemical Co. All reagents used in this study were analytical grade.

2.2. Extraction of gelatin from yellowfin tuna skin

The yellowfin tuna skin gelatin (YSG) was extracted based on modified methods of previous studies [3, 6, 7, 8, 9, 10, 11]. Skin of yellowfin tuna (*T. albacares*) stored at 20 °C. Before use, the skins were allowed to thaw below 10 °C and skins were scraped manually by a scalpel. A part of fat on those skins was removed by vigorous stirring in chill water and then de-greased by tumbling in warm water (70 °C) for 10 min.

Then the skin was cleaned in distilled water and cut into 2 × 2 cm squares using scissors. The cleaned skin was treated with a skin to solution ratio of 1:6 (v/w) of alkali solution (0.01 M; 0.03 M and 0.05 M NaOH) at room temperature for 12 h; 18 h and 24 h to remove the non-collagen protein and subcutaneous tissue after they were swollen. Alkaline treated skin was then washed with tap water until neutral or faintly basic pH (pH < 7.5) of wash water was obtained. To swell the collagenous material in the fish skin matrix, the alkaline treated skin was soaked in 0.075 M acetic acid with a skin to solution ratio of 1:6 (w/v) for 18 h at room temperature (25 °C). The treated skin was washed with distilled water until the wash water pH was greater than four. This required 6–7 washing cycles.

Gelatin was extracted from the bleached skin using distilled water (60 °C) for 4 h with a skin/water ratio of 1:4 (w/v) in a water bath shaker (Memmert, WB14, Germany). During extraction, the mixture was stirred continuously. Solubilized gelatin was separated from residual skin fragments by using two-layer filter cloth. The resultant filtrate was cabinet dried and the dry matter from cabinet dried process was ground and referred to as "gelatin powder".

2.3. Measurements of physicochemical characteristics

2.3.1. Determination of yield. The yields of the gelatins obtained were calculated on both weight and protein basis[13] as follows:

(i) % Yield (wet wt basis) = *(Dry wt of gelatin / wet wt of skin) × 100 %*
(ii) % Yield (dry wt basis) = *(Dry wt of gelatin) / (wet wt of skin - moisture content) ×100 %*
(iii) % Yield (protein basis) = *(Dry wt of gelatin / protein content of skin) ×100 %*

2.3.2. Determination of gel strength. Gel strength was determined using a Texture analyzer (TA.XT Plus, Stable Micro Systems Ltd., Surrey, England). Gelatin was dissolved with distilled water (6.67 %, w/v) and agitated at 60 °C, 150 strokes·min^{-1} for 30 min in a shaking water bath (SS40-D Shaking Bath, Grant Instruments Ltd., Cambridge, England) until completely dispersed. The gelatin solution was filled in glass measuring bottles (40.1-mm dia 52-mm height, flat bottom) and then kept at 9–10 °C for 17–18 h. After cool maturation, the gel strength expressed in Bloom value was measured while the sample was still at 9–10 °C. A plunger with a 12.7 mm diameter had a penetration depth of 4 mm into the gelatin gels and a penetration speed of 1 mm·s^{-1} was used. The force in g at this temperature is the Bloom strength [13].

2.3.3. Determination of viscosity. The viscoelastic studies were performed on a Brookfield DVIII + Rheometer (Model RY, Middleboro, MA, USA) using a 6.67 % of gelatin solution at 60 °C equipped with an SC4-21 spindle at 100 rpm. Dry gelatin was dissolved in Milli-Q water and gently stirred at 60 °C in a shaking water bath for about 20 min before the viscosity measurements (centipoises) were taken directly using the above viscometer coupled to a water bath at 60 °C [14].

2.3.4. Determination of melting point. Solutions containing 6.67 % (w/v) gelatin were prepared in screw cap test tubes. The test tubes were filled and kept in the refrigerator (7 °C) for 16–18 h, and then were transferred into a cold-water bath (4 °C). The measurement was performed at a heating rate of 0.2 °C·min^{-1}, using oil drop method. The amount of 0.5 % methyl-red chloroform was used as an indicator of melting point. The temperature, at which the gel melted, allowing the chloroform to start falling, was recorded as the melting point [15].

2.4. Statistical analysis
All the collected data were analyzed using Analysis of Variance (ANOVA) and Duncan's Multiple Test to determine the significance level among the means using the statistical program analyses package [17].

3. Results and Discussion

3.1. Yield of gelatin
Yields of yellowfin tuna skin gelatins (YSG) extracted under treatment conditions at different of alkaline (NaOH) concentration combined with treatment time is shown in table 1.

Table 1. Yields of yellowfin tuna skin gelatins (YSG) under various NaOH concentration and treatment time.

Treatments	Yield of YSG (%)
0.01 M; 12 h	12.49^{h}
0.03 M; 12 h	12.02^{f}
0.05 M; 12 h	8.74^{c}
0.01 M; 18 h	12.12^{g}
0.03 M; 18 h	11.81^{e}
0.05 M; 18 h	8.6^{b}
0.01 M; 24 h	12.1^{g}
0.03 M; 24 h	9.76^{d}
0.05 M; 24 h	8.4^{a}

*Values are the means of triplicates.
*Different superscripts in the same column indicate significant differences ($p < 0.05$).

The yields of YSG which used 0.01 M NaOH for the alkaline treatment combined with a treatment time of 12 h gave the highest total gelatin yield (12.49 % dry basis wt) compare to the other treatments. This suggested that the use of NaOH in a concentration of 0.01 M combined with a treatment time of 12 h provided the best conditioning treatment to remove non-collagen protein and subcutaneous tissue from tuna skin prior to gelatin extraction which resulted in better yield recovery.

IOP Conf. Series: Earth and Environmental Science **139** (2018) 012047 doi:10.1088/1755-1315/139/1/012047

Our findings are in line with previous work reported by previous researchers [11], who reported that the extraction conditions, including solvent, temperature and extraction time, might affect the yield of the extracted gelatin. Calcium hydroxide solution was used to remove impurities (proteoglycan, blood, mucins, sugars, fat, etc.) and change collagen to the optimal type for gelatin extraction [16]. Furthermore, using a prolonged extraction of whole cod skins, achieved a yield of gelatin between 11 % and 14 %, depending on the concentrations of the sodium hydroxide, sulfuric acid, and citric acid solutions used in the preliminary treatment of raw material [23].

3.2. Gelatin gel strength

Comparison of the gel strength of YSG extracted under various treatment conditions of different alkaline (NaOH) concentration combined with various treatment time is shown in table 2.

Table 2. Gel strength of yellowfin tuna skin gelatin (YSG) under various NaOH concentration and treatment time compared with bovine gelatin.

Treatments	Gel strength of YSG (g Bloom)
0.01 M; 12 h	291.73^{f}
0.03 M; 12 h	261.8897^{e}
0.05 M; 12 h	183.7049^{c}
0.01 M; 18 h	262.5762^{e}
0.03 M; 18 h	203.8145^{d}
0.05 M; 18 h	165.6546^{b}
0.01 M; 24 h	262.2446^{e}
0.03 M; 24 h	201.4912^{d}
0.05 M; 24 h	123.1543^{a}
Gel strength of bovine (g Bloom)	309.7401^{g}

*Values are the means of triplicates.
*Different superscripts in the same column indicate significant differences ($p < 0.05$).

The gel strength of YSG which used 0.01 M NaOH for the alkaline treatment combined with a treatment time of 12 h gave the highest gel strength (291.73 g Bloom) compare to the other treatments, however, this value was slightly lower than that of Commercial Bovine Gelatin (CBG) (309.7401 g Bloom). In accordance with the result for the best yield, the use of 0.01 M NaOH combined with a treatment time of 12 h shows the best pretreatment condition prior to gelatin extraction. This result suggested that the removal of collagen protein in pretreatment step (using NaOH) is important for the gelatin extraction process and it might further affect the gel strength of obtained gelatin.

Gel strength is the most important attribute of gelatin and determines the quality of produced gelatin [18]. Furthermore, the difference in bloom strength among species was possibly due to the different composition, particularly in terms of amino acid composition and size of protein chains [8]. It is likely that relatively limited amino acids and glycine content might have resulted in less organized triple helical structures and is largely responsible for lower bloom strength [24]. The difference in bloom strength among species was possibly due to the different composition, particularly in terms of amino acid composition and size of protein chains [25]. The low hydroxyproline content in fish skin gelatin was a major reason for the low gel strength of these gelatins. It is well established that hydrogen bonds between water molecules and free hydroxyl groups of amino acids in gelatin are essential for gel strength [26].

3.3. Gelatin viscosity

Comparison of the viscosity of YSG extracted under various treatment conditions of different alkaline (NaOH) concentration combined with various treatment time is shown in table 3.

Table 3. The viscosity of yellowfin tuna skin gelatin (YSG) under various NaOH concentration and treatment time compared with bovine gelatin.

Treatments	Viscosity of YSG (cPs)
0.01 M; 12 h	8.6^{j}
0.03 M; 12 h	4.5^{f}
0.05 M; 12 h	4.1^{c}
0.01 M; 18 h	8.5^{i}
0.03 M; 18 h	4.3^{e}
0.05 M; 18 h	3.9^{b}
0.01 M; 24 h	8.2^{h}
0.03 M; 24 h	4.2^{d}
0.05 M; 24 h	3.8^{a}
Viscosity of bovine (cPs)	5.6^{g}

*Values are the means of triplicates.
*Different superscripts in the same column indicate significant differences ($p < 0.05$).

The viscosity of YSG which used 0.01 M NaOH for the alkaline treatment combined with treatment time of 12 h gave the highest viscosity (8.6 cP) compare to the other treatments. This suggested that the use of 0.01 M NaOH combined with a treatment time of 12 h gave the best pretreatment condition for gelatin extraction. The viscosity of YSG solution (8.6 cP) was higher than that of CBG (5.6 cP) and it shows a significantly different ($p < 0.05$). Viscosity is the second most important commercial physical property of a gelatin [19]. The viscosity of gelatin solutions varies with sources in terms of molecular weight and molecular size distribution of proteins [20]. Viscosity did not appear to be affected by the protein yield but was clearly correlated with the molecular weight distribution of the extracted proteins. The gelatin-containing smaller ratio of large molecule portion like β-chains exhibited lower viscosity value [21]. The viscosity of gelatin solutions varies with sources in terms of concentration, pH [27] as well as temperature [28] of the gelatin solutions.

3.4. Gelatin melting point

Comparison of the melting point of YSG extracted under various treatment conditions of different alkaline (NaOH) concentration combined with various treatment time is shown in table 4.

Table 4. The melting point of yellowfin tuna skin gelatin (YSG) under various NaOH concentration and treatment time compared with bovine gelatin.

Treatments	Melting Point of YSG (°C)
0.01 M; 12 h	35^{d}
0.03 M; 12 h	33.5^{c}
0.05 M; 12 h	31^{a}
0.01 M; 18 h	33^{bc}
0.03 M; 18 h	33^{bc}
0.05 M; 18 h	32.5^{b}
0.01 M; 24 h	33^{bc}
0.03 M; 24 h	33^{b}
0.05 M; 24 h	32.5^{b}
Melting point of bovine (°C)	36.33^{e}

*Values are the means of triplicates.
*Different superscripts in the same column indicate significant differences ($p < 0.05$).

The melting point of YSG which used 0.01 M NaOH for the alkaline treatment combined with a treatment time of 12 h gave the highest melting point (35 °C) compare to the other treatments. The melting point of YSG (35 °C) was slightly lower than that of CBG (36.33 °C). An aqueous solution of a few percent gelatin forms thermally reversible gels with water and the gel melting temperature (< 35 °C) is below body temperature, which gives gelatin products unique organoleptic properties and flavor release [5]. The wide range of gelling temperatures is greatly influenced by the origin of the raw material used in the process [22].

4. Conclusion

The present study indicated that the optimal condition for YSG preparation was using 0.01 M NaOH for 12 h. The results show that the concentration of NaOH had a significant effect on yield, gel strength, viscosity and melting point.

References

[1] Ward A G and Courts A 1977 *The Science and Technology of Gelatin* (London: Academic Press)
[2] Schrieber R and Gareis H 2007 *Gelatine Handbook* (Weinhem Wiley-VCH GmbH and Co.)
[3] Go´mez-Guille´n M C, Turnay J, Ferna´ndez-Dı´az M D, Olmo N, Lizarbe M A and Montero P 2002 *Food Hydrocolloids* **16** 25–34
[4] Baziwane D and He Q 2003 *Food Reviews International* **19** 423–435
[5] Glicksman M 1969 *Gum Technology in the Food Industry* (New York: Academic Press)
[6] Kolodziejska I, Skierka E, Sadowska M, Kolodziejski W and Niecikowska C 2008 *Food Chemistry* **107** 700–706
[7] Grossman S and Bergman M 1992 *Process for the Production of Gelatin from Fish Skin* (United States: Patent No. 5,093,474)
[8] Muyonga J H, Cole C G B and Duodu K G 2004 *Food Hydrocolloids* **18** 581–592
[9] Zhou P and Regenstein J M 2005 *Journal of Food Science* **70** 392–396
[10] Aewsiri T, Benjakul S, Visessanguan W, Eun J B, Wierenga P A and Gruppen H 2009 *Food Chemistry* **117** 160–168
[11] Jongjareonrak A, Rawdkuen S, Chaijan M, Benjakul S, Osako K and Tanaka M 2010 *Food Science and Technology* **43** 161–165
[12] Jamilah B, Tan K W, Umi Hartina M R and Azizah A 2011 *Food Hydrocolloids* **25**(5) 1256–1260
[13] Wangtueai S and Noomhorm A 2009 *Food Science and Technology* **42** 825–834
[14] Mohtar N F, Perera C and Young Quek S 2010 *Food Chemistry* **122** 307–313
[15] Tabarestani H S, Maghsoudlou Y, Motamedzadegan A and Mahoonak A R S 2010 *Bioresource Technology* **101** 6207–6214
[16] Steel R G D 1991 *Principles and Procedures of Statistics* (New York: McGraw-Hill Book Co. Inc.)
[17] Johnston-Banks F A 1990 *Gelatin In P Harris* (Ed) *Food gels* (London: Elsevier) pp 233–289
[18] Kasankala L M, Xue Y, Weilong Y, Hong S D and He Q 2007 *Bioresource Technology* **98** 3338–3343
[19] Johnston-Bank F A 1983 *Journal of the Society of Leather Technologists and Chemists* **68** 141–145
[20] Sperling L H 1985 *Introduction to Physical Polymer Science* (New York: Wiley)
[21] Niu L, Zhou X, Yuan C, Bai Y, Lai K and Yang F 2013 *Food Hydrocolloids* **33** 336–341
[22] Karim A A and Bhat R 2009 *Food Hydrocolloids* **23** 563–576
[23] Gudmundsson M and Hafsteinsson H 1997 *Journal of Food Science* **62** 37–39
[24] Binsi P K, Shamasundar B A, Dileep A O, Badii F and Howell N K 2009 *Food Hydrocolloids* **23** 132–145
[25] Muyonga J H, Cole C G B and Duodu K G 2004 *Food Hydrocolloids* **18** 581–592
[26] Arnesen J A and Gildberg A 2002 *Bioresource Technology* **82** 191–194
[27] Cho S H, Jahncke M L, Chin K B and Eun J B 2006 *Food Hydrocolloids* **20** 810–816

[28] Arnesen J A and Gildberg A 2007 *Bioresource Technology* **98** 53–57

IOP Conf. Series: Earth and Environmental Science **139** (2018) 012048 doi:10.1088/1755-1315/139/1/012048

The effect of mimosa and syntan mixture on the quality of tanned red snapper leather

M Pratama[1], L Sahubawa[1*], A Pertiwiningrum[2], Y Rahmadian[2] and I D Puspita[1]

[1] Fisheries Sciences Departement, Faculty of Agriculture, Universitas Gadjah Mada, Jl. Flora Bulaksumur, Yogyakarta, Indonesia
[2] Faculty of Animal Sciences, Universitas Gadjah Mada, Jl. Fauna, Bulaksumur, Yogyakarta, Indonesia
*Email: Latifsahubawa2004@yahoo.com

Abstract. This study aimed to determine the effect of syntan and mimosa combination on the quality of tanned red snapper leather. The study was designed using complete randomized design (CRD) and the advanced test of Least Significant Difference (LSD) with three replications and three treatments of combined syntan and mimosa, namely: p1 (4 % syntan, 5 % mimosa), p2 (8 % syntan, 5 %mimosa), p3 (12 % syntan, 5 % mimosa). Data were analyzed using SPSS 18. The measured parameters were tensile strength ($N \cdot cm^{-2}$), elongation (%), tear strength ($N \cdot cm^{-1}$), enervation (mm), wrinkle temperature (°C), fat/oil content (%) and water content (%). The results indicate that the combined syntan and mimosa tanners gave no significant effect on the physical parameters (tensile strength, elongation, tear strength, enervation and wrinkle temperature) and chemical parameters (moisture and fat/oil). All treatments (p1, p2 and p3) met the Indonesian National Standard (SNI) 06-4586-1998 for chrome tanned freshwater snake leather for the parameters of tensile strength, tear strength, enervation, wrinkle temperature and moisture. The parameter of elongation and fat/oil content did not meet with the SNI 06-4586-1998. Among the three combinations of a tanner, syntan 4 % and mimosa 5 % treatment gave the best leather quality.

1. Introduction

Red snapper fish is one of the marine fishery commodities with high economic value with increased production volume from year to year in the last five years. Based on the Ministry of Marine, total volume of national capture fishery production in 2012 amounted to 5,435,632 tons, while the volume of red snapper production in 2008 to 2012 respectively: 109,299; 115,523; 123,794; 118,608 and 119,008 ton [1]. Along with the increase of the catch, it will certainly have an impact on the development of fishery processing industry in general, and especially the industry of fish snapper.

Snapper fillet industry in the production process produces a large amount of waste that has the potential to be used as raw material for food or feed products and other derivative products. Snapper fillet processing industry produces about 40% of raw material waste per ton consisting of leather (35 %), head (25 %), offal (25 %) and bone (15 %) [2].

Leather waste from the fish filet industry is generally used for the processing of leather cracker which has low economic value, whereas fish leather can be processed as raw material (leather tanned) and leather products that have high economic value [2, 3]. Along with advances in science and technology in the field of leather processing, the red snapper leather waste in particular and the leather of fish, in general, can be utilized into a variety of commercial leather products that have high economic value because it has a scarred scar on the leather surface that is very interesting and unique.

The tanning process produces tanned leather (finished leather) as raw material for commercial leather products (bags, wallets, shoes, jackets, HP covers, key chains, belts, backpacks, laptop bags and other accessories) [2]. The attractiveness of the leather products is strongly influenced by the quality of tanned leather, where the quality of the tanned leather is favorably affected by the tanning and tanning process used [3]. The tanner commonly used in the leather tanning industry is chrome because it produces high-quality tanned leather, but chromium is very harmful to human health and the environment so that alternate tanner alternatives are needed that can produce high quality tanned leather. One alternative replacement is the use of vegetable tannin or a combination of vegetable tannin with chemicals (mimosa and syntan or formaldehyde) and or with minerals.

The combination tanning is the use of two or more tanning materials in the tannery process [4]. The use of tanner mixtures can improve the quality of tanned leather as each tanner can cover each other's deficiencies through its own advantages [5]. Each tanner (syntan, formaldehyde, mimosa and chrome) has its own characteristics and advantages. Syntan has characters that do not give the effect of color on tanned leather, resulting in tanned leather that is solid, soft, and resistant to acid and base influences and can increase water resistance of tanned leather [4]. Mimosa has the ability to produce tanned leathers that contain, tighten and protect the leather from microbial activity and provide color to tanned leather, as a secondary effect of tannin compounds [6]. Research on the combination of tanners has been done by previous researchers [7], which used a mixture of 6 % mimosa and syntan 4 % for stingray leather tanning resulted in high tensile strength, tear strength, enervation and wrinkle temperature but low in water and fat content. The use of a mixture tanner is recommended in the process of leather tanning with the aim of improving and enhancing the quality of the raw material of leather [8]. Furthermore, loyal tanners (natural, mineral and chemical) have their respective advantages so that mixtures of two or more tanners will improve the quality of tanned skin [8].

2. Materials and Methods

2.1. Equipment and materials

Equipment used in processing tanning is knife, scales, measuring cup, plastic bucket, small bowl, paper pH, thermometer, boards, gloves, stacking boards, softness tester, tensile strength tester and wrinkle temperature tester. Materials used in the tanning process are red snapper leather, salt, water, sodium sulfide (Na_2S), lime ($Ca(OH)_2$), ammonium sulphate (ZA), palcobate, hustapol NID, formicacid, syntan, mimosa, chrome, sodium bicarbonate ($NaHCO_3$), sincal-MS, paint base, sulfonation oil, antifungus and katalix GS.

2.2. Methods

The method of processing fish leather as finished leather (raw material) products and leather goods are "tanning method", which consists of three stages: stage pre-tanning, tanning and finishing (figure 1) [9, 10]. The method used in the study is a completely randomized design (CRD) single factor, using single factor as the source of the treatment with three levels: p1 (mixture of 4 % syntan and mimosa 5 %), p2 (mixture of 8 % syntan and mimosa 5 %) and p3 (mixture of 12 % syntan and mimosa 5 %). Statistical analysis of the data processed by SPSS version 18, if the results of the analysis show a real effect then continued with the least significant difference test (BNT) at a significant level of 95 % to determine the differences between treatments. Quality parameters of tanned fish leather analyzed are, (1) physical (tensile strength, elongation, tear strength, enervation and wrinkle temperature) and (2) chemical (moisture and content oil/fat) [9, 10].

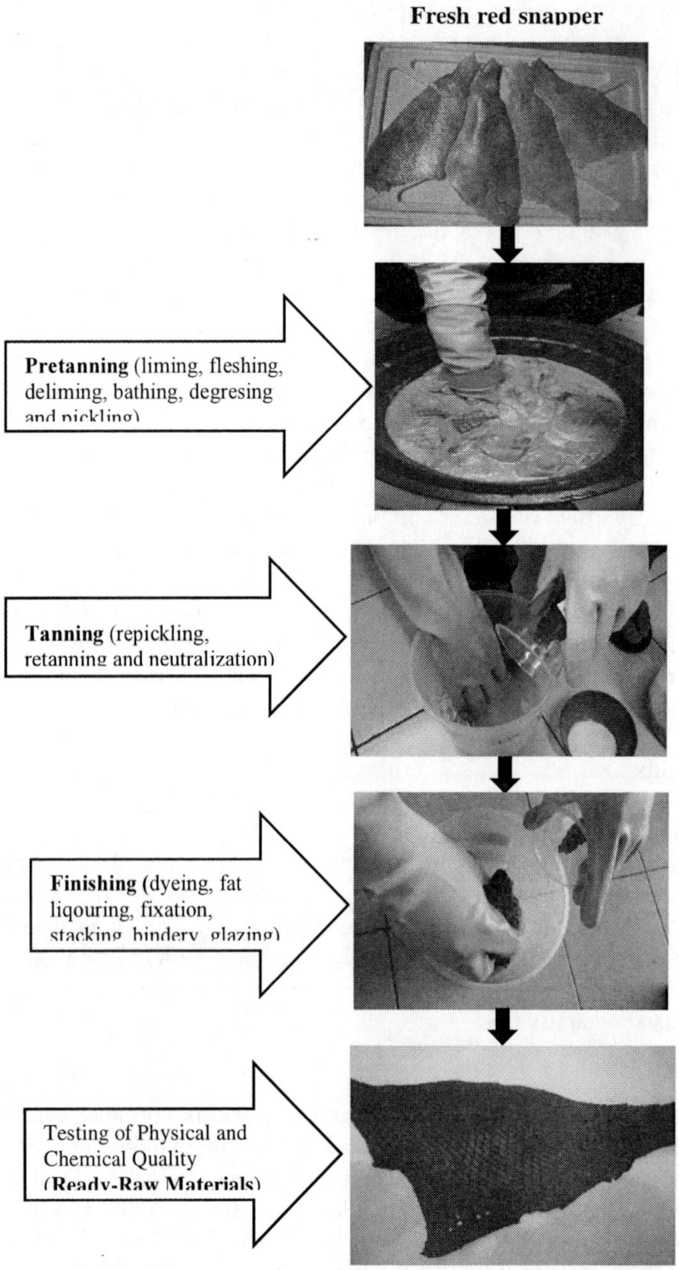

Figure 1. Tanning process of red snapper leather.

3. Results and Discussion

The study used mixed treatment of syntan and mimosa tanner as a single factor, consisting of three treatments that were p1 (a mixture of 4 % syntan and 5 % mimosa), p2 (a mixture of 8 % syntan and 5 % mimosa), p3 (a mixture of 12 % syntan and 5 % mimosa). Leather quality parameters that analyzed consisted of physical quality (tensile strength, elongation, tear strength, enervation and wrinkle temperature) and chemical quality (moisture and fat/oil content). The result of the physical and chemical quality of tanned red snapper leather was shown in table 1.

3.1. Tensile strength

The result of the variance analysis shows that the experimental treatments (a mixture of Syntan and mimosa) had no significant effect on the tensile strength of the tanned red snapper leather sample at a significant level of 95 %. The mean value of tensile strength of tanned red snapper leather sample from the treatment of p1, p2 and p3 were 1,339.19 N·cm^{-2}; 1,129.76 N·cm^{-2} and 1,075.78 N·cm^{-2}, respectively. The mean value of tensile strength of leather samples from each treatment tended to decrease with increasing concentration of tanners.

Table 1. The result of physical and chemical testing of tanned red snapper leather.

Kind of Test	Test Parameters	Treatment (Tanning Materials)			Standard
		p1	p2	p3	
Physical Quality	Tensile strength (N·cm^{-2})	1,339.19	1,129.76	1,075.78	> 1000 (SNI of freshwater snake leather)
	Elongation (%)	57.72	45.99	55.86	< 30 (SNI of snake leather)
	Tear strength (N·cm^{-1})	266.93	203.89	266.84	> 150 (SNI of fereshwater snake leather)
	Enervation (mm)	4.03	3.29	4.05	> 2 Burks's Bay TM (Nukom manufacturing)
	Wrinkle temperature (°C)	93.97	85.53	84.92	> 70 (SNI of stringray leather)
Chemical Quality	Moisture (%)	8.95	9.02	8.93	< 18 (SNI of freshwater snake leather)
	Oil content (%)	15.16	14.24	13.93	2 to 6 (SNI of freshwater snake leather)

Based on the analysis result, tensile strength value of sample tended to decrease but still higher than the resulting reported by previous researchers [11], using mimosa 20 % in red snapper leather with the value of 1,115 N·cm^{-2}; also higher than the result reported by previous researchers [12], using 3 % of chrome on red snapper with value of 1,150.314 N·cm^{-2}. This condition indicates that the use of mixed tanner concentration tends to be better compared with the use of single tanning material (without combination).

All the treatments in this study fulfill the requirements of BSN [9] concerning freshwater snake leather tanned by chrome, with the recommended minimum standard for tensile strength of 1,000 N·cm^{-2}. The resulting tensile strength was quite high because the syntan has a great ionization capacity in the tanning process so that the penetration process of the tanner material goes well and binds perfectly to the collagen protein [13]. Syntans have a complex active substance in the form of polyhydroxy benzoles that have large (positive and negative) hydroxyl groups in order to react perfectly with leather collagen forming a hydrogen bond to produce high-quality tanned leather [14].

3.2. Elongation

The results of variance analysis showed that the experiment did not significantly affect the elongation of tanned red snapper leather sample at a significant level of 95 % ($p > 0.05$). The average value of leather elongation of each treatment (p1, p2 and p3) was 57.72 %; 45.99 % and 55.86 %, respectively. The addition of syntan concentration has no significant effect on the change of sample elongation characteristic. The results of this study show the same characteristics with the result reported by previous researchers on red tilapia with a value of 56 % elongation at 10 % of syntan treatment [15]. In contrast, higher than the result reported by previous researchers who using 5 % chrome on red snapper with a value of 38.3 % elongation [12]. Based on these comparative results, the use of mixed syntan and mimosa on tanning process and the use of chrome on the re-tanning process obtained a better elongation value compared with the use of syntan alone as a single tanner. The mean value of elongation of all treatments did not fulfill BSN [9], on chrome-tanned freshwater snake leather with the requirement of 30 % maximum elongation quality. High leather elongation values were attributed to the sample containing a large amount of oil/fat [16].

3.3. *Tear strength*

The results of the variance analysis show that the experiment did not significantly affect the tear strength of the leather sample at a significant level of 95 %. Average value of tear strength of snapper leather sample from treatment p1, p2 and p3 was 266.93 N·cm^{-1}, 203.89 N·cm^{-1} and 266.84 N·cm^{-1}, respectively. The resulted quality score has fulfilled the requirements of BSN [9], on the chrome-tanned freshwater snake leather with a minimum tear strength standard of 150 N·cm^{-1}. The characteristic of the tear strength was almost the same as the tensile strength that was strongly influenced by the quality of the tanner. Syntan has a polyhydroxy benzoles having alarge amount of positive and negative charged hydroxyl groups that can react with collagen perfectly to form a hydrogen bond to produce a better quality of tanned leather [14].

The tear strength is affected by the histology of the leather of the fish (the extent/width of the leather and the constituent cells) affecting the leather-forming collagen fibers, where the more dense collagen fibers, the stronger the formed-leather collagen fibers the leather forms, the stronger the strength tensile and tear strength of leather sample [17]. The tear strength of this study was higher than results reported by previous researchers [18], using 10 % of mimosa on tilapia leather with a value of 137.40 N·cm^{-1}, but lower than results reported by previous researchers that using 5 % of chrome on the snapper leather with the same tear strength value of 289.29 N·cm^{-1} [12].

3.4. *Enervation*

The result of variance analysis showed that the experiment did not give any significant effect to enervation parameter of the tanned red snapper leather sample at a significant level of 95 % ($p > 0.05$). The enervation average value of red snapper leather from the treatments of p1, p2 and p3 were 4.03 mm, 3.29 mm and 4.05 mm, respectively. The standard parameters for the elongation were not mentioned in the BSN [9], on chrome-tanned freshwater snake leather, but they are usually compared to the Burks's Bay TM requirements by the Nukom industry on leather quality for superior leather and suede leather with a minimum quality standard of 2 mm. Thus, the enervation average value of leather sample produced had met the quality requirements. High enervation value is caused by active substances of tanners entering into leather fibers (collagen) leaving room in the leather to be filled by oil during the process of oil addition resulting in a flexible leather sample [19]. The value of the study results was higher than the results reported by previous researchers [18], using 10 % of mimosa (2.15 mm) and 10 % of chrome (2.37 mm) on the tilapia leather and that using 10 % of syntan on the tilapia leather with the value of 3.76 mm [15].

3.5. *Wrinkle temperature*

The results of variance analysis showed that the experiment did not significantly affect the wrinkle temperature of the snapper leather sample at a significant level of 95 %. Average of leather wrinkle values of snapper fish samples from treatment p1, p2 and p3 were 93.97 °C, 85.53 °C and 84.92 °C, respectively. The minimum quality requirement for wrinkle temperature based on BSN [10] about stingray leather for finished goods is 70 °C, hence, the mean value of wrinkle temperature in this experiment had fulfilled the quality requirement of tanned leather. This means that the samples of tanned leather with syntan and mimosa are very resistant to changes in ambient temperature.

3.6. *Water content*

The result of variance analysis showed that the treatments did not significantly affect the water content of red snapper samples at a significant level of 95 %. Average value of water content of snapper leather samples from treatment p1, p2 and p3 was 8.95 %, 9.02 % and 8.93 %, respectively. The higher the concentration of tanning materials, the more water tied up by the tanner so that the water content of the leather sample tends to decrease [7]. The mean value of water content of all treatments has fulfilled the requirements of BSN [10] on stingray leather for finished goods requiring maximum water content of 18 %. The water content of leather samples significantly influences the quality of stored leather products. Tanned fish leathers with low water content (according to ISO standard) are more resistant to microbial activity and changes in ambient temperature.

3.7. Oil content

All treatments did not significantly affect the fat content of red snapper samples at a significant level of 95 % ($p > 0.05$). The mean value of the fat content of the snapper leather samples from the treatment of p1, p2 and p3 was 15.16 %, 14.24 % and 13.93 %, respectively, tend to decrease along with the increase of tanner concentration. Mean of the fat content value of all treatments did not fulfill with BSN [9] about chrome-tanned freshwater snake leather which requires 2 to 6 %. Oil process occurs oil penetration efforts into the webbing of leather that uses to improve the leather enervation. Excessive concentration of oil given will result in high oil levels of red snapper leather. The oil/fat content will affect the preservation and consumers' convenience.

4. Conclusion

Treatment of the tanning mixture did not have a significant effect on the 95 % significance level on the physical and chemical quality of the red snapper samples. The parameters of tensile strength, tear strength, elongation, wrinkle temperature and moisture content had fulfilled the quality standard of BSN [9] about chrome-tanned freshwater snake leather respectively, whereas elongation parameters and oil content had not met the SNI standard. The leather sample obtained by treatment with p1 (mixture 4 % of syntan and mimosa 5 %) was the best (efficient) treatment.

References

[1] Ministry of Marine Affairs and Fisheries 2013 *Fishery Statistics of Marine Fishing Catch* http: //statistik.kkp.go.id. Reaccessed on 22 September 2015

[2] Sahubawa L and A Pertiwiningrum 2015 *Effect Concentration of Tanning Materials of Plant (Mimosa) on the Leather Quality of White Snapper Green Job-Type Fish and Its Application in Commercial Leather Products Processing* Research Report (not published) (Yogyakarta: Department of Fisheries Faculty of Agriculture Universitas Gadjah Mada)

[3] Haq N, Yunizal and Suherman 2000 *Fish Leather Preserving and Tanning Technology* (Jakarta: Research and Development Center for Sea Exploration and Fisheries)

[4] Widari R and Suparti 2013 *Magazine of Central Leather Rubber and Plastics (BBKKP) Yogyakarta* **29**(02) 99–104

[5] Sahubawa L, Pertiwiningrum A and Adityo T P 2011 *Leather, Rubber and Plastic Magazines* **27**(01) 38–45

[6] Anonyms 2011 *Report on the Development of E-Learning Learning Grants* (Yogyakarta: Learning Development Center Universitas Gadjah Mada)

[7] Asgha B 2010 *The Effect of Mimosa Tanner Material and Syntan on Quality of Tanned Stingray Leather* Research Report (Yogyakarta: Faculty of Agriculture Universitas Gadjah Mada)

[8] Sahubawa L 2013 *Grant of Community Service* (Yogyakarta: Universitas Gadjah Mada)

[9] BSN 1998 SNI 06-4586-1998 *Indonesian National Standard of Finished Leather From Freshwater Snakes Leather Tanned Chrome* (Jakarta: National Standardization Agency)

[10] BSN 1999 SNI 06-6121-1999 *Indonesian National Standard for Stingray Leather Goods* (Jakarta: National Standardization Agency)

[11] Susanti M 2006 *Study of Use of Tanning Materials of Plant (Mimosa) Toward Physical Quality of Tanned Red Snapper Leather* (Yogyakarta: Research Report Faculty of Agriculture Universitas Gadjah Mada)

[12] Budiyanto R A 2010 *The Effect of Chromium (Cr_2O_3) on Tanned Snapper Leather Quality* (*Lutjanus* sp.) (Bogor: Research Report Faculty of Fisheries and Marine Science Bogor Agricultural Institute)

[13] Untari S 2001 *Application of Leather Tanning Technology for Bag Material with Quick Tanning System* (Yogyakarta: Central Research and Development of Leather, Rubber and Platics (BBKKP))
[14] Triatmojo S and Mohammad Z A 2014 *Eco-friendly Tannery* (Yogyakarta: Gadjah Mada University Press)
[15] Hayati E 2009 *Chemical and Natural Tanning Effectiveness on Red Tilapia Fish Quality* (Yogyakarta: Research Report Agriculture Faculty Universitas Gadjah Mada)
[16] Untari S, Setyowati D and Jatmikowati E S 2004 *Livestock Bulletin Diponegoro University Semarang* **28**(2) 87–95
[17] Untari S, Emiliana S, Wrd, Sutyasmi S and Susila J 2009 *Intensive Program of State Ministry of Research and Technology* (Yogyakarta: Central Rubber and Plastic Leather)
[18] Astrida M 2008 Minithesis Department of Fisheries Faculty of Agriculture Universitas Gadjah MadaYogyakarta
[19] Mustakim U S, Widati and Purnaningtyas L 2007 *Journal of Animal Product Science and Technology* **2**(1) 26–34

IOP Conf. Series: Earth and Environmental Science **139** (2018) 012049 doi:10.1088/1755-1315/139/1/012049

Enhancing the economic value and consumer preferences of commercial *mondol* stingray (*Himantura gerardi*) leather creative products

L Sahubawa[1*], A Pertiwiningrum[2] and Y Rahmadian[2]

[1] Departement of Fisheries, Faculty of Agriculture, Universitas Gadjah Mada, Jl. Flora, Bulaksumur, Yogyakarta, Indonesia
[2] Faculty of Animal Sciences, Universitas Gadjah Mada, Jl. Fauna, Bulaksumur Yogyakarta, Indonesia
*Email: Latifsahubawa2004@yahoo.com

Abstract. The research objectives were to design, assess the economic value and consumer preference level of stingray leather products. The research method included a product design, analysis of economic value and consumer preferences. Mondol stingray (*Himantura gerardi*) leather, with a length of 50 cm and width of 30 cm, were processed into ID card wallet, man and women's wallet and key holder. The number of respondents involved to analyze the preference level is 75 respondents (students, lecturers and employees of Universitas Gadjah Mada). Indicators of consumer preferences were model, color, price and purchasing power. The price of ID card wallet is Rp. 450,000; women wallet is Rp. 650,000 and a key holder is Rp. 300,000. Consumer preferences on ID card wallet were as follow: 84 % stated very interesting model; 83 % stated very interesting color; 61 % stated cheap and 53 % had enough. Consumer preferences of women's wallet were as follow: 81 % stated very interesting model; 84 % stated very interesting color; 56 % stated cheap and 57 % had enough. Consumer preferences on key holder were as follow: 49 % stated interesting model; 72 % stated very interesting color; 61 % stated cheap and 57 % had enough.

1. Introduction

Ministry of Maritime Affairs and Fisheries, The Republic of Indonesia, has given an attention in the implementation of the concept of the blue economy by the utilization of fishery industries by-products/wastes into a variety of creative products with high economic potential [1]. Creative products are innovative products with high economic value that derived from the development of creativity, skills and innovation of individuals or groups, that potentially capable of generating creativity, intellectual property and employment [2]. One of an example of creative product derived from the fishery industry by-products is stingray leather. Stingrays are usually obtained as a by-catch product of catching operation using gill net, purse seine and hand poles. Nationwide, the production of stingrays until the end of 2014 reached 70,447 tons, including species of bird stingrays, shovel noses, *kekeh*, bats and flower/tiger [3]. Stingrays belong to economically important fish groups but are less favored by consumers because of its pungent meat (due to high ammonia content). In some areas that have considerable fisheries potential, they are often thrown back into the sea. Stingrays are generally sold fresh for ± Rp. 4,000–5,000 per kg [4, 5]. With the advancement of innovation as well as science, technology and art in the leather creative industry, stingray leather can be utilized for the development of commercial leather creative products and accessories with high economic value, meanwhile, its meat can be processed into commercial food products (jerky and steak) that are free from ammonia stench [6]. Tanned stingray leather (a type of *mondol* stingray) with a size of 8 inches and a price of Rp. 75,000/sheet can be processed into one ladies wallet (main product, 2-in-1 type) with the price of Rp. 350,000 and two pieces of bracelet/accessories (derived product) with the price of Rp. 300,000 [7]. This shows that one sheet of stingray tanned skin potentially gives the total economic value of Rp. 650,000 with the value added of 866 %.

At the time of the catching season, the stingrays are landed in considerable quantities and are only sold fresh. Fresh stingray leather can be processed into tanned leather as raw material on the processing of leather creative product at a price ranged between Rp. 100,000–125,000 per sheet. The price range of *mondol* stingrays leather (per sheet) of small, medium and large size was Rp. 60,000–75,000; Rp. 75,000–100,000 and Rp. 100,000–125,000, respectively [8, 9], while other types of stingray leather, such as *pariduri*, have higher selling prices because of the pearls/beads along the center of the skin [7, 8].

The purposes of the research were to design, process and determine the level of consumer demand for commercial stingray creative products. This research provided information on processing technology as well as the economic value of stingray creative products to the community (especially coastal communities and small islands) as well as leather craftsmen and investors of fish-based goods and products.

2. Materials and Methods

2.1. Materials and equipment

The main raw materials used were stingray leather (*mondol* type) as well as supporting materials (cow leather, satin fabric, paperboard, glue, paint/dye and other materials). Stingray leather that was used has a length of 50 to 60 cm (size 8 inches) and the width of more than 30 cm, purchased from Kub Zazmi Leather, Moyudan, Sleman, Yogyakarta (figure 1). Important equipment used were a leather machine, grinding machine, dye machine, sewing machine, scissors and cutter.

2.2. The processing of stingray leather creative products

The processing of stingray leather creative products was performed by the following stages: design of product patterns; preparation and cutting of paperboard, stingray leather and satin cloth; skin coloration; forming of stingray sewing lines; stingray rays, cowhide and satin fabrics; gluing leather and fabric sheets; suturing; installation of accessories (buttons and rings); finishing (removal of dust and glue/glue remnants), shine and packaging. The processing of stingray creative products using pattern design method: cutting, coloring, gluing, sewing and finishing. The stingray creative products consisted of the main and derivative products. Main products were produced from the center part of the leather, while derivative products were made from the remaining leather from the main product processing. Main products were identity card wallet (2-in-1 type) or ladies/women wallet (2-in-1 type); while the derivative product was accessories (keyholders).

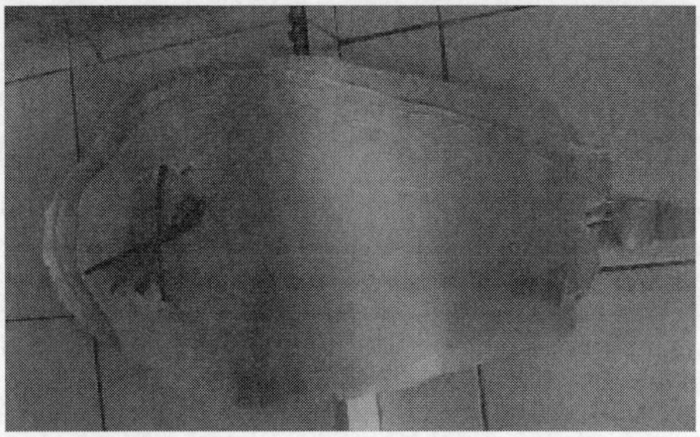

Figure 1. Tanned *mondol* stingray leather sample.

2.3. Analyze of consumer preferences

The consumer acceptances toward the stingray leather creative products were analyzed using consumer preferences/demand method [7]. The number of respondents used for the analysis of the level of consumer preference was 50 women and 25 men (total 75 people) representing three market segments (students, lecturers and employees) at Universitas Gadjah Mada, Yogyakarta. Consumer demand indicators consisted of: (1) product model (very interesting, interesting, interesting enough and not attractive); (2) color (very interesting, interesting, interesting enough and not interesting); (3) price (very expensive, expensive, cheap enough and very cheap) and (4) purchasing power (very high, high, low and very low).

3. Results and Discussion

3.1. Types, specifications and prices of stingray creative products (main products)

The types of stingray leather creative products chosen to be produced in this experiment were the products with the attractive shapes and colors as well as having high competitive selling points. There were two types of products, main product and derivative product. The main product was (1) 2-in-1 type identity card wallet (for passport and card) or (2) 2-in-1 type women's wallet, with specifications respectively as shown in figure 2 and figure 3. In addition to the main products, we also produced derivative products, accessories (key holder), which use raw materials from the remaining leather from the main product. Specifications of derivative products were shown in figure 4.

Identity card wallet (2-in-1 type) was one of the new products manufactured from the stingray leather. The similar product in the market had been made from cow leather. This type of wallet is used to facilitate the storage of medium to large personal identity cards, such as passport and credit card, which is often used in travel and banking transactions. The size of identity card wallet produced from stingray leather in this experiment was larger than the standard product. This product was not only designed to facilitate the storage of important personal identities documents but was also designed to add its functionality as a unique and exciting commercial accessories product. With the dimension of 17 cm × 11 cm × 2 cm, the identity card wallet was priced at Rp. 450,000 per specimen. The price of this product was more expensive than similar products made from another type of leather (Rp 250,000–300,000), but it offered attractive patterns and unique motifs, moreover, this kind of product was difficult to be found in outlets or commercial leather products store [7].

The women's wallet (grasp model, 2-in-1 type) was one of the newly engineered products of similar products made from stingray leather [10], but this new product has three interior parts and thicker than other women's wallet produced previously. This product was designed to facilitate the storage of medium to a large personal identity card, banknotes, bills, business cards and other securities. The design of this product was not only facilitating important personal documents but was also designed to add its functionality as a unique and exciting commercial accessories product. With

IOP Conf. Series: Earth and Environmental Science **139** (2018) 012049 doi:10.1088/1755-1315/139/1/012049

the dimension of 20 cm × 11 cm × 2.5 cm, the women's wallet was priced at a range of Rp. 600,000–650,000 per specimen. The price of this product was more expensive than similar products from stingray leather (Rp. 350,000–400,000), but it offer and attractive looks, unique motifs and luxurious style, moreover, this product was still difficult to be found by the consumers (especially for the collectors of genuine leather products).

The key holders (type 2-in-1) were one of the newly engineered products of similar products made from the stingray leather [10]. This new product had a larger and more attractive size than similar products produced by other three manufacturers in Yogyakarta. This product was designed to facilitate the storage of small personal identity cards, such as Motor Vehicle Identity Letters and personal business cards. This product was not only aimed to facilitate the storage of identity document of motor vehicles (cars and motorcycles), but its functionality was also added as a commercial accessories product. With the dimension of 11 cm × 5 cm × 1.5 cm, the key holder was priced at Rp. 150,000 per specimen. The price of this product was higher than similar products (Rp. 125,000) that were sold by other producers in Yogyakarta, but this product offered an interesting and luxurious impression.

Figure 2. ID Card wallet (2-in-1 type, light green color) (specification: length 17 cm, width 11 cm, thickness 2 cm and price Rp. 450,000).

Figure 3. Women's wallet (2in-1 type, light green color) (specification: length 20 cm, width 11 cm, thickness 2.5 cm and price Rp. 650,000).

Figure 4. Key holders (processed from the rest of the leather of the main product) (specification: length 11 cm, width 5 cm, thickness 1.5 cm and price Rp. 150,000).

3.2. The level of consumer preference

3.2.1. ID card wallets (2-in-1 type). The level of consumer preference on the ID card wallet made from stingray leather was shown in table 1. Based on the results of consumer preference analysis of 75 respondents, it was known that the average of 84 % consumers stated that the model of identity card wallet was "very interesting" (highest on employees) and 16 % stated "interesting" (highest on lecturer). The consumer preference on color was 83 % stated "very interesting" (highest on the students) and 17 % stated "interesting" (highest on lecturers). The consumer preference on price was 39 % stated "expensive" (high on the employee) and 61 % stated "cheap enough" (highest on lecturers). The level of purchasing power was shown by 36 % of consumers was categorized as having "high purchasing power" (highest in lecturers), 53 % having "low/enough purchasing power" (highest on lecturers) and only 11 % having "very low purchasing power" (not affordable) with the highest score on employees.

Table 1. Level of consumer preference on ID card wallet as the stingray leather creative product.

Consumer Demand Indicators	Level of Consumer Preference (Quality and Quantity)			
Models	very interesting 84 %	interesting 16 %	interesting enough 0 %	not attractive 0 %
Color	very interesting 83 %	interesting 17 %	interesting enough 0 %	not interesting 0 %
Prices	very expensive 0 %	expensive 39 %	cheap enough 61 %	very cheap 0 %
Purchasing power	very high 0 %	high 36 %	Low/enough 53 %	very low 11 %

3.2.2. Woman's wallet (2-in-1 type). The level of consumer preference on the women's wallet made from stingray leather was shown in table 2. Based on the results of consumer preference analysis of 75 respondents, it was known that 81 % of consumers stated that the model of women's wallet (2-in-1 type) was "very interesting" (highest on employees) and 19 % stated "interesting" (highest on students). The color preference analysis showed that 84 % of consumers stated the color of the product was "very interesting" (highest on lecturers and students) and 16 % stated "interesting" (highest in students). The consumer preference on price was shown by 44 % of consumers stated "expensive" (highest on the employee) and 56 % stated "cheap enough" (highest on lecturer). For the consumer purchasing power, amount of 28 % consumers was categorized having "high purchasing power" (high on lecturers), 57 % having "low/enough purchasing power" (the highest in lecturers) and only 15 % having "very low purchasing power" (not affordable) with the highest score found on employees.

Table 2. Level of consumer preference on women's wallet (2-in-1 type) as the stingray leather creative product.

Consumer Demand Indicators	Level of Consumer Preference (Quality and Quantity)			
Models	very interesting 81 %	interesting 19 %	interesting enough 0 %	not attractive 0 %
Color	very interesting 84 %	interesting 16 %	interesting enough 0 %	not interesting 0 %
Prices	very expensive 0 %	expensive 44 %	cheap enough 56 %	very cheap 0 %
Purchasing power	very high 0 %	high 28 %	low/enough 57 %	very low 15 %

3.2.3. Accessories (key holders). The level of consumer preference on the key holders made from the stingray leather was shown in table 3. From the analysis of consumer preference, it was known that 48 % of consumers stated that the product/accessories model was "very interesting" (highest on the students); 49 % stated "interesting" (highest on lecturers) and only 3 % stated "interesting enough" (highest in students). The consumer preference on color was shown by 72 % of consumers stated that the color of the product was "very interesting" (highest on the student) and 28 % stated "interesting" (highest on lecturer). For the price, amount of 40 % consumers stated "expensive" (highest on the employee) and 60 % stated "cheap enough" (highest on lecturer). The purchasing power of 31 % consumers was categorized as having "high purchasing power" (highest on lecturers), 57 % having "low/enough purchasing power" (highest on lecturers) and only 12 % showing "very low purchasing power" (not affordable) with the highest value encountered on the employee.

Table 3. Level of consumer preference on key holders as the stingray leather creative product.

Consumer Demand Indicators	Level of Consumer Preference (Quality and Quantity)			
Models	very interesting 48 %	interesting 49 %	interesting enough 3 %	not attractive 0 %
Color	very interesting 72 %	interesting 28 %	interesting enough 0 %	not interesting 0 %
Prices	very expensive 0 %	expensive 40 %	cheap enough 60 %	very cheap 0 %
Purchasing power	very high 0 %	high 31 %	Low/enough 57 %	very low 12 %

Based on the results of the above analysis, it was known that lecturers, employees and students have the same perception on the "model and color" of stingray leather creative products. The perception of lecturers on "price and purchasing power" was higher than employees and students. This is because lecturers have a steady income and higher than employees and students. When compared to employees (who had a steady income), it turned out that students had a higher desire to buy stingray creative products, even though they did not have a steady income. This is because the creative products are unique, attractive and luxurious, moreover, it can be used as accessories on the important occasions so it would be needed by the women. Stingray creative products have a high economic value because they are interesting, unique and rarely found in the market, therefore they are able to be used as souvenir products or special gifts (birthdays and weddings) [7]. The uniqueness of stingray creative product is in its beads and pearls that spread on the surface of the leather and looks more attractive if given any color (black, brown, green, red, maroon, natural, pink and other colors) [11]. A stingray leather creative product with high selling prices was considered as a very special leather product and they were often purchased by local and state officials.

Overall, the level of demand and purchasing power of lecturers on stingray creative products was very high. This was due to the higher levels of income, education level, experience and desire compared to employees and students. Factors affecting consumers' purchasing power are the level of education, income, needs, habits and prices and goods model [12, 13]. In addition, geographical environmental factors, consumer behavior and product marketing strategies significantly influence the choice and purchase of new product types [14].

The higher the person's education level, the higher the need to be fulfilled. The higher consumer's income, the higher purchasing power and the various needs to be achieved. The level of demand for each consumer is varied, whereas consumers living in urban areas have higher purchasing power when compared to consumers living in rural areas. The lifestyle of people in the modern era shows a higher consumption rate of products/goods and services (consumerism). If the price of the products/goods or services increases, the consumer's purchasing power tends to decrease, otherwise if the price of the products/goods and services decreases, it is assured that consumer's purchasing power will increase. The new model of goods also will stimulate consumers to buy the goods in large enough quantities. Consumers are always striving for maximum satisfaction by meeting the various needs on desired products/goods and services [13]. This research indicates that consumers had a good acceptance of stingray leather creative products and general consumer preference was on their attractive model and color, as well as their affordable price.

4. Summary and Recommendation

4.1. Summary

Stingrays were utilized into creative leather products with high economic value, i.e. identity card wallet (a type of 2-in-1), women's wallet (a type of 2-in-1) and key holders. Consumer preferences on identity card wallet were as follow: (1) model: 84 % stated very interestingand 16 % stated interesting; (2) color: 83 % stated very interesting and 17 % stated interesting; (3) price: 39 % stated expensive and 61 % stated cheap; and (4) purchasing power: 36 % was high, 53 % was enough/low and 11 % was very low. Consumer preferences on women's wallet were as follow: (1) model: 81 % stated very interesting and 19 % stated interesting; (2) color: 84 % stated very interesting and 16 % stated interesting; (3) price: 44 % stated expensive and 56 % stated cheap; and (4) purchasing power: 28 % was high, 57 % was enough/low, and 15 % was very low. Consumer preferences on key holder were as follow: (1) model: 48 % stated very interesting, 49 % stated interesting and 3 % stated interesting enough; (2) color: 72 % stated very interesting and 28 % stated interesting; (3) price: 39 % stated expensive and 61 % stated cheap; and (4) purchasing power: 31 % was high, 57 % was enough/low, and 12 % was very low.

4.2. Recommendation

To increase the economic value of stingray creative products, it is necessary to create innovation and development of new products that are interesting or trendy. In addition, it is necessary to use the remaining leather from the main product processing for the production of other attractive commercial accessories with high economic value.

References

[1] Anonimous 2013 *Report from the Minister of Marine Affairs and Fisheries of Republic Indonesia at the Opening Ceremony of Quality Management Training of Marine and Fisheries Resources* (Jakarta)

[2] Ministry of Trade of The Republic of Indonesia 2009 *Development of the Blue Economy Concept in the Marine Affair and Fisheries Sector* Message from the Minister of Marine Affairs and Fisheries of Indonesia Republic (Jakarta: Opening Ceremony of Quality Management Training of Marine and Fisheries Resources)

[3] Anonymous 2015 *Data on Production Statistics and Volume of National Fisheries Production* (Jakarta: Directorate General of Capture Fisheries Ministry of Marine Affairs and Fisheries Republic of Indonesia)

[4] Sahubawa L 2008 *Vucer Grant for Community Service* (Yogyakarta: Universitas Gadjah Mada)

[5] Sahubawa L and Pertiwiningrum A 2010 *Community Service Handing Grant* (Yogyakarta: Universitas Gajdah Mada)

[6] Sahubawa L 2013 *Grant of Community Service* (Yogyakarta: Universitas Gadjah Mada)

[7] Sahubawa L and Pertiwiningrum A 2016 *Presentation Material of International Conference on Integrated Coastal Management and Marine Biotechnology (ICM-MBT)* in 29th - 30th November 2016 (Bogor: International IPB Conference Center (IICC))

[8] Sahubawa L 2011 *Proceedings of Leather, Rubber and Plastics R and D Workshop* ISBN: 878–979–8378–14–0 (Yogyakarta)

[9] Sahubawa L, Pertiwiningrum A and Pamungkas 2011 *Leather, Rubber and Plastics Magazine* ISSN: 1829–6971 (Yogyakarta)

[10] Sahubawa L Pertiwingrum A, Lelana I Y B and Meylinda 2010 *Proceedings of the VII Annual Seminar* (Yogyakarta: Departement of Fisheries Faculty of Agriculture Universitas Gadjah Mada)

[11] Sahubawa L 2017 *Material Presented* (Bogor: 3rd EMBRIO International Symposium)

[12] Anonimous 2017a http://dokumen.tips/documents/beberapa-factor-what-is-using-the-power-buy-the-community.html Accessed on 18th July 2017 at: 09.08 WIB

[13] Anonimous. 2017b https://prezi.com/0scppa5st92y/daya-beli-konsumen/ Accessed on 18th July 2017 at 09.20 WIB

[14] Anonimous. 2017c http://dkey0303.blogspot.co.id/2012/03/makalah-consumer-behavior.html Accessed on 18th July 2017 at 09:26 GMT

IOP Conf. Series: Earth and Environmental Science **139** (2018) 012050 doi:10.1088/1755-1315/139/1/012050

The changes of astaxanthin content and chemical characteristics of tiger prawn (*Penaeus monodon*) due to processing: boiling, smoking and frying

F Swastawati

Fisheries Technology Study Program, Department of Fisheries, Faculty of Fisheries and Marine Sciences, Diponegoro University, Jl. Prof. Soedarto, S.H., Tembalang, Semarang, Central Java, Indonesia
*Email: fronthea_thp@yahoo.co.id

Abstract. Food processing using high temperatures can cause changes in pigment color and chemical characteristics in food stuffs, including prawn. The aim of this research was to evaluate the changes in pigment and chemical characteristics of tiger prawn caused by boiling, smoking and frying. Ten kg of tiger prawn was boiled, smoked and fried at the temperature of ± 100 °C for ± 10 min. The results showed that boiling, smoking and frying gave a significant effect ($P < 0.05$) on the astaxanthin pigment, pH, moisture, protein, salt content, Aw and color. The content of astaxanthin pigments in fresh prawn, boiled prawn, smoked prawn and fried prawn was: 132.79 ± 1.5 $\mu g \cdot g^{-1}$; 82.89 ± 0.92 $\mu g \cdot g^{-1}$; 78.28 ± 0.1 $\mu g \cdot g^{-1}$ and 91.35 ± 2.59 $\mu g \cdot g^{-1}$, respectively. The value of °Hue on fresh prawn, boiled prawn, smoked prawn and fried prawn was: 87.85°; 52.5°; 55.94° and 53.98°. The tiger prawn processed by the smoking method has preferable by panelist rather than processed by boiling and frying.

1. Introduction

Tiger prawn is a popular seafood in many countries and it becomes a favorite seafood choice by consumers because of its specific taste and aroma. Although it has a tangy smell in its raw condition, after processing, distinct flavor and specific delicious taste will arise. Based on data from the Ministry of Marine Affairs and Fisheries in 2013, Indonesia exported shrimp to various destination countries reaching a total volume of 162,000 tons. Production of shrimp aquaculture during the last five years (2010–2014) has a tendency to continue rising. The average increase of shrimp production during the last five years is 13.83 % per year. In 2012 shrimp production of 415,703 tons was reached while in 2013 amounted to 645,955 tons [1].

Food processing by heating which is also known as the cooking process is the process of heating food at a temperature of 100 °C or more. The main purpose is to get a better taste, better aroma and softer texture; moreover, it also kills microbes and deactivates all enzymes. Cooking can be done by boiling, smoking and frying. The use of heat in the cooking process is very influential on the nutritional value of the foods including boiling, smoking and frying. The goals of the salt-boiled process for shrimp are to reduce a load of microorganisms to an acceptable level and to improve the flavor of shrimp. A suitable boiling process is important to get a suitable condition to minimize loss of muscle protein and maintain the red color. Smoking by using liquid smoke can produce smoked fish that is higher in quality when compared with the traditional smoking process. Smoking by using liquid smoke may affect the color and sensory quality of smoked fish products [2]. Frying is a fast and simple culinary technique employed worldwide that consists of cooking food above the boiling point of water, by means of partial (shallow-frying) or total (deep-frying) immersion in liquid edible oils or fats at frying temperature. Although fried foods are highly appreciated for their sensory characteristics, they must also satisfy health-related consumer concerns.

High-temperature processing can also affect the change of pigment and chemical characteristics in foodstuffs, including prawn. Prawn color turns red because it contains astaxanthin, a pigment that gives the characteristic of orange to red. Changes in pigment can also indicate that there has been a change in chemical characteristics of prawn. The purpose of this research was to evaluate the effect of processing treatments on the changes of astaxanthin content and chemical characteristics in tiger prawn.

2. Materials and Methods

2.1. Materials

The material used in this study was ± 10 kg of tiger prawn with a size of ± 20 cm in length and an average weight of ± 150 g. Cooking oil, the food grade liquid smoke made of coconut shell, boiling water and salt as a flavor enhancer.

2.2. Methods

This research was conducted experimentally using completely randomized design with duplication. After washed thoroughly, the prawns were dipped into 5 % liquid smoke and 5 % of brine solution. Then they were processed by boiling (at ± 100 °C), smoking (at ± 100 °C) and frying (at ± 100 °C) for ± 10 min [3].

2.3. Measurement of astaxanthin content

Astaxanthin content was measured following Becerra et al. [4], a total of 10 g samples was extracted with 40 mL 0.05 % BHT solution (in acetone) with a digital homogenizer for 2 min. The samples were cooled during the homogenization process to prevent the sample from heating up. After extraction, the sample was centrifuged at 4,000 rpm, temperature of 40 °C for 5 min. To separate the water-soluble compound, the acetone extract from the sample was transferred to a 250 mL separating funnel with 40 mL n-hexane. Next, 100 mL of aquadest containing 0.5 % (w/v) of sodium chloride was added to the mixture. After continuous shaking, phase separation was achieved and then the top layer was separated and transferred into 50 mL measuring flask. The absorbance spectrum of a water-soluble compound was measured at a wavelength of 472 nm using a spectrophotometer. The astaxanthin stock solution (0.2 mg·mL^{-1}) of 0.5; 1.0; 1.5; 2.0 and 3.0 mL were respectively diluted with 10 mL n-hexane. Then the absorbance spectrum of each solution was recorded with a wavelength of 472 nm. The standard curve was obtained by plotting the concentration on absorbance.

2.4. Measurement of pH

The sample was ground and weighed as much as 1 g in a cup glass. Then, 10 mL of aquadest was added and stirred. Subsequently, pH level of the sample in the container was measured by using a calibrated pH meter with a buffer solution of pH 4 and pH 7. The pH value was obtained based on the reading at pH meter until the digital number indicated the constant number [5].

2.5. Measurement of moisture content

The determination of moisture content was based on the different weight of samples before and after drying. The method used was a drying technique for approximately 24 h at 105 °C until the sample reached a stable weight. Percentage of moisture contents (wet weight) was calculated by using the following formula [5]:

$$Moisture\ Content\ (\%) = \frac{B-C}{B-A} x\ 100\ \%$$

Information:
A: empty vial weight (g)
B: cup weight + initial sample (g)
C: cup weight + dry sample (g)

2.6. Measurement of protein content
Determination of protein content was done based on AOAC [5]. Basically, it could be divided into three stages namely the process of destruction, distillation and titration. In the process of destruction the sample was heated with concentrated H_2SO_4 thus it broke down into their respective elements. The destruction process was completed when the solution became clear. The distillation stage resulted in the broken down of ammonium sulfate into ammonia by adding alkalis NaOH and heated. The formed ammonia which was accommodated in concentrated H_3BO_3 added with BCG and methyl red indicator. The amount of H_3BO_3 reacted with ammonia could be known by its reaction with 0.02 M HCl. The end of the titration was marked by the change of solution color from green to pink. Blank treatment was performed to determine the nitrogen from the reagents being used.

2.7. Measurement of A_w
The instrument was calibrated by inserting the $BaCl_2$ $2H_2O$ liquid and closed for 3 min to reach the reading scale of 0.9. Then, samples were inserted into A_w meter for 3 min and A_w scale was recorded. If the temperature scale reached above 20 °C, the scale reading was added as much as the excess temperature multiplied by the correction factor of 0.002 °C, as well as the temperature below 20 °C [6].

3. Results and Discussion

3.1. Astaxanthin content

Table 1. Astaxanthin content tiger prawn before and after the heating process.

Treatment	Astaxanthin ($\mu g \cdot g^{-1}$)
Fresh	132.79 ± 1.5^a
Boiled	82.89 ± 0.92^b
Smoked	78.28 ± 0.1^c
Fried	91.35 ± 2.59^d

Note: Value from average of duplicate ± standard deviation.
*Value with different lower case notation shows a significant difference ($p < 0.05$).

Table 1 shows that the content of astaxanthin in tiger prawn decreased after heat processing. The heating processes led to the reduction of astaxanthin levels as the proteins protecting astaxanthin had been denatured by heat, resulting in the decomposition of astaxanthin. In boiling treatment, this reduction could be related to denaturation and partial solubilization of the carotene protein complex. In this regard, a significant decrease in proteins content (myofibrillar, sarcoplasmic and stroma) in boiled shrimp, suggesting that part of these proteins could have been dissolved in the salt solution [7]. The reduction of astaxanthin due to heat is also reported in drying treatment. Astaxanthin content in cooked shrimp decreased by 75 % after four days of direct sun drying, reaching the concentration of 25.0 ± 2.7 $\mu g \cdot g^{-1}$ [4]. The authors noted that the decline in astaxanthin content in shrimp during sun drying mainly affects the color of the product.

3.2. Moisture content

Table 2. Moisture content, protein content and water activity of tiger prawn before and after the heating process.

Treatment	Moisture content (%)	Protein content (%)	A_w
Fresh	88.72 ± 0.54[a]	27.6 ± 1.5[a]	0.93 ± 0.0007[a]
Boiled	72.7 ± 0.32[b]	14.17 ± 1.4[b]	0.83 ± 0.0007[b]
Smoked	59.84 ± 0.59[c]	22.57 ± 0.23[c]	0.72 ± 0.0014[c]
Fried	57.6 ± 1.17[d]	14.3 ± 0.68[d]	0.62 ± 0.002[d]

Note: Value from average of duplicate ± standard deviation.
*Value with different lower case notation shows a significant difference ($p < 0.05$).

The lowest moisture content was found in fried prawn at 57.6 ± 1.17 % with a reduction of 35.076 %, followed by the smoked prawns with a reduction of 32.55 %, and boiled prawn with a reduction of 18.05 %. The reduction in water content as a consequence of processing related to the temperature being used, the higher the temperature the higher the water content decreases. It happened because of the process of steaming and boiling used temperature of 90–100 °C, while the frying process used a temperature of 200–205 °C. As a comparison study by previous researchers [8], showed that the largest reduction in water content occurred in soybean cake (20.13 %) followed by chicken (16.74 %), mackerel (16.05 %) and the smallest was in tofu (7.66 %).

Moisture content is related to the presence of microorganisms that grow in the body of prawn since water is a medium that supports the growth of microorganisms. The lower the water content, the lower microorganisms decompose. Water is the largest content in fish [9]. Water is a means of microorganisms to thrive. The moisture content reduction in fish can extend the shelf life of the products. Salting process is capable oflosing water in fish because salt has a higher concentration than water, so the water will come out of the tissue or called plasmolysis. The decrease of the moisture content was the most prominent change in fish fillet after drying. Protein, fat and ash content increased significantly in the processed sample [10].

3.3. Protein content

Protein content in prawn decreased after treated with boiling, smoking or frying process. Protein content in both boiled and fried prawn had almost the same value of 14.17 ± 1.4 % and 14.3 ± 0.68 %, respectively, with a decrease of 48.19 %. The lowest reduction in protein content occurred in smoked prawn (18.22 %) with a protein content of 22.57 ± 0.23 %. The reduction of protein content was directly proportional to the used processing temperature, meanwhile, in the process of frying and boiling the used temperature was more than 100 °C hence the protein was damaged by heat. Other researchers reported that the reduction of protein content in boiled bloated fish was 3.12 % whereas fried bloated fish was 5.24 % [8]. During cooking, heat induces protein denaturation and aggregation, causing actin, myosin and collagen shrinkage, which leads to water expulsion from muscle cells and the consequent water loss. During boiling, protein content also decreased. Water and protein losses are called cooking loss [7].

Processing with high temperatures can lead to proteins damage because the protein bonds have undergone changes and breaks. It also deals with the solubility of proteins, water-soluble proteins and faster salts lost due to heat. Protein solubility decreased with increasing temperature, it indicated changes in protein solubility [3]. Denaturation of proteins related to protein solubility, as a result of the heat, the secondary or tertiary structures of the protein was damaged, hence it transformed into a primary structure. Salt-soluble and water-soluble proteins were more easily lost in the primary structure rather than those in secondary or tertiary structures. Boiling, smoking and frying maintained and concentrated the crude protein content. After storage, there was a reduction in crude protein content of the fried fish rather [11].

3.4. Water activity

A_w of fresh prawn was 0.93. The lowest A_w occurred in fried prawn with a value of 0.62. While the A_w of smoked and boiled prawns were 0.72 and 0.83, respectively. Processing with high temperatures could decrease the value of A_w. It was in line with the decrease in water content. The relationship of water content with water activity (A_w) was indicated by the tendency that the higher the water content the higher the A_w value would be and vice versa [12].

Table 3. The value of pH and salt content of tiger prawn before and after the heating process.

Treatment	pH	Salt content (%)
Fresh	7.6 ± 0.14^a	0.78 ± 0.04^a
Boiled	7.55 ± 0.07^b	1.4 ± 0.16^b
Smoked	6.45 ± 0.07^c	4.2 ± 0.12^c
Fried	7.3 ± 0.14^d	3.4 ± 0.07^d

Note: Value from average of duplicate ± standard deviation.
*Value with different lower case notation shows a significant difference ($p < 0.05$).

3.5. pH value

The pH value of all samples decreased from the initial pH of fresh prawn with a value of 7.6. The value of pH on boiled prawn and fried prawn were 7.55 and 7.3, respectively. The lowest pH value occurred in smoked prawns with a value of 6.45. The pH value in smoked prawn was acidic, whereas the pH value of other treatments indicated that the pH was neutral. The low pH of the smoked prawns occurred as the effect of the addition of liquid smoke prior to processing. This acidic pH trend was caused by the absorption of the liquid organic liquids into the sample. The longer the immersion of liquid smoke, the more organic acids such as acetic acid, formic acid would enter into the fish meat during smoking process and resulted in a reduction of pH value. The smoke attached to the fish meat caused the fish meat to be more acidic therefore the pH value tends to be lower [13].

3.6. Salt content

Samples were dipped into 5 % of brine before heat processing. Salt content in all samples decreased significantly after processing. The decrease in salt content was caused by salt dissolution into processing media such as oil and water. The highest salt content was found in smoked prawns with a value of 3.4 %, while the lowest was found in boiled prawn with a value of 1.4 %. The salt content of smoked catfish ranges from 2.62 to 5.83 % [3]. In this study, cooking loss observed was 32 %, which is higher than losses reported by previous researchers [7] for boiled shrimp in a salt solution (4 %, w/v) for 7 min. The greater cooking loss we found can be attributed to the longer boiling time (15 min vs. 7 min). On the other hand, the NaCl content in shrimp showed a significant increase after boiling in brine.

3.7. Colour analysis

Processing affected the color change in prawn. The value of °HUE of boiled prawn reduced 35.35 °HUE into the yellow-redcolor category, the smoked prawn reduced 31.91 °HUE and fried prawn reduced 33.87 °HUE into the red color category (table 4). Changes in astaxanthin are marked in red color on prawn. Colour changes from reddish pink to orange in prawn are formed by the release of astaxanthin from protein-bound due to proteolysis [14].

Table 4. Colour analysis of tiger prawn before and after the heating process.

Treatment	Colour			°HUE
	L (*Lightness*)	+a* (*Redness*)	+b* (*Yellowness*)	
Fresh	35.64 ± 0.7[a]	0.2 ± 0.01[a]	5.32 ± 0.66[a]	87.85[d]
Boiled	46.57 ± 0.49[c]	14.77 ± 0.68[c]	19.23 ± 0.71[c]	52.5[b]
Smoked	48.24 ± 0.22[e]	15.63 ± 0.44[e]	23.12 ± 0.08[e]	55.94[a]
Fried	41.32 ± 0.33[b]	12.25 ± 0.45[a]	16.85 ± 1.75[b]	53.98[f]

Note: Value from average of duplicate ± standard deviation.
*Value with different lower case notation shows a significant difference ($p < 0.05$).

3.8. Sensory value of fresh prawn

The results of thesensory value of fresh prawn showthat appearance, odor, and texture covered average values of 8.73 ± 0.45; 8 ± 0.51 and 8.8 ± 0.41, respectively. Fresh prawn had a specific prawn's texture of elastic, compact and solid. Based on these specifications, the prawns used in this research were still in the fresh category. The sensory of fresh prawn fulfill requirements of sensory of fresh fish based on Indonesian National Standard for fresh fish which minimal sensory value was seven.

3.9. The prawn preferences test

Table 5. Preferences test of tiger prawn after heat processing treatment.

Treatment	Hedonic parameter			
	Appearance	Flavour	Odor	Texture
Boiled prawn	6.5 ± 0.51[a]	6.6 ± 0.56[a]	7.2 ± 0.48[a]	6.23 ± 0.5[a]
Smoked prawn	8.57 ± 0.5[c]	8.1 ± 0.31[b]	8 ± 0.49[b]	8.57 ± 0.5[c]
Fried prawn	7.57 ± 0.5[b]	8.1 ± 0.42[b]	7.3 ± 0.48[a]	7.87 ± 0.57[b]

Note: Value from anaverage of 10 panelists ± standard deviation.
*Value with different lowercase notation shows a significant difference ($p < 0.05$).

3.9.1. The appearance. The panelists gave the highest score for the appearance of the smoked prawn. The results of the appearance were as follows: surface quality prawn was bright, shiny, invisible dirt in the form of dried blood and the specific color of prawn. Chemical reactions among substances contained in the smoke could occur in smoked fishery products. The chemical reaction between phenol and oxygen could change the color of the smoke product to be a brownish yellow. Phenol, the major carbonyl component in smoke reacting with proteins and fats in fish, has an important role in affecting the appearance of smoked fish [9].

3.9.2. Flavors. The panelist's preference score in the flavor of both smoked and fried prawn was the highest with a similar average value of 8.1. The taste specifications on fried prawn were savory on the contrary the smoked prawn had a distinctive taste of smoke products. In addition, boiled prawn had a slightly bland taste when compared with the other two processing treatments. The acceptance of panelists on taste parameters was based on not only the maturity level but also taste of "umami" in the processed prawn. Processing techniques with heating were able to produce products that have extraordinary taste compared with other techniques [15].

3.9.3. Odor. The highest panelists preference score of the odor was in the smoked prawn based on the specification that they had the distinctive aroma of smoke. It was the effect of the increase of phenol levels due to heating. The increase of phenol level occurred because of temperature and heating time. The higher the temperature and curing time resulted to the higher the phenol production [9]. In fact, the aroma of boiled prawn and fried prawn was still included in a favorite category with a value of 7.2

± 0.48 and 7.3 ± 0.48. The aroma of a food that undergone high-temperature processing was caused by the volatile compounds. The aroma produced in food after processing affected by the compound in food and the way of cooking [15].

3.9.4. Texture. The highest score for texture preference was in the smoked prawns with a value of 8.57 ± 0.5. The texture of smoked prawn was solid, compact and not fragile. The addition of liquid smoke might affect the texture of prawn result from the reaction of liquid smoke components such as phenol. The main carbonyl component in the smoke played an important role. The smell, taste and texture of smoked fish were the results of the reaction of carbonyl groups in the smoke reacting with proteins and fats in fish. Smoke plays an important role in the formation of color, texture and taste [9].

4. Conclusion

The different prawn processing by boiling, smoking or frying affected pigment concentration changes and chemical characteristics in tiger prawn. Further research on the influence of various types of oil in frying, various types of liquid smoke and different boiling temperature on the changes in pigment concentration and chemical characteristics of prawn needs to be conducted in the future.

References

[1] Ministry of Marine Affairs and Fisheries 2014 *Marine and Fisheries in Figures* (Jakarta: Centre of Data Statistic and Information)
[2] Swastawati F, Surti T, Agustini T W and Riyadi P H 2015 *Kne.Life Science* **1** 156–161
[3] Swastawati F, Baarri A N A, Agustini T W, Dewi E N, Wijayanti I, Prasetyo D Y B and Khan D 2016 *UTM Press* **78**(4) 55–59
[4] Becerra, J A H, Flores A A O, Alfaro G V, Rodriguez I S, Estrada M T R and García H S 2014 *Food Chemistry* **14** 832–839
[5] Association of Offcial Analytical Chemists (AOAC) 2006 *Offcial Methods of Analysis* 18th ed. (Maryland: AOAC Int. Gaithersburg MD)
[6] *Manual Procedure of Novasina Humetic Water Activitymeter type EEJA-3*
[7] Niamnuy C, Devahastin, S and Soponronnarit S 2008b *Food Chemistry* **108** 165–175
[8] Sundari, Almasyhurin D and Lamid A 2015 *Media Litbangkes*. **25**(4) 235–242
[9] Swastawati F, Surti T, Agustini T W and Riyadi P H 2013 *Jurnal Aplikasi Teknologi Pangan* **2**(3) 126–131
[10] Wu T and Mao L 2008 *Food Chemistry* **110** 647–653
[11] Holma, Ayinsa K and Maalekuu K 2013 *American Journal of Food and Nutrition* 2157–2167
[12] Legowo A M, Mulyani S and Kusrahayu 2012 *Teknologi Pengolahan Susu* (Semarang: Universitas Diponegoro)
[13] Swastawati F, Siagian W D L and Wijayanto D 2014 *Jurnal Pengolahan dan Bioteknologi Hasil Perikanan* **4**(1) 21–39
[14] Chaijan M and Panpipat W 2012 *Asian Journal Food and Agro-Indusry* **5**(2) 163–171
[15] Winarno 2004 *Food and Nutrition Chemicals* (Jakarta: Gramedia Pustaka Utama)

IOP Conf. Series: Earth and Environmental Science **139** (2018) 012051 doi:10.1088/1755-1315/139/1/012051

Identification of chitinolytic bacteria isolated from shrimp pond sediment and characterization of their chitinase encoding gene

A U Triwijayani[1], I D Puspita[1], Murwantoko[1] and Ustadi[1]

[1] Department of Fisheries, Faculty of Agriculture, Universitas Gadjah Mada, Jl. Flora Gd. A4, Bulaksumur, Yogyakarta, Indonesia

Abstract. Chitinolytic bacteria are a group of bacteria owning enzymes that able to hydrolyze chitin. Previously, we isolated chitinolytic bacteria from shrimp pond sediment in Bantul, Yogyakarta, and obtained five isolates showing high chitinolytic index named as isolate PT1, PT2, PT5, PT6 and PB2. The aims of this study were to identify chitinolytic bacteria isolated from shrimp pond sediment and to characterize the chitinase encoding gene from each isolate. The molecular technique was performed by amplification of 16S rDNA, amplification of chitinase encoding gene and sequence analysis. Two chitinolytic bacteria of PT1 and PT2 were similar to *Aeromonas bivalvium* strain D15, PT5 to *Pseudomonas stutzeri* strain BD-2.2.1, PT6 to *Serratia marcescens* strain FZSF02 and PB2 to *Streptomyces misionensis* strain OsiRt-1. The comparison of chitinase encoding gene between three isolates with those in Gen Bank shows that PT1 had similar sequences with the *chi1* gene in *Aeromonas* sp. 17m, PT2 with *chi1* gene in *A. caviae* (CB101) and PT6 with *chiB* gene in *S. Marcescens* (BJL200).

1. Introduction

Chitinolytic bacteria are microorganisms enable to degrade chitin using chitinase enzymes. Chitinolytic bacteria can be found in the various type of environment, such as rhizosphere, phyllosphere, soil and water environment includes oceans, lakes and shrimp ponds [1]. Shrimp pond sediment is a distinct environment for chitinolytic bacteria bioprospecting, as it contains high chitinous substrate, as well as other organic residues, resulted from the degradation of crustacean shells during an intensive culture. Chitinolytic bacteria isolated from shrimp pond sediment had been reported, such as *Vibrio alginolytic* [2], *Bacillus licheniformis* [3] and *Chitinimonas taiwanensis* [4]. Previously we isolated 46 chitinolytic bacteria from the pond sediment of local shrimp aquaculture plant of PT. Indokor Bangun Desa located in Bantul, Yogyakarta, in which five of them showed high chitinolytic index [5].

Chitinase, an enzyme that catalyzes the breakdown of β-1,4-glycosidic bond in chitin polymer, is produced by bacteria to hydrolyze chitin into its monomer or dimer that will be further used as a nutrient for growth [6]. Chitinase is also used to produce chitin hydrolyzate products that have been widely applied in the fields of health, food and agriculture [1]. Therefore, the productions of bacterial chitinase gain much interest and have been explored by many researchers. Chitinase encoding genes, called the chitinase gene (chi gene), from various bacteria, have been studied and used to construct recombinant enzyme to understand the characteristics of the enzyme. Three chitinase encoding gene *chiA*, *chiB* and *chiC* from *Serratia marcescens* was reported previously [7]. The chitinase encoding gene *chiA* in *Aeromonas caviae* was also reported [8]. This study aimed to identify chitinolytic bacteria from shrimp pond sediments and characterize the chitinase encoding genes of each isolate.

IOP Conf. Series: Earth and Environmental Science **139** (2018) 012051 doi:10.1088/1755-1315/139/1/012051

2. Materials and Methods

2.1. Bacterial cell preparation
Five isolates of chitinolytic bacteria were obtained from the collection of Laboratory of Quality and Safety of Fisheries Product, Universitas Gadjah Mada. The isolates were previously isolated from shrimp pond sediment in PT. Indokor Bangun Desa located in Bantul, Yogyakarta [5]. One loop full of bacteria suspension in glycerol stock was transferred into 7 mL of chitin medium and incubated at 37 °C for 24 h.

2.2. DNA isolation
As much as 1 mL of cell culture from the previous step was inserted into a 1.5 mL microtube and centrifuged at 14,000 × g for 1 min. The supernatant was removed and cell pellet was used for genomic DNA isolation using Presto™ Mini gDNA Bacteria Kit (Geneaid) following manufacturer's protocol. The cell pellet was added with 200 µL of Gram + buffer solution that was previously mixed with lysozyme 4 $mg \cdot mL^{-1}$. The mixture was homogenized and incubated for 30 min at 37 °C (inverted every 10 min). Then, 20 µL Proteinase K solution was added to the mixture, homogenate and incubated for 10 min at 60 °C (inverted every 3 min). A 200 mL GB Buffer solution was added to the sample solution and homogenized for 10 sec. The sample solution was incubated at 70 °C for 10 min (inverted every 3 min). Then, 200 µL of ethanol absolute was added to the sample solution and mixed gently. The sample was moved into the GD column then centrifuged at 14,000 × g for 2 min. Collection tubes containing liquids were discarded. The GD column was assembled into a new 2 mL collection tube. A 400 µL W1 buffer solution was added to the GD column, centrifuged at 14,000 × g for 30 sec, discarded the liquid in the collection tube and placed back the GD column into the collection tube. Then, wash buffer (600 µL) solution was added to the GD column, centrifuged at 14.000 × g for 30 sec, discarded the liquid and continued to centrifuge the column for 3 min at a rate of 14.000 × g. The GD column was moved into a clean 1.5 mL microtube. The pre-heated Elution Buffer solution of 100 µL was added to the GD column and allowed to stand for 3 min until the solution was absorbed. The sample was centrifuged at 14.000 × g for 30 sec to obtain the purified DNA.

2.3. Amplification of 16S rDNA fragment
The purified DNA from the previous step was amplified by means of PCR using the Kapa Biosystem® PCR kit. The PCR mixture consisted of 6 µL ready PCR mix (consisted of Taq polymerase, dNTP, Mg^{2+} and loading dye), 5 µL ddH_2O, 0.5 µL forward primer, 0.5 µL reverse primer and 0.5 µL DNA template. The universal primer of 27F (5'-AGAGTTTGATCMTGGCTCAG-3') and 1492R (5'-CGGTTACCTTGTTACGACTT-3') was used in this step. The mixture was inserted into the thermocycler (T100™ Biorad) following the process of pre-denaturation (95 °C for 3 min), denaturation (95 °C for 30 sec), annealing (55 °C for 30 sec), elongation (72 °C for 1.5 min), final extension (72 °C for 5 min) and infinite hold (until 12 °C). The process of denaturation, annealing and elongation were repeated for 30 cycles.

2.4. Electrophoresis
The size of resulting amplicon was confirmed by electrophoresis using 1 % agarose gel in Tris Boric EDTA (TBE) solution added with 0.75 µL fluorosafe. The formed agarose gel is transferred to mini-gel electrophoresis. A 3 µL of the marker and 5 µL of amplicon were incorporated into the well. The process was held for 15 min at 100 Volts. The band formed on the gel was visualized using UV Transilluminator. The size of the band produced from the sample was obtained by comparing with the size of the marker.

2.5. Molecular identification of the 16S rDNA sequence

The 16S rDNA fragment was sent to First Base Laboratory Singapore for sequence analysis and matched with the nucleotide database available at Gen Bank using BLAST tool in NCBI and further processed using MEGA version 5. The identified bacterial species provided basic information for the design process of primers.

2.6. Molecular identification of the putative chitinase encoding gene

The specific primers were designed for amplification of the putative chitinase encoding gene of each isolate. The nucleotide sequences of a chitinase gene from five reference bacteria of the same genus were obtained from Gen Bank. The sequence was analyzed using BLASTX (NCBI) to identify the active site of each reference gene to be included for the consideration for primer design. Then the chitinase genes of references were aligned using BioEdit software and the primers were designed based on the conserved domain. Primers characteristic was analyzed by Oligocalculator software to identify the GC content and melting temperature (T_m). All designed primers were confirmed not to have self-complementary and self-annealing sequence that potential to form a hairpin loop. The design of primers was sent to First Base Laboratory Singapore to be assembled.

Table 1. The primers used for amplification of the putative chitinase encoding gene.

Genus	Primer	T_m (°C)	% GC
Aeromonas	1218F (5'-GAGAACAACAAGCCCTACGC-3 ')	53.8	55
	2450R (5'-CCTCGTGCATGGCGTTG-3')	51.9	65
Serratia	4F (5'-TCCRMACGYAAAGCSGTTATTGG-3')	53.5–58.8	43–57
	1475R (5'-TCATTAYGCYASRCGGCCCACYTTCAGCC-3')	62.9–68.8	52–66

The amplification of the putative chitinase encoding gene was performed by means of PCR using the Kapa Biosystem® PCR kit following the manufacturer's protocol as mentioned previously. Genomic DNA was used as the template and specific primers (table 1) was used for each isolate. The thermal cycle used for amplification was as follow: pre-denaturation (95 °C, 3 min), denaturation (95 °C, 30 sec), annealing (56–57 °C for *Aeromonas* and 58 °C for *Serratia*, 30 sec), elongation (72 °C, 1.5 min), final extension (72 °C, 5 min) and infinite hold (12 °C). The process of denaturation, annealing and elongation were repeated for 30 cycles. The size of the amplicon was confirmed by electrophoresis on 1 % agarose gel and sent to First Base Laboratory Singapore for sequence analysis.

2.7. Characterization of the putative chitinase encoding gene

The sequence of putative chitinase encoding gene from each isolate was matched with the reference gene database in BLAST (NCBI) for identification. Further, the characteristic of putative chitinase encoding gene was also determined using BLATSX.

3. Results and Discussion

3.1. Identification of chitinolytic bacteria with 16S rDNA

Five chitinolytic bacteria showing high chitinolytic index value from the previous experiment was identified [5]. The visualization of PCR product from 16S rDNA fragment of all isolates on agarose showed a single band of 1,500 bp. Molecular identification based on BLAST results on 16S rDNA sequences in Gen Bank showed that PT1 and PT2 were similar to *Aeromonas bivalvium* strain D15, PT5 to *Pseudomonas stutzeri* strain BD-2.2.1, PT6 to *Serratia marcescens* strain FZSF02 and PB2 to *Streptomyces misionensis* strain OsiRt-1 (table 2).

An isolate having a sequence similarity of 16S rDNA greater than 97 % with the reference sequence represent the same species, meanwhile a sequence similarity between 93–97 % may represent the similar identity at the genus level but differ at the species level [9]. All five bacteria isolates identified in this research showed a similarity of 99 % or more with the reference, however further morphological and biochemical identification is necessary to support the molecular identification results.

Table 2. Molecular identification of five chitinolytic bacteria isolated from pond sediment.

Isolates	Species reference	Identity	Accession number
PT1	*Aeromonas bivalvium* strain D15	99 %	KF500920.1
PT2	*A. bivalvium* strain D15	99 %	KF500920.1
PT5	*Pseudomonas stutzeri* strain BD-2.2.1	100 %	LC125170.1
PT6	*Serratia marcescens* strain FZSF02	99 %	KU145144.1
PB2	*Streptomyces misionensis* strain OsiRt-1	99 %	KU321340.1

3.2. Identification of the putative chitinase encoding gene

The primers for the amplification of putative chitinase encoding gene were designed based on the conserved domain in chitinase genes from five reference bacteria of the same genus. In this research, we only observed the putative chitinase encoding gene of three isolates, namely PT1, PT2 and PT6, because the sequence of a chitinase gene from reference bacteria of other two isolates was not available by the time when the research was conducted. Figure 1 shows that the target gene was amplified. The length of the targeted putative chitinase encoding gene from PT1 and PT2 was 1,157 bp, and PT6 was 1,500 bp (figure 1).

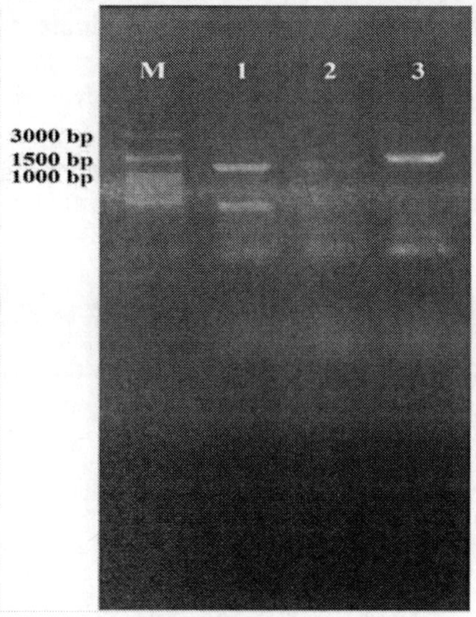

Figure 1. Agarose gel electrophoresis of the chitinase encoding gene. Lane M: DNA Marker, lane 1: PT1 strain, lane 2: PT2 and lane 3: PT6.

Bioinformatics analysis of chitinase encoding genes of PT1, PT2 and PT6 is shown in table 3. The chitinase encoding gene of isolate PT1 had a similarity of 85 % to *chi1* gene of *Aeromonas* sp. 17m and PT2 showed 84 % similarity to *chi1* gene of *A. caviae* CB101. The low percentage of similarity indicates that chitinase encoding gene possessed by PT1 and PT2 was not similar to chitinase gene of the reference bacteria. Previously, PT1 and PT2 were identified as *A. bivalvium*. This result indicates that the variation of chitinase encoding gene might be presented at the species level. By the time of this research was conducted, the reference of chitinase encoding gene from *A. bivalvium* was not available in Gen Bank.

The chitinase encoding gene of isolate PT6 showed the similarity of 99 % to *chiB* gene of *Serratia marcescens* BJL200. This result was in accordance with the result of molecular identification in which PT6 was identified as *Serratia marcescens*. The *ChiB* of *S. marcescens* is an exochitinase catalyzing the hydrolysis of chitin into GlcNAc and $(GlcNAc)_2$ [7]. *S. marcescens* possesses three types of chitinase, namely *ChiA* (exochitinase), *ChiB* (exochitinase) and *ChiC* (endochitinase), in which each of them has a specific characteristic and exhibits different mechanisms to hydrolyze chitin [7]. In the chitinolytic system of *S. marcescens*, these three types of chitinase work synergistically to produce the final product of GlcNAc from the hydrolysis of chitin.

Table 3. Identification the putative chitinase encoding gene of PT1, PT2 and PT6.

Isolates	Chitinase Gene	Identity	Accession Number
PT1	*Aeromonas* sp. 17m *chi1* gene	85 %	JN987186.1
PT2	*A. caviae* CB101 *chi1* gene	84 %	AJ534329.1
PT6	*S. marcescens* BJL200 *chiB* gene	99 %	Z36295.1

Characteristics of each putative chitinase encoding gene were further analyzed by comparing the deduced amino acid with the amino acid sequence of the reference chitinase protein provided from BLASTX (table 4). The deduced amino acid from PT1 and PT2 putative chitinase encoding gene were similar to GH18 chitinase from *Aeromonas molluscorum*, however, the similarity of the deduced amino acid sequence from PT1 was higher (93 %) than PT2 (86 %). Meanwhile, the deduced amino acid from PT6 putative chitinase encoding gene had a 93 % similarity to Glyco18 of chitinase B from *S. marcescens*. The deduced amino acid sequence from all isolates showed 10 active sites with a different position between PT6 and other two isolates, however, they had the identical amino acid residues in the active sites. The amino acid residues located at the active sites were tyrosine, phenylalanine, aspartic acid, glutamic acid, methionine and tryptophan. The identical amino acid residues at the active sites indicate that they may have a similar mechanism of hydrolyzing chitin polymer.

Table 4. Characteristics of the putative chitinase encoding gene from three chitinolytic bacteria isolated from pond sediment.

Isolates	Region name chitinase	Identity	Active site	Amino acid residues	Accession number
PT1	GH18_chitinase (*Aeromonas molluscorum*)	93 %	163, 191, 311, 313, 315, 388, 390, 391, 444, 539	Y, F, D, D, E, M, Y, D, Y, W	WP_005896739.1
PT2	GH18_chitinase (*A. molluscorum*)	86 %	163, 191, 311, 313, 315, 388, 390, 391, 444, 539	Y, F, D, D, E, M, Y, D, Y, W	WP_005896739.1
PT6	Glyco_18 (Chitinase B *Serratia marcescens*)	93 %	10, 51, 140, 142, 144, 212, 214, 215, 292, 403	Y, F, D, D, E, M, Y, D, Y, W	AIP98318.1

4. Conclusion

Five chitinolytic bacteria obtained from shrimp sediments were identified as *Aeromonas bivalvium* (PT1 and PT2), *Pseudomonas stutzeri* (PT5), *Serratia marcescens* (PT6) and *Streptomyces misionensis* (PB2). The comparison of the putative chitinase encoding gene between three isolates with chitinase gene in Gen Bank shows that PT1 had similar sequences with the *chi1* gene in *Aeromonas* sp. 17m, PT2 with *chi1* gene in *A. caviae* (CB101) and PT6 with *chiB* gene in *S.*

marcescens (BJL200). The deduced amino acid from PT1 and PT2 putative chitinase encoding gene was similar to GH18 chitinase from *Aeromonas molluscorum* and PT6 was similar to Glyco18 of chitinase B from *S. marcescens*. The deduced amino acid sequence showed that they had the identical amino acid residues in the active sites.

Acknowledgment
Cahya Kurnia Fusianto for conducting primer design and molecular work.

References

[1] Herdyastuti N, Raharjo T J, Mudasir and Matsjeh S 2009 *Indonesia Journal of Chemistry* **9**(1) 37–47
[2] Vincy V, Shoba M V, Viveka S, Vijaya T M and Rani G J B 2014 *European Journal of Experimental Biology* **4**(3) 78–82
[3] Prabowo Y Y and Herdyastuti N 2014 *Proceedings of the National Seminar on Chemistry* 75–79
[4] Shu-Chen C, Wang J T,Vandamme P, Hwang J H 2014 *Syst. Appl. Microbiol.* **27**(1) 43–49
[5] Kholifah A 2015 *Isolation and Identification of Chitinolytic Bacteria from Shrimp Pond Sediments* (Yogyakarta: Faculty of Agriculture UGM)
[6] Artimo P, Jonnalagedda M, Arnold K, Baratin D, Csardi G, de Castro E, Duvaud S, Flegel V, Fortier A, Gasteiger E, Grosdidier A, Hernandez C, Ioannidis V, Kuznetsov D, Liechti R, Moretti S, Mostaguir K, Redaschi N, Rossier G, Xenarios I and Stockinger H 2012 *Nucleic Acids Res.* **40**(W1) W597-W603
[7] Brurberg M B, Synstad B, Klemsdal S S, Aalten D M F V, Sundheim L and Eijsink V G H 2000 *Recent Research Development in Microbiology* Manuscript
[8] Sitrit Y, Vorgias C E, Chet I and Oppenheim A B 1995 *Bacteriological Journal* **177**(14) 4187–4189
[9] Hagstörm A, Pinhassi J and Zweifel U L 2000 *Aquatic Microbial Ecology* **21** 231–244

Optimization of alginate alkaline extraction technology from *Sargassum polycystum* and its antioxidant properties

E Yudiati[1*], G W Santosa[1], M R Tontowi[1], S Sedjati[1], E Supriyantini[1] and M Khakimah[1]

[1] Marine Science Department, Faculty of Fisheries and Marine Science, Diponegoro University, Jl. Prof. Soedarto, S.H., Tembalang, Semarang 50275
*Email: eyudiati@gmail.com

Abstract. The quality of alginate may differ considerably depending on several aspects. The aims of this research were to identify, characterize the alginate, examine the quality and determine the antioxidant properties. Identification was done phenotypically. Characterization was done by FT-IR spectroscopy by comparing the samples to standard alginate (Sigma, USA). The extraction was done in two treatments: soaked with 5 % NaOCl, extraction of Na_2CO_3 (3.0; 5.0; 7.0 and 9.0 %)/EDTA, KCl and precipitated with ethanol (T1) and dried. T2 was pretreated with KOH, HCl, boiled by Na_2CO_3 (3.0; 5.0; 7.0 and 9.0 %), depigmented, HCl and NaOH addition precipitated with ethanol absolute and then sundried. Antioxidant properties test were done by DPPH and NBT radical scavenging assays. Phenotypic identification showed that the species was *Sargassum polycystum*. There were similarities in signal vibration between the samples and the standard. The highest yield was produced from 7.0 % Na_2CO_3, while the average yield of T1 (37.56 %) was significantly higher ($P \leq 0.05$) than T2 (21.77 %). The level of dynamic viscosity was correlated with alkali concentration. Indication in T1 results showed that the higher alkali concentration, the better yield. The best yield was the lowest viscosity, produced the strongest antioxidant activity.

1. Introduction

Alginates are a group of hydrocolloids, localized in the cell wall and matrix of brown seaweeds and play an important structural role in providing mechanical resistance and flexibility to seaweeds against currents [1]. This natural biomaterial has some excellent properties, such as non toxicity, biodegradability, reproducibility, and biocompatibility. It has a wide range of applications in many fields can be applicated in cosmeceutical, foods application [2, 3, 4], nutraceutical and drugs as well as immunostimulants in marine culture [5, 6].

Antioxidants decrease the oxidative damage mediated by free radicals [7]. Due to the severe side effects of many synthesized chemicals which act as a free radical scavenger [8], nowadays, researchers are trying to use antioxidants with the natural origin, including alginates. Researchers have also shown that alginate polymer has antioxidant properties and this property increases by breaking the polymer chain into oligosaccharide [9, 10, 11]. Antioxidants have widespread applications in medical and food industry [12].

The chemical composition and molecular mass of the alginate may differ considerably depending on several factors, such as the algal species, ecological and seasonal variation, or on the age and type of tissue sampled [13, 14]. On the other hand, the extraction and further processing conditions may greatly influence the structure and the composition of the isolated alginate and subsequently its considerable practical utility. The main step in the extraction protocol used in the alginate industry is the alkaline treatment. Based on our previous study, compared to acid and calcium alginate, the alkaline alginate produced the highest yield [6].

The brown algae *Sargassum polycystum*is abundant in North Sea of Java, Indonesia and still un exploited. The simple and cheap extraction methods of this local tropical alga will pursue some new information regarding methods of extraction with different alkaline concentration, yield, alginates characterization, viscosity and molecular weight as well as its superoxide radical scavenger and DPPH activity.

2. Materials and Methods

2.1. Identification of brown seaweed

In fact, Sargassum genus from Phaeophyta is more than 400 species all over the world [15], including Indonesia [16]. The key of phenotypic characters of brown algae was determined by the main axis, vesicle, receptacle and thallus [17].

2.2. Preparation of sodium alginate of S. polycystum

S. polycystum was collected from the seabed of Panjang Island, Jepara Regency, Central Java, Indonesia. The sodium alginate extract was prepared by two treatments: treatment 1 (T1) [18] and treatment 2 (T2) [19]. The sodium alginate from T1 was collected by overnight magnetic stirrer extraction with 3.0; 5.0; 7.0; and 9.0 % Na_2CO_3/50 μM EDTA. Prior to this, the dried alga was soaked with 5 % NaOCl for 30 min. The pellet was filtered and 0.13 M KCl was added and followed by 96 % ethanol in 1:1 volume and stirred well. Centrifugation was performed at 3,500 rpm for 5 min. Finally, the Na-alginate was collected and then dried overnight in the oven at 60 °C.

The sodium alginate of T2 was prepared by immersing the algae with 0.7 % KOH (1:30) w/v, rinsed with tap water and then soaked with HCl 5 % for one night. After tap water rinsing, the alga were then boiled with Na_2CO_3 (3.0 %; 5.0 %; 7.0 % and 9.0 %) (1:20) w/v for 2 h, filtered and discoloured by 4 % NaOCl. The extraction was then added with 10 % HCl (pH 2.8–3.2), followed by NaOH 10 % addition (pH7), precipitated with 96 % ethanol (1:20) w/v and then centrifugated. The Na-alginate was collected and then sun-dried. The two treatments of alginate yield were then determined by comparing the dried weight before and after extraction (%).

2.3. Physiochemical properties of sodium alginate

2.3.1. Water content. The water content was done following AOAC [20]. The samples of alginate were dried at 105 °C.

2.3.2. Ash content. The ash content was obtained following AOAC [20]. This was done by burning the sample in a hot air oven at 200 °C.

2.3.3. FT-IR spectroscopy. The characterizations of alginates were determined spectrophotometrically by signal vibration using Fourier Transformed-Infrared. The preparation was done by mixing the samples with KBr in pellets formation (10 % w/w). The signal was recorded in the 400–500 cm^{-1} region using a Thermo Nicolet 380 FTIR (Germany).

2.4. Viscosity

The viscosity examination was done with a simple device by comparing the flow times of two liquids of known density [21].

2.5. Antioxidant activity

2.5.1. Superoxide radical-scavenging property. This assay is based on the ability of the alginate to inhibit the photochemical reduction of Nitro Blue Tetrazolium (NBT), in the riboflavin-light-NBT system [22]. The alginate solutions were prepared in 50 mM potassium dihydrogen phosphate buffer (pH 7.8). In brief, reaction mixture composed of 13 mM methionine, 2 μM riboflavin, 100 μM EDTA, 75 μM NBT and various concentrations of samples (0.05; 1:2 w/v). The reaction mixtures were kept in front of a fluorescent light for 40 min [23], then absorbance was measured using 96-well plates at 560 nm by microplate spectrophotometer (R-Biopharm Well Reader, Germany). Unilluminated identical tubes containing the reaction mixture were used as blanks (control). The relative superoxide radical-scavenging activity was calculated using the equation:

Relative activity (%) $= As/Ac \times 100$

Where: Ac is the absorbance of the control and As is the absorbance of the sample.

2.5.2. DPPH radical scavenging activity assay. The concentration of samples was one w/v. An aliquot of each sample (100 μL) was mixed with 100 μL of 0.1 mM DPPH (prepared with ethanol) and then followed by incubation for 30 min. [24]. The absorbance of each sample was read at 515 nm using a microplate reader (R-Biopharm Well Reader, Germany). The percentage of scavenged DPPH was calculated using the following equation:

DPPH scavenging activity (%) $= [(Ac /As) /Ac] \times 100$

Where: Ac is the absorbance of the control (100 μL of ethanol with 100 μL of the DPPH solution) and As is the absorbance of the sample. BHA was used as a positive reference.

2.6. Statistical analysis
All data were subjected to one-way and two-way analysis of variance (ANOVA) at the level of significance of 0.01 and 0.05. A multiple comparison (LSD) test was used to examine significant differences among treatments using IBM SPSS Statistics 20 computer software.

3. Results and Discussion

3.1. Brown algae identification
The morphology of this species are thallus of variable size from 10 cm in shallow exposed habitats to 1.5 m indeeper areas; holdfast discoidal or conical; main axis cylindrical and warty, bearing up to 10 stolon-like branches, 0.5–1 cm long; stolon-like branches cylindrical or flattened, smooth or with few spines, exhibiting sometimes short leaf-like lateral or secondary haptera; secondary axes cylindrical, distichous and bearing numerous “y” shaped spine-like protuberances, up to 0.2 cm in diameter; leaves linear to lanceolate or oblong and straight in lateral view, apex acute or obtuse, margins irregularly serrate, petioles short, basecuneate, midribs percurrent; cryptostomata small, abundant and scattered irregularly on the leaf surface, up to 3 cm long and 0.8 cm wide; vesicles spherical or slightly oval, smooth or bearing prominent cryptostomata and a short spine-like mucron, up to 0.25 cm in diameter, petioles cylindrical and shorter than the vesicles; receptacles unisexual; female receptacles simple or branched, cylindrical or slightly flattened, lanceolate, warty, bearing coarse spines; male receptacles simple or bifid, cylindrical, lanceolate, smooth or bearing a few fine spines, possibly mixed with vesicles. Based on the phenotypic characters, it was concluded that the species of this brown seaweed is *Sargassum polycystum*.

The classification is subgenus Sargassum, section Polycystae, type locality from Sunda Strait, Indonesia. Geographical distribution is worldwide, especially Indo-Pacific region and lives in shallow reef flats and rocky bottom habitat.

3.2. Alginate extraction and yield

The yield of *S. polycystum* sodium alginate of T1 and T2 is showed in figure 1. This *S. polycystum* was locally from northern waters area of Java, Indonesia. Based on the result, T1 treatment was much simple and easy method when compared to T2. In fact, the yield obtained from T1 alginates was higher than T2 alginate. The EDTA addition in T1 extraction reached the highest yield at 7.0 % and 9.0 % alkaline extraction. EDTA (*ethylenediaminetetraacetic acid)* has been known as a chelating agent and this, consequently, was improved the extract yield. Similar data were reported from experiment conducting with *S. siliquosum* from the southern sea of Java, Indonesia [6] and the *Sargassum* sp. extract from Madagascar [25]. The other experiment found that in absence of EDTA, the alginate yield is lower (10–13 %) [26].

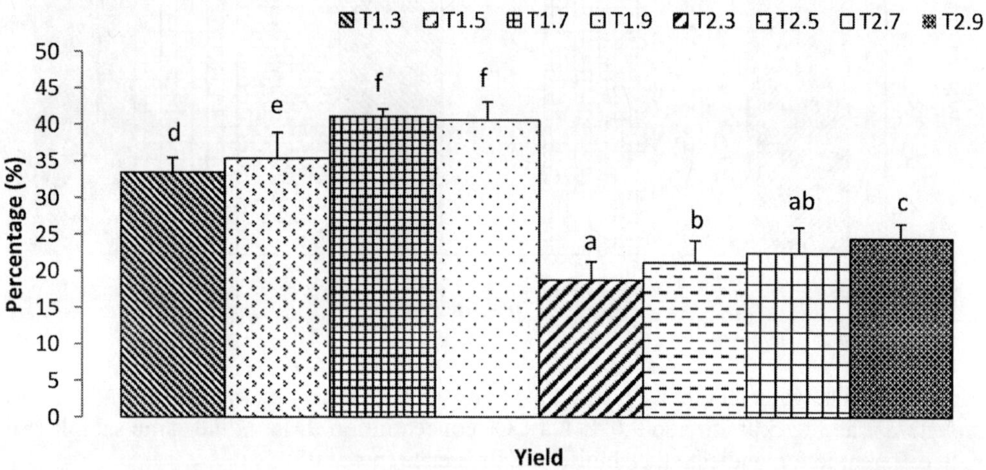

Figure 1. The yield of sodium alginate in different methods (T1 and T2) of *S. polycystum* at 3.0; 5.0; 7.0 and 9.0 % Na_2CO_3 concentration. Data at the same sampling time with different letters indicate the significant difference ($p < 0.05$).

Application of Na_2CO_3 during extraction enabling the separation of cellulose from alginate which contained in alginofit cell walls. Therefore, this process increased the yield [27, 28]. The higher Na_2CO_3 concentration, the higher yield was obtained. We used 96 % ethanol for precipitation. This influenced the yield due to its ability to bind water from alginate solution, therefore the sodium alginate was separated.

The condition of bleaching in T1 treatment was administered in the initial ground algae, while T2 treatment was administered after alkaline extraction. The T1 resulted in better yield than T2, supported by the similar data by another researcher [26]. Moreover, instead of low alginate yield, the T2 treatment might lead the occurrence of chain degradation.

Some researchers found that *Sargassum* sp. is a potential source of alginate. The best yield of alginate from *Sargassum tenerrimum* from Tuticorin Coast was reported to obtain 32.57 % yield [13]. Similar yield (31 %) of *Sargassum* sp. was reported by previous study [25]. Up to now, the best alginate yield (40 %) was obtained by *Sargassum siliquosum* from the south coast of Java [6] and *Sargassum polycystum* from this research, which originated from the north coast of Java. Both species were the local species of Indonesia.

3.3. *Water content*

The water content was influenced by precipitation of ethanol. Ethanol has an ability to bind water from alginate solution, which reduces the water content [29]. This present study showed the similar water content of T1 (12.13 %) and T2 (13 %) with the earlier report by other researchers [13, 29]. The water content that permitted by Food Chemical Codex (FCC) was less than 15 % [30]. The average of water content for *S. polycystum* in this research was shownin figure 2. This water content has already fulfilled the permitted condition by FCC.

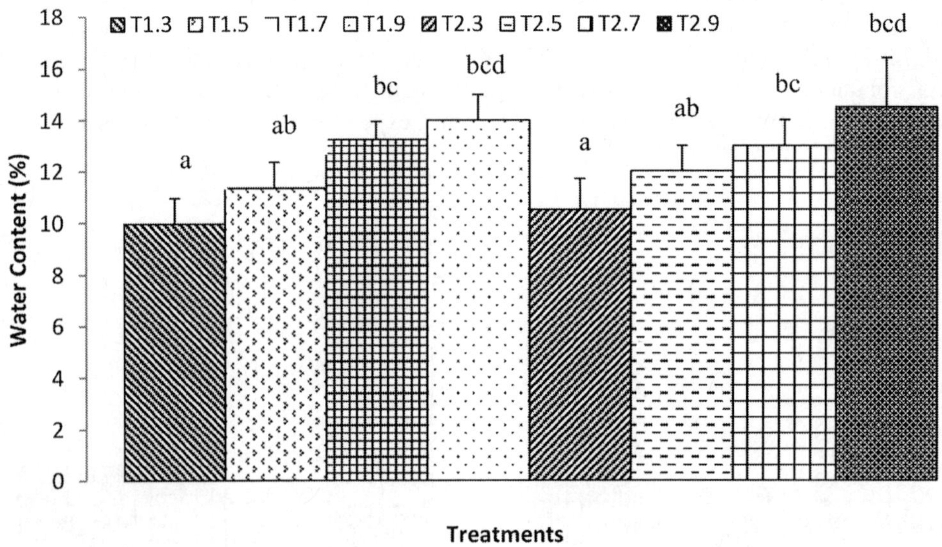

Figure 2. The water content of sodium alginate in different methods (T1 and T2) of *S. polycystum* at 3.0; 5.0; 7.0 and 9.0 % Na_2CO_3 concentration. Data at the same sampling time with different letters indicate the significant difference ($p < 0.05$).

3.4. *Ash content*

The ash content shows the mineral salt. The composition of alginates directly depends on the season and on the ecological conditions of algae growth [13, 14]. Mineral salt could be found on the surface and in thallus. Conditions of hydrology and hydrochemistry on the habitat also influence the ash content [29]. The average of ash content for *S. polycystum* in this research was shown in figure 3.

In native state, alginates exist as an insoluble salt form of mixed counter ions found in seawater, especially Na^+, Mg^{2+} and Ca^{2+} [31]. The addition of Na_2CO_3 in different concentration in T1 and T2 treatments have proven that the higher concentration, the higher ash content. This indicates that the extraction process of alkaline was succeeded. The extraction of alginates is based on the conversion from the insoluble to their soluble form using the Na^+ counter ions slightly basic conditions [26]. The ash content in this research was 16.75 % at a minimum level and 25.56 % at maximum level which was similar to 21.43 % recorded ash content in *S. tenerrimum* [13] and around 20 % in *S. polycystum* in the southern waters area of Java [29].

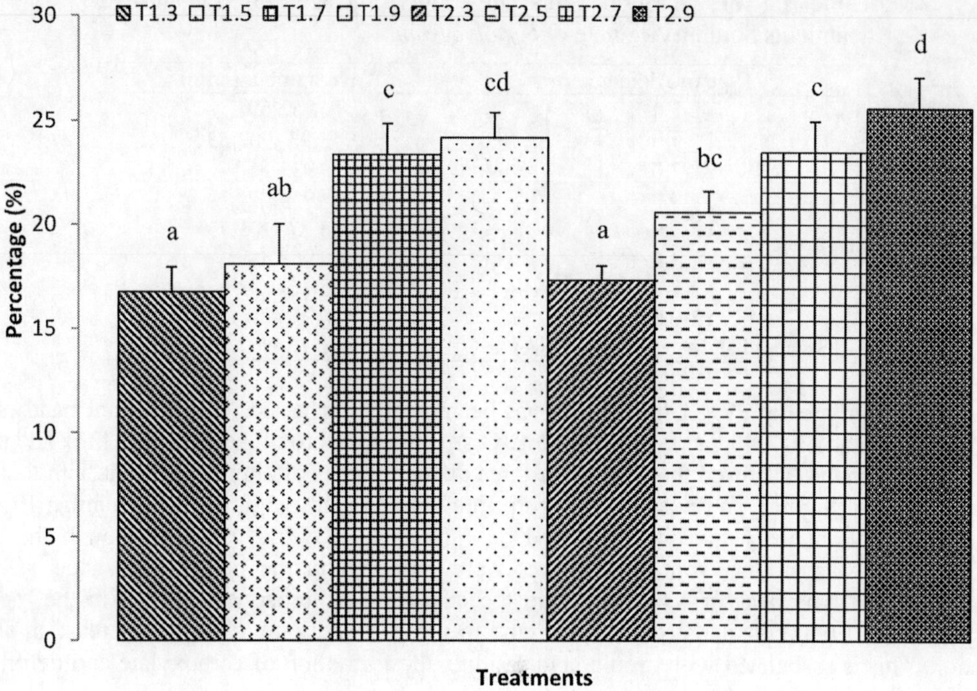

Figure 3. Ash content of sodium alginate in different methods (T1 and T2) of *S. polycystum* at 3.0; 5.0; 7.0 and 9.0 % Na_2CO_3 concentration. Data at the same sampling time with different letters indicate the significant difference ($p < 0.05$).

3.5. FT-IR spectroscopic analysis

The FT-IR spectra from T1 and T2 in 3.0; 5.0; 7.0 and 9.0 % of the Na_2CO_3 concentration of alginate compared to the standard alginate (Sigma, USA) can be seen in Figure 4 a, b, c and d the vibration signal is shown in table 1. There was a wide band at 3,400 cm^{-1} shows the signal of O-H stretching vibration, while the signal at 2,900 and 1,600 cm^{-1} was interacted with C-H stretching vibration and O-C-O carboxylate bound asymmetrically. The absorbance around 1,401 cm^{-1} is correlated to the deformation vibration of C-OH, which is the contribution of O-C-O symmetrically stretching vibration from carboxylate group [32, 33]. It is shown that the overall spectral pattern was not change by different alginate alkaline methods. The additional bands did not appear.

The observed band at around 1,300 cm^{-1} was predicted from deformation of C-C-H (and O-C-H) attributes. Furthermore, 1,095 bands was the stretching from C-O vibration at pyranose ring. The stretching formation from C-C vibration was measured at 1,033 cm^{-1}. The indication of uronic acid which formatted by the C-O group was observed at 946 cm^{-1} wavelength number [34, 35]. Moreover, the recorded signal at around 900 cm^{-1} shows the existence of asymmetric α-L-gulopyranuronate vibration ring [32, 34].

IOP Conf. Series: Earth and Environmental Science **139** (2018) 012052 doi:10.1088/1755-1315/139/1/012052

Table 1. The vibration signal on fingerprint area of T1 and T2 treatments Sodium Alginate of *S. polycystum.*

Type of Alginate	Wave number (cm^{-1})
T1-3.0 %	952.59
T1-5.0 %	952.93–902,53
T1-7.0 %	933.55
T1-9.0 %	954.02
T2-3.0 %	954.37–901.75
T2-5.0 %	953.44
T2-7.0 %	953.36–902.65
T2-9.0 %	953.26
Standard alginate (Sigma, USA)	941.26–902.96

The fingerprint area at 950–750 cm^{-1} [32] has been mostly discussed. The spectrum band of three types of alginate at 930–940cm^{-1} is referred to C-O stretching of uronic acid residue [36].The present study of FT-IR analysis showed that spectra of three different types of alginate were fit with that of the standard alginate (Sigma), and positively fingerprinted at a specific alginate wave number (950–750 cm^{-1}), though the intensity was varied. Based on figure 4a, b, c and d below, though the overall spectral pattern was similar, some differences were observed in the height and shape of certain absorption bands. The peak at 1,600 cm^{-1} and the characteristic broad absorbance in the region of fingerprint area, 1,000 cm^{-1} and around 1,300 cm^{-1} increased remarkably with the increase in alkaline extraction. This is associated with uronic acid residue, the formation of carboxylate and deformation of C-C-H and O-C-H attributes.

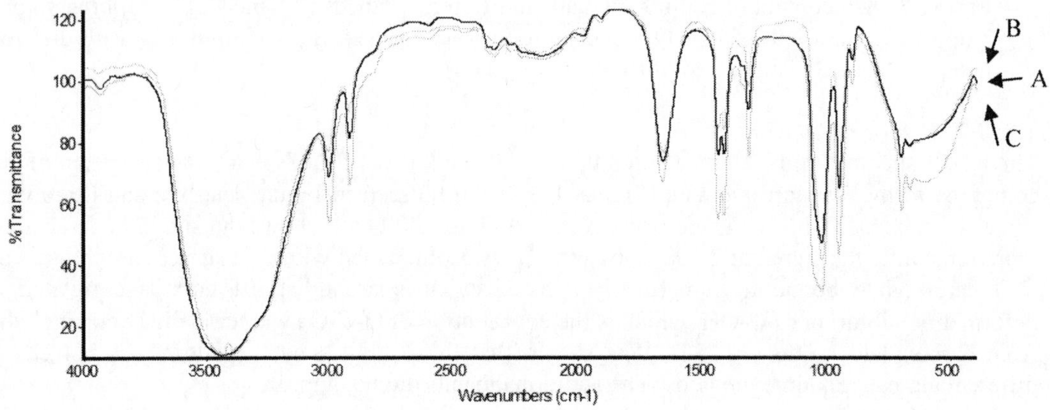

Figure 4a. The FT-IR spectra of Standard (A), T1-3.0 % (B), T2-3.0 % (C) sodium alginate of *Sargassum polycystum.*

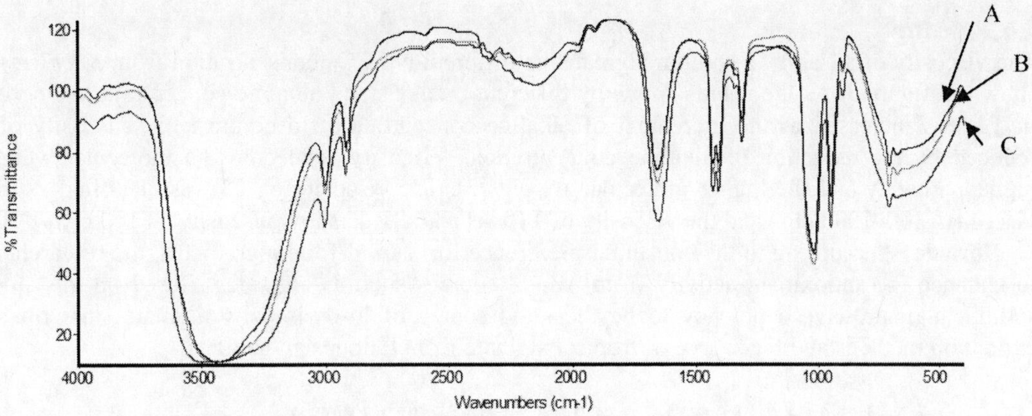

Figure 4b. The FT-IR spectra of Standard (A), T1-5.0 % (B), T2-5.0 % (C) sodium alginate of *Sargassum polycystum.*

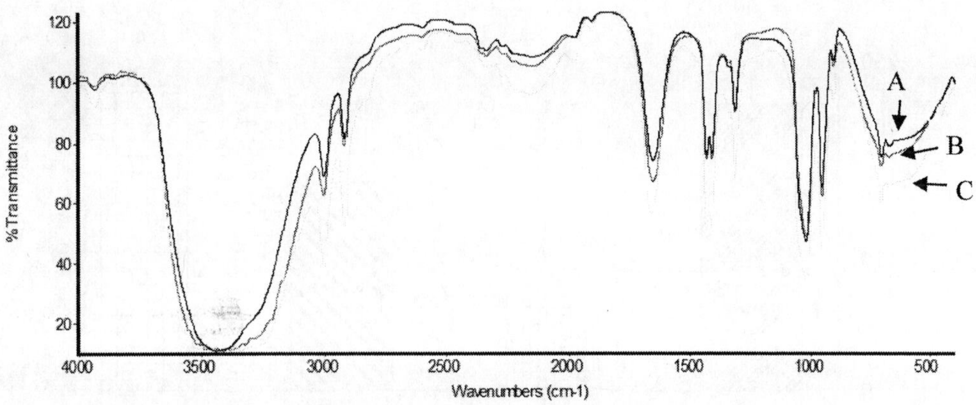

Figure 4c. The FT-IR spectra of Standard (A), T1-7.0 % (B), T2-7.0 % (C) sodium alginate of *Sargassum polycystum.*

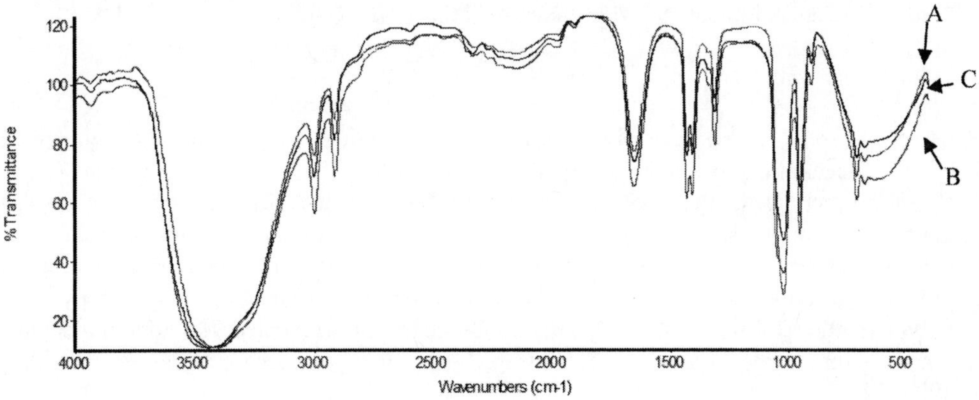

Figure 4d. The FT-IR spectra of Standard (A), T1-9.0 % (B), T2-9.0 % (C) sodium alginate of *Sargassum polycystum.*

3.6. Viscosity

The viscosity of T1 and T2 sodium alginates in different Na_2CO_3 concentration from *S. polycystum* is shown in figure 5. As the results show, by different methods and alkaline concentration, the viscosity in T2 was increased as the increment of alkaline concentration. In contrast, the viscosity of T1 is reduced by the reduction of alkaline concentration. Viscosity related to the molecular weight, the higher viscosity and the higher molecular weight [7]. The viscosity of T2 was classified as medium viscosity (300–1000 cPs) and the viscosity of T1 was classified as low viscosity (< 100 cPs) [37].

Nowadays, according to the consumer preference for natural food ingredients, the researchers tend to enhance the antioxidant activity from natural sources such as alginates. Based on this study, T1 sodium alginate have a potency to be the good source of low viscosity alginates in terms of the extraction methods and the source of tropical alginate from Indonesian habitat.

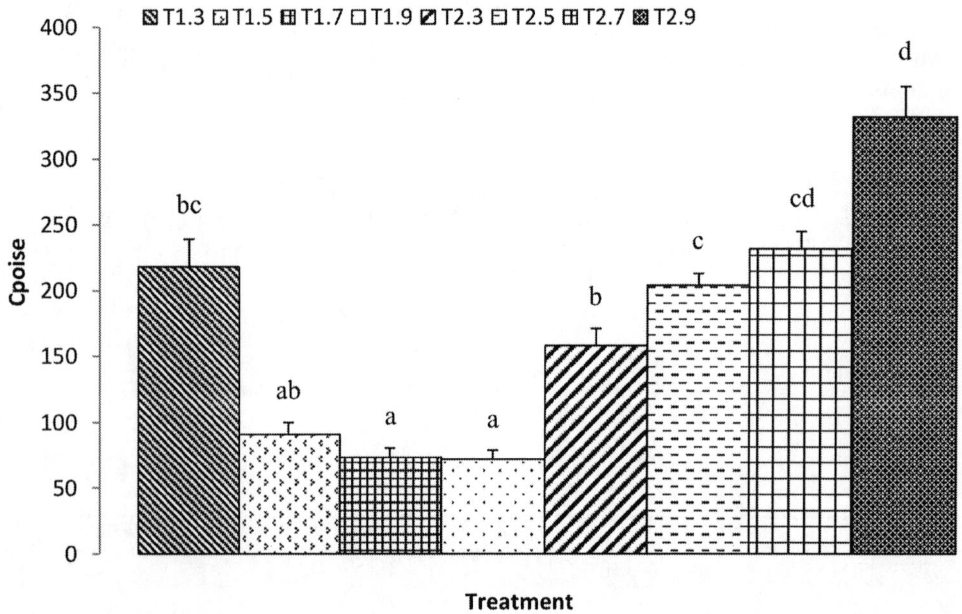

Figure 5. The viscosity of sodium alginate in different methods (T1 and T2) of *S. polysystum* at 3.0; 5.0; 7.0 and 9.0 % Na_2CO_3 concentration. Data at the same sampling time with different letters indicate the significant difference ($p < 0.05$).

3.7. Antioxidant activity

3.7.1. Superoxide radical-scavenging property. The ability of T1 and T2 alginates in different Na_2CO_3 concentration to scavenge O_2^- radical was studied by superoxide radical (O_2^-) scavenging assay in the presence of the riboflavin-light-NBT system (figure 6). As the results show, by applying alkaline extraction, superoxide radical (O_2^-) scavenging activity increased in a concentration and time-dependent manner. These results confirmed that T1-7.0 % alginate in three concentrations had better antioxidant activity compared with others. The increase in antioxidant activity caused by the depolymerization of these polysaccharides. Some studies revealed the low molecular of sodium alginate increased the bioactivity, including the immunity of fish [38] as well as the antioxidant activity [7].

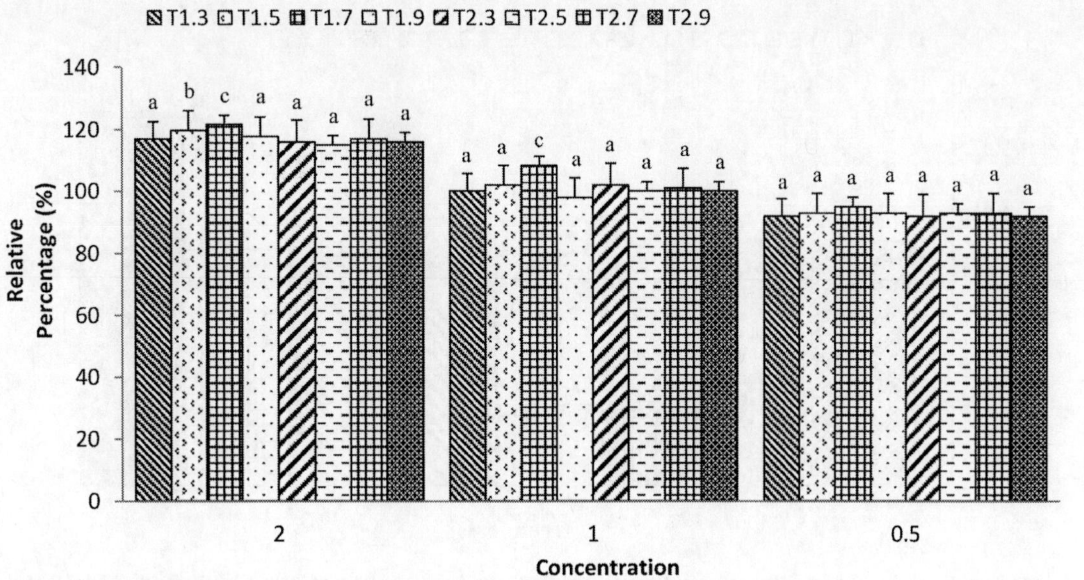

Figure 6. Three concentrations (0.05; 1:2 w/v) of relative percentage superoxide radical scavenging activity of sodium alginate in different methods (T1 and T2) of *S. polycystum* at 3.0; 5.0; 7.0 and 9.0 % Na_2CO_3 concentration. Data at the same sampling time with different letters indicate the significant difference ($p < 0.05$).

3.7.2. DPPH radical scavenging activity assay. DPPH has been extensively used as a free radical to evaluate antioxidant substances that reduce DPPH by donating hydrogen to form the non-radical DPPH-H. The DPPH radical scavenging activity of T1 and T2 sodium alginates in different Na_2CO_3 concentration from *S. polycystum* and BHT as positive control are shown in figure 6. The employment of different treatment and Na_2CO_3 concentration produced alginates with diverse DPPH scavenging activities. The results exhibited a concentration-dependent antiradical activity. The order of DPPH assay measurements was according to that of NBT assay.

The highest antiradical activity (17.48 %) was reached by T1 at 7.0 % Na_2CO_3 concentration. The ability to scavenge free radical is presumably because the donate electrons or hydrogen atoms to inactivate this radical action [9]. In general, the antiradical activity of T2 was lower than T1. This due to the fact that T2 drying technique was sundried and this allowed the degradation of the hydrochemical compound. The drying technique of T1 used the oven at stable temperature (60 °C), therefore the water content was lower than T2. Furthermore, the administration of bleaching reagent after Na_2CO_3 extraction caused chain degradation, though promote the yield of extraction, slightly [9]. The antiradical activity of these treatments was much lower than BHT. Based on DPPH and superoxide scavenging assay, T1-7.0 % $Na_2CO_3^-$ treatedalginate displayed best antioxidant activity.

Further efforts have to explore to enhance this antioxidant activity. Researchers have also shown that alginate polymer has antioxidant properties and this property increases by breaking the polymer chain [7, 9, 10, 11]. These include the purification of sodium alginate, producing the oligosaccharide alginate by different techniques i.e. irradiation [11, 39], enzyme application [1, 9, 10] or simply by heating the alginate powder in certain temperature [7].

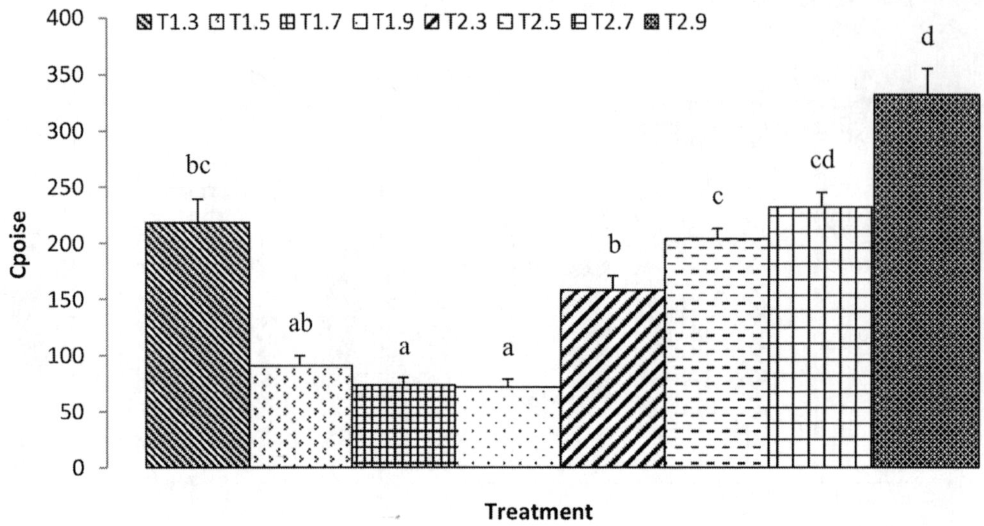

Figure 7. Percentage inhibition of sodium alginate in different methods (T1 and T2) of *S. polycystum* at 3.0; 5.0; 7.0 and 9.0 % Na_2CO_3 concentration and BHT as a positive control. Data at the same sampling time with different letters indicate the significant difference ($p < 0.05$).

There was a strong relationship between superoxide radical scavenging activity, DPPH activity and viscosity. Low viscosity indicated the low molecular weight. Similarly to other researchers [7, 38], antioxidant measurements confirmed antioxidant activity of alginate increased upon a decrease in molecular weight.

4. Conclusion

The *S. polycystum* from Panjang Island Coast of Jepara, Indonesia had a high yield of sodium alginate (41.08 %), low viscosity and high antioxidant potency. The FT-IR analysis shows that all types of alginate had a similar character compared with standard Alginate (Sigma, USA). The high yield of alginate from Indonesia promising good opportunities concerning the application of alginate in food, biomedical as well as immunostimulants in marine culture.

Acknowledgement

Part of this study was financially supported by a grant from the Faculty of Fisheries and Marine Sciences, Diponegoro University, contract no. 36/UN7.5.10/HK/2017. The authors would like to express the appreciation to those who helped support this research at Integrated Laboratory, Diponegoro University and Brackishwater Aquaculture Development Center, Jepara.

References

[1] Borazjani N F, Tabarsa M, Yo S G and Rezaei M 2017 *Int. J. Biol. Macromol* **101** 703–711
[2] Lee K Y and Mooney D J 2012 *Prog. Polym. Sci.* **37**(1) 106–126
[3] Pawar S N and Edgar K J 2012 *Biomaterials* **33**(11) 3279–3305
[4] Yang J S, Xie Y J and He W 2011 *Carbohydr. Polym.* **84**(1) 33–39
[5] Isnansetyo A, Irpani H M, Wulansari T A and Kasanah N 2014 *Aquacultura Indonesiana* **15** 73–80
[6] Yudiati E and Isnansetyo A 2017 *Ilmu Kelautan* **22**(1) 7–14
[7] Kelishomi Z H, Goliaei B, Mahdavi H, Nikoofar A, Rahimi M, Moosavi-Movahedi A A, Mamashli F and Bigdeli B 2016 *Food Chem.* **196** 897–902
[8] Melo-Silveira R, Fidelis G, Viana R, Soeiro V, Silva R, Machado D and Oliveira R H 2014

Molecules **19**(4) 5360–5378
[9] Falkeborg M, Cheong L-Z, Gianfico C, Sztukiel K M, Kristensen K, Glasius M, Xu Xand Guo Z 2014 *FoodChem.* **164** 185–194
[10] Zhao X, Li B, Xue C and Sun L 2012 *J. Appl. Phycol.* **24**(2) 295–300
[11] Sen M and Atik H 2012 *Radiat Phys. and Chem.* **81**(7) 816–822
[12] Sindhi V, Gupta V, Sharma K, Bhatnagar S, Kumari R and Dhaka N 2013 *Journal of Pharmacy Research* **7**(9) 828–835
[13] Parthiban C, Parameswari K, Saranya C, Hemalath A and Anantharaman P 2012 *Asian Pac. J. Trop. Biomed.* 1–4
[14] Vauchel P, Leroux K, Kaas R, Arhaliass A, Baron R and Legrand J 2009 *Bioresour. Technol.* **100** (3)1291–129
[15] Nguyen V T, Le N H, Lin S M, Steen F and De Clerck O 2013 *Botanica Marina* **56** 207–227
[16] Rachmat R 1999 *Prosiding Pra Kipnas VII Forum Komunikasi Ikatan Fikologi Indonesia* pp. 31–35
[17] Nguyen V T 2015 PhD Thesis Ghent University Belgium
[18] Kim D S, Lim D J, Moon S H, Suh H H and Park Y I 2004 *Kor. J. Microbiol. Biotechnol.* **32** 362–365
[19] Ayu P M T, Rudiyansyah and Harlia 2013 *Jurnal Kimia Khatulistiwa* **2**(2) 112–117
[20] AOAC 1990 *Official Methods of Analysis of the Association of Official Analytical Chemists* 15th ed (Washington) pp 456–579
[21] Celik E, Keskin I, Kayatekin I, Ak Azem F and Özkan E 2013 *Mater. Charact.* **58** 349–357
[22] Köse L P, Gülçin I, Gören A C, Namiesnik J, Martinez-Ayala A L and Gorinstein S 2015 *Ind. Crops. Prod.* **74** 712–721
[23] Ak T and Gülçin I 2008 *Chem. Biol. Interac.* **174**(1) 27–37
[24] Banerjee A, Dasgupta N and De B 2005 *J. Food Chem.* **90** 727–733
[25] Rahelivao M R, Andriamanantoanina H, Heyraud A and Rinaudo M 2013 *Food Hydrocol.* **32** 143–146
[26] Andriamanantoanina H and Rinaudo M 2010 *Carbohydr. Polym.* **82** 555–560
[27] Truss V, Taure D and Grasdalen H 2001 *Proc. Estonian Acad. Sci. Chem.* **50** 95–103
[28] Mushollaeni W 2007 *Young Lecturer Experiment Paper* 20–33
[29] Mushollaeni W 2011 *African Journal of Food Science* **5**(6) 349–352
[30] FCC 1993 *Food Chemical Codex National* (Washington: Academy Press) p 434
[31] Fourest E and Volesky B 1997 *Appl. Biochem. Biotechnol.* **67** 215–226
[32] Mathlouthi M and Koenig J L 1986 *Adv. Carbohydr. Chem. Biochem.* **44** 7–66
[33] Silverstein R M and Webster F X 1991 *Spectrometric Identification of Organic Compounds* (USA: Wil)
[34] Chandia N P, Matsuhiro BandVa´squez A E 2001 *Carbohydr. Polym.* **46** 81–87
[35] Chandia N P, Matsuhiro B, Mejias E and Moenne A 2004 *J. App. Phycol.* **16** 127–133
[36] Leal D, Matsuhiro B, Rossi M and Caruso F 2008 *Carbohydr. Res.* **343** 308–316
[37] Sigma 2008 *Biochemical Reagents for Life* Science Research (Sigma Aldrich Pte, Lt) pp 706
[38] Doan H V, Tapingkae W, Moonmanee T and Seepai A 2016 *Fish & Shellfish Immunol.* **55** 186–194
[39] Sen M 2011 *Appl. Radiat. Isot.* **69** 126–12